Optimization Methods Applied to Power Systems

Optimization Methods Applied to Power Systems

Volume 1

Special Issue Editors

Francisco G. Montoya
Raúl Baños Navarro

MDPI • Basel • Beijing • Wuhan • Barcelona • Belgrade

Special Issue Editors

Francisco G. Montoya
University of Almería
Spain

Raúl Baños Navarro
University of Almería
Spain

Editorial Office
MDPI
St. Alban-Anlage 66
4052 Basel, Switzerland

This is a reprint of articles from the Special Issue published online in the open access journal *Energies* (ISSN 1996-1073) from 2018 to 2019 (available at: https://www.mdpi.com/journal/energies/special_issues/optimization).

For citation purposes, cite each article independently as indicated on the article page online and as indicated below:

LastName, A.A.; LastName, B.B.; LastName, C.C. Article Title. *Journal Name* **Year**, *Article Number*, Page Range.

Volume 1
ISBN 978-3-03921-130-2 (Pbk)
ISBN 978-3-03921-131-9 (PDF)

Volume 1-2
ISBN 978-3-03897-154-8 (Pbk)
ISBN 978-3-03897-155-5 (PDF)

Contents

About the Special Issue Editors

Francisco G. Montoya, professor at the Engineering Department and the Electrical Engineering Section in the University of Almeria (Spain), received his M.S. from the University of Malaga and his Ph.D. from the University of Granada (Spain). He has published over 70 papers in JCR journals and is the author or coauthor of books published by MDPI, RA-MA, and others. His main interests are power quality, smart metering, smart grids and evolutionary optimization applied to power systems, and renewable energy. Recently, he has become passionately interested in Geometric Algebra as applied to Power Theory.

Raúl Baños Navarro is an associate professor at the Department of Engineering, University of Almeria (Spain). He received his first Bachelor's degree in Computer Science at the University of Almeria and his second Bachelor's degree in Economics by the National University of Distance Education (UNED). He wrote his Ph.D. dissertation on computational methods applied to optimization of energy distribution in power networks. His research activity includes computational optimization, power systems, renewable energy systems, and engineering economics. The research is being carried out at Napier University (Edinburgh, UK) and at the Universidade do Algarve (Portugal). As a result of his research, he has published more than 150 papers in peer-reviewed journals, books, and conference proceedings.

Preface to "Optimization Methods Applied to Power Systems"

Power systems are made up of extensive complex networks governed by physical laws in which unexpected and uncontrolled events can occur. This complexity has increased considerably in recent years due to the increase in distributed generation associated with increased generation capacity from renewable energy sources. Therefore, the analysis, design, and operation of current and future electrical systems require an efficient approach to different problems such as load flow, parameters and position finding, filter designing, fault location, contingency analysis, system restoration after blackout, islanding detection, economic dispatch, unit commitment, etc. The evolution is so frenetic that it is necessary for engineers to have sufficiently updated material to face the new challenges involved in the management of new generation networks (smart grids).

Given the complexity of these problems, the efficient management of electrical systems requires the application of advanced optimization methods for decision-making processes. Electrical power systems have so greatly benefited from scientific and engineering advancements in the use of optimization techniques to the point that these advanced optimization methods are required to manage the analysis, design, and operation of electrical systems. Considering the high complexity of large-scale electrical systems, efficient network planning, operation, or maintenance requires the use of advanced techniques. Accordingly, besides classical optimization techniques such as Linear and Nonlinear Programming or Integer and Mixed-Integer Programming, other advanced techniques have been applied to great effect in the study of electrical systems. Specifically, bio-inspired meta-heuristics have allowed scientists to consider the optimization of problems of great importance and obtain quality solutions in reduced response times thanks to the increasing calculation power of the current computers.

Therefore, this book includes recent advances in the application optimization techniques that directly apply to electrical power systems so that readers may familiarize themselves with new methodologies directly explained by experts in the field.

Francisco G. Montoya, Raúl Baños Navarro
Special Issue Editors

Editorial

Optimization Methods Applied to Power Systems

Francisco G. Montoya *, Raúl Baños, Alfredo Alcayde and Francisco Manzano-Agugliaro

Department of Engineering, University of Almeria, ceiA3, 04120 Almeria, Spain; rbanos@ual.es (R.B.);
aalcayde@ual.es (A.A.); fmanzano@ual.es (F.M.-A.)
* Correspondence: pagilm@ual.es; Tel.: +34-950-015791; Fax: +34-950-015491

Received: 6 May 2019; Accepted: 13 June 2019; Published: 17 June 2019

1. Introduction

Continuous advances in computer hardware and software are enabling researchers to address optimization solutions using computational resources, as can be seen in the large number of optimization approaches that have been applied to the energy field.

Power systems are made up of extensive complex networks governed by physical laws in which unexpected and uncontrolled events can occur. This complexity has increased considerably in recent years due to the increase in distributed generation associated with increased generation capacity from renewable energy sources. Therefore, the analysis, design, and operation of current and future electrical systems require an efficient approach to different problems (like load flow, parameters and position finding, filter design, fault location, contingency analysis, system restoration after blackout, islanding detection of distributed generation, economic dispatch, unit commitment, etc.). Given the complexity of these problems, the efficient management of electrical systems requires the application of advanced optimization methods that take advantage of high-performance computer clusters.

This special issue belongs to the section "Electrical Power and Energy System". The topics of interest in this special issue include different optimization methods applied to any field related to power systems, such as conventional and renewable energy generation, distributed generation, transport and distribution of electrical energy, electrical machines and power electronics, intelligent systems, advances in electric mobility, etc. The optimization methods of interest for publication include, but are not limited to:

- Expert Systems
- Artificial Neural Networks
- Fuzzy Logic
- Genetic Algorithms
- Evolutionary Algorithms
- Simulated Annealing
- Tabu Search
- Ant Colony Optimization
- Particle Swarm Optimization
- Multi-Objective Optimization
- Parallel Computing
- Linear and Nonlinear Programming
- Integer and Mixed-Integer Programming
- Dynamic Programming
- Interior Point Methods
- Lagrangian Relaxation and Benders Decomposition-Based Methods
- General Stochastic Techniques.

2. Statistics of the Special Issue

The statistics of the call for papers for this special issue related to published or rejected items were: Total submissions (113), published (36; 31.8%), and rejected (77; 68.3%).

The authors' geographical distribution by countries for published papers is shown in Table 1, where it is possible to observe 144 authors from 19 different countries. Note that it is usual for an article to be signed by more than one author, and for authors to collaborate with others of different affiliation.

Table 1. Geographic distribution by countries of authors.

Country	Number of Authors
China	80
Spain	11
South Korea	9
Cameroon	5
Malaysia	5
United States	5
Taiwan	4
Thailand	4
Viet Nam	4
Brazil	3
Egypt	3
Algeria	2
France	2
Russian Federation	2
Chile	1
Germany	1
Mexico	1
New Zealand	1
Singapore	1
Total	**144**

3. Authors of this Special Issue

The authors of this special issue and their main bibliometric indicators are summarized in Table 2, where they have been ordered from the highest to the lowest H-index. The novel authors, those considered with an H-index equal to zero are 29, and those of H-index equal to 1 are 27. On the other hand, the internationally recognized authors, those considered with an H-index of 10 or higher, are 31. It is remarkable that these authors (H-index ≥10), on average, have more than 123 co-authors, more than 110 documents published, and more than 1069 citations.

Table 2. Affiliations and bibliometric indicators for the authors.

Author	Affiliation
Jurado F.	Universidad de Jaen
Watson N.	University of Canterbury
Trentesaux D.	University of Valenciennes et du Hainaut-Cambresis
Liu N.	North China Electric Power University
Premrudeepreechacharn S.	Chiang Mai University
Sun Y.	Hohai University
Gu W.	Southeast University
Aguado, J.A.	Universidad de Málaga
Baños R.	Universidad de Almeria
Montoya F.	Universidad de Almeria
Maciel P.	Universidade Federal de Pernambuco
Liu M.	South China University of Technology
Zhang C.	Shandong University
Liu Z.	North China Electric Power University

Table 2. *Cont.*

Author	Affiliation
Wu Z.	Southeast University
Miao S.	Huazhong University of Science and Technology
Yu J.	Chongqing University
Ferreira J.	Universidade de Pernambuco
Won D.	Inha University, Incheon
Bai L.	The University of North Carolina at Charlotte
Hu Y.	Hohai University
Yao L.	National Taipei University of Technology
Lim W.	UCSI University
Yang F.	Chongqing University
Sun H.	Hebei University of Technology
Callou G.	Universidade Federal Rural de Pernambuco
Lee J.	University of Louisiana at Lafayette
Zhao D.	North China Electric Power University
Zhang X.	Shantou University
Li Y.	Zhejiang University City College
Gutiérrez-Alcaraz G.	Tecnológico Nacional de México / I.T.
Huang N.	Northeast Electric Power University
Xiang J.	Zhejiang University
Morshed M.	University of Louisiana at Lafayette
Sun B.	Shandong University
Bekrar A.	University of Valenciennes et du Hainaut-Cambresis
Rhee S.	Yeungnam University
Kamel S.	Aswan University
Xie M.	South China University of Technology
Tutsch D.	Bergische Universitat Wuppertal
Sidorov D.	Melentiev Energy Systems Institute of Siberian Branch of the Russian Academy of Sciences
Zhang X.	Nanyang Technological University
Zhou B.	China Southern Power Grid
Perng J.	National Sun Yat-Sen University Taiwan
Panasetsky D.	Melentiev Energy Systems Institute of Siberian Branch of the Russian Academy of Sciences
Zheng T.	Tsinghua University
Li J.	Northeast China Institute of Electric Power Engineering
Hinojosa V.	Universidad Técnica Federico Santa María
Siritaratiwat A.	Khon Kaen University
Hua D.	South China University of Technology
Hamouda A.	Université Ferhat Abbas de Sétif
Zhang L.	Tianjin University of Commerce
Alcayde A.	Universidad de Almeria
Ge W.	State Grid Liaoning Electric Power Supply Co., Ltd.
Zhang L.	Chongqing University
Zhang C.	Hunan University
Wu J.	Beihang University
Wang Y.	North China Electric Power University
Febrero-Garrido L.	Defense University Center
Chambers T.	University of Louisiana at Lafayette
Truong A.	HCMC University of Technology and Education
Nganhou J.	University of Yaoundé
Li Y.	Huazhong University of Science and Technology
Lin L.	Jilin Institute of Chemical Technology
Jiang T.	North China Electric Power University
Ebeed M.	Sohag University
Chatthaworn R.	Khon Kaen University
Duong T.	Industrial University of Ho Chi Minh City
Hamandjoda O.	University of Yaoundé

Table 2. *Cont.*

Author	Affiliation
Chun Y.	Hongik University
Ye C.	Huazhong University of Science and Technology
Mei S.	Qinghai University
Nguyen T.	Industrial University of Ho Chi Minh City
Mao T.	China Southern Power Grid
Wang Y.	Hohai University
Arrabal-Campos F.	Universidad de Almeria
Tiang S.	UCSI University
Hmida J.	University of Louisiana at Lafayette
Tan T.	UCSI University
Chen S.	Anqing Teachers College
Sahli Z.	Université Ferhat Abbas de Sétif
Kim C.	Yeungnam University
Li F.	Shandong University
Meva'a L.	University of Yaoundé
Wadood A.	Yeungnam University
Le Y.	State Grid Zhejiang Electric Power Corporation
Khunkitti S.	Khon Kaen University
Hong Wong C.	UCSI University
Shim M.	Inha University, Incheon
Dong X.	North China Electric Power University
Du Y.	State Grid Ganzhou Electric Power Supply Company
Xie L.	China Electric Power Research Institute
Li L.	Huazhong University of Science and Technology
Du X.	Southeast University
Fang C.	State Grid Shanghai Municipal Electric Power Company
Ndzana B.	University of Yaoundé
Yew Pang J.	Heriot-Watt University, Malaysia
Hu Z.	Zhejiang Electric Power CorporationWenzhou Power Supply Company
Chen Y.	Zhejiang University
Liu J.	State Grid Shanghai Municipal Electric Power Company
Xue L.	Northeast China Institute of Electric Power Engineering
Yimen N.	University of Yaoundé
Khurshiad T.	Yeungnam University
Kim N.	Hyosung Group
Shao B.	State Grid Liaoning Electric Power Company Limited Electric Power Research Institute
Guo B.	Jilin University
Li K.	Beihang University
Kuang J.	Shandong University
Yu J.	Anyang Institute of Technology
Sun J.	Beihang University
Ling P.	State Grid Shanghai Municipal Electric Power Company
Guo B.	North China Electric Power University
Li C.	Huazhong University of Science and Technology
Leiva, J	Universidad de Malaga
Li J.	Electric Power Research Institute of State Grid Liaoning Electric Power Co. Ltd.
Kuo Y.	Taiwan Power Company
Yang X.	Chongqing University
Yu L.	Tianjin University of Commerce
Zhang Y.	Zhoushan Power Company of State Grid
Niu F.	Zhejiang University
Ogando-Martínez A.	Universidad de Vigo
Han X.	State Grid Sichuan Electric Power Company
Ren X.	Tianjin University of Commerce
Gan C.	Zhoushan Power Company of State Grid

Table 2. *Cont.*

Author	Affiliation
Xiao L.	Tianjin University of Commerce
Fan C.	State Grid Sichuan Electric Power Research Institute
Ton T.	Thu Duc College of Technology
Zhang J.	Northeast Electric Power University
Chen H.	Tsinghua University
Zhou H.	Northeast Electric Power University
López-Gómez J.	Universidad de Vigo
Jiang S.	Anqing Teachers College
Lu S.	Taiwan Power Company
Sun G.	South China University of Technology
Cheng P.	Guangzhou Power Supply Bureau Co., Ltd.
Li X.	North China Electric Power University
Cheng W.	Shenzhen Power Supply Bureau Co., Ltd.
Cheng R.	Shenzhen Power Supply Bureau Co., Ltd.
Lee H.	Korea Electrotechnology Research Institute
Chen Z.	State Grid Sichuan Electric Power Research Institute
Shi J.	Shenzhen Power Supply Bureau Co., Ltd.
Abdo M.	Aswan University
Carmona R.	Universidad de Malaga
Wei W.	South China University of Technology

4. Brief Overview of the Contributions to this Special Issue

4.1. Keyword Analysis

The analysis of the keywords identifies or summarizes the work of the researchers. This section analyses the keywords obtained from the 36 manuscripts published in this special issue [1–36]. The keyword analysis of the papers of this special issue shows a wide variety of terms, reaching 135 different keywords. Figure 1 shows a cloud of words using author keywords. The most used and highlighted keywords are: Optimal power flow, genetic algorithm, optimization, particle swarm optimization, demand response, energy management, metaheuristic, and wind power. If we split the author keywords in simple words, it is possible to get Figure 2, where the highlighted words are now: Optimal, power, energy, system, and algorithm.

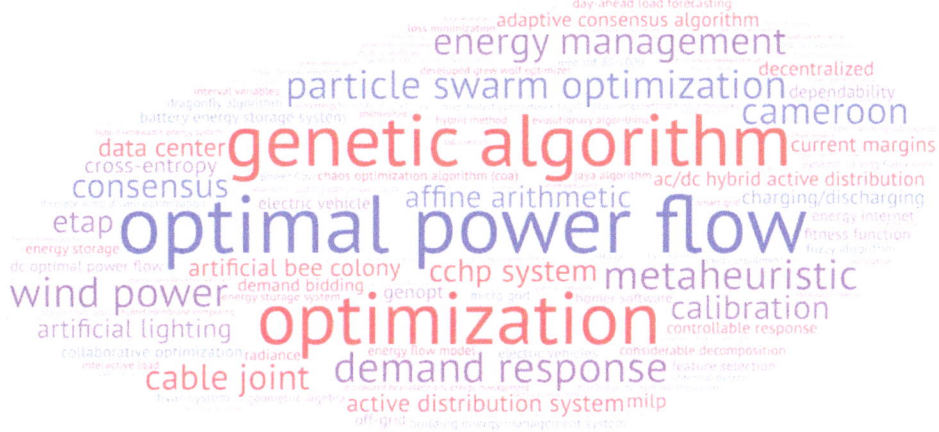

Figure 1. Cloud word of the author keywords related to the special issue.

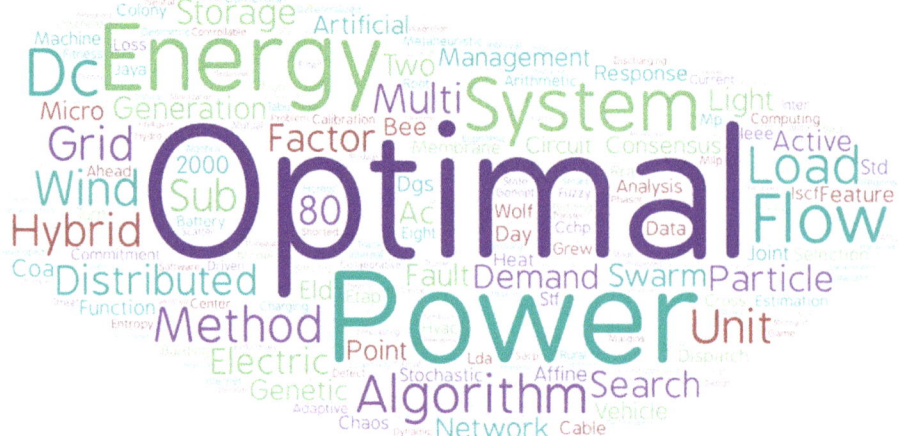

Figure 2. Cloud word for split author keywords related to the special issue.

4.2. Analysis of Author Relationship

Figure 3 shows a graph with the authors of this special issue. Each author is a node and a different color indicates their affiliation country. If an author collaborates with another one, then a link highlights the relationship between them. The larger the size of the node, the larger the H-index of this author. As expected, there is no relationship between authors of the different manuscripts, unless they are authors who have contributed to more than one, but they were exactly the same authors. What does attract attention is that there are at least nine papers with international collaboration, i.e., between authors from different countries, and two of them are collaborations between authors from at least three different countries.

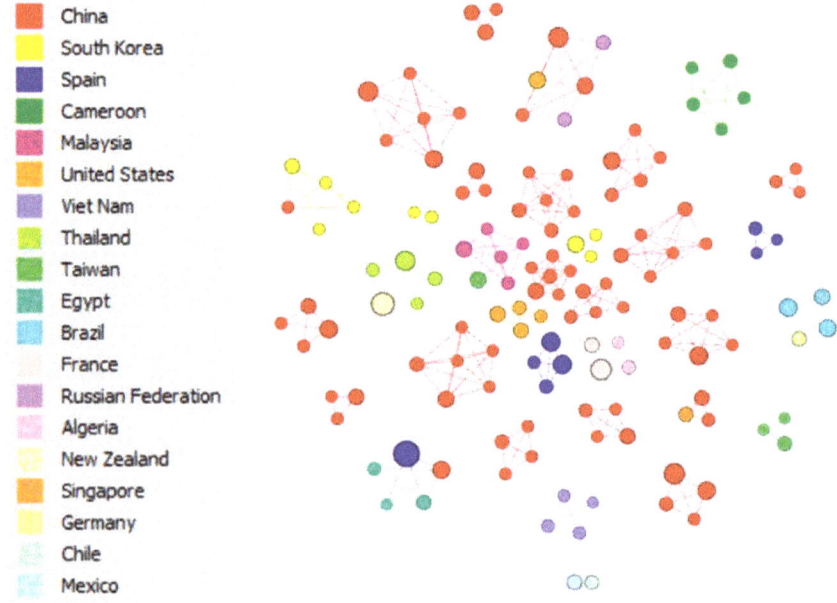

Figure 3. International interconnection between authors.

Conflicts of Interest: The authors declare no conflict of interest

References

1. Leiva, J.; Carmona Pardo, R.; Aguado, J.A. Data Analytics-Based Multi-Objective Particle Swarm Optimization for Determination of Congestion Thresholds in LV Networks. *Energies* **2019**, *12*, 1295. [CrossRef]
2. Alcayde, A.; Baños, R.; Arrabal Campos, F.M.; Montoya, F.G. Optimization of the Contracted Electric Power by Means of Genetic Algorithms. *Energies* **2019**, *12*, 1270. [CrossRef]
3. Montoya, F.G.; Alcayde, A.; Arrabal Campos, F.M.; Baños, R. Quadrature Current Compensation in Non-Sinusoidal Circuits Using Geometric Algebra and Evolutionary Algorithms. *Energies* **2019**, *12*, 692. [CrossRef]
4. Chen, Z.; Han, X.; Fan, C.; Zheng, T.; Mei, S. A Two-Stage Feature Selection Method for Power System Transient Stability Status Prediction. *Energies* **2019**, *12*, 689. [CrossRef]
5. Xie, M.; Du, Y.; Cheng, P.; Wei, W.; Liu, M. A Cross-Entropy-Based Hybrid Membrane Computing Method for Power System Unit Commitment Problems. *Energies* **2019**, *12*, 486. [CrossRef]
6. Chen, S.; Chen, H.; Jiang, S. Optimal Decision-Making to Charge Electric Vehicles in Heterogeneous Networks: Stackelberg Game Approach. *Energies* **2019**, *12*, 325. [CrossRef]
7. Mao, T.; Zhang, X.; Zhou, B. Intelligent Energy Management Algorithms for EV-charging Scheduling with Consideration of Multiple EV Charging Modes. *Energies* **2019**, *12*, 265. [CrossRef]
8. Xiao, L.; Sun, H.; Zhang, L.; Niu, F.; Yu, L.; Ren, X. Applications of a Strong Track Filter and LDA for On-Line Identification of a Switched Reluctance Machine Stator Inter-Turn Shorted-Circuit Fault. *Energies* **2019**, *12*, 134. [CrossRef]
9. Viet Truong, A.; Ngoc Ton, T.; Thanh Nguyen, T.; Duong, T. Two States for Optimal Position and Capacity of Distributed Generators Considering Network Reconfiguration for Power Loss Minimization Based on Runner Root Algorithm. *Energies* **2019**, *12*, 106. [CrossRef]
10. Perng, J.W.; Kuo, Y.C.; Lu, S.P. Grounding System Cost Analysis Using Optimization Algorithms. *Energies* **2018**, *11*, 3484. [CrossRef]
11. Li, X.; Zhao, D.; Guo, B. Decentralized and Collaborative Scheduling Approach for Active Distribution Network with Multiple Virtual Power Plants. *Energies* **2018**, *11*, 3208. [CrossRef]
12. Cheng, W.; Cheng, R.; Shi, J.; Zhang, C.; Sun, G.; Hua, D. Interval Power Flow Analysis Considering Interval Output of Wind Farms through Affine Arithmetic and Optimizing-Scenarios Method. *Energies* **2018**, *11*, 3176. [CrossRef]
13. Chen, Y.; Xiang, J.; Li, Y. SOCP Relaxations of Optimal Power Flow Problem Considering Current Margins in Radial Networks. *Energies* **2018**, *11*, 3164. [CrossRef]
14. Li, J.J.; Shao, B.Z.; Li, J.H.; Ge, W.C.; Zhang, J.H.; Zhou, H.Y. Intelligent Regulation Method for a Controllable Load Used for Improving Wind Power Integration. *Energies* **2018**, *11*, 3085. [CrossRef]
15. Wu, J.; Li, K.; Sun, J.; Xie, L. A Novel Integrated Method to Diagnose Faults in Power Transformers. *Energies* **2018**, *11*, 3041. [CrossRef]
16. Ben Hmida, J.; Javad Morshed, M.; Lee, J.; Chambers, T. Hybrid Imperialist Competitive and Grey Wolf Algorithm to Solve Multiobjective Optimal Power Flow with Wind and Solar Units. *Energies* **2018**, *11*, 2891. [CrossRef]
17. Ye, C.; Miao, S.; Li, Y.; Li, C.; Li, L. Hierarchical Scheduling Scheme for AC/DC Hybrid Active Distribution Network Based on Multi-Stakeholders. *Energies* **2018**, *11*, 2830. [CrossRef]
18. Ferreira, J.; Callou, G.; Tutsch, D.; Maciel, P. PLDAD—An Algorihm to Reduce Data Center Energy Consumption. *Energies* **2018**, *11*, 2821. [CrossRef]
19. Kim, N.K.; Shim, M.H.; Won, D. Building Energy Management Strategy Using an HVAC System and Energy Storage System. *Energies* **2018**, *11*, 2690. [CrossRef]
20. Yimen, N.; Hamandjoda, O.; Meva'a, L.; Ndzana, B.; Nganhou, J. Analyzing of a photovoltaic/wind/biogas/pumped-hydro off-grid hybrid system for rural electrification in Sub-Saharan Africa—Case study of Djoundé in Northern Cameroon. *Energies* **2018**, *11*, 2644. [CrossRef]
21. Yao, L.; Lim, W.; Tiang, S.; Tan, T.; Wong, C.; Pang, J. Demand bidding optimization for an aggregator with a Genetic Algorithm. *Energies* **2018**, *11*, 2498. [CrossRef]

22. Lee, H.L.; Chun, Y.H. Using Piecewise Linearization Method to PCS Input/Output-Efficiency Curve for a Stand-Alone Microgrid Unit Commitment. *Energies* **2018**, *11*, 2468. [CrossRef]
23. Kuang, J.; Zhang, C.; Li, F.; Sun, B. Dynamic Optimization of Combined Cooling, Heating, and Power Systems with Energy Storage Units. *Energies* **2018**, *11*, 2288. [CrossRef]
24. Khunkitti, S.; Siritaratiwat, A.; Premrudeepreechacharn, S.; Chatthaworn, R.; Watson, N. A Hybrid DA-PSO Optimization Algorithm for Multiobjective Optimal Power Flow Problems. *Energies* **2018**, *11*, 2270. [CrossRef]
25. Dong, X.; Zhang, X.; Jiang, T. Adaptive Consensus Algorithm for Distributed Heat-Electricity Energy Management of an Islanded Microgrid. *Energies* **2018**, *11*, 2236. [CrossRef]
26. Gutierrez Alcaraz, G.; Hinojosa, V. Using Generalized Generation Distribution Factors in a MILP Model to Solve the Transmission-Constrained Unit Commitment Problem. *Energies* **2018**, *11*, 2232. [CrossRef]
27. Ogando Martínez, A.; López Gómez, J.; Febrero-Garrido, L. Maintenance Factor Identification in Outdoor Lighting Installations Using Simulation and Optimization Techniques. *Energies* **2018**, *11*, 2169. [CrossRef]
28. Sahli, Z.; Hamouda, A.; Bekrar, A.; Trentesaux, D. Reactive Power Dispatch Optimization with Voltage Profile Improvement Using an Efficient Hybrid Algorithm. *Energies* **2018**, *11*, 2134. [CrossRef]
29. Sun, Y.; Wang, Y.; Bai, L.; Hu, Y.; Sidorov, D.; Panasetsky, D. Parameter Estimation of Electromechanical Oscillation Based on a Constrained EKF with C&I-PSO. *Energies* **2018**, *11*, 2059.
30. Yu, J.; Kim, C.H.; Wadood, A.; Khurshiad, T.; Rhee, S.B. A novel multi-population based chaotic JAYA algorithm with application in solving economic load dispatch problems. *Energies* **2018**, *11*, 1946. [CrossRef]
31. Wu, Z.; Du, X.; Gu, W.; Ling, P.; Liu, J.; Fang, C. Optimal Micro-PMU Placement Using Mutual Information Theory in Distribution Networks. *Energies* **2018**, *11*, 1917. [CrossRef]
32. Lin, L.; Xue, L.; Hu, Z.; Huang, N. Modular predictor for day-ahead load forecasting and feature selection for different hours. *Energies* **2018**, *11*, 1899. [CrossRef]
33. Liu, N.; Guo, B.; Liu, Z.; Wang, Y. Distributed Energy Sharing for PVT-HP Prosumers in Community Energy Internet: A Consensus Approach. *Energies* **2018**, *11*, 1891. [CrossRef]
34. Abdo, M.; Kamel, S.; Ebeed, M.; Yu, J.; Jurado, F. Solving Non-Smooth Optimal Power Flow Problems Using a Developed Grey Wolf Optimizer. *Energies* **2018**, *11*, 1692. [CrossRef]
35. Zhang, L.; LuoYang, X.; Le, Y.; Yang, F.; Gan, C.; Zhang, Y. A Thermal Probability Density–Based Method to Detect the Internal Defects of Power Cable Joints. *Energies* **2018**, *11*, 1674. [CrossRef]
36. Bravo Rodríguez, J.C.; del Pino López, J.C.; Cruz Romero, P. A Survey on Optimization Techniques Applied to Magnetic Field Mitigation in Power Systems. *Energies* **2019**, *12*, 1332. [CrossRef]

Article

A Thermal Probability Density–Based Method to Detect the Internal Defects of Power Cable Joints

Li Zhang [1], Xiyue LuoYang [1,*], Yanjie Le [2], Fan Yang [1], Chun Gan [2] and Yinxian Zhang [2]

[1] State Key Laboratory of Power Transmission Equipment & System Security and New Technology, Chongqing University, Chongqing 400044, China; zldy02@cqu.edu.cn (L.Z.); yangfancqu@gmail.com (F.Y.)

[2] Zhoushan Power Company of State Grid, Zhoushan 316021, China; fylhfut@126.com (Y.L.); czxvbnmcz@126.com (C.G.); matthew920619@gmail.com (Y.Z.)

* Correspondence: luoyangxiyue@163.com; Tel.: +86-187-2584-8089

Received: 22 May 2018; Accepted: 22 June 2018; Published: 27 June 2018

Abstract: Internal defects inside power cable joints due to unqualified construction is the main issue of power cable failures, hence in this paper a method based on thermal probability density function to detect the internal defects of power cable joints is presented. First, the model to calculate the thermal distribution of power cable joints is set up and the thermal distribution is calculated. Then a thermal probability density (TPD)-based method that gives the statistics of isothermal points is presented. The TPD characteristics of normal power cable joints and those with internal defects, including insulation eccentricity and unqualified connection of conductors, are analyzed. The results indicate that TPD differs with the internal state of cable joints. Finally, experiments were conducted in which surface thermal distribution was measured by FLIR SC7000, and the corresponding TPDs are discussed.

Keywords: Cable joint; internal defect; thermal probability density

---

1. Introduction

Unqualified construction and external destruction are the main issues in internal defects of power cable joints. The statistics show that more than 70% of defects occurred in cable joints during the past decade [1]. Internal defects of power cables will cause an increase of electromagnetic loss, insulation aging, and surface temperature changes. Excessive contact resistance due to unqualified connections of conductors and eccentricity of the core are common internal defects of cable joints.

At present, many researchers concentrate on calculating and measuring power cable temperature characteristics, because the working conditions of cable joints can be derived from the surface temperature. Many measuring techniques have been proposed, including temperature sensors, optical fibers, infrared thermal imagers, and so on [2–4]. Due to the advantages of their noncontact, secure, and real-time characteristics [5,6], infrared thermal imagers are widely used in fault monitoring and diagnosing [7,8].

At present, researchers concentrate on thermal analysis to check the faults and ampacity of power cables. In [9], a method to invert the temperature of conductors in cable joints was proposed, which was composed of two parts, radial-direction temperature inversion (RDTI) in the cable and axial-direction temperature inversion (ADTI) in the conductor. Reference [10] stated that the failure of cables and their joints can be classified by estimating or measuring ambient temperature and other parameters, because the temperature of cable insulation is a function of both ambient temperature and thermal resistivity of the ground. Reference [11] applied thermographic analysis to analyze associated regions with high surface temperature and proposed a method to diagnose faulty connections of parallel conductors. In [12], an equivalent Laplace thermal model of single-core cable was developed with lumped parameter methods based on the thermal circuit model. Reference [13] found that the partial discharge activity of power cables can be used to reflect the temperature cycling caused by load variation. Insulation eccentricity and unqualified connections of conductors are common internal defects of cable. Insulation eccentricity of cable causes not only a huge waste of the material but also electrical property problems [14]. Excess contact resistance due to unqualified connection of conductors is the main contributor to overheating and can accelerate insulation aging [15,16]. At present, the common methods to evaluate the degree of insulation eccentricity are x-ray, photoelectromagnetic, and eddy current [17,18].

Based on current research, this paper presents a new method to detect internal defects of cable joints by using thermal probability density (TPD). First, a three-dimensional (3D) electromagnetic-thermal coupling model of power cable is established and thermal distribution is calculated. Then, the distributions of TPD under different insulation eccentricity conditions are analyzed. According to the characteristics of TPD, the insulation eccentricity of power cable joints can be judged accurately. A platform is built to verify the accuracy of the proposed method. Finally, applying this method to excess contact resistance, the contact coefficient K can also be determined.

2. Model for Thermal Distribution of Power Cable Joints

The XLPE (crosslinked polyethylene) power cable (8.7/15 kV YJV 1 × 400) is taken as an example, and an axial cross-section model of the cable joint is shown in Figure 1.

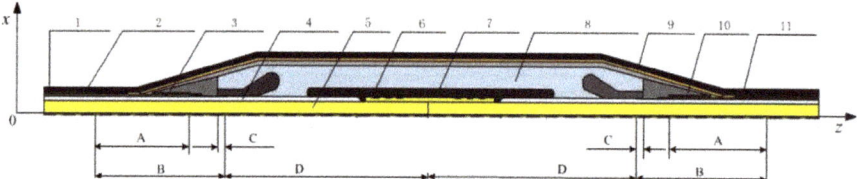

1-Cable sheath; 2-External semi-conductive layer; 3-Cable shielding layer; 4-XLPE insulation; 5-Conductor; 6-Connection tube; 7-Semi-conductive band; 8-Cold-shrinkable joint; 9-copper mesh belt; 10-Sealant; 11-PVC (polyvinyl chloride) band

Figure 1. Axial cross-section model of cable joint.

The parameters of the cable joint are shown in Table 1.
The lengths of different parts of the cable joint (as shown in Figure 1) are listed in Table 2.
The parameters required for calculation in the temperature field are shown in Table 3.
For a single cable joint laid in the air, the laying parameters are given in Table 4.

Table 1. Parameters of cable joint.

Conductor diameter	23.8 mm
Insulation thickness	4.5 mm
Shielding layer thickness	0.5 mm
Sheath thickness	2.5 mm
External diameter of cable	41 mm
Conductor cross-section area	400 mm²

Table 2. Length parameters of cable joint (mm).

A	B	C	D	E
90	140	25	175	120

Table 3. Material physical parameters used for temperature field calculation.

Material	Thermal Conductivity/(W·(m·°C)$^{-1}$)	Density/(kg·m^{-3})	Specific Heat Capacity/(J·(kg·°C)$^{-1}$)
Conductor	400	8920	385
Semiconductor	0.48	1350	1470
Insulation	0.286	1200	2250
Sheath	0.167	1380	2100

Table 4. Simulation parameters.

Ambient Temperature	Convection Heat Transfer Coefficient h	Current
20 °C	5.6 W/(m^2·K)	1000 A

3. Thermal Probability Density Distribution–Based Method

Let T_{max} and T_{min} represent the maximum and minimum temperature, respectively. C_i is the count of T_i, where $T_{min} < T_i < T_{max}$.

Set $C_T = \sum C_i$, $P_i = C_i/C_T$, which is known. $0 \leq P_i < 1$, and $\sum P_i = 1$.

It is obvious that when faults arise in high-voltage equipment, the thermal distribution changes, hence the curve of P_i will change, which can be used to determine internal faults. This is the thermal probability density (TPD)–based method.

To use the TPD method in practice, infrared imaging technology can be used, which is widely used to analyze the operating state of electrical equipment and the contamination level of insulators. Two infrared images of low-voltage bushing under normal and fault conditions are shown in Figure 2.

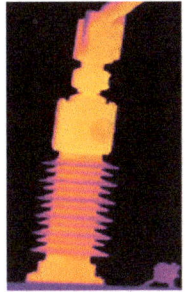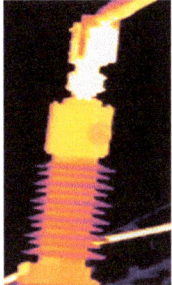

Figure 2. Infrared images.

The surface temperature distribution and the temperature span (the difference between T_{max} and T_{min}) will change with the operating conditions [19]. From the perspective of thermodynamic entropy, regarding each set of temperature data in the infrared image as a state, an infrared image contains a lot of temperature data, and a statistical method is chosen to analyze the data.

Probability density functions are often used to represent the distribution of data samples. As the cable surface temperature distribution is unknown, the nonparametric kernel density estimation method is used to calculate the surface temperature distribution [20,21]. The gray scale is used in infrared images to record the temperature data. Scattering the infrared image and treating temperature as a discontinuous physical quantity, the temperature matrix can be obtained, as shown in Figure 3. There are N × M temperature values in Figure 3, and each temperature value corresponds to a temperature state.

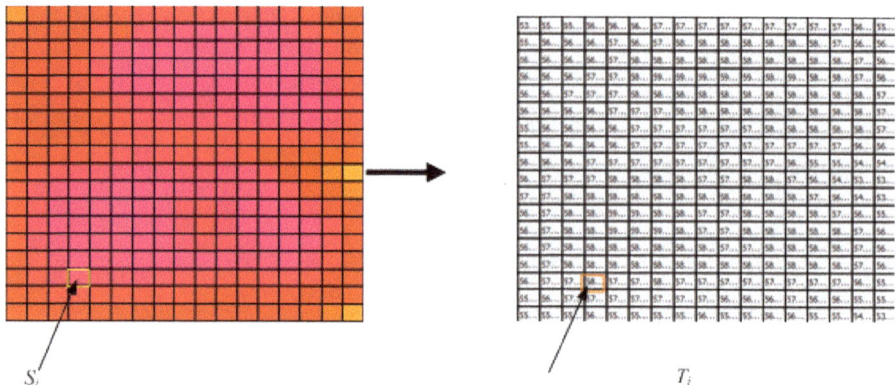

Figure 3. Illustration of gray level distribution corresponding to temperature image.

Dividing the entire temperature range into several small subintervals and regarding the temperatures located in the same subinterval as isothermal points T_i ($T_{min} < T_i < T_{max}$), the quantity of T_i can be determined using the statistical method, and then TPD can be drawn. The probability density of any temperature T_i is calculated according to Equation (1):

$$\hat{y}(x) = \frac{1}{nh} \sum_{j=1}^{n} K\left(\frac{x - x_i}{h}\right) \tag{1}$$

where $K(u)$ is a kernel function and h is the window width.

In order to verify the accuracy of the calculation results, the asymptotic mean integrated square error (*AMISE*) is often used to detect the accuracy of $\hat{y}(x)$. The expression is as follows:

$$AMISE(h) = \frac{1}{nh}R(K) + \frac{1}{4}h^4[\mu_2(K)]^2R(y'') \tag{2}$$

where $R(K) = \int K(z)^2 dz$, $\mu_2(K) = \int z^2 K(z)dz$, $R(y'') = \int [y''(x)]^2 dz$.

When $\frac{d}{dh}[AMISE(h)] = 0$, the best window width value ($h_{optimal}$) can be calculated using Equation (3):

$$h_{optimal} = \left\{\frac{R(K)}{[\mu_2(K)]^2 R(y'')n}\right\}^{\frac{1}{5}} \tag{3}$$

The calculation results of $h_{optimal}$ and *AMISE* under different kernel functions are shown in Table 5. The comparison results show that the Gaussian kernel function has the smallest error. The Gaussian kernel is shown in Equation (4):

$$K(x) = \frac{1}{\sqrt{2\pi}} \exp(-\frac{1}{2}x^2) \tag{4}$$

Table 5. Results of several kernel functions. *AMISE*, asymptotic mean integrated square error.

Parameter	Uniform Kernel	Triangular Kernel	Gaussian Kernel
$h_{optimal}$	0.436806	0.618528	0.5548
AMISE (10^{-5})	7.7542	6.1584	5.7456

The following characteristics are used to characterize TPD of the cable joint:

(1) Variance *s*: represents the element difference within an array. The formula is as follows:

$$s = \sqrt{\frac{1}{n-1}\sum_{i=1}^{n}(T_i - \overline{T})^2} \tag{5}$$

where $\overline{T}$ is the average temperature.

(2) Peak-peak difference *P*: represents the difference between the peaks of high temperature and low temperature. The formula is as follows:

$$P = P_2 - P_1 \tag{6}$$

where *P* is the peak-peak difference, P_2 is the peak value of the high temperature, and P_1 is the peak value of the low temperature

The details of the process are as follows: First, the infrared camera is used to get the surface temperature of the cable joint; then, TPD is obtained according to Equations (1)–(4), as shown in Figure 4. Finally, the defect type and degree of cable are judged based on the characteristic of TPD.

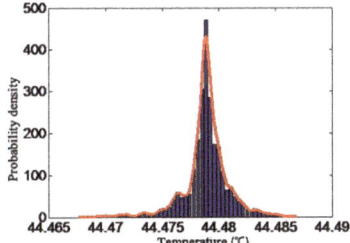

Figure 4. Thermal probability density (TPD) under normal conditions.

4. Simulation and Results

4.1. Cable Eccentricity

4.1.1. Measurement Precision with Resistor

A cross-section of the cable joint is shown in Figure 5, and the degree of insulation eccentricity is defined as $D = \frac{D_1 - D_1'}{2}$, where D_1, D_1' represents the insulation thickness.

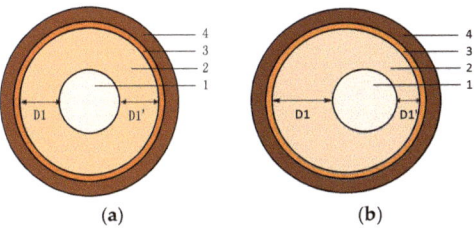

Figure 5. Cable cross-section diagram: (**a**) normal, (**b**) eccentricity.

Based on the model of cable joint shown in Figure 1, the temperature distribution was calculated when *D* = 0 mm, 2 mm, 3 mm, 4 mm, 5 mm, and 6 mm, and the results when *D* = 3 mm are shown in Figure 6. The temperature distribution is not uniform when the cable joint is eccentric, which is the basis for the detection of cable eccentricity. TPDs under normal and insulation eccentricity are shown in Figure 7.

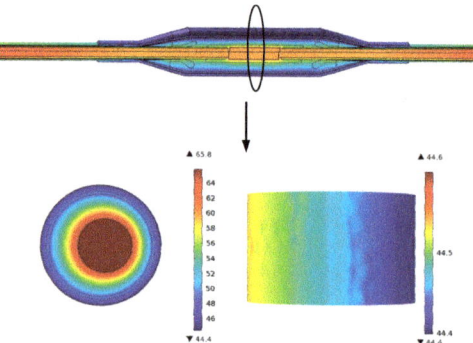

Figure 6. Temperature distribution when $D = 3$ mm.

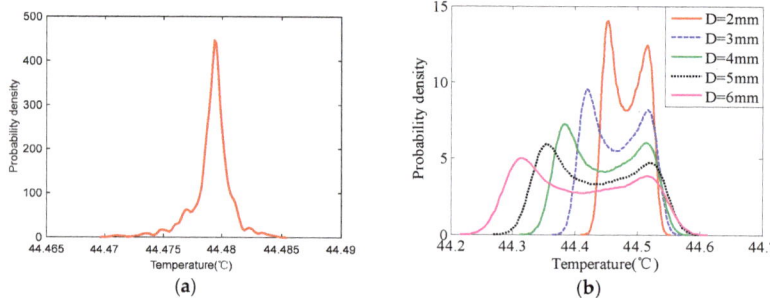

Figure 7. TPDs of cable under (**a**) normal and (**b**) insulation eccentricity conditions.

Comparing Figure 7a,b, it can be seen that under normal conditions, the distribution of cable surface temperature is uniform and the temperature is concentrated at 44.479 °C. When the cable is eccentric, its TPD changes from a single peak to a bimodal wave. In addition, the peak-peak difference increases as the degree of eccentricity increases, and when D changes from 2 mm to 6 mm, the corresponding peak-peak difference increases from 0.06 °C to 0.20 °C.

Table 6. Characteristics of TPD under different eccentricities.

D	Variance	Peak-Peak Difference
0 mm (normal)	2.68×10^{-6}	0
2 mm	0.0018	0.06
3 mm	0.0037	0.10
4 mm	0.0063	0.13
5 mm	0.0096	0.16
6 mm	0.0136	0.20

The change rule of the characteristic parameters under different eccentricity is shown in Table 6. The variance increases in the form of a quadratic function with increased D. When D increases from 2 to 6 mm, the variance increases from 0.0018 to 0.0136, and the change rule is shown in Figure 8a. The rule can be expressed with the function $s = 0.0035D^2 - 0.00015D + 0.0001$, where s is the variance. In addition, the peak-peak difference increases in the form of the first-order function, which is expressed as $P = 0.034D - 0.006$, where P is the peak-peak difference. When D increases from 2 to 6 mm, the peak-peak difference changes from 0.06 to 0.20, as shown in Figure 8b.

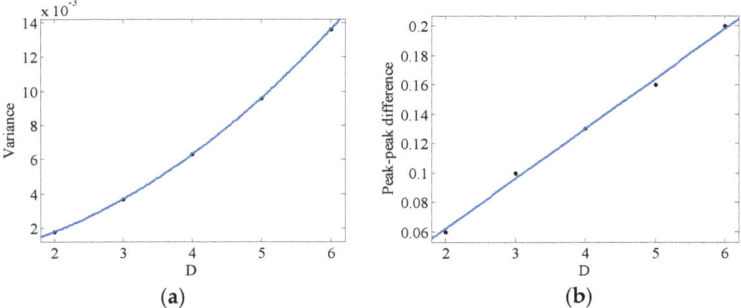

Figure 8. Relationship between characteristics of TPD and cable eccentricity: (**a**) variance, (**b**) peak-peak difference.

4.1.2. Experimental Verification

Based on the principle of equal heat source, a surface temperature measurement platform was built. Using graphite rods as a core conductor to simulate the actual operation of large loads not only overcomes the problem of imposing a high load current of 400 A and above in laboratory conditions, but also reduces the cost of the experiment. Figure 9 shows the structure of the analog cable. The resistance of each graphite rod is about 1 Ω.

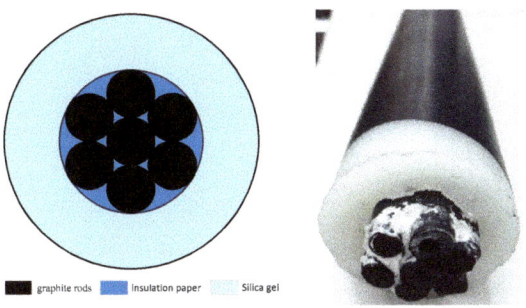

Figure 9. Structure of the cable model.

Different degrees of insulation eccentricity were simulated in different parallel ways using the graphite rods, as shown in Figure 10. A 12 V/30 A adjustable constant-current source was used to supply power, and the output current could be adjusted in the range of 5 A to 30 A to ensure the same internal heat. The related data are shown in Table 7. Temperature was measured by an FLIR SC7000 infrared camera, whose accuracy is 0.1 °C. The outer side of the cable and its support parts were painted black so that the radiation coefficient was close to 1. The infrared camera was placed at the same level as the cable and the steady-state temperature data were recorded. The experimental platform is shown in Figure 11.

Table 7. Internal heat of cable.

Case	Resistance	Current	Energy
Normal	1 Ω	10 A	100 J
Case 1	1 Ω	10 A	100 J
Case 2	0.5 Ω	14 A	98 J
Case 3	0.33 Ω	17 A	95.37 J

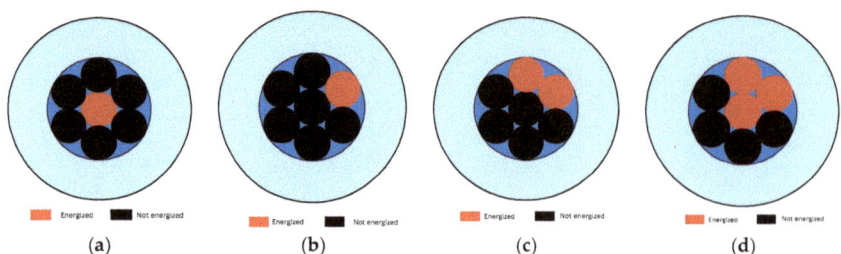

Figure 10. Structure of the insulation eccentricity cables: (**a**) normal, (**b**) case 1, (**c**) case 2, (**d**) case 3.

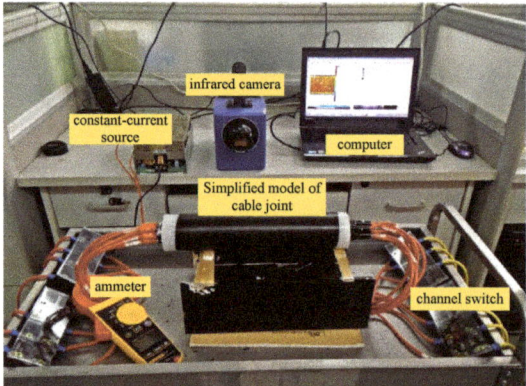

Figure 11. Schematic diagram of the experimental scheme.

The surface temperature distribution recorded by the infrared thermal imager is shown in Figure 12 and TPDs are shown in Figure 13. Under normal conditions, the surface temperature distribution is uniform and the temperatures are concentrated at 33 °C. When the cable is eccentric, the surface temperature distribution is not uniform. The waveform is distorted from a single peak wave to a bimodal one, and the tendency of the related parameters of waveform is consistent with that obtained in simulation, which proves the feasibility of the proposed method.

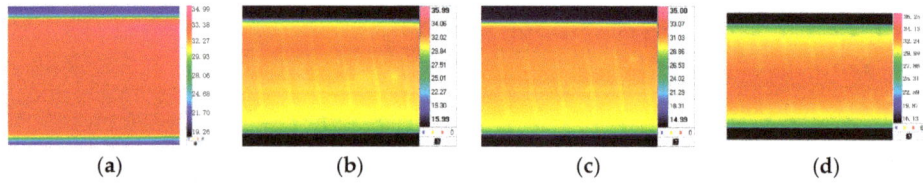

Figure 12. Infrared images of cable surface temperature: (**a**) normal, (**b**) case 1, (**c**) case 2, (**d**) case 3.

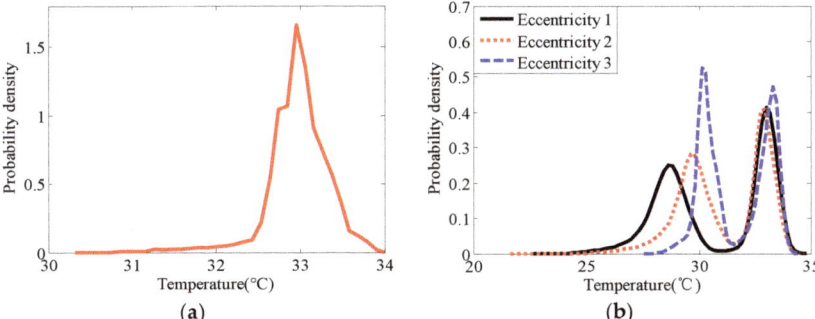

Figure 13. TPDs in experiment under different insulation eccentricities: (**a**) normal, (**b**) eccentricity.

4.2. Contact Resistance

This method can not only be applied to cable eccentricity testing but also be used to analyze the degree of excess contact resistance.

Due to crimping process defects, the thermal loss of cable joint increases, resulting in a surface temperature distribution difference. In order to quantitatively characterize the influence of contact resistance, the contact coefficient K is defined as $K = R_1/R_2$, where $R_1 = \frac{1}{\sigma_2}\frac{1}{\pi r_2^2}$ and $R_2 = \frac{1}{\sigma_1}\frac{1}{\pi r_1^2}$. R_1 is the resistance of the connection portion and R_2 is the conductor resistance of a cable body of the same length. A schematic diagram of contact coefficient is shown in Figure 14, and the formula is expressed as Equation (7).

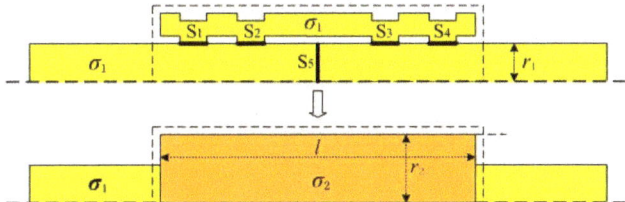

Figure 14. Structure and equivalent model of cable conductor connection.

$$K = \frac{\frac{1}{\sigma_2}\frac{1}{\pi r_2^2}}{\frac{1}{\sigma_1}\frac{1}{\pi r_1^2}} = \frac{\sigma_1}{\sigma_2}\left(\frac{r_1}{r_2}\right)^2 \tag{7}$$

Based on the model of cable joint shown in Figure 1, the temperature distribution was calculated when $K = 1, 3, 5, 7, 11$, and the results when $K = 5$ are shown in Figure 15. Surface temperatures of the cable joint and the cable body are different, so the contact coefficient can be determined by the temperature difference between them.

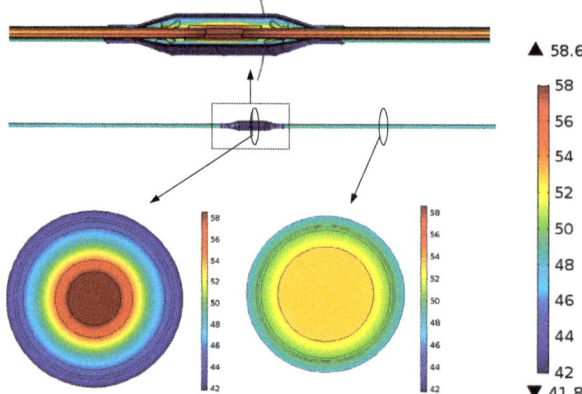

Figure 15. Temperature distribution of $K = 5$.

The TPDs of the cable with different K values are shown in Figure 16. The peak-peak difference decreases gradually with the increase of K, and changes gradually from a bimodal wave to a unimodal wave ($K = 9$, $K = 11$). The variance values and peak-peak differences with different K are shown in Table 8.

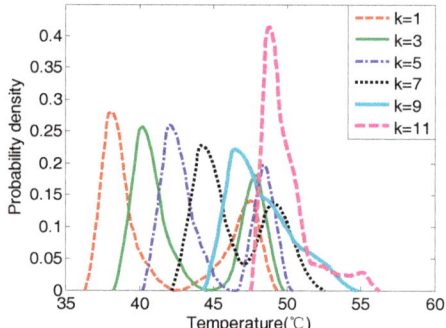

Figure 16. TPDs of cable with different K values.

Table 8. Characteristics of TPDs under different K values.

K	Variance	P_1	P_2	Peak-Peak Difference
$K = 1$	17.30	38.02	47.49	9.47
$K = 3$	12.23	40.21	47.88	7.67
$K = 5$	8.46	42.04	48.44	6.40
$K = 7$	5.70	44.25	48.89	4.64
$K = 9$	3.87	46.44	48.90	2.46
$K = 11$	2.60	48.76	48.76	0

Figure 17a shows that the variance decreases in the form of a quadratic function as K increases. When contact coefficient K increases from 1 to 11, variance is reduced from 17.30 to 2.60. The relationship between variance and contact coefficient can be described with $s = 0.12K^2 - 2.88K + 19.95$, where s is variance. The peak-peak difference also decreases with the increase of K, and if the contact coefficient K continues to increase, the peak-peak difference will become negative. The function that represents the relationship between the peak-peak difference and the contact coefficient is $P = -0.9258K + 10.66$, where P is the peak-peak difference. The change trend is shown in Figure 17b.

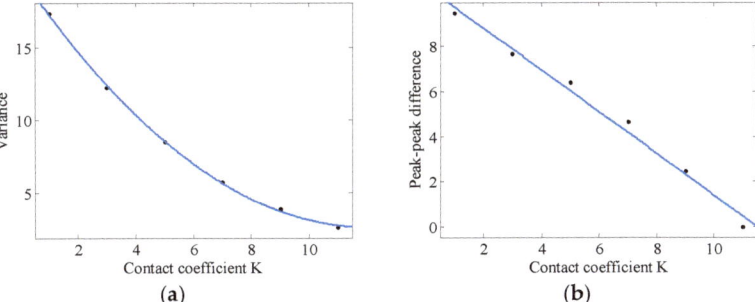

Figure 17. Relationship between characteristics of TPD and K: (**a**) variance, (**b**) peak-peak difference.

To further analyze the reason for the curve distribution in Figure 18, we analyzed the temperature distribution of the cable surface axis with different contact coefficient K, as shown in Figure 18.

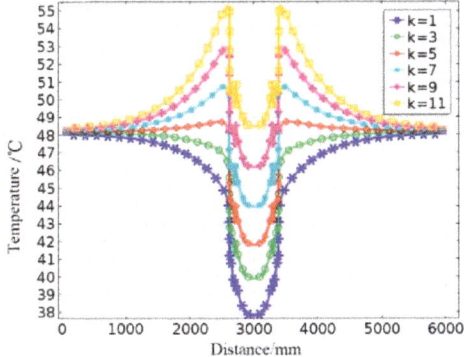

Figure 18. Surface axial temperature distribution curve of cable with different K values.

Figure 18 shows that when K is small, the temperature at the cable joint is lower than that at the cable body, because the cable joint has a greater heat dissipation area. When K is small, thermal convection plays a major role in the cable joint temperature being significant lower than the body temperature and a low temperature peak, shown in Figure 16. With increased K, the heat yield of cable joint increases gradually, therefore the joint temperature increases gradually and the peak-peak difference reduces gradually. Since contact resistance only changes the heat generation rate and heat conduction in the axial direction becomes weak when the distance from the center of the cable joint is more than 2.5 m, the cable body temperature 2.5 m from the center does not change with K.

5. Conclusions

This paper proposes a method of using infrared temperature measurement and analyzes the regularities of TPD to estimate the type and degree of internal faults of cable, based on a three-dimensional electromagnetic-thermal multiphysics model of power cable. When cable internal faults occur, the distributions of surface temperature probability density curves are different. Combining the characteristic of TPD, a comprehensive judgment can be made to determine the type and degree of cable defects accurately. In addition, an experimental platform was built to verify the method proposed in this paper, and the experimental results are consistent with the simulation results, which verifies the feasibility of the method.

Author Contributions: This paper is a result of the collaboration of all co-authors. L.Z. conceived and designed the study. X.L. was responsible for the modeling results, experiment and wrote most of the article. F.Y. provided the theory for the modeling and established the model. Y.L., C.G. and Y.Z. supervised the project and helped with most of the correction.

Funding: This research was funded by [National Key R&D Program of China] grant number (2017YFB0902703).

Acknowledgments: This work was supported by the State Grid Science and Technology Project (Research on Temperature Field Detection Technology of Cable Joint). We are thankful to all our lab fellows for providing support during research experiments and for valuable suggestions.

Conflicts of Interest: The authors declare no conflict of interest.

Nomenclature

TPD	thermal probability density
T_{max}	maximum temperature
T_{min}	minimum temperature
T_i	$T_{min} < T_i < T_{max}$
C_i	the count of T_i
AMISE	asymptotic mean integrated square error
$h_{optimal}$	best window width value
s	variance
$\overline{T}$	average temperature
P	peak-peak difference
P_2	peak value of high temperature
P_1	peak value of low temperature
D	degree of insulation eccentricity
K	contact coefficient
R_1	resistance of connection portion
R_2	conductor resistance of cable body

References

1. Shaker, Y.O.; EI-Hag, A.H.; Patel, U.; Jayaram, S.H. Thermal modeling of medium voltage cable terminations under square pulses. *IEEE Trans. Dielectr. Electr. Insul.* **2014**, *21*, 932–939. [CrossRef]
2. Xiang, X.; Tu, P.; Zhao, J. Application of fiber Bragg grating sensor in temperature monitoring of power cable joints. In Proceedings of the 2011 International Conference on Electronics, Communications and Control (ICECC), Ningbo, China, 9–11 September 2011; pp. 755–757.
3. Gan, W.; Wang, Y. Application of the distributed optical fiber grating temperature sensing technology in high-voltage cable. In Proceedings of the 2011 International Conference on Electronic & Mechanical Engineering and Information Technology, Harbin, China, 12–14 August 2011; Volume 9, pp. 4538–4541.
4. Cui, H.; Xu, Y.; Zeng, J.; Tang, Z. The methods in infrared thermal imaging diagnosis technology of power equipment. In Proceedings of the 2013 IEEE 4th International Conference on Electronics Information and Emergency Communication, Beijing, China, 15–17 November 2013; pp. 246–251.
5. He, H.; Lee, W.-J.; Luo, D.; Cao, Y. Insulator Infrared Image Denoising Method Based on Wavelet Generic Gaussian Distribution and Map Estimation. In Proceedings of the 2016 IEEE Industry Applications Society Annual Meeting, Portland, OR, USA, 2–6 October 2016.
6. Chaturvedi, D.K.; Iqbal, M.S.; Pratap, M. Intelligent health monitoring system for three phase induction motor using infrared thermal image. In Proceedings of the 2015 International Conference on Energy Economics and Environment (ICEEE), Noida, India, 27–28 March 2015; pp. 1–6.
7. Tang, Q.; Liu, J.; Dai, J.; Yu, Z. Theoretical and experimental study on thermal barrier coating (TBC) uneven thickness detection using pulsed infrared thermography technology. *Appl. Therm. Eng.* **2017**, *114*, 770–775. [CrossRef]

8. Tang, Q.; Dai, J.; Bu, C.; Qi, L.; Li, D. Application of infrared thermography for predictive/preventive maintenance of thermal defect in electrical equipment. *Appl. Therm. Eng.* **2013**, *61*, 220–227.

9. Ruan, J.; Liu, C.; Huang, D.; Zhan, Q.; Tang, L. Hot spot temperature inversion for the single-core power cable joint. *Appl. Therm. Eng.* **2013**, *104*, 146–152. [CrossRef]

10. Sturchio, A.; Fioriti, G.; Pompili, M.; Cauzillo, B. Failure rates reduction in SmartGrid MV underground distribution cables: Influence of temperature. In Proceedings of the 2014 AEIT Annual Conference—From Research to Industry: The Need for a More Effective Technology Transfer (AEIT), Trieste, Italy, 18–19 September 2014; pp. 1–6.

11. Mendes, M.A.; Tonini, L.G.R.; Muniz, P.R.; Donadel, C.B. Thermographic analysis of parallelly cables: A method to avoid misdiagnosis. *Appl. Therm. Eng.* **2013**, *104*, 231–236. [CrossRef]

12. Lei, M.; Liu, G.; Lai, Y.; Li, J.; Li, W.; Liu, Y. Study on thermal model of dynamic temperature calculation of single-core cable based on Laplace calculation method. In Proceedings of the 2010 IEEE International Symposium on Electrical Insulation, San Diego, CA, USA, 6–9 June 2010; pp. 1–7.

13. Steennis, F.; Wagenaars, P.; van der Wielen, P.; Wouters, P.; Li, Y.; Broersma, T.; Harmsen, D.; Bleeker, P. Guarding MV cables on-line: With travelling wave based temperature monitoring, fault location, PD location and PD related remaining life aspects. *IEEE Trans. Dielectr. Electr. Insul.* **2016**, *23*, 1562–1569. [CrossRef]

14. Fedorov, M.E. Optical Laser Diffraction Transducer for Measuring Single-Wire Electric Cable Eccentricity. *IOP Conf. Ser. Mater. Sci. Eng.* **2015**, *81*, 012074. [CrossRef]

15. Ruzlin, M.M.M.; Shafi, A.H.H.; Basri, A.G.A. Study of cable crimping factors affecting contact resistance of medium voltage cable ferrule and lug. In Proceedings of the 22nd International Conference and Exhibition on Electricity Distribution (CIRED 2013), Stockholm, Sweden, 10–13 June 2013; pp. 1–4.

16. Luo, H.; Cheng, P.; Liu, H.; Kang, K.; Yang, F.; Yang, Q. Investigation of contact resistance influence on power cable joint temperature based on 3-D coupling model. In Proceedings of the 2016 IEEE 11th Conference on Industrial Electronics and Applications (ICIEA), Hefei, China, 5–7 June 2016; pp. 2265–2268.

17. Robinson, A.P.; Lewin, P.L.; Sutton, S.J.; Swingler, S.G. Inspection of high voltage cables using X-ray techniques. *IEEE Int. Symp. Electr. Insul.* **2004**, *19*, 372–375.

18. Yang, F.; Cheng, P.; Luo, H.; Yang, Y.; Liu, H.; Kang, K. 3-D thermal analysis and contact resistance evaluation of power cable joint. *Appl. Therm. Eng.* **2016**, *93*, 1183–1192. [CrossRef]

19. Perpiñà, X.; Castellazzi, A.; Piton, M.; Mermet-Guyennet, M.; Millán, J. Failure-relevant abnormal events in power inverters considering measured IGBT module temperature inhomogeneities. *Microelectron. Reliab.* **2007**, *47*, 1784–1789. [CrossRef]

20. Comaniciu, D.; Ramesh, V.; Meer, P. Kernel-Based Object Tracking. *IEEE Trans. Pattern Anal. Mach. Intell.* **2003**, *40*, 564–575. [CrossRef]

21. Pană, C.; Severi, S.; de Abreu, G.T.F. An adaptive approach to non-parametric estimation of dynamic probability density functions. In Proceedings of the 2016 13th Workshop on Positioning, Navigation and Communications (WPNC), Bremen, Germany, 19–20 October 2016; pp. 1–4.

Article

Solving Non-Smooth Optimal Power Flow Problems Using a Developed Grey Wolf Optimizer

Mostafa Abdo [1], Salah Kamel [1,2], Mohamed Ebeed [3], Juan Yu [2] and Francisco Jurado [4,*]

[1] Department of Electrical Engineering, Faculty of Engineering, Aswan University, Aswan 81542, Egypt; mostafaabdu777@yahoo.com (M.A.); skamel@aswu.edu.eg (S.K.)
[2] State Key Laboratory of Power Transmission Equipment and System Security and New Technology, Chongqing University, Chongqing 400030, China; dianlixi@cqu.edu.cn
[3] Department of Electrical Engineering, Faculty of Engineering, Sohag University, Sohag 82524, Egypt; mebeed@eng.sohag.com
[4] Department of Electrical Engineering, University of Jaén, EPS Linares, 23700 Jaén, Spain
* Correspondence: fjurado@ujaen.es; Tel.: +34-953-648518; Fax: +34-953-648586

Received: 13 June 2018; Accepted: 26 June 2018; Published: 28 June 2018

Abstract: The optimal power flow (OPF) problem is a non-linear and non-smooth optimization problem. OPF problem is a complicated optimization problem, especially when considering the system constraints. This paper proposes a new enhanced version for the grey wolf optimization technique called Developed Grey Wolf Optimizer (DGWO) to solve the optimal power flow (OPF) problem by an efficient way. Although the GWO is an efficient technique, it may be prone to stagnate at local optima for some cases due to the insufficient diversity of wolves, hence the DGWO algorithm is proposed for improving the search capabilities of this optimizer. The DGWO is based on enhancing the exploration process by applying a random mutation to increase the diversity of population, while an exploitation process is enhanced by updating the position of populations in spiral path around the best solution. An adaptive operator is employed in DGWO to find a balance between the exploration and exploitation phases during the iterative process. The considered objective functions are quadratic fuel cost minimization, piecewise quadratic cost minimization, and quadratic fuel cost minimization considering the valve point effect. The DGWO is validated using the standard IEEE 30-bus test system. The obtained results showed the effectiveness and superiority of DGWO for solving the OPF problem compared with the other well-known meta-heuristic techniques.

Keywords: power system optimization; optimal power flow; developed grew wolf optimizer

1. Introduction

Recently, OPF problems have become a strenuous task for optimal operation of the power systems. The main objective of OPF is finding the best operation, security and economic settings of electrical power systems. In this study, the operating variables of systems are determined optimally for different objective functions such as fuel cost minimization, power loss minimization, emission and voltage deviation minimization, etc., while in addition, enhancing system stability, loadability and voltage profiles. Practically, the solution of OPF problem must satisfy the equality and inequality system constraints [1,2].

OPF is a non-smooth and non-linear optimization problem that is considered a complicated problem. This problem becomes especially more difficult when the equality and inequality operating system constraints are considered. Thus, solving the OPF problem needs more efficient and developed meta-heuristic optimization algorithms. Many conventional methods have been developed in order to solve the OPF problem such as NLP [3], LP [4], QP [5], Newton's Method [6], IP [7]. However, these methods face some problems in solving nonlinear or non-convex objective functions. In addition, these methods may fall into local minima, hence new optimization algorithms have been proposed to avoid the shortcomings of these methods. From these methods; GA [8,9], MFO [10], DE [11,12], PSO [13], MSA [14], EP [15,16], ABC [17], GSA [18], BBO [19], SFLA [20], forced initialized differential evolution algorithm [21], TS [22], MDE [23], SOS [24], BSA [25] and TLBO [26], decentralized decision-making algorithm [27]. The thermal generation units have multiple valves to control the output generated power. As the valves of thermal generation units are opened in case of steam admission, a sudden increase in losses is observed which leads to ripples in the cost function curve (known as the valve-point loading effect). Several optimization techniques have been employed for solving the OPF considering the valve-point loading effect such as ABC [17], GSA [18], SFLA [20], SOS [24], BSA [25] and Hybrid Particle Swarm Optimization and Differential Evolution [28].

The conventional and some meta-heuristics methods could not efficiently solve the OPF problem, thus several new or modified versions of optimization techniques have been proposed. The GWO algorithm is considered a new optimization technique that proposed by Mirjalili [29]. GWO simulates the grey wolves' social hierarchy and hunting behavior. The main phases of gray wolf hunting are the approaching, encircling and attacking the prey by the grey wolves [29,30]. It should point out that the conventional GWO technique updates its hunters towards the prey based on the condition of leader wolves. However, the population of GWO is still inclined to stall in local optima in some cases. In addition, the GWO technique is not capable of performing a seamless transition from the exploration to exploitation phases. In this paper, a new developed version of GWO is proposed to effectively solve the OPF problem. The DGWO is based on enhancing the exploration phase by applying a random mutation in order to enhance the searching process and avoid the stagnation at local optima. The exploitation process is improved by updating the populations of GWO in spiral path around the best solution to focus on the most promising regions. DGWO is applied for minimizing the quadratic fuel cost, fuel cost considering the valve loading. The obtained simulation results by the DGWO are compared with those obtained by the classical GWO and other well-known techniques to demonstrate the effectiveness of the proposed algorithm.

The rest of paper is organized as follows: Section 2 presents the optimal power flow problem formulation. Section 3 presents the mathematical formulation of GWO and DGWO techniques. Section 4 presents the numerical results. Finally, the conclusions presented in Section 5.

2. Optimal Power Flow Formulation

Solution of OPF problem aims to achieve certain objective functions by adjustment some control variables with satisfying different operating constraints. Generally, the optimization problem can be mathematically represented as:

$$Min \quad F(x,u) \tag{1}$$

Subject to:

$$g_j(x,u) = 0 \quad j = 1,2,\ldots,m \tag{2}$$

$$h_j(x,u) \leq 0 \quad j = 1,2,\ldots,p \tag{3}$$

where, F is a certain objective function, x are the state variables, u is the control variables vector, g_j and h_j are equality and inequality operating constraints, respectively. m and p are the number of the equality and inequality operating constraints, respectively. The state variables vector (x) can be given as:

$$x = [P_{G1}, V_{L1} \ldots V_{LNPQ}, Q_{G1} \ldots Q_{GNPV}, S_{TL1} \ldots S_{TLNTL}] \tag{4}$$

where, P_{G1} is the generated power of slack bus, V_L is the load bus voltage, Q_G is the generated reactive power, S_{TL} is the power flow in the line, NPQ is the load buses number, NPV is the generated buses number and NTL is the lines number. The independent variables u can be given as:

$$u = [P_{G2} \ldots P_{GNG}, V_{G1} \ldots V_{GNG}, Q_{C1} \ldots Q_{CNC}, T_1 \ldots T_{NT}] \tag{5}$$

where, P_G is the generated active power, V_G is the generated voltage, Q_C is the shunt compensator injected reactive power, T is the transformer tap setting, NG is the generators number, NC is the shunt compensator units and NT is the transformers number.

2.1. Objective Functions

2.1.1. Quadratic Fuel Cost

The first objective function is the quadratic equation of total generation fuel cost which formulated as follows:

$$F_1 = \sum_{i=1}^{NPV} F_i(P_{Gi}) = \sum_{i=1}^{NPV} \left(a_i + b_i P_{Gi} + c_i P_{Gi}^2 \right) \tag{6}$$

where, F_i is the fuel cost. a_i, b_i and c_i are the cost coefficients.

2.1.2. Quadratic Cost with Valve-Point Effect and Prohibited Zones

Practically, the effect of valve point loading for thermal power plants should be considered. This effect occurred as a result of the rippling influence on the unit's cost curve which produced from each steam admission in the turbine as shown in Figure 1.

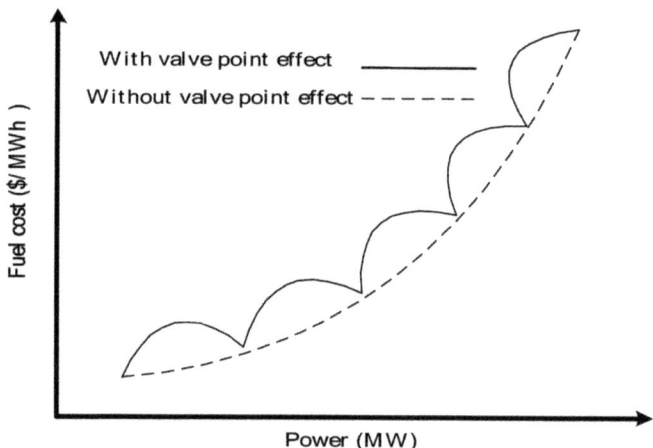

Figure 1. Cost function with and without valve point effect.

The valve point loading effect is considered by adding a sine term to the fuel cost as:

$$F(x,u) = \sum_{i=1}^{NPV} F_i(P_{Gi}) = \sum_{i=1}^{NPV} \left(a_i + b_i P_{Gi} + c_i P^2{}_{Gi} \right) + \left| d_i \sin \left(e_i \left(P_{Gi}^{min} - P_{Gi} \right) \right) \right| \tag{7}$$

where, d_i and e_i are the fuel cost coefficients considering the valve-point effects.

2.1.3. Piecewise Quadratic Cost Functions

Due to the different fuel sources (coal, natural gas and oil), their fuel cost functions can be considered as a non-convex problem which is given as:

$$F(P_{Gi}) = \begin{cases} a_{i1} + b_{i1}P_{Gi} + c_{i1}P_{Gi}^2 & P_{Gi}^{min} \leq P_{Gi} \leq P_{G1} \\ a_{i2} + b_iP_{Gi} + c_iP_{Gi}^2 & P_{G1} \leq P_{Gi} \leq P_{G2} \\ \cdots \\ a_{ik} + b_{ik}P_{Gi} + c_{ik}P_{Gi}^2 & P_{Gi\,k-1} \leq P_{Gi} \leq P_{Gi}^{max} \end{cases} \tag{8}$$

where, a_{ik}, b_{ik} and c_{ik} are cost coefficients of the ith generator for fuel type k.

2.2. Operating Constraints

2.2.1. Equality Operating Constraints

The operating equality constrains can be represented as:

$$P_{Gi} - P_{Di} = |V_i| \sum_{j=1}^{NB} |V_j| \left(G_{ij} \cos \delta_{ij} + B_{ij} \sin \delta_{ij} \right) \tag{9}$$

$$Q_{Gi} - Q_{Di} = |V_i| \sum_{j=1}^{NB} |V_j| \left(G_{ij} \cos \delta_{ij} + B_{ij} \sin \delta_{ij} \right) \tag{10}$$

where, P_{Gi} and Q_{Gi} are the generated power at bus i. P_{Di} and Q_{Di} are load demand at bus i. G_{ij} and B_{ij} are the real and imaginary parts of admittance between bus i and bus j, respectively.

2.2.2. Inequality Operating Constrains

The inequality operating constrains can be given as:

$$P_{Gi}^{min} \leq P_{Gi} \leq P_{Gi}^{max} \quad i = 1, 2, \ldots, NG \tag{11}$$

$$V_{Gi}^{min} \leq V_{Gi} \leq V_{Gi}^{max} \quad i = 1, 2, \ldots, NG \tag{12}$$

$$Q_{Gi}^{min} \leq Q_{Gi} \leq Q_{Gi}^{max} \quad i = 1, 2, \ldots, NG \tag{13}$$

$$T_i^{min} \leq T_i \leq T_i^{max} \quad i = 1, 2, \ldots, NT \tag{14}$$

$$Q_{Ci}^{min} \leq Q_{Ci} \leq Q_{Ci}^{max} \quad i = 1, 2, \ldots, NC \tag{15}$$

$$S_{Li} \leq S_{Li}^{min} \quad i = 1, 2, \ldots, NTL \tag{16}$$

$$V_{Li}^{min} \leq V_{Li} \leq V_{Li}^{max} \quad i = 1, 2, \ldots, NPQ \tag{17}$$

where, P_{Gi}^{min} and P_{Gi}^{max} are the minimum and maximum generated active power limits of ith generator, respectively. V_{Gi}^{min} and V_{Gi}^{max} are the lower and upper output voltage limits of ith generator, respectively. Q_{Gi}^{min} and Q_{Gi}^{max} are the minimum and maximum generated reactive power limits of ith generator, respectively. T_i^{min} and T_i^{max} are the lower and upper limits of regulating transformer i. Q_{Ci}^{min} and Q_{Ci}^{max} are the minimum and maximum injected VAR of ith shunt compensation unit. S_{Li} is the apparent power flow in ith line while S_{Li}^{min} is the maximum apparent power flow of this line. V_{Li}^{min} and V_{Li}^{max} are the lower and upper limits of voltage magnitude load bus i, respectively.

The dependent state variables can be considered in OPF solution using the quadratic penalty formulation as:

$$F_g(x,u) = F_i(x,u) + K_G\left(P_{G1} - P_{G1}^{lim}\right)^2 + K_Q \sum_{i=1}^{NPV}\left(Q_{Gi} - Q_{Gi}^{lim}\right)^2 + K_V \sum_{i=1}^{NPQ}\left(V_{Li} - V_{Li}^{lim}\right)^2$$
$$+ K_S \sum_{i=1}^{NTL}\left(S_{Li} - S_{Li}^{lim}\right)^2 \qquad (18)$$

where, K_G, K_Q, K_V, K_S and K_S are the penalty factors. x^{lim} is the limit value that can be given as:

$$x^{lim} = \begin{cases} x^{max}; & x > x^{max} \\ x^{min}; & x < x^{min} \end{cases} \qquad (19)$$

where, x^{max} and x^{min} are the upper and lower limits of the dependent variables, respectively.

3. Developed Grey Wolf Optimizer

3.1. Grey Wolf Optimizer

GWO is a robust swarm-based optimizer inspired by the social hierarchy of grey wolves [27]. The pack of grey wolves has a special social hierarchy where the leadership in the pack can be divided into four levels; alpha, beta, omega and delta. Alpha wolf (α) is the first level in the social hierarchy hence it is the leader that guides the pack and the other wolves follow its orders. Beta wolf (β) is being in the second level of leadership that helps the alpha wolf directly for the activities of the pack. Delta (δ) wolves come in the third level of hierarchy where, they follow α and β wolves. The rest of wolves are the omegas (ω) that always have to submit to all the other dominant wolves. Figure 2 illustrates the social hierarchy ranking of wolves in GWO. In the mathematical model of GWO, the fittest solution is considered as the alpha (α), where, the second and third best solutions are called beta (β) and delta (δ), respectively. Finally, omega (ω) are considered the rest of the candidate solutions. However, the GWO based on three steps:

A.   Encircling prey.
B.   Hunting the prey.
C.   Attacking the prey.

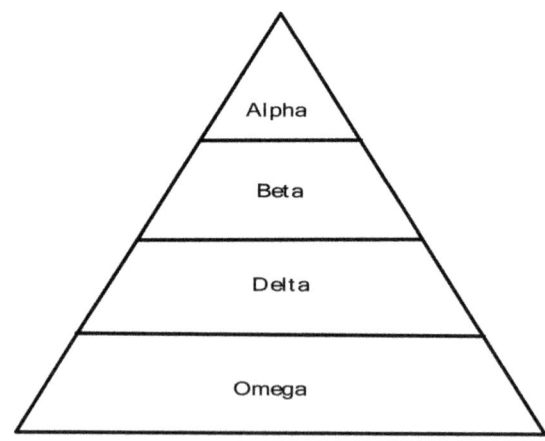

Figure 2. Social hierarchy of wolves in GWO.

3.1.1. Encircling Prey

The grey wolves encircle the prey in hunting process that can be mathematically modeled as:

$$D = \left| C \times X_{p_{(i,j)}}(t) - X_{(i,j)}(t) \right| \tag{20}$$

$$X_{(i,j)}(t+1) = X_{p_{(i,j)}}(t) - A \times D \tag{21}$$

where, t is the current iteration, X_p is the prey position vector, and X indicates the position vector of a grey wolf. A and C are coefficient vectors that can be calculated as:

$$A = 2a \times r_1 - a \tag{22}$$

$$C = 2 \times r_2 \tag{23}$$

where, a is a value can be decreased linearly from 2 to 0 with iterations. r_1 and r_2 are random numbers in range [0, 1].

3.1.2. Hunting the Prey

In hunting process, the pack is affected by α, β and δ. Hence, the first three best solutions are saved as best agents (α, β, δ) and the other search agents are updated their positions according to the best agents as:

$$D = \left| C \times X_{p_{(i,j)}}(t) - X_{(i,j)}(t) \right| \tag{24}$$

$$D_\alpha = \left| C_1 \times X_{\alpha_{(i,j)}} - X_{(i,j)} \right| \tag{25}$$

$$D_\beta = \left| C_2 \times X_{\beta_{(i,j)}} - X_{(i,j)} \right| \tag{26}$$

$$D_\delta = \left| C_3 \times X_{\delta_{(i,j)}} - X_{(i,j)} \right| \tag{27}$$

$$X1_{(i,j)} = X_{\alpha_{(i,j)}} - A_1 \times (D_\alpha) \tag{28}$$

$$X2_{(i,j)} = X_{\beta_{(i,j)}} - A_2 \times (D_\beta) \tag{29}$$

$$X3_{(i,j)} = X_{\beta_{(i,j)}} - A_3 \times (D_\delta) \tag{30}$$

$$X_{(i,j)}(t+1) = \frac{X1_{(i,j)} + X2_{(i,j)} + X3_{(i,j)}}{3} \tag{31}$$

where, i is number of populations (vectors) and j is number of variables (individuals). A_1, A_2 and A_3 are random vectors. The step size of the ω wolves is expressed in Equations (25)–(27), respectively. The final location of the ω wolves is formulated in Equations (28)–(31).

3.1.3. Attacking the Prey

The last stage in hunting is attacking the prey when the prey stopped. This can be achieved mathematically by reducing the value of a gradually from 2 to 0, consequently, A is varied randomly in range [−1, 1].

3.2. Developed Grey Wolf Optimizer

DGWO technique is presented as a new version for the conventional GWO. In this technique, the exploration and exploitation processes of GWO is enhanced. The exploration process is enhanced by integration a random mutation to find new searching regions to avoid the local minimum problem. The random mutation is applied as follows:

$$X_{(i,j)}^{new} = L_{(i,j)} + R\left(U_{(i,j)} - L_{(i,j)}\right) \tag{32}$$

where, R is a random number over [0, 1]. $X_{(i,j)}^{new}$ is a new generated vector. L and U are the lower and upper limits of control variables, respectively. In the exploitation of DGWO, the search process is focusing on the promising area by updating the search agents around the best solution $(X_{\alpha_{(i,j)}})$ in logarithmic spiral function as:

$$X_{(i,j)}^{new} = \left| X_{(i,j)}(t) - X_{\alpha_{(i,j)}}(t) \right| \times e^{bt} \cos(2\pi q) + X_{\alpha_{(i,j)}}(t) \tag{33}$$

where:

$X_{\alpha_{(i,j)}}$: the best position (alpha wolf position).
b: is a constant value for defining the logarithmic spiral shape.
q: is a random number $[-1, 1]$.

For balancing the exploration during the initial searching process and exploitation in the final stages of the search process, an adaptive operator is used which changed dynamically as:

$$K(t) = K_{min} + \frac{K_{max} - K_{min}}{T_{max}} \times t \tag{34}$$

The procedures of DGWO algorithm for solving the OPF problem can be summarized as follows:

(1) Initialize maximum number of iterations (T_{max}) and search agents (N).
(2) Read the input system data.
(3) Initialize grey wolf population X as:

$$X_n = x_n^{min} + rand(0,1)\left(x_n^{max} - x_n^{min}\right) \tag{35}$$

where, $n = 1, 2, 3 \ldots, j$, x_n^{min} and x_n^{max} are the minimum and maximum limits of control variables which are predefined values. *rand* is a random number in range [0, 1].

(4) Calculate the objective function for all grey wolf population using Newton Raphson load flow method.
(5) Determine $X_{\alpha_{(i,j)}}$, $X_{\beta_{(i,j)}}$, $X_{\delta_{(i,j)}}$ (first, second, and third best search agent).
(6) Update the location of each search agent according Equations (24)–(31) and calculate the objective function using Newton Raphson load flow for the updated agents.
(7) Update the values of a [2:0], A and C according Equations (22) and (23).
(8) Update the adaptive operator, K according to Equation (34)
(9) **IF** $K < rand$, update the position of search agent based on random mutation according to Equation (32) **ELSE IF** $K > rand$, update the position of search agent locally in spiral path using Equation (33) **END IF** Fitness $(X_{(i,j)}^{new}) <$ Fitness $(X_{(i,j)})$

$$X_{(i,j)} = X_{(i,j)}^{new}$$

ELSE, END where, Fitness $\left(X_{(i,j)}\right)$ is the objective function of the position vector n while Fitness $(X_{(i,j)}^{new})$ is the objective function of the updated position vector j.

(10) Repeat steps from (4) to (9) until the iteration number equals to its maximum value.

(11) Find the best vector ($X_{\alpha_{(i,j)}}$) which include the system control variables and its related fitness function.

However, the OPF solution process using the DGWO is shown in Figure 3.

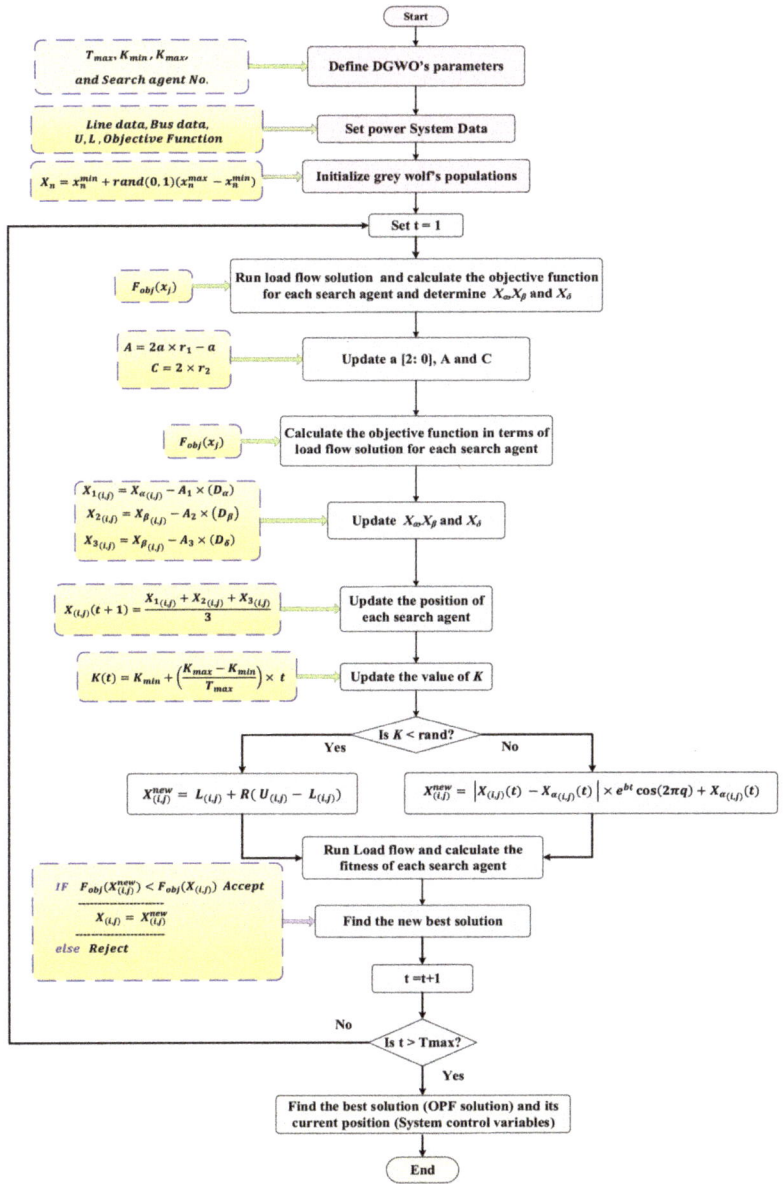

Figure 3. The solution process of OPF problem using DGWO.

4. Simulation Results

The DGWO is validated using the IEEE 30-bus test system. More details about this system can be found in [31]. The developed code has been written using MATLAB 2015 and the simulation run on a PC equipped with a core i5 processor, 2.50 GHz and 4 GB RAM. The upper and lower operating ranges and coefficients of generators are given in Table 1. The upper and lower limits of the load bus voltage are 1.05 p.u. and 0.95 p.u., respectively. The upper and lower limits of VAR compensation units are 0.00 p.u. and 0.05 p.u., respectively. The working voltage ranges of PV buses is [0.95, 1.1] p.u while the allowable range of transformer taps is [0.9, 1.1].The limits of transmission line power flows are given in [24]. The parameters of DGWO technique are selected as; number of populations = 50, maximum iteration = 100, $b = 1$, $K_{min} = 0.00001$ and $K_{max} = 0.1$. In this study, 100 runs have been performed for all the test cases to calculate the best cost, the worst cost and the average cost.

Table 1. Generator data coefficients.

Bus No.	P_G^{max} (MW)	P_G^{min} (MW)	Q_G^{min} (MVar)	Cost Coefficients			Prohibited Zones
				a	b	c	
1	250	50	−20	0	2.0	0.00375	(55–66), (80–120)
2	80	20	−20	0	1.75	0.0175	(21–24), (45–55)
5	50	15	−15	0	1.0	0.0625	(30–36)
8	35	10	−15	0	3.25	0.00834	(25–30)
11	30	10	−10	0	3.00	0.025	(25–28)
13	40	12	−15	0	3.00	0.025	(24–30)

4.1. Case1: OPF Solution without Considering the Valve Point Effects

In this case, the quadratic fuel cost effect is taken as an objective function to be minimized as given in Equation (6). The generator data for this case are listed in Table 1. The optimal control variables for this case obtained by GWO and DGWO techniques are listed in 4th and 5th columns of Table 2, respectively. The obtained fuel cost using GWO and DGWO are 801.259 $/h and 800.433 $/h, respectively. Table 3 gives the fuel costs obtained by GWO, DGWO and other optimization techniques. From Table 3, it can be observed that the obtained results using DGWO are better than those obtained by the others reported optimization techniques in terms of the best, the worst and the average fuel costs. The convergence characteristics of GWO and DGWO for this case are shown in Figure 4. It is clear that DGWO has stable and rapid convergence characteristic.

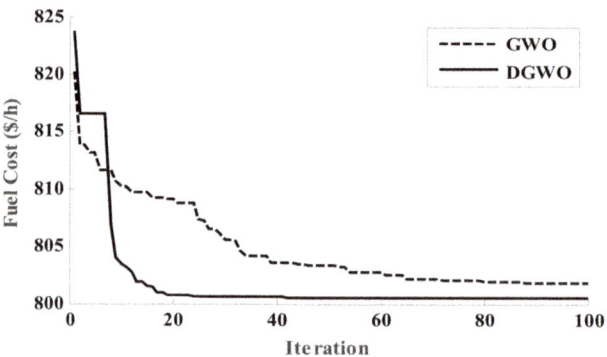

Figure 4. Convergence characteristics of fuel cost (Case 1).

Table 2. Optimal control variables for different cases obtained by GWO and DGWO.

Variables	Limit		Case 1		Case 2		Case 3	
	Min.	Max.	GWO	DGWO	GWO	DGWO	GWO	DGWO
P1 (MW)	50	250	171.094	176.949	212.633	219.801	140.00	140.00
P2 (MW)	20	80	48.615	48.519	25.684	28.358	54.992	55.000
P5 (MW)	15	50	21.123	21.326	17.612	15.047	34.930	24.105
P8 (MW)	10	35	22.068	21.571	14.185	10.000	25.008	35.000
P11 (MW)	10	30	15.479	12.026	10.651	10.000	16.934	18.239
P13 (MW)	12	40	13.665	12.001	13.751	12.000	18.223	17.664
V1 (p.u)	0.95	1.1	1.080	1.083	1.087	1.090	1.077	1.073
V2 (p.u)	0.95	1.1	1.062	1.063	1.062	1.065	1.064	1.060
V5 (p.u)	0.95	1.1	1.030	1.031	1.023	1.032	1.035	1.032
V8 (p.u)	0.95	1.1	1.036	1.035	1.035	1.035	1.044	1.040
V11 (p.u)	0.95	1.1	1.080	1.060	1.051	1.099	1.062	1.049
V13 (p.u)	0.95	1.1	1.054	1.050	1.060	1.037	1.036	1.060
T11	0.90	1.1	0.982	0.977	1.0128	0.948	1.023	0.994
T12	0.90	1.1	1.026	1.013	0.908	1.025	1.008	0.978
T15	0.90	1.1	0.989	0.934	0.986	0.970	1.019	0.971
T36	0.90	1.1	0.981	0.975	0.976	0.981	0.959	0.975
Q10 (MVar)	0.00	5.00	2.144	1.695	3.170	3.277	0.986	1.251
Q12 (MVar)	0.00	5.00	2.929	3.394	2.143	2.367	3.996	3.157
Q15 (MVar)	0.00	5.00	1.400	4.777	1.959	1.228	2.978	2.433
Q17 (MVar)	0.00	5.00	3.526	4.153	1.126	4.660	2.148	4.831
Q20 (MVar)	0.00	5.00	2.954	3.738	2.369	3.585	4.139	4.462
Q21 (MVar)	0.00	5.00	3.588	4.941	2.016	3.603	2.878	4.653
Q23 (MVar)	0.00	5.00	2.974	3.567	1.532	3.560	3.603	3.043
Q24 (MVar)	0.00	5.00	3.688	4.996	1.675	4.603	1.377	4.467
Q29 (MVar)	0.00	5.00	3.259	2.200	2.378	3.232	3.628	2.439
PLoss(MW)	NA	NA	8.6428	8.9921	11.1151	11.805	6.6860	6.6079
VD (p.u)	NA	NA	0.7285	0.8784	0.7055	0.8589	0.6170	0.8825
Lmax (p.u)	NA	NA	0.1299	0.1279	0.1328	0.1281	0.1307	0.1280
Fuelcost ($/h)	NA	NA	801.259	800.433	830.028	824.132	646.426	645.913
Computational time (s)	NA	NA	53.6	37.8	41.70	41.5	52.4	47.2

PLoss: Power losses, Lmax: Voltage stability index, VD: Summation voltage deviations.

Table 3. Simulation results of Case 1.

Algorithm	Best Cost	Average Cost	Worst Cost
DGWO	800.433	800.4674	800.4989
GWO	801.259	802.663	804.898
MSA [14]	800.5099	NA	NA
SOS [24]	801.5733	801.7251	801.8821
ABC [17]	800.6600	800.8715	801.8674
TS [22]	802.290	NA	NA
MDE [23]	802.376	802.382	802.404
IEP [15]	802.465	802.521	802.581
TS [15]	802.502	802.632	802.746
EP [16]	802.62	803.51	805.61
TS/SA [15]	802.788	803.032	803.291
EP [15]	802.907	803.232	803.474
ITS [15]	804.556	805.812	806.856
GA [9]	805.937	NA	NA

4.2. Case 2: OPF Solution Considering the Valve Point Effects

In this case, the OPF problem is solved considering the valve point effect as given in Equation (7). The optimal control variables obtained by the DGWO are given in 6th and 7th columns of Table 2, respectively. The minimum fuel costs obtained by GWO and DGWO are 830.028 $/h and 824.132 $/h, respectively. Table 4 gives the fuel costs obtained by DGWO, GWO, and other techniques under the same conditions (control variable boundaries, dependent variables limits and system constraints).

From Table 4, it can be observed that the obtained results from DGWO are better than those obtained by GWO and the other techniques. Figure 5 shows the convergence characteristics of the minimum fuel cost of the GWO and DGWO. From this figures, it can be observed that the DGWO is converged faster than GWO.

Table 2 gives the active power losses, voltage stability index and summation of voltage deviations. From this table, it can be observed that some values are increased for DGWO compared with GWO, this due to these values are not considered as objective functions. As it is well known that the optimization of single objective function probably not lead to enhance the other functions.

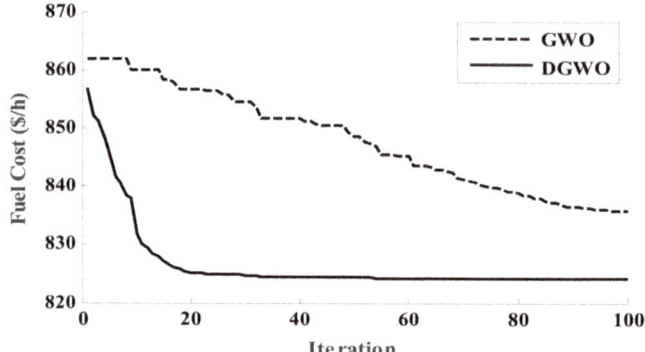

Figure 5. Convergence characteristics of fuel cost (Case 2).

Table 4. Comparison of the simulation results of Case 2.

Algorithm	Best Cost	Average Cost	Worst Cost
DGWO	824.132	824.295	824.663
GWO	830.028	844.639	852.388
SOS [24]	825.2985	825.4039	825.5275
BSA [25]	825.23	827.69	830.15
SFLA-SA [20]	825.6921	NA	NA
SFLA [20]	825.9906	NA	NA
PSO [20]	826.5897	NA	NA
SA [20]	827.8262	NA	NA

4.3. Case 3: OPF Solution Considering Piecewise Quadratic Fuel Cost Function

In this case, piecewise fuel cost function is taken as an objective function as given in Equation (8). In this case, two generation units at buses 1 and 2 are represented by piecewise quadratic cost functions [16]. The generated active power and the generation unit coefficients for this case are given in Table 5. The optimal control variables obtained by GWO and DGWO are listed in 8th and 9th columns of Table 2, respectively. The minimum piecewise fuel costs obtained by GWO and DGWO are 646.426 $/h and 645.913 $/h, respectively. The piecewise fuel costs obtained by DGWO, GWO, and other techniques given in Table 6. From Table 6, it can be observed that the obtained results from DGWO are better than those obtained by GWO and the other techniques in terms of the best, the worst and the average piecewise fuel costs. Figure 6 shows the convergence characteristics of the minimum fuel cost of the GWO and DGWO for this case. It is clear that DGWO has fast and stable convergence characteristic compared with GWO.

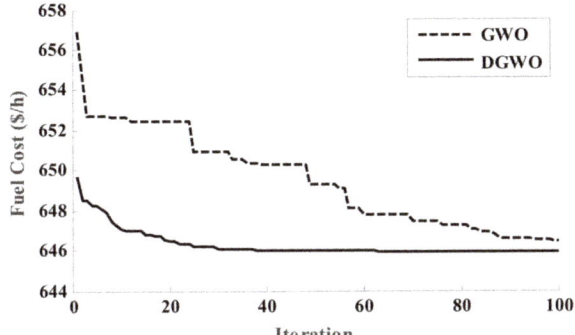

Figure 6. Convergence characteristics of fuel cost (Case 3).

Table 5. Cost coefficients of generators (Case 3).

Bus No.	Output Power Limit (MW)		Cost Coefficients		
	Min.	Max.	a	b	c
1	50	140	55.0	0.70	0.0050
	140	200	82.5	1.05	0.0075
2	20	55	40.0	0.30	0.0100
	55	80	80.0	0.60	0.0200

Table 6. Comparison of the simulation results of Case 3.

Algorithm	Best Cost	Average Cost	Worst Cost
DGWO	645.9132	645.993	646.095
GWO	646.426	647.432	648.681
GSA [18]	646.8480	646.8962	646.9381
Lévy LTLBO [26]	647.4315	647.4725	647.8638
PSO [13]	647.69	647.73	647.87
BBO [19]	647.7437	647.7645	647.7928
TLBO [26]	647.8125	647.8335	647.8415
MDE [23]	647.846	648.356	650.664
ABC [17]	649.0855	654.0784	659.7708
EP [16]	650.206	654.501	657.120
TS [15]	651.246	654.087	658.911
TS/SA [15]	654.378	658.234	662.616
ITS [15]	654.874	664.473	675.035

5. Conclusions

In this paper, DGWO has been proposed to efficiently solve the OPF problem and avoid the stagnation problems of the traditional GWO. This technique is based on modifying the grey wolf optimizer by employing a random mutation for enhancing its exploration process. This modification provides a flexibility to search in new areas. Moreover, the new generated populations are updated around the best solution in a spiral path to enhance the exploitation process and focus on the most promising areas. In the proposed technique, two equations should be added to the traditional GWO, the first equation is related to the random mutation and the second one for the spiral path updating process. The results obtained by the proposed algorithm have been compared with those obtained by the conventional GWO and other well-known optimization techniques. From the results obtained, it can be concluded that:

Energies **2018**, *11*, 1692

- The proposed technique has successfully performed to find the optimal settings of the control variables of test system.
- Different objective functions (quadratic fuel cost minimization, piecewise quadratic cost minimization, and quadratic fuel cost minimization considering the valve point effect) have been achieved using the proposed algorithm.
- The superiority of DGWO compared with the conventional GWO and other well-known optimization techniques has been proved.
- DGWO has a fast and stable convergence characteristic compared with the conventional GWO.

In the future work, the proposed algorithm will be applied in other planning and expansion studies in power systems with thermal and renewable generation units considering the uncertainties of load.

Author Contributions: M.A., S.K. and M.E. proposed the idea, obtained the results, and wrote the paper. J.Y. and F.J. contributed by drafting and critical revisions. All authors together organized and refined the manuscript in the present form.

Funding: This work was supported in part by the Project Supported by the National Key Research and Development Program of China under grant 2017YFB0902200, the Basic and Frontier Research Project of Chongqing under grant cstc2017jcyjBX0056, the Academician Lead Science and Technology Innovation Guide Project of Chongqing under grant cstc2017jcyj-yszxX0011 and the Fundamental Research Funds for the Central Universities 2018CDQYDQ0006.

Conflicts of Interest: The authors declare no conflict of interest.

Nomenclature

ABC	Artificial bee colony algorithm
BSA	Backtracking search algorithm
DGWO	Developed grey wolf optimizer
GA	Genetic algorithm
GWO	Grey wolf optimizer
LP	Linear programming
MSA	Moth swarm algorithm
OPF	Optimal power flow
QP	Quadratic programming
TS	Tabu search
MFO	Moth-flame algorithm
ITS	Improved Tabu Search
A_1, A_2, A_3	Random vectors
x	The state variables vector
L, U	The lower and upper boundary of control variables
Q_G	The reactive power output of generators
t	The current iteration
T_{max}	The maximum number of iterations
P_{Di}, Q_{Di}	The active and reactive load demand at bus i
δ_{ij}	Phase difference of voltages
V_L	The voltage of load bus
V_G	The voltage of generation bus
NPQ	Number of load buses
d_i, e_i	The fuel cost coefficients of the ith generator unit with valve-point effects
NTL	Number of transmission lines
R	Random number

rand	Random value
G	Transmission line conductance
B	Transmission line susceptance
X_p	The prey position vector
k	Adaptive operator
b	Constant value
K_G, K_Q, K_V, K_S, K_S	Penalty factors
$X_\alpha, X_\beta, X_\delta$	First, second, and third best search agents
max, min	Superscript refers to maximum and minimum values
BBO	Biogeography-based optimization
DE	Differential evolution
EP	Evolutionary programming
GSA	Gravitational search algorithm
MDE	Modified differentia evolution
NLP	Nonlinear programming
PSO	Particle swarm optimization
SFLA	Shuffle frog leaping algorithm
SOS	Symbiotic organisms search
TLBO	Teaching–learning-based optimization
IP	Interior point
F	The objective function
g_i, h_j	The equality and inequality constraints
u	The control variables vector
m, p	Number of equality and inequality constraints
Q_C	The injected reactive power of shunt compensator
P_{G1}	The generated power of slack bus
P_G	The output active power of generator
S_L	The apparent power flow in transmission line
T	Tap setting of transformer
NG	Number of generators
NC	Number of shunt compensator
NT	Number of transformers
NPV	Number of generators PV buses
a_i, b_i, c_i	The cost coefficients of ith generator.
NPV	Number of generation buses
I	Current
V	Magnitude of node voltage
R, X, Z	Resistance, reactance, impedance
P, Q, S	Active, reactive, apparent powers
X	The location of the present solution
q	A random number
X^{new}	New generated vector
$\alpha, \beta, \delta, \omega$	Alpha, beta, delta, omega fittest solutions
C, C_1, C_2, C_3	Random vectors

References

1. Lee, K.; Park, Y.; Ortiz, J. A united approach to optimal real and reactive power dispatch. *IEEE Trans. Power Appar. Syst.* **1985**, *PER-5*, 1147–1153. [CrossRef]
2. Zhu, J. *Optimization of Power System Operation*; John Wiley & Sons: Hoboken, NJ, USA, 2015; Volume 47.
3. Dommel, H.W.; Tinney, W.F. Optimal power flow solutions. *IEEE Trans. Power Appar. Syst.* **1968**, *PAS-87*, 1866–1876. [CrossRef]
4. Mota-Palomino, R.; Quintana, V. Sparse reactive power scheduling by a penalty function-linear programming technique. *IEEE Trans. Power Syst.* **1986**, *1*, 31–39. [CrossRef]

5. Burchett, R.; Happ, H.; Vierath, D. Quadratically convergent optimal power flow. *IEEE Trans. Power Appar. Syst.* **1984**, *PAS-103*, 3267–3275. [CrossRef]
6. Santos, A.J.; da Costa, G. Optimal-power-flow solution by Newton's method applied to an augmented Lagrangian function. *IEE Proc. Gener. Transm. Distrib.* **1995**, *142*, 33–36. [CrossRef]
7. Yan, X.; Quintana, V.H. Improving an interior-point-based OPF by dynamic adjustments of step sizes and tolerances. *IEEE Trans. Power Syst.* **1999**, *14*, 709–717.
8. Chung, T.; Li, Y. A hybrid GA approach for OPF with consideration of FACTS devices. *IEEE Power Eng. Rev.* **2000**, *20*, 54–57. [CrossRef]
9. Paranjothi, S.; Anburaja, K. Optimal power flow using refined genetic algorithm. *Electr. Power Compon. Syst.* **2002**, *30*, 1055–1063. [CrossRef]
10. Ebeed, M.; Kamel, S.; Youssef, H. Optimal setting of STATCOM based on voltage stability improvement and power loss minimization using Moth-Flame algorithm. In Proceedings of the 2016 Eighteenth International Middle East Power Systems Conference (MEPCON), Cairo, Egypt, 27–29 December 2016; pp. 815–820.
11. Varadarajan, M.; Swarup, K.S. Solving multi-objective optimal power flow using differential evolution. *IET Gener. Transm. Distrib.* **2008**, *2*, 720–730. [CrossRef]
12. Abou El Ela, A.A.; Abido, M.; Spea, S. Optimal power flow using differential evolution algorithm. *Electr. Power Syst. Res.* **2010**, *80*, 878–885. [CrossRef]
13. Abido, M. Optimal power flow using particle swarm optimization. *Int. J. Electr. Power Energy Syst.* **2002**, *24*, 563–571. [CrossRef]
14. Mohamed, A.-A.A.; Mohamed, Y.S.; El-Gaafary, A.A.; Hemeida, A.M. Optimal power flow using moth swarm algorithm. *Electr. Power Syst. Res.* **2017**, *142*, 190–206. [CrossRef]
15. Ongsakul, W.; Tantimaporn, T. Optimal power flow by improved evolutionary programming. *Electr. Power Compon. Syst.* **2006**, *34*, 79–95. [CrossRef]
16. Yuryevich, J.; Wong, K.P. Evolutionary programming based optimal power flow algorithm. *IEEE Trans. Power Syst.* **1999**, *14*, 1245–1250. [CrossRef]
17. Adaryani, M.R.; Karami, A. Artificial bee colony algorithm for solving multi-objective optimal power flow problem. *Int. J. Electr. Power Energy Syst.* **2013**, *53*, 219–230. [CrossRef]
18. Duman, S.; Güvenç, U.; Sönmez, Y.; Yörükeren, N. Optimal power flow using gravitational search algorithm. *Energy Convers. Manag.* **2012**, *59*, 86–95. [CrossRef]
19. Bhattacharya, A.; Chattopadhyay, P. Application of biogeography-based optimisation to solve different optimal power flow problems. *IET Gener. Transm. Distrib.* **2011**, *5*, 70–80. [CrossRef]
20. Niknam, T.; Narimani, M.R.; Azizipanah-Abarghooee, R. A new hybrid algorithm for optimal power flow considering prohibited zones and valve point effect. *Energy Convers. Manag.* **2012**, *58*, 197–206. [CrossRef]
21. Shaheen, A.M.; El-Sehiemy, R.A.; Farrag, S.M. Solving multi-objective optimal power flow problem via forced initialised differential evolution algorithm. *IET Gener. Transm. Distrib.* **2016**, *10*, 1634–1647. [CrossRef]
22. Abido, M. Optimal power flow using tabu search algorithm. *Electr. Power Compon. Syst.* **2002**, *30*, 469–483. [CrossRef]
23. Sayah, S.; Zehar, K. Modified differential evolution algorithm for optimal power flow with non-smooth cost functions. *Energy Convers. Manag.* **2008**, *49*, 3036–3042. [CrossRef]
24. Duman, S. Symbiotic organisms search algorithm for optimal power flow problem based on valve-point effect and prohibited zones. *Neural Comput. Appl.* **2017**, *28*, 3571–3585. [CrossRef]
25. Kılıç, U. Backtracking search algorithm-based optimal power flow with valve point effect and prohibited zones. *Electr. Eng.* **2015**, *97*, 101–110. [CrossRef]
26. Ghasemi, M.; Ghavidel, S.; Gitizadeh, M.; Akbari, E. An improved teaching–learning-based optimization algorithm using Lévy mutation strategy for non-smooth optimal power flow. *Int. J. Electr. Power Energy Syst.* **2015**, *65*, 375–384. [CrossRef]
27. Mohammadi, A.; Mehrtash, M.; Kargarian, A. Diagonal quadratic approximation for decentralized collaborative TSO+ DSO optimal power flow. *IEEE Trans. Smart Grid* **2018**. [CrossRef]
28. Naderi, E.; Azizivahed, A.; Narimani, H.; Fathi, M.; Narimani, M.R. A comprehensive study of practical economic dispatch problems by a new hybrid evolutionary algorithm. *Appl. Soft Comput.* **2017**, *61*, 1186–1206. [CrossRef]
29. Mirjalili, S.; Mirjalili, S.M.; Lewis, A. Grey wolf optimizer. *Adv. Eng. Softw.* **2014**, *69*, 46–61. [CrossRef]

30. Azizivahed, A.; Narimani, H.; Fathi, M.; Naderi, E.; Safarpour, H.R.; Narimani, M.R. Multi-objective dynamic distribution feeder reconfiguration in automated distribution systems. *Energy* **2018**, *147*, 896–914. [CrossRef]

31. IEEE 30-Bus Test System Data. Available online: https://www.ee.washington.edu/research/pstca/pf57/pg_tca30bus.hm (accessed on 26 June 2018).

Article

Distributed Energy Sharing for PVT-HP Prosumers in Community Energy Internet: A Consensus Approach

Nian Liu [1], Bin Guo [1], Zifa Liu [1,*] and Yongli Wang [2]

[1] State Key Laboratory of Alternate Electrical Power System with Renewable Energy Sources,
    North China Electric Power University, Beijing 102206, China; nianliu@ncepu.edu.cn (N.L.);
    binguo@ncepu.edu.cn (B.G.)
[2] School of Economics and Management, North China Electric Power University, Beijing 102206, China;
    wyl@ncepu.edu.cn
*   Correspondence: zifaliu@ncepu.edu.cn; Tel.: +86-010-6177-1623

Received: 27 June 2017; Accepted: 17 July 2018; Published: 20 July 2018

Abstract: Community Energy Internet (CEI) integrates electric network and thermal network based on combined heat and power (CHP) to improve the economy of energy system in Smart Community. In the CEI, an energy sharing framework for prosumers equipped with photovoltaic-thermal (PVT) system and heat pump (HP) is introduced. Supporting by the PVT and HP, the prosumer has four role attributes with either heat or electricity producer/consumer. A social welfare maximization model is built for the CEI, including PVT-HP prosumers, CHP system, and utility grid. Considering there are multiply participants in the local market of CEI, the social welfare maximization problem is decoupled by using Lagrange multiplier method. Moreover, a consensus-based fully distributed algorithm is designed to solve the problem. Finally, six residential buildings are selected as the case study to validate the effectiveness of the proposed method.

Keywords: energy internet; prosumer; energy management; consensus; demand response

1. Introduction

Energy Internet (EI) was proposed [1] to improve the utilization of renewable energy and meet the growing demand for energy. EI is a highly intelligent system which integrates distributed energy resources (DERs) and advanced Internet technology with existing Smart Grid [2]. A variety of energy, especially the renewable energy including photovoltaic (PV), wind turbine (WT), can be absorbed in a dynamic means for the distributed topology of the power network [3]. Furthermore, the power loss reduction, energy utilization efficiency improvement, and energy demand allocation optimization can be achieved under the EI [4]. From the perspective of energy policy, in China, the government has developed a series of energy policies, i.e., changing fuel from coal to natural gas and renewable energy. As a core component of the EI, the combined heat and power (CHP), PV and heat pump are bound to realize the goal and enhance the energy transaction efficiency.

Until recently, the research of the EI has drawn wide attention, and its topics may include architecture system [5–7], coordination control [8,9] and energy management [10–12]. A future electric power distribution system was proposed in [5] to suit for the plug-and-play of DERs and distributed storage devices. The main features of the future electric power distribution system is discussed in [6]. Based on the EI, a three-phase cascaded power electronic transformer was designed to connect with high voltage directly [7]. Energy hub is an effective means to realize reliable control of the EI. A decentralized model predictive control strategy was proposed in [8] to improve the operation of the coupled electricity and gas network system by considering predicted behavior and operational constraints. A residential energy hub is designed to coordinate solar energy with load demand and determine the scheduling of electric vehicles [9]. Particularly, there are multiple participants inside an EI, how to coordinate and optimize the interests of each participant is addressed in [10–13]. An optimization model of energy sharing among smart building is developed using non-cooperative game theory [10]. To improve the economy and reliability, a distributed energy management is proposed for interconnected CHP-based Microgrids with demand response (DR) [11]. Based on the feed-in-tariff, an energy sharing model is formulated among peer-to-peer PV prosumers [12].

It is worth noting that the electric efficiency of a PV panel is less than 20% in practical operation while the rest 80% of solar radiation is wasted into other forms of energy [14]. Hence, solar photovoltaic-thermal (PVT) hybrid system is proposed to improve both the electrical efficiency and the thermal efficiency [15]. Furthermore, the study in [16] shows that the idea of PVT system is economically feasible. However, the hot water produced by PVT system cannot meet the application requirements of temperature, from [17–19], it is demonstrated that working with a heat pump, PVT system will significantly increase system efficiency and provides both electricity and thermal energy for end users. Fuzzy Logic control has been applied to optimize the energy consumption of the PVT system [17]. A thermodynamics model is built for a refrigerant-based PVT assisted heat pump water heater system in [18]. A practical residential application of heat pump coupled with a PVT system is addressed in [19], the system can provide space heating and domestic hot water for a single-family dwelling located in North East Italy. PV panels of PVT combined with heat pump system can produce more additional power compared with the uncooled one [20]. In this paper, the end user equipped with PVT system and heat pump is named as a PVT-HP prosumer.

Distributed energy management has been widely used in Smart Grid for privacy protection [21], high efficiency [22], and global fairness [23]. Among them, consensus algorithm is considered to be an important solution for Smart Grid in a multi-agent system. For the economic dispatch problem, the traditional centralized economic dispatch problem was solved in a distributed way using consensus algorithm [24–26]. A strict analysis of convergence and optimality for the consensus algorithm under different topologies is addressed in [24]. The economic dispatch problem is solved in distributed way which consist of two stages in [25]. A flooding-based consensus algorithm is proposed in the first stage, and in the second stage, a nondeterministic method is used for solving the economic dispatch problem in parallel. The consensus algorithm in [26] enables generators to collectively learn the mismatch between demand and total amount of power generation as a feedback mechanism to adjust its own power generation. Furthermore, the transmission line losses and generator constraints are considered in the distributed economic dispatch using consensus algorithm [27,28]. The proposed approach is based on two consensus algorithms running in parallel and can handle networks of various size and topology [27]. A non-convex social welfare maximization problem by considering the transmission losses is formulated in [28]. The renewable energy or storage system are taken into consideration in [29,30]. An optimal DERs coordination problem over multiple time periods is proposed in [29] and consensus algorithm is used to coordinate distributed generators with multiple/single storages automatically and dynamically. Energy storage devices are incorporated into the economic dispatch problem in [30] for both iter-temporal energy arbitrage and providing spinning reserve. DR is applied to realize the social welfare maximization in Smart Grid [31,32]. A distributed approach is proposed to deal with energy management in the smart grid under dispatchable distributed generators and responsive loads using real-time pricing (RTP) in [31] and consensus networks is applied to maximize the social welfare. The problem of distributed energy management is addressed by formulating the economic dispatch and demand response in a united framework [32]. However, these studies cannot be directly applied to the Community Energy Internet (CEI) with PVT-HP prosumers. The reasons may include two aspects. First, the CEI actually consists of two different physical networks, one for electricity, the other for heat. Second, the heat and electricity are coupled on both sides of energy sources and end users. The electricity and heat can be generated simultaneously by the CHP system, while the end users may produce or consume both electricity and heat if they are PVT-HP prosumers.

To this end, the focus of this paper is on the distributed energy sharing of PVT-HP prosumers in a CEI. The contributions are as follows:

(1)   The CEI is constructed as a heat and electricity coupled network, in which the PVT-HP prosumers are modelled with four role attributes and heat-electricity DR ability.

(2)   A social welfare maximization model is built for the CEI, including PVT-HP prosumers, CHP system, and utility grid. By using Lagrange multiplier method, the problem is further decoupled into three sub-optimization problems correspondingly.

(3)   A consensus algorithm is designed to solve the optimization problem of the CEI, which can be fully-distributed solved by each participant in the local market.

2. Structure and Function of CEI

The system architecture of the CEI is shown in Figure 1. There are three entities: PVT-HP prosumer, CHP operator and utility grid. Each prosumer is equipped with PVT, heat pump and user energy management system (UEMS) [13]. UEMS is in charge of communicating with other entities, adjusting local load demand and determine the output of heat pump. The heat pump uses the water from the tank of PVT system as the low temperature heat resource to produce high temperature hot water for end users. Excess electric and thermal energy can be shared among prosumers and the electricity can be sold to utility grid as well. CHP operator guarantees thermal supply through recovering the waste heat of micro-gas turbine when the heat pump cannot meet the users' thermal demand and conducts trading with utility grid and prosumers inside the community. Its energy management system is CHP operator energy management system (CO-EMS). After using the self-produce and CHP power, if the CEI is still in lack of electricity, the insufficient can be balanced by the utility grid.

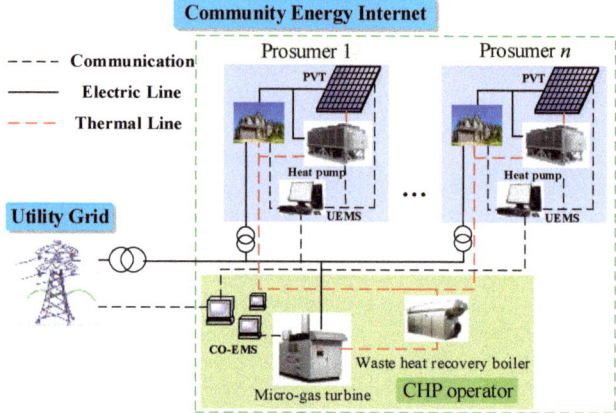

Figure 1. Structure of Community Energy Internet (CEI) with photovoltaic-thermal and heat pump (PVT-HP) prosumers.

3. Basic Knowledge of Consensus Algorithm

In this section, the notations of graph theory and two consensus protocols are presented.

3.1. Graph Theory

Consider a CEI with N PVT-HP prosumers, a CHP operator and utility grid. A directed connected graph $G = \{V, E\}$ is used to represent the communication topology of the CEI, where V is the node set and $E \subset \{V \times V\}$ is the edge set. The prosumers set, CHP operator set and utility grid set are expressed as V_P, V_{CHP} and V_G, respectively. A directed edge from i to j is denoted by an ordered pair $(i, j) \in E$, which means that if and only if (iff) node j can receive messages from node i. The in-neighbors of the ith node is denoted by $N_i^+ = \{j \in V | (j, i) \in E\}$ and $|N_i^+|$ is the cardinality of in-neighbor set of node i. Similarly, the out-neighbors of the ith node is denoted by $N_i^- = \{j \in V | (i, j) \in E\}$ and

$|N_i^-|$ is the cardinality of out-neighbor set of node i. Since it is obvious that node i can obtain its own state information, each node belongs to both its in-neighbor and out-neighbor set, i.e., $i \in N_i^+$ as well as $i \in N_i^-$ [24]. It is worth noting that the communication network is strongly connected, i.e., any two nodes have directed path between them.

3.2. Consensus Protocols

CHP can produce electric and thermal power while PVT-HP prosumer can act as a producer or a consumer according to the load level, output power of PV and heat pump. Besides, the deviated electric power is balanced by the utility grid. Thus, the network is divided into electric network and thermal network, i.e., the electric network set V_e and the thermal network set V_t and note that $V_t \subset V_e$. Let us define two stochastic matrices named row stochastic matrix $R = [r_{ij}]$ and column stochastic matrix $S = [s_{ij}]$ associated with the electric network of N nodes as follows:

$$r_{i,j} = \left\{ \begin{array}{ll} \frac{1}{|N_i^+|}, & j \in N_i^+ \\ 0, & j \notin N_i^+ \end{array} \right. \quad s_{i,j} = \left\{ \begin{array}{ll} \frac{1}{|N_j^-|}, & i \in N_j^- \\ 0, & i \notin N_j^- \end{array} \right. \tag{1}$$

Similarly, $X = [x_{ij}]$ and $Y = [y_{ij}]$ are the stochastic matrices of thermal network of M nodes:

$$x_{i,j} = \left\{ \begin{array}{ll} \frac{1}{|M_i^+|}, & j \in M_i^+ \\ 0, & j \notin M_i^+ \end{array} \right. \quad y_{i,j} = \left\{ \begin{array}{ll} \frac{1}{|M_j^-|}, & i \in M_j^- \\ 0, & i \notin M_j^- \end{array} \right. \tag{2}$$

For simplicity, the electric network is taken as an example for the following part. It is assumed that $P(k) = [P_1(k), P_2(k), ..., P_N(k)]$ denotes the state vector of all nodes in the electric network at iteration k. For the initial value $P(0)$, the following two lemmas will be helpful for the design of consensus algorithm [24].

Lemma 1. *(Consensus): If the network is strongly connected, for the discrete consensus algorithm $P(k+1) = RP(k)$, there holds that $\lim\limits_{k \to \infty} P_i(k) = c, \forall i \in V_e$, where c is a constant.*

Lemma 2. *(Ratio Consensus): If the network is strongly connected, for the discrete consensus algorithm $P(k+1) = SP(k)$, there holds that $\lim\limits_{k \to \infty} P_i(k) = \kappa_i \sum\limits_{i=1}^{N} P_i(0), \forall i \in V_e$, where κ_i is the ith element of the unit eigenvector corresponding to eigenvalue 1 of the matrix S.*

4. System Model

4.1. Profit of Utility Grid

When the CEI is in lack of electric power, utility grid purchases the electricity from the power plant (i.e., coal-fired generation) to meet the demand of the end users. On the contrary, if there is plenty of solar energy, CEI can feed the excess power back to the utility grid to get income. Hence, the cost of utility grid at time t can be denoted as:

$$C_i(t) = \left\{ \begin{array}{ll} a_i P_i(t)^2 + b_i P_i(t) + c_i, & P_i(t) > 0 \\ \tau P_i(t), & P_i(t) < 0 \end{array} \right. \tag{3}$$

where $P_i(t)$ is the net electric power of the CEI at time t, a_i, b_i, c_i are the cost coefficients of the power plant, τ is the unit price of PV energy selling to the utility grid.

It's assumed that the cost function of utility grid should be continuous, here c_i is set as 0 in Equation (3). For the convenience of calculations, the piecewise function of utility grid can be approximated by a quadratic function by referring to [10]:

$$C_i(t) = a_i{}' P_i(t)^2 + b_i{}' P_i(t) \tag{4}$$

Therefore, for the utility grid i, $i \in V_G$, the profit can be expressed as:

$$P_G(t) = p_e(t) P_i(t) - C_i(t) \tag{5}$$

where $P_G(t)$ is the profit of utility grid, $p_e(t)$ is the electric price at time slot t in CEI, $P_i(t)$ is the electricity selling to the CEI.

4.2. Utility of Prosumer

Generally, a PVT-HP prosumer can get revenue through trading energy with utility grid, CHP operator and other prosumers, which means that the utility function of a prosumer should consider the profit of selling energy and the cost of purchasing energy firstly. For prosumer i, $i \in V_P$, its profit or cost at each time slot is defined as:

$$P_P(t) = p_e(t)(PV_i(t) - P_i(t) - P_i^{hp}(t)) + p_h(t)(H_i^{hp}(t) - H_i(t)) \tag{6}$$

where $P_P(t)$ is the profit (positive) or cost (negative) of a prosumer at time slot t, $p_h(t)$ is the thermal price at time slot t, $P_i(t)$ is the load demand of electric power, $P_i^{hp}(t)$ is the electric power consumed by the heat pump, PV_i is the predicted electric power of PVT system, H_i is the load demand of thermal power, H_i^{hp} is the thermal power produced by the heat pump.

For the heat pump, it uses the water from the tank of PVT system as the low temperature heat resource to produce high temperature hot water. The thermal power produced by the heat pump is denoted by:

$$H_i^{hp}(t) = P_i^{hp}(t)COP \tag{7}$$

where COP is the coefficient of performance of the heat pump.

Moreover, from the perspective of DR, the prosumers adjust their usage of energy motivated by the prices, and then get financially benefit. Simultaneously, adjusting the prosumer's load profiles would cause uncomfortable [33] or inconvenience [10] impact. Thus, the equivalent negative cost on the utility of prosumers can be defined as follows.

$$I_i(t) = v_i(P_i(t) - P_i^0(t))^2 + \omega_i(H_i(t) - H_i^0(t))^2 \tag{8}$$

where v_i, ω_i are the coefficients of inconvenience and uncomfortable, respectively; $P_i^0(t)$ and $H_i^0(t)$ are the initial electric power and thermal power consumption, respectively.

From Equation (8), the negative impact of DR increases with the deviation of power consumption, both on electricity and heat. Hence, for prosumer i, $i \in V_P$, its profit function at each time slot can be updated as follows:

$$P_P(t) = p_e(t)(PV_i(t) - P_i(t) - P_i^{hp}(t)) - p_h(t)(P_i^{hp}(t)COP - H_i(t)) - I_i(t) \tag{9}$$

4.3. Profit of CHP

CHP produces electric power as well as thermal power leading to a high overall efficiency, its profit function is represented as [11]:

$$P_{CHP}(t) = p_e(t)P_i(t) + p_h(t)H_i(t) - F_i(P_i(t), H_i(t)) \tag{10}$$

$$F_i(P_i(t), H_i(t)) = \phi + \beta P_i(t) + \gamma P_i(t)^2 + \delta H_i(t) + \theta H_i(t)^2 + \varphi P_i(t)H_i(t) \tag{11}$$

where ϕ, β, γ, δ, θ, φ are the cost coefficients of CHP system.

The heat-to-electric rate of CHP may be different due to the variable load conditions. CHP can operate either in the Following Thermal Load (FTL) mode or in the Following Electric Load (FEL) mode. For CHP i, $i \in V_{CHP}$, the coupling model of thermal and electric is denoted by [34]:

$$P_i(t) = K_{CHP}H_i(t) = \frac{\eta_e}{(1 - \eta_e)\eta_r}H_i(t) \tag{12}$$

where η_e is the electric efficiency, η_r is the heat recovery rate of heat recovery boiler, and K_{CHP} is the coupling coefficient between heat and electricity.

Here, we can rewrite Equations (10) and (11) when the cost of CHP is denoted by H_i:

$$P_{CHP}^H(t) = p_e(t)P_i(t) + p_h(t)H_i(t) - F_i(H_i(t)) \tag{13}$$

$$F_i(H_i(t)) = \phi + \beta(K_{CHP}H_i(t)) + \gamma(K_{CHP}H_i(t))^2 + \delta H_i(t) + \theta H_i(t)^2 + \varphi H_i(t)(K_{CHP}H_i(t)) \tag{14}$$

When the load rate of CHP is less than 30%, its electric efficient will be much lower, so we only start the CHP unit when its load rate is greater than 30%:

$$P_i(t) = \begin{cases} P_i(t), & P_i(t) \geq 0.3Cap \\ 0, & P_i(t) < 0.3Cap \end{cases} \tag{15}$$

where Cap is the rated power of CHP unit and it is assumed that the electric efficient is a constant when the load rate of CHP is over 30%.

5. Problem Formulation and Algorithm

5.1. Optimization Problem

In the energy market of the CEI, prosumers, utility grid and CHP operator compete with each other to maximize its own interests for their selfishness. To guarantee both efficiency and fairness, social welfare maximization is introduced and widely used. Social welfare maximization ensures the overall interests and maximizes each individual interest simultaneously. Here, we add up all the profits or utilities of the participants as the social welfare:

$$\begin{aligned} W(t) &= P_G(t) + P_P(t) + P_{CHP}^H(t) \\ &= \sum_{i \in V_G} [p_e(t)P_i(t) - C_i(t)] \\ &+ \sum_{i \in V_P} [p_e(t)(PV_i(t) - P_i(t) - P_i^{hp}(t)) + p_h(t)(P_i^{hp}(t)COP - H_i(t)) - I_i(t)] \\ &+ \sum_{i \in V_{CHP}} [p_e(t)P_i(t) + p_h(t)H_i(t) - F_i(H_i(t))] \end{aligned} \tag{16}$$

At any time, the electric power and thermal power should be balanced in practical operation.

$$\sum_{i \in V_G} P_i(t) + \sum_{i \in V_{CHP}} P_i(t) = \sum_{i \in V_P} (P_i(t) + P_i^{hp}(t) - PV_i(t)) \tag{17}$$

$$\sum_{i \in V_{CHP}} H_i(t) = \sum_{i \in V_P} (H_i(t) - P_i^{hp}(t)COP) \tag{18}$$

Now, we can rewrite Equation (16) by using Equations (17) and (18):

$$max \quad \sum_{i \in V_G} -C_i(t) + \sum_{i \in V_P} -I_i(t) + \sum_{i \in V_{CHP}} -F_i(H_i(t)) \tag{19}$$

Based on the previous description, energy management of the CEI can be formulated as a convex optimization problem:

$$\min \quad \sum_{i \in V_G} C_i(t) + \sum_{i \in V_P} I_i(t) + \sum_{i \in V_{CHP}} F_i(H_i(t))$$

$$s.t. \quad \sum_{i \in V_G} P_i(t) + \sum_{i \in V_{CHP}} P_i(t) = \sum_{i \in V_P} (P_i(t) + P_i^{hp}(t) - PV_i(t))$$

$$\sum_{i \in V_{CHP}} H_i(t) = \sum_{i \in V_P} (H_i(t) - P_i^{hp}(t)COP)$$

$$P_i^{\min} \le P_i^{hp}(t) \le P_i^{\max}, \; i \in V_P$$

$$P_i^{\min} \le P_i(t) \le P_i^{\max}, \; i \in V_P$$

$$H_i^{\min} \le H_i(t) \le H_i^{\max}, \; i \in V_P$$

$$P_i^{\min} \le P_i(t) \le P_i^{\max}, \; i \in V_G$$

$$P_i^{\min} \le P_i(t) \le P_i^{\max}, \; i \in V_{CHP}$$

$$H_i^{\min} \le H_i(t) \le H_i^{\max}, \; i \in V_{CHP}$$

where the equality constraints are global power constraints which describes the electric power balance and thermal power balance, the inequality constraints are local power constraints for all the participants, P_i^{min} and P_i^{max} are the lower and upper limit of electric power, H_i^{min} and H_i^{max} are the lower and upper limit of thermal power, respectively.

5.2. Problem Decoupling

To resolve the convex optimization problem with consensus algorithm, we need to decouple the global constraints. The Lagrangian multiplier method is a typical approach to transfer the equality constraints to the objective function. The corresponding Lagrangian function of original problem is given by:

$$
\begin{aligned}
&\Gamma(P(t), P^{hp}(t), H(t), \lambda(t), \mu(t)) \\
&= \sum_{i \in V_G} C_i(t) + \sum_{i \in V_P} I_i(t) + \sum_{i \in V_{CHP}} F_i(H_i(t)) \\
&+ \lambda(t)[\sum_{i \in V_P} (P_i(t) + P_i^{hp}(t) - PV_i(t)) - (\sum_{i \in V_G} P_i(t) + \sum_{i \in V_{CHP}} P_i(t))] \\
&+ \mu(t)[\sum_{i \in V_P} (H_i(t) - P_i^{hp}(t)COP) - \sum_{i \in V_{CHP}} H_i(t)]
\end{aligned}
\tag{20}
$$

where $P(t) = [P_1(t), ..., P_N(t)]'$, $H(t) = [H_1(t), ..., H_M(t)]'$ and $P^{hp}(t) = [P_1^{hp}(t), ..., P_N^{hp}(t)]'$ and $\lambda(t)$, $\mu(t)$ are the Lagrange multipliers at time t that are introduced to decouple the global power constraints.

There exist different global constraints between electric network and thermal network and leading to the separation of them. We define $\lambda_i(t)$ as incremental cost (incremental utility) for the energy sources (the demand) i in electric network as follows:

$$
\lambda_i(t) = \begin{cases} C_i'(t), \; i \in V_G \\ 0, \; i \in V_{CHP} \\ \frac{\partial I_i(t)}{\partial P_i}, \; i \in V_P \end{cases}
\tag{21}
$$

Similarly, the incremental cost (incremental utility) $\mu_i(t)$ for the energy sources (the demand) i of thermal network is given by:

$$
\mu_i = \begin{cases} F'(H_i(t)) - \lambda_i(t)K_{CHP}, i \in V_{CHP} \\ \frac{\partial I_i(t)}{\partial H_i}, i \in V_P \end{cases}
\tag{22}
$$

From Equations (20)–(22), we can decouple the primal problem into 3 sub-optimization problems only with local constraints.

Prosumer subproblem ($i \in V_P$):

$$
\begin{aligned}
\min \quad &I_i(t) + \lambda_i(t)(P_i(t) + P_i^{hp}(t) - PV_i(t)) \\
&+ \mu_i(t)(H_i(t) - P_i^{hp}(t)COP)
\end{aligned}
\tag{23}
$$

$$s.t. \quad P_i^{\min} \leq P_i^{hp}(t) \leq P_i^{\max}, \; i \in V_P$$
$$P_i^{\min} \leq P_i(t) \leq P_i^{\max}, \; i \in V_P \tag{24}$$
$$H_i^{\min} \leq H_i(t) \leq H_i^{\max}, \; i \in V_P$$

Utility grid subproblem ($i \in V_G$):

$$min \quad C_i(t) - \lambda_i(t)P_i(t) \tag{25}$$

$$s.t. \quad P_i^{\min} \leq P_i \leq P_i^{\max} \tag{26}$$

CHP subproblem ($i \in V_{CHP}$):

$$minF_i(H_i(t)) - \lambda_i(t)P_i(t) - \mu_i(t)H_i(t) \tag{27}$$

$$s.t. \quad P_i^{\min} \leq P_i(t) \leq P_i^{\max} \tag{28}$$
$$H_i^{\min} \leq H_i(t) \leq H_i^{\max}$$

Now, all the subproblems only have local constraints to be solved, which are appropriate to use consensus algorithm in a directed communication network.

5.3. Design of Algorithm

The local electric and thermal power mismatches are denoted as $\xi_i^E(t), i \in V_e$ and $\xi_i^H(t), i \in V_t$, respectively. $\rho_e(t)$ and $\rho_t(t)$ represent the gain parameters. The final errors $\varepsilon_{\xi^E}(t)$, $\varepsilon_{\xi^H}(t)$ and $\varepsilon_{\lambda^E}(t)$, $\varepsilon_{\lambda^H}(t)$ are for the power mismatch and incremental cost (utility). To simplify the description of algorithm, t is omitted in the following part, i.e., ξ_i^E denotes $\xi_i^E(t)$. The detail of the algorithm can be found in Algorithms 1 and 2.

Algorithm 1 Algorithm for participants in the CEI

1: Set parameters: a', b', v_i, ω_i, COP, ϕ, β, γ, δ, θ, φ
2: **For** $t = 1$
3: Initialization: Set $\lambda_i(0)$, $P_i(0)$, $\xi_i^E(0)$, $\mu_i(0)$, $H_i(0)$ and $\xi_i^H(0)$ as follows:

$$H_i(0) = 0, i \in V_t \tag{29}$$

$$P_i(0) = 0, i \in V_e \tag{30}$$

$$\mu_i(0) = \begin{cases} (F'(H_i)) - \lambda_i(0)K_{CHP}, i \in V_{CHP} \\ \frac{\partial I_i(H_i)}{\partial H_i}, i \in V_P \end{cases} \tag{31}$$

$$\lambda_i(0) = \begin{cases} 2a_i'P_i + b_i', i \in V_G \\ \frac{\partial I_i(P_i)}{\partial P_i}, i \in V_P \\ 0, i \in V_{CHP} \end{cases} \tag{32}$$

$$\xi_i^H(0) = 0, i \in V_t \tag{33}$$

$$\xi_i^E(0) = \begin{cases} 0, i \in V_G \cup V_{CHP} \\ -PV_i, i \in V_P \end{cases} \tag{34}$$

4: Execute **Algorithm 2**: Calculate the optimal power and incremental costs (utilities), i.e., $\lambda_i(t)$, $P_i(t)$, $i \in V_e$, $\mu_i(t)$, $H_i(t)$, $i \in V_t$.
5: **IF** $t > 24$
   End For
6: **Else** $t = t + 1$

Algorithm 2 Iteration Process

While true

1: $k = 1$
2: Update incremental cost (utility): 1) Update μ_i according to

$$\mu_i(k+1) = \sum_{j \in V_t} x_{ij}\mu_j(k) + \rho_t \zeta_i^H(k), i \in V_t \tag{35}$$

2) Update λ_i according to

$$\lambda_i(k+1) = \sum_{j \in V_e} r_{ij}\lambda_j(k) + \rho_e \zeta_i^E(k), i \in V_e \tag{36}$$

3: Update thermal power H_i according to: when $i \in V_{CHP}$,

$$H_i(k+1) = \arg\min[F_i(H_i(k)) - \lambda_i(k+1)K_{CHP} - \mu_i(k+1)H_i(k)], H_i(k) \in [H_i^{min}, H_i^{max}] \tag{37}$$

$$H_i(k+1) = \begin{cases} H_i(k+1), & H_i(k+1)K_{CHP} \geq 0.3Cap \\ 0 & , H_i(k+1)K_{CHP} < 0.3Cap \end{cases} \tag{38}$$

when $i \in V_P$,

$$[H_i(k+1), P_i(k+1), P_i^{hp}(k+1)] = \arg\min[I_i(k) + \lambda_i(k+1)(P_i(k) + P_i^{hp}(k) - PV_i) \\ + \mu_i(k+1)(H_i(k) - P_i^{hp}(k)COP)], \\ H_i(k) \in [H_i^{min}, H_i^{max}], P_i(k) \in [P_i^{min}, P_i^{max}], P_i^{hp}(k) \in [P_i^{min}, P_i^{max}] \tag{39}$$

$$H_i(k+1) = H_i(k+1) - P_i^{hp}(k+1)COP \tag{40}$$

4: Update electric power P_i according to: when $i \in V_G$,

$$P_i(k+1) = \arg\min[C_i(P_i(k)) - \lambda_i(k+1)P_i(k)], P_i(k) \in [P_i^{min}, P_i^{max}] \tag{41}$$

when $i \in V_{CHP}$,

$$P_i(k+1) = H_i(k+1)K_{CHP}, P_i(k) \in [P_i^{min}, P_i^{max}] \tag{42}$$

when $i \in V_P$,

$$P_i(k+1) = P_i(k+1) + P_i^{hp}(k+1) \tag{43}$$

5: Update thermal power mismatch ζ_i^H according to:

$$\zeta_i^H(k+1) = \sum_{j \in V_i} y_{ij}(k)\zeta_i^H(k) + [(H_i(k+1) - H_i(k))|i \in V_P] + [(H_i(k) - H_i(k+1))|i \in V_{CHP}] \tag{44}$$

6: Update electric power mismatch ζ_i^E according to:

$$\zeta_i^E(k+1) = \sum_{j \in V_e} s_{ij}(k)\zeta_i^E(k) + [(P_i(k+1) - P_i(k))|i \in V_P] + [(P_i(k) - P_i(k+1))|i \in V_G \cup V_{CHP}] \tag{45}$$

7: **If** $|\zeta_i^E(k)| \leq \varepsilon_{\zeta E}, \forall i \in V_e$, $|\lambda_i(k) - \lambda_i(k-1)| \leq \varepsilon_{\lambda E}, \forall i \in V_e$, $|\zeta_i^H(k)| \leq \varepsilon_{\zeta H}, \forall i \in V_t$ and $|\mu_i(k) - \mu_i(k-1)| \leq \varepsilon_{\lambda H}, \forall i \in V_t$, Output: $\lambda_i(k), P_i(k), i \in V_e, \mu_i(k), H_i(k), i \in V_t$. **Break.**
8: $k = k+1$

For Algorithm 1, in Initialization, the value of $\lambda_i(0)$, $P_i(0)$, $\mu_i(0)$, $H_i(0)$ and $\zeta_i^H(0)$ can be set to any valid value. Please note that the initial value of $\zeta_i^E(0)$ should be $-PV_i$ when $i \in V_P$ for the PV energy is preferentially self-consumed.

For Algorithm 2, first, the convergence of both incremental cost (utility) and local power mismatch are guaranteed by the update rules Equations (35) and (36) which are derived from Lemma 1. Second, the thermal and electric power are calculated based on the updated $\lambda_i(k+1), \mu_i(k+1)$ according to Equations (37)–(43). Third, each participant updates its own local power mismatch based on the updated power $H_i(k+1), P_i(k+1)$ according to Equations (44) and (45) that come from Lemma 2. Fourth, the iteration breaks until all the λ_i, μ_i approach to the same value and the total power mismatch get close to 0. By choosing a small enough value for ρ_e, ρ_t, the iterative procedure finally converges to the global optimum.

6. Case Study

6.1. Basic Data

In this paper, a CEI comprises 6 residential buildings is chosen as the study case and each building represents a PVT-HP prosumer. The parameters of utility grid, CHP and PVT-HP prosumer are listed in Table 1. All the load data are collected from the smart residential buildings of demonstration projects in Beijing [33]. The settings of the case study are consistent with these projects to make the method capable of the practical applications. The daily curves of electric, thermal load, and electric net power are shown in Figure 2 and the range of load adjustment is set between −20% and +20%. Figure 2 shows the daily load profiles and PV energy curve of the CEI. Figure 2a is the daily electric load profiles of six prosumers, the peak loads appear at 20:00. Figure 2b is the daily thermal load profiles of six prosumers. With the change of temperature, end users need more heat at night. By equipping with PVT systems, end users utilize the solar energy to meet part of the electric load and thermal load from 8:00 to 20:00. Therefore, the electric net power has shown negative values from 12:00 to 16:00. $\rho_e, \rho_t, \varepsilon_{\xi^E}$ and ε_{λ^H} are set as 0.000595, 0.000555, 0.01 and 0.01, respectively.

Table 1. Parameters of participants.

Participant	Parameters	Value
Utility grid	Cost coefficients	$a = 0.00059, b = 0.302, c = 0$
	Capacity	$[-500, 1000]$ (kW)
CHP	Cost coefficients	$\phi = 0.03395, \beta = 4.6425, \gamma = 0.00442$
		$\delta = 1.345, \theta = 0.00384, \varphi = 0.004$
	Efficiency	$\eta_e = 0.3441, \eta_r = 0.80$
	Capacity	50 (kW)
PVT-HP prosumer	COP	3
	Coefficients of comfort	$v_i \in (0.03\text{–}0.055), \omega_i \in (0.025\text{–}0.1)$
	Capacity of heat pump	15 (kW)

6.2. Results of Simulation

6.2.1. Convergence and Optimality of Consensus Algorithm

In this section, we apply the proposed model in the CEI and use MATLAB (2014a, The MathWorks, Inc, Natick, MA, USA) to programme for testing the convergence and optimality of Algorithm 2. The iterative optimization processes of incremental cost (incremental utility) μ_i, the local power of each participator H_i and the local thermal power mismatch ξ_i^H at 1:00 are shown in Figure 3. As shown in Figure 3a, the incremental cost (incremental utility) converges to its final value $\mu^* = 0.2701$ CNY/kWh with iterations, while all the local thermal power mismatch ξ_i^H approach to 0.0098 as shown in Figure 3b. Moreover, from Figure 3c, the results show that the output power of CHP is 0 kW at the beginning and then gradually approaches to the convergence value $H_i^* = 17.01$ kW. The net thermal power of prosumer 1 and prosumer 2 are negative which means that they share excess thermal energy with other prosumers. More importantly, all the prosumers and CHP operator achieve the same incremental cost (incremental utility), under which all the participants achieve their own goal, i.e., maximize the individual welfare.

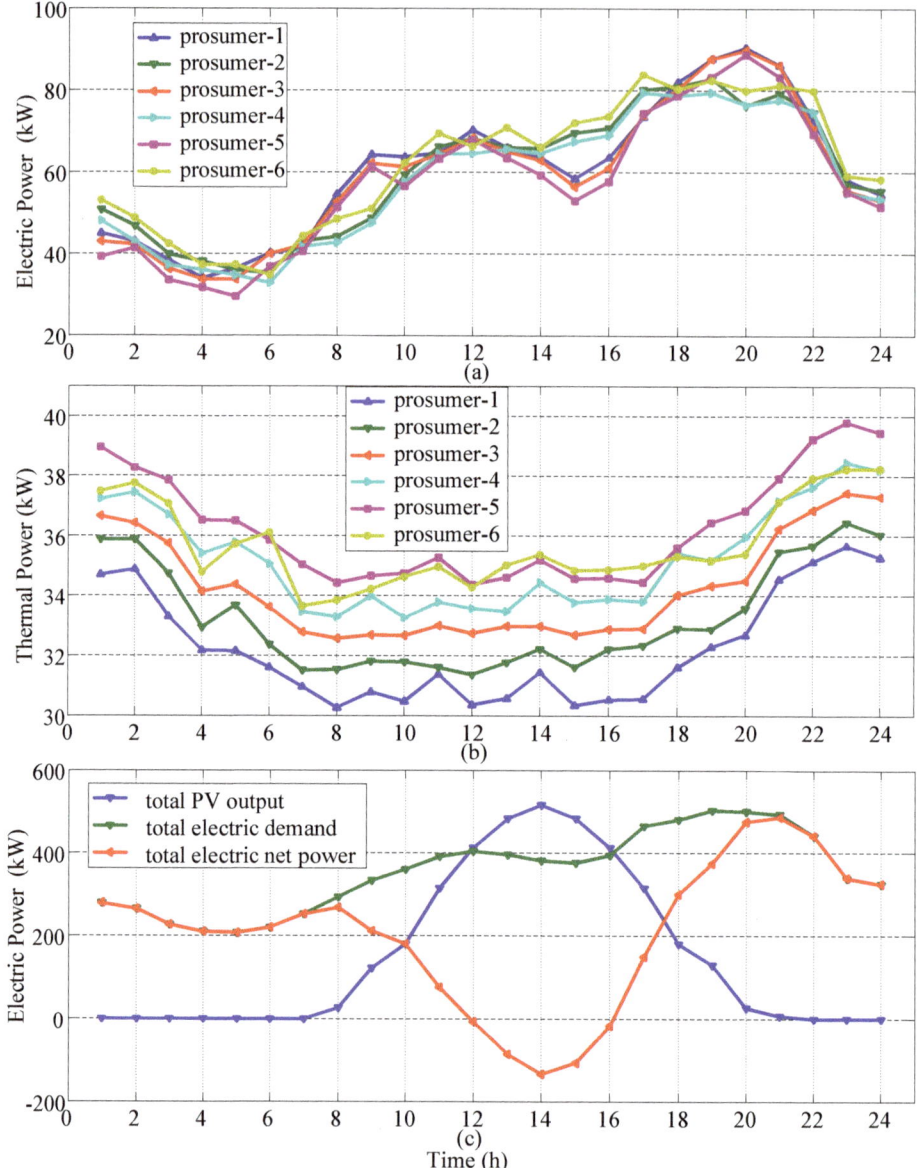

Figure 2. Daily load curves of prosumers: (**a**) daily electric load curves; (**b**) daily thermal load curves; (**c**) daily electric net power curves.

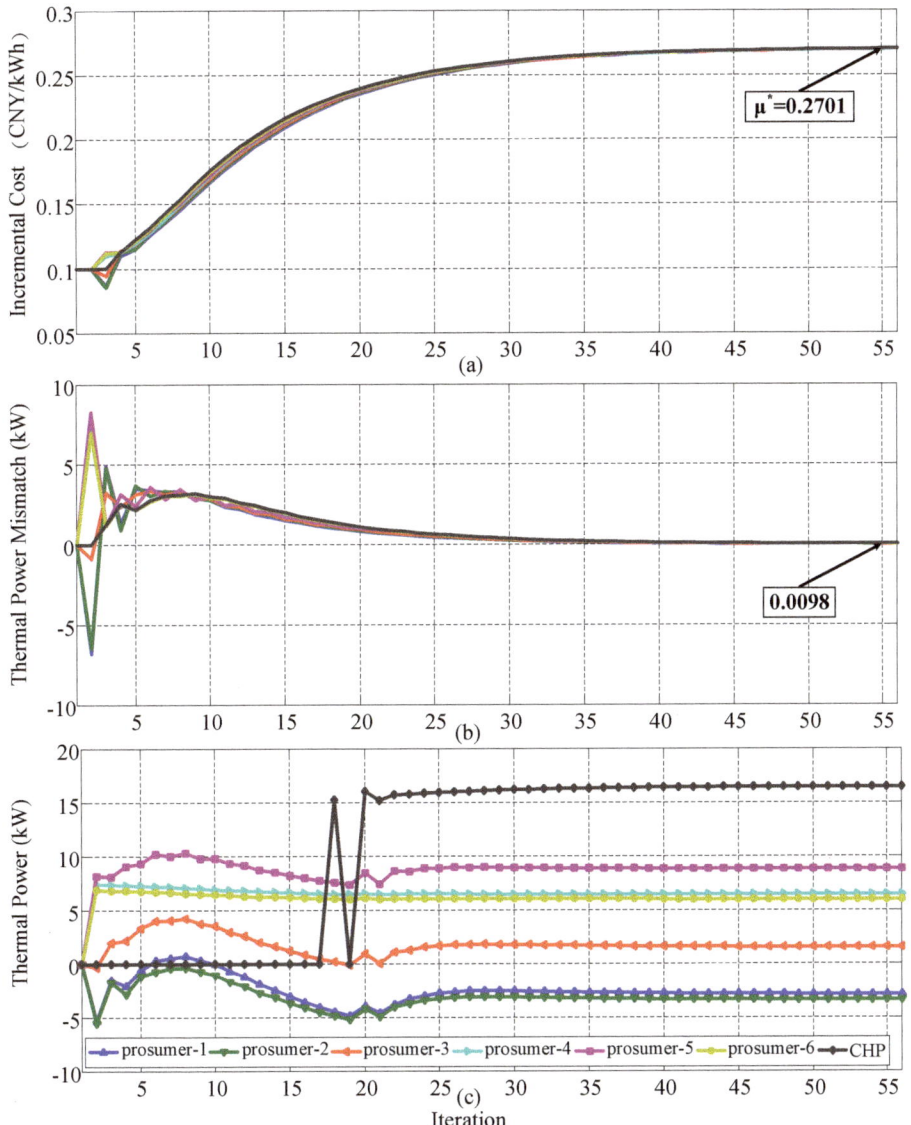

Figure 3. Convergence results of thermal network: (**a**) incremental cost (utility); (**b**) local thermal power mismatch; (**c**) local generated or consumed thermal power.

Figure 4 shows the convergence results of electric network. In Figure 4a,b, all the participants convergence to the same incremental cost (utility), i.e., $\lambda_i^* = 0.740$ CNY/kWh, while the local power mismatch get close to 0.0027. Moreover, Figure 4c shows that the variation tendency of CHP is the same as the results in Figure 3c. In this time slot, there is no solar energy, the prosumers has to buy 373.46 kWh from the utility grid to meet the electric demand of load and heat pump. The total social welfare of the CEI approaches to its convergence result at 1555.0 CNY, as shown in Figure 5.

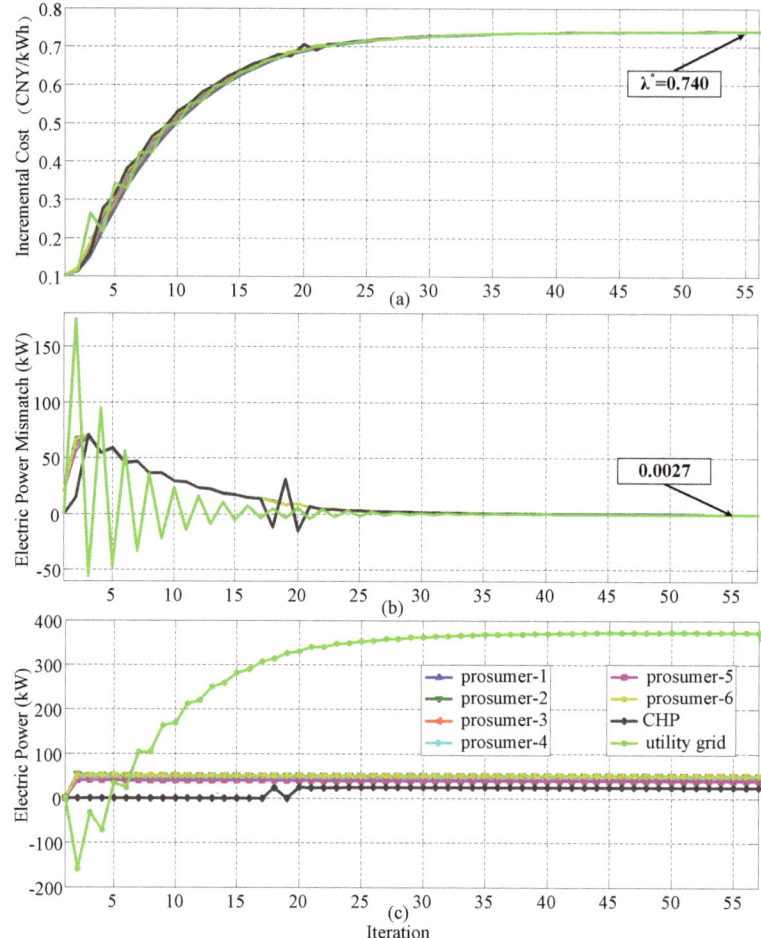

Figure 4. Convergence results of electric network: (**a**) incremental cost (utility); (**b**) local electric power mismatch; (**c**) local generated and consumed electric power.

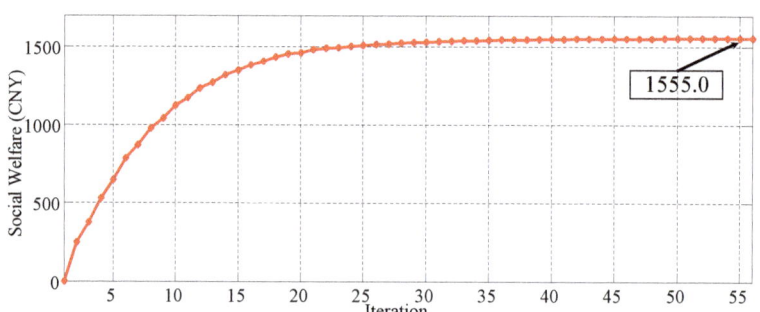

Figure 5. Convergence results of total social welfare.

6.2.2. Results of Price and Net Power

By using the basic data of six prosumers, CHP and utility grid, consensus-based energy sharing of the CEI on each time slot of a day can be solved by executing Algorithm 1. The price of electric power and thermal power during the daytime are shown in Figure 6. From 0:00 to 8:00 and 20:00 to 24:00, the electric price is high because there is no solar energy and the end users have to purchase energy from the utility grid. The electric price changes with the variety of PV output from 8:00 to 20:00 and gets its minimal value 0.1973 CYN/kWh at 14:00. After 20:00, the electric price increases for the reason that there is a peak electric load from 19:00 to 22:00. The variation tendency of the thermal price is consistent with the thermal load curve. At night, the higher thermal load level leads to the higher prices, i.e., 0.25–0.3 CNY/kWh. The thermal price decreases with the thermal load reduction and increases gradually with the increasing demand during the day.

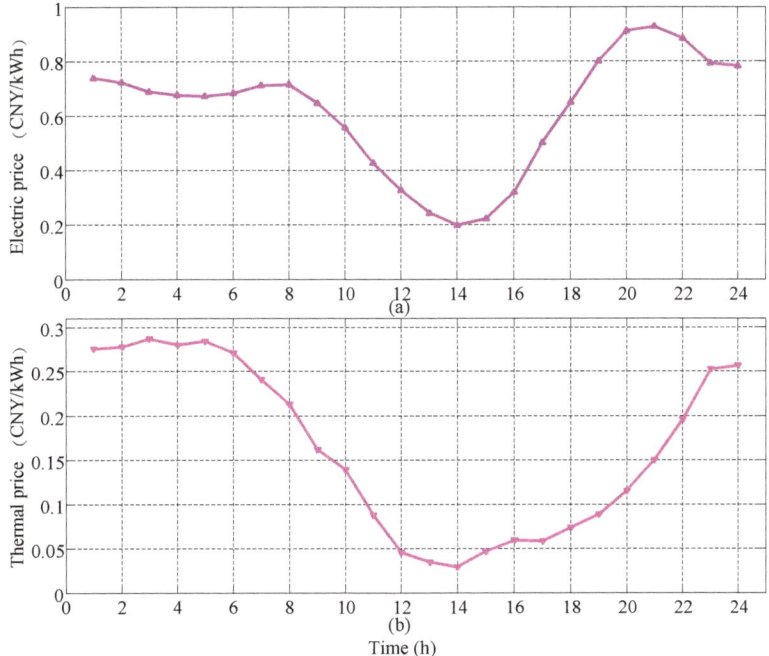

Figure 6. (**a**) Electric price and (**b**) thermal price during the daytime.

The net power of each prosumer, CHP and utility grid can be seen in Figure 7. As shown in Figure 7a, the net thermal power of prosumers 4–6 are positive which means they are always a thermal consumer during the day and prosumers 1–3 can act as a thermal producer or a thermal consumer alternatively. Especially, from 10:00 to 17:00, as a thermal producer, prosumers 1–3 share their excess thermal energy to prosumers 4–6. During the time, the heat pump of each prosumer makes full use of the abundant solar energy to produce high temperature thermal energy to meet the load demand. The CHP is stopped during 9:00–17:00 to increase the economic effectiveness because the load rate is less than 30%. Moreover, Figure 7b shows the net electric power of each prosumer. From 12:00 to 16:00, there are plenty of solar energy and all the prosumers act as a electric producer to sell electric energy to utility grid.

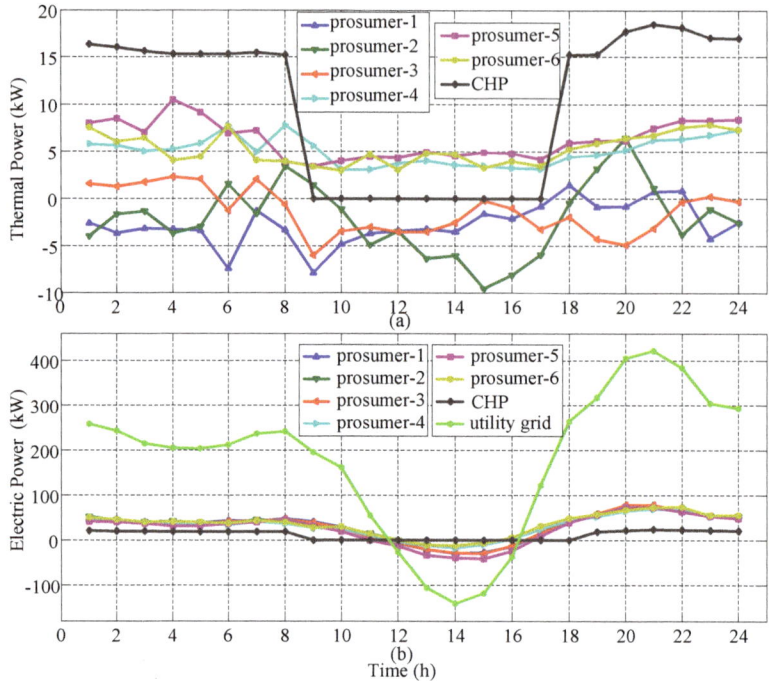

Figure 7. Net power during the day. (**a**) Net thermal power; (**b**) Net electric power.

By using the results of each time slot, an interesting result can be found that there are four types of role attributes for PVT-HP prosumers. That is, in one time slot, a thermal producer can be an electric consumer, and a thermal consumer can also be acted as an electric producer, which can be further shown in Figure 8. The horizontal axis represents electric net power while the vertical axis represents the thermal net power.

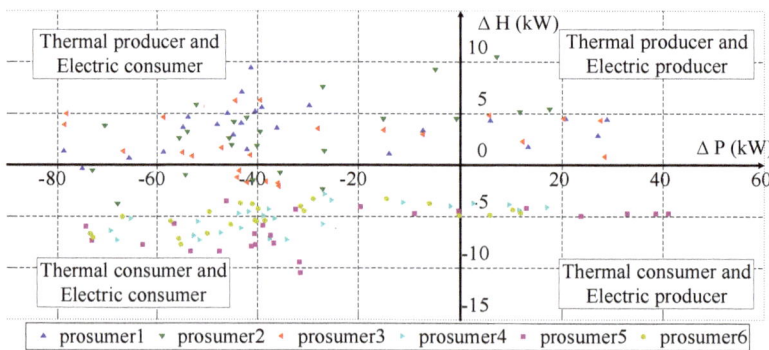

Figure 8. Roles of prosumer during the day.

The results show that the number of prosumer roles acted as electric producer is much less than the number of electric consumer. This is mainly due to the fact that the solar electric power is only excess during the periods 12:00–16:00 of daytime. In contrast, the roles of thermal producers and consumers are distributed evenly in the vertical axis, because the capacity of heat pump is matched with the load demand.

6.3. Analysis of Computation Time and Applications

With the development of smart grid, the user side has gradually been equipped with smart meters, building energy management system, as well as high speed Internet connections, which lay the hardware foundation of the consensus algorithm applications. Figures 3 and 4 show the Algorithm 2 can approach the optimum solution in 55 iterations. According to the designed Algorithm 2 in Section 5.3, only increment cost (utility) and local power mismatch are exchanged between participants at each iteration which is presented in Equations (35), (36), (44) and (45). Each participant broadcasts its value of last iteration while receives data from neighboring participants. Therefore, there are about 2 Bytes data broadcasted and 14 Bytes data received in one iteration for a participant, the total data exchange for the hour-ahead optimization is less than 110 Bytes (broadcasted) and 770 Bytes (received). If we use the LTE 230 MHZ VPN network for data exchange, the latency of each message is less than 3 s (average 2 s). Furthermore, we use a computer with Intel Core i5-4570 CPU 3.2 GHz (Intel, Santa Clara, CA, USA), 8G memory (Samsung Corporation, Seoul, Korea), and MATLAB 2014a as the time cost testing environment for the algorithm. The average computation time is 0.037 s for one iteration, combined with the communication latency, the maximum time cost of the hour-ahead optimization is about 2.78 min (average 1.87 min). Thus, the optimal scheduling of the CEI can be start at 5 min prior to the energy sharing. As mentioned above, the consensus algorithm is bound to be a good distributed algorithm in future energy management system.

7. Conclusions

In this paper, we have proposed a consensus-based energy sharing method for solving social welfare problem of the CEI. From the results of case study, we have shown that all the increment costs (utilities) convergence to a same value while the power balance is satisfied, which means that all participants achieve their own maximal profits and the social welfare maximization problem has been solved. Furthermore, if the capacity of PVT and heat pumps are properly configured and the capacity of heat pump can meet the demand of thermal load, which means the heat pump can make full use of the waste heat of PVT, the prosumers can make full use of solar energy to guarantee the thermal power balance and reduces the installed capacity of CHP. It is an interesting result that a thermal producer can be an electric consumer and a thermal consumer can also act as an electric producer in the market of the CEI. In practical work, the sun's illumination has a greater impact on the system. The energy storage devices are to be an useful tool to solve the problem. Therefor, we will take the storage system into account in our future work.

Author Contributions: The paper was a collaborative effort between the authors. N.L., B.G., Z.L., Y.W. contributed collectively to the theoretical analysis, modeling, simulation and manuscript preparation.

Funding: This work was supported in part by the National Natural Science Foundation of China under Grant 51577059 and in part by the Fundamental Research Funds for the Central Universities under Grant 2018ZD13.

Conflicts of Interest: The authors declare no conflict of interest.

References

1. Rifkin, J. The third industrial revolution: How lateral power is transforming energy, the economy, and the world. *Survival* **2011**, *2*, 67–68. [CrossRef]
2. Wang, K.; Yu, J.; Yu, Y.; Qian, Y.; Zeng, D.; Guo, S.; Xiang, Y.; Wu, J. A Survey on Energy Internet: Architecture, Approach, and Emerging Technologies. *IEEE Syst. J.* **2017**. [CrossRef]
3. Sabbah, A.I.; El-Mougy, A.; Ibnkahla, M. A Survey of Networking Challenges and Routing Protocols in Smart Grids. *IEEE Trans. Ind. Inform.* **2013**, *10*, 210–221. [CrossRef]

4.   Bui, N.; Castellani, A.P.; Casari, P.; Zorzi, M. The internet of energy: a web-enabled smart grid system. *IEEE Netw.* **2012**, *26*, 39–45. [CrossRef]

5.   Huang, A.Q.; Crow, M.L.; Heydt, G.T.; Zheng, J.P.; Dale, S.J. The Future Renewable Electric Energy Delivery and Management (FREEDM) System: The Energy Internet. *Proc. IEEE* **2010**, *99*, 133–148. [CrossRef]

6.   Huang, A. FREEDM system—A vision for the future grid. In Proceedings of the 2010 IEEE Power and Energy Society General Meeting, Providence, RI, USA, 25–29 July 2010; pp. 1–4.

7.   Ji, Z.; Sun, Y.; Wang, S.; Zhao, J. Design of a three-phase cascaded Power Electronic Transformer based on energy Internet. In Proceedings of the International Conference on Sustainable Power Generation and Supply, Hangzhou, China, 8–9 September 2012; pp. 1–6.

8.   Arnold, M.; Negenborn, R.R.; Andersson, G.; Schutter, B.D. *Distributed Predictive Control for Energy Hub Coordination in Coupled Electricity and Gas Networks*; Springer: Berlin, Germany, 2010; pp. 235–273.

9.   Rastegar, M.; Fotuhi-Firuzabad, M. Load Management in a Residential Energy Hub with Renewable Distributed Energy Resources. *Energy Build.* **2015**, *107*, 234–242. [CrossRef]

10. Ma, L.; Liu, N.; Wang, L.; Zhang, J.; Lei, J.; Zeng, Z.; Wang, C.; Cheng, M. Multi-party energy management for smart building cluster with PV systems using automatic demand response. *Energy Build.* **2016**, *121*, 11–21. [CrossRef]

11. Liu, N.; Wang, J.; Wang, L. Distributed energy management for interconnected operation of combined heat and power-based microgrids with demand response. *J. Mod. Power Syst. Clean Energy* **2017**, *5*, 478–488. [CrossRef]

12. Liu, N.; Yu, X.; Wang, C.; Li, C.; Ma, L.; Lei, J. Energy-Sharing Model With Price-Based Demand Response for Microgrids of Peer-to-Peer Prosumers. *IEEE Trans. Power Syst.* **2017**, *32*, 3569–3583. [CrossRef]

13. Liu, N.; Chen, Q.; Liu, J.; Lu, X.; Li, P.; Lei, J.; Zhang, J. A Heuristic Operation Strategy for Commercial Building Microgrids Containing EVs and PV System. *IEEE Trans. Ind. Electron.* **2015**, *62*, 2560–2570. [CrossRef]

14. Radziemska, E. The effect of temperature on the power drop in crystalline silicon solar cells. *Renew. Energy* **2003**, *28*, 1–12. [CrossRef]

15. He, W.; Chow, T.T.; Ji, J.; Lu, J.; Pei, G.; Chan, L.S. Hybrid photovoltaic and thermal solar-collector designed for natural circulation of water. *Appl. Energy* **2006**, *83*, 199–210. [CrossRef]

16. Huang, B.J.; Lin, T.H.; Hung, W.C.; Sun, F.S. Performance evaluation of solar photovoltaic/thermal systems. *Sol. Energy* **2014**, *70*, 443–448. [CrossRef]

17. Putrayudha, S.A.; Kang, E.C.; Evgueniy, E.; Libing, Y.; Lee, E.J. A study of photovoltaic/thermal (PVT)-ground source heat pump hybrid system by using fuzzy logic control. *Appl. Therm. Eng.* **2015**, *89*, 578–586. [CrossRef]

18. Tsai, H.L. Modeling and validation of refrigerant-based PVT-assisted heat pump water heating system. *Sol. Energy* **2015**, *122*, 36–47. [CrossRef]

19. Emmi, G.; Tisato, C.; Zarrella, A.; Carli, M.D. Multi-Source Heat Pump Coupled with a Photovoltaic Thermal (PVT) Hybrid Solar Collectors Technology: a Case Study in Residential Application. *Int. J. Energy Prod. Manag.* **2016**, *1*, 382–392.

20. Bertram, E.; Glembin, J.; Rockendorf, G. Unglazed PVT collectors as additional heat source in heat pump systems with borehole heat exchanger. *Energy Procedia* **2012**, *30*, 414–423. [CrossRef]

21. Liang, Y.; Liu, F.; Wang, C.; Mei, S. Distributed demand-side energy management scheme in residential smart grids: An ordinal state-based potential game approach. *Appl. Energy* **2017**, *206*, 991–1008. [CrossRef]

22. Pipattanasomporn, M.; Feroze, H.; Rahman, S. Multi-agent systems in a distributed smart grid: Design and implementation. In Proceedings of the 2009 IEEE/PES Power Systems Conference and Exposition (PSCE '09), Seattle, WA, USA, 15–18 March 2009; pp. 1–8.

23. Shi, W.; Xie, X.; Chu, C.C.; Gadh, R. Distributed Optimal Energy Management in Microgrids. *IEEE Trans. Smart Grid* **2015**, *6*, 1137–1146. [CrossRef]

24. Zhang, Z.; Chow, M.Y. Convergence Analysis of the Incremental Cost Consensus Algorithm Under Different Communication Network Topologies in a Smart Grid. *IEEE Trans. Power Syst.* **2012**, *27*, 1761–1768. [CrossRef]

25. Elsayed, W.T.; El-Saadany, E.F. A Fully Decentralized Approach for Solving the Economic Dispatch Problem. *IEEE Trans. Power Syst.* **2015**, *30*, 2179–2189. [CrossRef]

26. Yang, S.; Tan, S.; Xu, J.X. Consensus Based Approach for Economic Dispatch Problem in a Smart Grid. *IEEE Trans. Power Syst.* **2013**, *28*, 4416–4426. [CrossRef]

27. Binetti, G.; Davoudi, A.; Lewis, F.L.; Naso, D.; Turchiano, B. Distributed Consensus-Based Economic Dispatch With Transmission Losses. *IEEE Trans. Power Syst.* **2014**, *29*, 1711–1720. [CrossRef]

28. Zhao, C.; He, J.; Cheng, P.; Chen, J. Consensus-Based Energy Management in Smart Grid With Transmission Losses and Directed Communication. *IEEE Trans. Smart Grid* **2017**, *8*, 2049–2061. [CrossRef]

29. Wu, D.; Yang, T.; Stoorvogel, A.A.; Stoustrup, J. Distributed Optimal Coordination for Distributed Energy Resources in Power Systems. *IEEE Trans. Autom. Sci. Eng.* **2017**, *14*, 414–424. [CrossRef]

30. Xing, H.; Lin, Z.; Fu, M.; Hobbs, B.F. Distributed algorithm for dynamic economic power dispatch with energy storage in smart grids. *IET Control Theory Appl.* **2017**, *11*, 1813–1821. [CrossRef]

31. Asr, N.R.; Zhang, Z.; Chow, M.Y. Consensus-based distributed energy management with real-time pricing. In Proceedings of the 2013 IEEE Power and Energy Society General Meeting, Vancouver, BC, Canada, 21–25 July 2013; pp. 1–5.

32. Meng, W.; Wang, X. Distributed Energy Management in Smart Grid With Wind Power and Temporally Coupled Constraints. *IEEE Trans. Ind. Electron.* **2017**, *64*, 6052–6062. [CrossRef]

33. Liu, N.; He, L.; Yu, X.; Ma, L. Multiparty Energy Management for Grid-Connected Microgrids With Heat- and Electricity-Coupled Demand Response. *IEEE Trans. Ind. Inform.* **2018**, *14*, 1887–1897. [CrossRef]

34. Wang, J.; Sui, J.; Jin, H. An improved operation strategy of combined cooling heating andpower system following electrical load. *Energy* **2015**, *85*, 654–666. [CrossRef]

Article

Modular Predictor for Day-Ahead Load Forecasting and Feature Selection for Different Hours

Lin Lin [1],*, Lin Xue [2], Zhiqiang Hu [3] and Nantian Huang [2]

[1] College of Information and Control Engineering, Jilin Institute of Chemical Technology, Jilin 132022, China
[2] School of Electrical Engineering, Northeast Electric Power University, Jilin 132013, China;
    xuelin1428@163.com (L.X.); huangnantian@126.com (N.H.)
[3] Zhejiang Electric Power Corporation Wenzhou Power Supply Company, Wenzhou 325000, China;
    huzhiqiang19910714@163.com
* Correspondence: jllinlin@126.com; Tel.: +86-139-4420-8142

Received: 25 June 2018; Accepted: 12 July 2018; Published: 20 July 2018

Abstract: To improve the accuracy of the day-ahead load forecasting predictions of a single model, a novel modular parallel forecasting model with feature selection was proposed. First, load features were extracted from a historic load with a horizon from the previous 24 h to the previous 168 h considering the calendar feature. Second, a feature selection combined with a predictor process was carried out to select the optimal feature for building a reliable predictor with respect to each hour. The final modular model consisted of 24 predictors with a respective optimal feature subset for day-ahead load forecasting. New England and Singapore load data were used to evaluate the effectiveness of the proposed method. The results indicated that the accuracy of the proposed modular model was higher than that of the traditional method. Furthermore, conducting a feature selection step when building a predictor improved the accuracy of load forecasting.

Keywords: day-ahead load forecasting; modular predictor; feature selection

---

1. Introduction

The main idea of short-term load forecasting (STLF) is to predict future loads with horizons of a few hours to several days. Accurate STLF predictions play a vital role in electrical department load dispatch, unit commitment, and electricity market trading [1]. With the permeation of renewable resources in grids and the technological innovation of electric vehicles, load components become more complex and make STLF difficult; therefore, strict requirements of stability and accuracy are needed [2–6].

STLF is an old but worthy theme for research. General forecasting methods can be divided into two branches: the statistical method and the artificial intelligence method. Statistical methods such as regression analysis, exponential smoothing, Kalman filter, and autoregressive integrated moving average (ARIMA) are easy to apply but modeling is difficult for complex loads [7–9]. Artificial intelligence methods show better performance than statistical methods in load forecasting and include fuzzy logic, the artificial neural network (ANN), the support vector machine (SVM), Gaussian process regression (GPR), and random forest (RF) [10–17]. The relationship of input and output is confirmed by a list of rules by fuzzy logic. However, the prior knowledge required to select the parameters in the membership function and the rules makes the modeling process complex [18]. The artificial neural network method is applied to the STLF of power systems owing to its self-learning ability and robustness to data noise. However, shortcomings such as the difficulty in determining initial network parameters and over-fitting still exist [19]. By adopting a structural risk minimization principle, the complexity and the learning ability of an SVM can be balanced. With low-dimension conditions and few samples, the SVM can maintain its generalization ability. Compared to the artificial neural network, the SVM has many advantages. The parameters of the SVM should be determined through a computational optimization by algorithm such as the genetic algorithm or the particle swarm optimization algorithm [20,21]. GPR is a kernel-function-based algorithm whose transcendental function is established in the form of probability distribution, and the posterior function can be acquired by Bayesian logic. The parameter of kernel function in GPR is obtained automatically in the process of training [22]. RF is a type of integrated machine-learning algorithm based on a decision tree. The main advantages of RF are immunity to noise and insensitivity to its parameters [23].

In addition to the forecasting method, input feature selection is a vital factor that influences the accuracy and efficiency of load forecasting. A model using a few features has difficulty analyzing the effect of external conditions on the load. However, as the complexity of a model increases, the accuracy and efficiency will be influenced. Feature selection is a process of selecting a subset of variables from an original high-dimensionality variable set that retains the most efficient variables while reducing the effects of the irrelevant variables [24]. Feature selection methods can be classified as wrapper, filter, and embedded [25]. In the wrapper method, the performance of a predictor is chosen as the criterion for feature selection. An exhaustive search is performed to identify the optimal feature subset from numerous combinations of features at which the predictor performs best. However, the wrapper method needs to evaluate 2N subsets which leads to an NP-hard problem with too many features [26]. Therefore, evolutionary algorithms such as the memetic algorithm [27], the genetic algorithm [28], and the particle swarm optimization algorithm [29] can reduce the complexity of computation. Filter methods, such as mutual information (MI) and RreliefF, are ranking methods that evaluate features by analyzing the relationship between the inputs and outputs and a feature score or weight is given to each feature for ranking. To acquire an optimal feature subset, the accuracy of the predictor is used as the criterion [30]. Compared to wrapper methods, filter methods do not rely on other learning algorithms and the computational cost is light [31–33]. Embedded methods, such as the classification and regression tree (CART) and RF, which combine feature selection with a learning algorithm, analyze and compute the importance value of features in a training process [25]. Experiments need to be performed according to a specific forecasting case that considers the advantages and disadvantages of different kinds of feature selection methods, the size of training sets, and the performance of a predictor to determine the most-accurate forecasting method.

Although the performance of a predictor can be provided by feature selection, it should be noted that the load time series presents a day-cycle characteristic, which means the load characteristics at the same time on different days are similar [34]. In addition, the load at different hours of a day is affected by consumption behavior and leads to significantly different feature responses. A single predictor with a feature selection for forecasting all future load periods may not reach the load requirement of different hours, and the accuracy of the total forecast result will decrease. Therefore, a modular model that consists of several single predictors used for forecasting the load of different hours is needed. The relation of the load at different hours to be forecast and a feature could be analyzed by a modular predictor with a feature selection for a specific hour of load, and thus the accuracy can be improved [35]. In addition, in electric power dispatching, for different electric power departments, the demand of the time of submission of the STLF result is different. Therefore, when constructing a candidate feature set for STLF, the time factor should be considered.

Considering the construction of a feature set, feature selection, and modeling objects, a novel modular parallel forecasting model with feature selection for day-ahead load forecasting was proposed. First, to meet the requirement of the dispatch department and electricity market, the load time series which records every hour according to different forecasting moments was reconstructed to a different load sub-time series. Second, the candidate feature set included 173 features extracted from historic load and calendar. Then, five feature selection methods—MI, conditional mutual information (CMI), RreliefF, CART, and RF—were used to analyze the importance between each feature and different prediction targets and to rank the features in descending order. Third, combined with various predictors, the sequential forward-selection algorithm and a decision criterion based on the mean absolute percentage error (MAPE) were utilized to obtain optimal feature subsets corresponding to different prediction targets. Finally, the optimal modular predictor including several optimal sub-predictors with optimal feature subsets for different forecasting periods was built. The optimal combination method was determined by comparing the forecast results. The proposed method was tested through a day-ahead load forecasting experiment using actual load data from New England and Singapore.

2. Feature Selection

The input feature (variable), as one of the key factors in a predictor build, has a significant influence on the accuracy of the predictor in day-ahead load forecasting. In this study, the filter method and embedded method were adopted for feature selection before building the predictor.

2.1. Filter Method of Feature Selection

The filter method is a feature ranking method that computes a feature's numerical value to evaluate its importance. Therefore, the estimation of a feature is important to the feature selection result. MI, CMI, and RreliefF methods were used as filters in this study.

2.1.1. Mutual Information

The Mutual Information (MI) method measures the common information between two random variables. For two random variables X and Y, the MI between X and Y can be estimated as:

$$I(X,Y) = \sum_{X,Y} P(x,y) \log \frac{P(x,y)}{P(x)P(y)} \tag{1}$$

where $P(x)$ and $P(y)$ are the marginal density functions corresponding to X and Y, respectively. $P(x,y)$ is the joint probability density function. In load forecasting, the feature is defined as X, the target variable is defined as Y, and $I(X,Y)$ represents their strength of relevance. The larger $I(X,Y)$ is, the more dependent X is. If $I(X,Y)$ is zero, X and Y are independent. The MI method can measure the relevance between a feature and a target variable effectively; however, the redundancy is analyzed differently.

2.1.2. Conditional Mutual Information

The Conditional Mutual Information (CMI) method measures the relevance of two variables when the variable Z is known. In the feature selection of load forecasting, let us suppose the selected feature set is S and the CMI between feature X_i and target Y is defined as:

$$I(Y;X_i|S) = I(Y;S|X_i) - I(Y;S) \tag{2}$$

where $I(Y;X_i|S)$ represents the new information that X_i supplies to S. The larger $I(Y;X_i|S)$ is, the more information X_i can supply, and the less is the redundancy to S. Compared to the MI method, the redundancy among features can be reduced by CMI.

2.1.3. RreliefF

RreliefF is the extended version of relief for regression [36]. By evaluating the feature weight, the feature quality is measured. Relief works by randomly selecting an instance and searching the nearest neighbor from the same class and from a different class. The weight $W[X_i]$ of feature X_i estimated by relief is an approximation of the difference of probabilities:

$$\begin{aligned} W[X_i] \quad &= \quad P(\text{diff, value of } X_i | \text{nearest inst. from diff. class}) \\ &- \quad P(\text{diff, value of } X_i | \text{nearest inst. from same. class}) \end{aligned} \tag{3}$$

For RreliefF, the probability of two instances belonging to different classes can be evaluated by their relative distances for classification. However, for STLF, the predicted value is continuous; therefore, Equation (3) should be reformulated. By using Bayes' theorem, $W[X_i]$ can be obtained as:

$$W[X_i] = \frac{P_{diffC|diffX_i}P_{diffX_i}}{P_{diffC}} - \frac{(1 - P_{diffC|diffX_i})P_{diffX_i}}{1 - P_{diffC}} \tag{4}$$

2.2. Embedded Method for Feature Selection

In the embedded method, feature selection is performed during the training process where the contribution of the feature combination is efficiently evaluated. The embedded method can be directly applied to STLF and can collaborate with other feature selection methods according to their estimated importance.

2.2.1. Classification and Regression Tree

The Classification and Regression Tree (CART) method uses a binary recursive partitioning algorithm [37]. By splitting the current samples into two sub-samples, a father node generates two child nodes. The final model of CART is a simple binary tree.

The generation of the CART can be divided into two steps:

Step one: first, the root node is split. A best feature X^{bset} chosen from the feature set serves as the criterion for node splitting. To select the best feature, the minimum variance of child nodes is the objective function. The variance of the child node of X_i is defined as:

$$\text{var}(q) = \sum_{X_i \in q} (y_i - \bar{y}_q)^2 \tag{5}$$

where $\bar{y}_q$ is the average of observation values y_i at node q. The importance of feature X_i according to the variance is defined as:

$$V_C(X_i) = \frac{1}{\sum\limits_{X_i \in q} (y_i - \bar{y}_q)^2} \tag{6}$$

Step two: for each child node, repeat Step one until the CART grows completely. The predictive model can be expressed as $t(x, T)$, where $T = (x_i, y_i)$, $i = 1,2,\ldots,n$ and $x \in R$ is the training set. For STLF, the forecasting value of load $\hat{y}$ is obtained when inputting the new $\hat{x}$.

$$\hat{y} = t(\hat{x}, T) \tag{7}$$

2.2.2. Random Forest

Random Forest (RF) is a machine-learning algorithm that uses a combination of CART with a bootstrap sample for classification and regression [38]. For a training set T with n samples, the bootstrap sample means randomly selecting n samples from T replacements. The probability that each sample selected is $1/n$, means one sample may appear several times. After a complete bootstrap sample, the samples that were not sampled form the out-of-bag (OOB) dataset. Different from CART, the feature for node splitting in RF is selected from m features which are chosen from the original feature set. The basis of selecting the best feature for node splitting is Equation (5). The predictive output of RF is obtained by averaging the results of the trees:

$$\hat{y} = \frac{1}{N_t} \sum_{i=1}^{N_t} t(\hat{x}, T^i) \tag{8}$$

where N_t is the number of trees.

In addition, the OOB error and the importance of each feature are computed in the process of modeling. Each tree has an OOB dataset, and the OOB error is evaluated by predicting the OOB dataset using the tree model corresponding to the OOB dataset. The OOB error is defined as:

$$e = \frac{1}{N_t} \sum_{i=1}^{N_t} (y_i - \hat{y}_i)^2 \tag{9}$$

A feature's importance is estimated by permutating the feature and averaging the difference of OOB errors before and after the permutation of all trees. For instance, for the ith tree whose OOB data is OOB_i and OOB error is e_i, after permutation, the new OOB data will be OOB_i' and the OOB error will be e_i'. The feature's importance in this tree is computed as:

$$VI_i = e_i' - e_i \tag{10}$$

3. The Short-Term Load Forecasting (STLF) Predictor

Selecting an appropriate predictor is key to improving the accuracy of STLF. Five state-of-the-art predictors were applied in this study: support vector regression (SVR), back-propagation neural network (BPNN), CART, GPR, and RF. The SVR, BPNN, and GPR are introduced briefly in this section. The detailed mathematical theories of these algorithms are shown in the references [39–41].

3.1. Support Vector Regression

By using the non-sensitive loss function, an Support Vector Regression (SVM), which is used only for classification, is extended for regression to be applied for load forecasting in power systems and is called support vector regression (SVR).

Given a training set T, the model for the load that decreases the difference between the predictive value $f(x)$ and the true load y as much as possible is expected to be:

$$f(x) = \omega^T x + b \tag{11}$$

In SVR, the maximum difference that can be tolerated between $f(x)$ and y is ε. The mathematical model can be expressed as:

$$\begin{cases} \max\limits_{\alpha,\alpha^*} \left[-\frac{1}{2} \sum\limits_{i=1}^{n} \sum\limits_{j=1}^{n} (\alpha_i - \alpha_i^*)(\alpha_j - \alpha_j^*) K(x_i, x_j) - \sum\limits_{j=1}^{n} (\alpha_i + \alpha_i^*)\varepsilon + \sum\limits_{i=1}^{n} (\alpha_i - \alpha_i^*)y_i \right] \\ s.t. \begin{cases} \sum\limits_{i=1}^{n} (\alpha_i - \alpha_i^*) = 0 \\ 0 \le \alpha_i, \alpha_i^* \le C \end{cases} \end{cases} \tag{12}$$

where C is the regularization parameter, $K(x_i, x_j) = \varphi(x_i)\varphi(x_j)$ is the kernel function, and α_i, α_i^* are Lagrange factors.

The radial basis function selected in this study is expressed as:

$$K(x_i, x_j) = \exp\left(-\frac{\|x_i - x_j\|^2}{2\sigma^2} \right) \tag{13}$$

where σ^2 is the kernel width.

The SVR model is obtained by solving Equation (12):

$$f(x) = \sum_{i=1}^{n} (\alpha_i - \alpha_i^*) K(x_i, x) + b \tag{14}$$

where b is the bias value.

3.2. Back-Propagation Neural Network

The Back-Propagation Neural Network (BPNN) is a type of artificial neural network consisting of an input layer, a hidden layer, and an outer layer trained by a back-propagation algorithm with the mean squared error (MSE) as the objective function. The main idea of the BPNN is to deliver the output-layer error from back to front by which the error of the hidden layer is computed. The learning process of BPNN is divided into two steps:

Step 1: The output of each neural unit in the input and hidden layers is estimated.

Step 2: By using the output error, the error of each neural unit which is used for updating the former layer weight is computed.

The objective function of the gradient minimization is based on:

$$e_f = \frac{1}{2} \sum_i (y_i - \hat{y}_i)^2 \tag{15}$$

where y_i is the actual value of neural unit i and $\hat{y}_i$ is the predictive value. To compute the minimum value of e_f, a modification value is needed to correct the weight. The modification value is defined as:

$$\Delta w_{ij}(t) = -\eta \frac{\partial e}{\partial w_{ij}} + \alpha \Delta w_{ij}(t-1) = -\eta \frac{\partial e}{\partial net_i} \frac{\partial net_i}{\partial w_{ij}} + \alpha \Delta w_{ij}(t-1) = -\eta \delta_i O_j + \alpha \Delta w_{ij}(t-1) \tag{16}$$

$$net_i = \sum_j w_{ij} O_j \tag{17}$$

$$O_i = \frac{1}{1 + e^{-net_i}} \tag{18}$$

where η is the learning rate, net_i is the input of neuron i, O_i is the output of neuron i, and α is the momentum factor.

The modified weight is:

$$w_{ij}(t+1) = w_{ij}(t) + \Delta w_{ij} \tag{19}$$

The final output $\hat{y}_i$ of neuron i can be estimated by the iteration of weight w_{ij} when meeting precision requirements.

3.3. Gaussian Process Regression (GPR)

Gaussian Process Regression (GPR) is a random process in which the random variables obey the Gaussian distribution and is used to establish the input and output maps. For STLF, the load data collected is polluted by noise. Assuming that the noise follows a normal distribution $\varepsilon \sim N(0, \sigma_n^2)$, then the joint prior distribution of observation y and the predictive value f^* are defined as:

$$\begin{bmatrix} y \\ f^* \end{bmatrix} \sim N \left(0, \begin{bmatrix} K(X,X) + \sigma_n^2 I_n & K(X, x^*) \\ K(x^*, X) & k(x^*, x^*) \end{bmatrix} \right) \tag{20}$$

where n is the number of training samples, $K(X, X)$ is the covariance matrix, and I_n is the unit matrix.

The posterior distribution of f^* is defined as:

$$f^* | X, y, x^* \sim N \left[\overline{f}^*, \text{cov}(f^*) \right] \tag{21}$$

where $\overline{f}^*$ is the mean value of f^* and $\text{cov}(f^*)$ is the variance of f^*.

Then, $\overline{f}^*$ and $\text{cov}(f^*)$ can be computed as:

$$\begin{cases} \overline{f}^* = K(x, X) \left[K(X, X) + \sigma_n^2 I_n \right]^{-1} y \\ \text{cov}(f^*) = k(x^*, x^*) - K(x^*, X) \times \left[K(X, X) + \sigma_n^2 I_n \right]^{-1} K(X, x^*) \end{cases} \tag{22}$$

The covariance function of GPR is the squared exponential function:

$$k(x, x\prime) = \sigma_f^2 exp\left[-\frac{1}{2}(x - x\prime)M^{-1}(x - x\prime)\right] \qquad (23)$$

where $\theta = \left\{M, \sigma_f^2, \sigma_n^2\right\}$ is a hyper-parameter that can be solved by the maximum likelihood method [41].

4. Data Analysis

4.1. Load Analysis

Affected by different factors, load sequence appears as a type of complicated non-linear time series. To construct a reasonable original feature set and achieve better forecasting for a region, the load characteristics and other factors should be analyzed.

Figure 1 shows the power load of New England in different time lengths. By observing Figure 1a,b, the load patterns from 2011 to 2013 are similar. Influenced by climate, load patterns differ by season. In Figure 1c, the load curves of two continuous weeks in four seasons are presented (the first day is Monday). It is easy to see that the weekday and weekend load demands differ, and the load demand presented a cycling mode with a period of seven days. The Tuesday load curves of the different seasons shown in Figure 1d shows that the Tuesday load pattern of different weeks is similar. The load increased rapidly from 6:00 am to 11:00 am, which corresponds to the beginning of work, and reached the first peak load. The second peak load occurred from 19:00 pm to 20:00 pm.

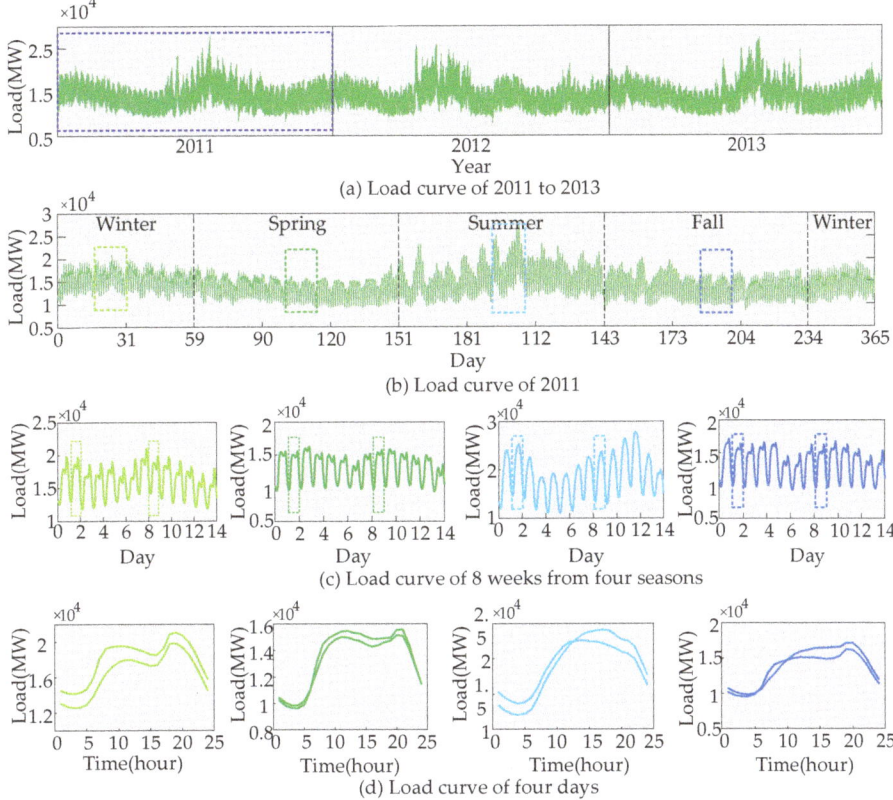

Figure 1. The power load of New England.

As analyzed above, the load characteristics can be summarized as

(1) The same-day load patterns are similar and represent the week-cycle of the load.

(2) The weekday and weekend load patterns were similar respectively and represent the day-cycle of the load.

4.2. Candidate Feature Set

An appropriate feature set plays a significant role in modeling an uncomplicated but outstanding predictor. However, a candidate feature set that contains sufficient information must be found to ensure that effective features can be selected by the feature selection method. The two main feature types are the endogenous predictor (load feature) and the exogenous predictor (calendar feature).

The time interval before the predictive moment before submission of the dispatch department's forecasting result should be considered when extracting features. To ensure the universality of the original feature set, we used the interval time $p = 24$. A feature set consisting of 145 internal historic load features (from lag 24 to lag 168) from a one-week data window was chosen as a part of the candidate feature set. The maximum, minimum, and mean loads were also included. Except for the load feature, calendar features such as hour of day, day type, working day, and non-working day were also considered. The candidate feature set with 173 features was formed as shown in Table 1.

Table 1. The feature information.

Feature Type	Feature Name	Feature Number
Endogenous predictor	$F_{L(t-i)}$, $i = 24, 25, \ldots , 168$	145
	$F_{L(\max,d-k)}$, $F_{L(\min,d-k)}$, $F_{L(\text{mean},d-k)}$, $k = 2, 3, 4, 5, 6, 7$	18
Exogenous predictor	F_D^W, $W = 1, 2, 3, 4, 5, 6, 7$	7
	F_W	2
	F_{Hour}	1

Feature explanation:

Endogenous predictor:

$F_{L(\max,d-k)}$ is the maximum power load k days before, $k = 2, 3, 4, 5, 6, 7$.

$F_{L(\min,d-k)}$ is the minimum power load k days before, $k = 2, 3, 4, 5, 6, 7$.

$F_{L(\text{mean},d-k)}$ is the average power load k days before, $k = 2, 3, 4, 5, 6, 7$.

$F_{L(t-i)}$ is the historic power load i hours before the forecasting hour t, and $i = 24, 25, 26, \ldots , 168$.

Exogenous predictor:

F_D^W is the day of week, which is signed by 0 or 1 (W = 1, 2, 3, 4, 5, 6, 7 represents Monday to Sunday).

F_W is work day or non-work day (0 is a work day and 1 is a non-work day).

F_{Hour} is the moment of hour (1 to 24).

5. Experimental Setup

5.1. Proposed STLF Process with Feature Selection

Figure 2 provides an overview of the proposed method which covers the construction of the feature set, the dataset separation, and the feature selection for the load with respect to the different hours and the modeling for different hours. Figure 2a shows the one-day structure of a sample. The inputs include 173 features, and the output is the predicted load.

The diagram of the proposed method is displayed in Figure 2b. The training set was separated into 24 training subsets corresponding to each hour. The features in each training subset were ranked in descending order according to their feature scores as computed by the feature selection method. Then, the optimal feature subset was selected using the predictor and the MAPE-based criteria. Finally, the modular predictor was constructed based on 24 predictors with the obtained optimal subsets.

The process of selecting the optimal feature subset in modeling is shown in Figure 2c. According to the ranked feature order, the predictor was used to test the feature subset consisting of the top i features, and the criteria based on the MAPE was used to select the optimal feature subset.

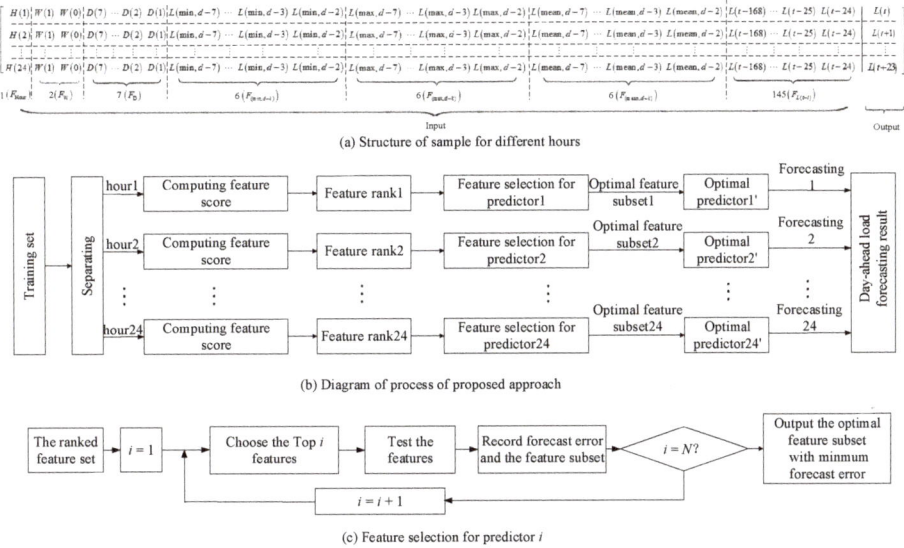

(a) Structure of sample for different hours

(b) Diagram of process of proposed approach

(c) Feature selection for predictor *i*

Figure 2. Overview of the proposed method.

5.2. Dataset Split

The data used in this study were from New England [42] and Singapore [43]. The New England data were recorded every hour from 2011 to 2013 for a total of 26,304 data points. The Singapore data were recorded every half hour from 2014 to 2015 for a total number of 35,040 data points. To apply the proposed method, the hourly load of Singapore was extracted to form a new load time series. The data used for training and testing the predictor consisted of the feature set (173 features) and the predictive object (the load corresponding to different hours) as shown in Figure 2.

Each dataset was split into three parts: a training set (14,616 New England samples, 11,712 Singapore samples), a validation set (2928 New England samples, 2094 Singapore samples), and a test set (8760 New England samples, 2904 Singapore samples). The training and the validation sets were used to build the predictor and to select an optimal feature subset. The test set was used to examine the performance of the feature subset and the predictor. Detailed information about the datasets is shown in Table 2.

Table 2. Experimental data description.

Data Set	Detail Information of Experimental Data (New England)			Detail Information of Experimental Data (Singapore)	
	2011	**2012**	**2013**	**2014**	**2015**
Training set	Jan., Feb., Mar., Apr., May, Jun., Jul., Aug., Sept., Oct., Nov., Dec.	Jan., Feb., Apr., Jun., Jul., Aug., Oct., Dec.	-	Jan., Feb., Mar., Apr., May, Jun., Jul., Aug., Sept., Oct., Nov., Dec.	Jan., Apr., Aug., Dec.
Validation set	-	Mar., May, Sept., Nov.	-	-	Feb., May, Jul., Oct.
Test set	-	-	Jan., Feb., Mar., Apr., May, Jun., Jul., Aug., Sept., Oct., Nov., Dec.	-	Mar., Jun., Sept., Nov.

5.3. Evaluation Criterion

To evaluate the performance of the proposed method, three criteria, the MAPE, the mean absolute error (MAE), and the root mean square error (RMSE) were used as follows:

$$MAPE = \frac{1}{n}\sum_{i=1}^{n}\left|\frac{y_i - \hat{y}_i}{y_i}\right| \times 100\% \tag{24}$$

$$MAE = \frac{1}{n}|y_i - \hat{y}_i| \tag{25}$$

$$RMSE = \sqrt{\frac{1}{n}(y_i - \hat{y}_i)^2} \tag{26}$$

where y_i is the actual load, $\hat{y}_i$ is the predictive load, and n is the number of predictive loads.

6. Results

The software used were MATLAB 2016b (Version 9.1.0.441655, Mathworks Inc., Natick, MA, USA) and Rx64 3.3.2 (Version 3.3.2, GUN Project, developed at Bell Laboratories). It is noted that the CART algorithm in the rpart package in R identifies part of the features whose total importance value is 100. The parameter of each predictor was set by:

BPNN: the number of neurons in the hidden layer was $N_{neu} = 2 \times N_{feature} + 1$, iteration $T = 1000$ [44].

SVR: the regularization parameter $C = 1$, the non-sensitive loss function $\varepsilon = 0.1$, the kernel width $\delta^2 = 2$ [15].

RF: $m = N_{feature}/3$ and $N_{Tree} = 500$ [16,23].

CART: no pruning parameter was set because the tree grows completely.

GPR: the parameter of GPR was tuned by learning the training data.

6.1. Load Forecasting for New England

6.1.1. Feature Selection for Different-Hour Loads

Feature Score for Feature Analysis

Feature selection methods rate the importance of a feature by assigning a numerical value to represent the relation between the feature and the target. For example, the value of a feature computed by MI is called an MI value, while that computed by RF and CART is called its permutation importance. The feature score is used for easy description. Parts of normalized feature score curves computed by different feature selection methods are shown in Figure 3. The feature score curves of typical hours (hour 5, hour 6, hour 10, and hour 11 when the valley and peak loads appear) were chosen for analysis. The feature score curves that used the same feature score calculation method were different at various hours. For example, the MI curves were much different for hour 5, hour 6, hour 10, and hour 11, and the features with the highest scores were different from each other (marked by a red circle).

The feature score shows the importance between the feature and the target variable. When selecting a feature subset, the feature with the highest score should be retained and one with the lowest should be eliminated.

The top 10 features after ranking are shown in Table 3, where it is clear that the top 10 features for the same hour were similar. For example, for hour 5, the same top 10 features were selected by the various methods such as $F_{L(t-24)}$, $F_{L(t-25)}$, $F_{L(t-26)}$, and $F_{L(t-27)}$ and similar features such as $F_{L(t-28)}$, $F_{L(t-29)}$, $F_{L(t-30)}$, and $F_{L(t-31)}$. However, there was an obvious difference in the features of hour 5 and hour 6 which may have been caused by the different load characteristics shown in Figure 1d.

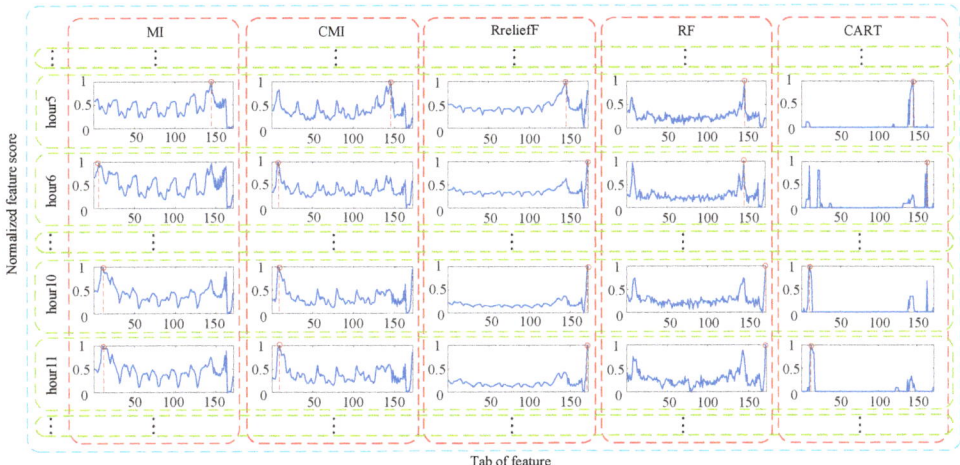

Figure 3. Normalized feature score of features evaluated by kinds of feature selection methods.

Table 3. Top 10 features of ranked of feature by different feature selection corresponding to Figure 3.

	MI	CMI	RreliefF	RF	CART
Hour 5	$F_{L(t-24)}, F_{L(t-25)},$ $F_{L(t-26)}, F_{L(t-27)},$ $F_{L(t-28)}, F_{L(t-29)},$ $F_{L(min,d-2)}, F_{L(t-30)},$ $F_{L(mean,d-2)}, F_{L(t-44)}$	$F_{L(t-24)}, F_{L(t-25)},$ $F_{L(t-29)}, F_{L(t-28)},$ $F_{L(t-160)}, F_{L(t-26)},$ $F_{L(t-161)}, F_{L(t-162)},$ $F_{L(t-27)}, F_{L(max,d-2)}$	$F_{L(t-24)}, F_{L(t-25)},$ $F_{L(t-26)}, F_{L(t-27)},$ $F_{L(t-28)}, F_W^0, F_W^1,$ $F_{L(t-28)}, F_{L(max,d-2)},$ $F_{L(t-31)}$	$F_{L(t-24)}, F_{L(t-25)},$ $F_{L(t-163)}, F_{L(t-162)},$ $F_{L(t-26)}, F_{L(t-164)},$ $F_{L(t-30)}, F_{L(t-29)},$ $F_{L(t-160)}, F_{L(t-27)}$	$F_{L(t-24)}, F_{L(t-25)},$ $F_{L(t-26)}, F_{L(t-27)},$ $F_{L(t-28)}, F_{L(t-30)},$ $F_{L(t-163)}, F_{L(t-160)},$ $F_{L(t-161)}, F_{L(t-162)}$
Hour 6	$F_{L(t-160)}, F_{L(t-162)},$ $F_{L(t-161)}, F_{L(t-24)},$ $F_{L(t-164)},$ $F_{L(mean,d-7)},$ $F_{L(t-163)}, F_{L(t-159)},$ $F_{L(t-28)}, F_{L(t-29)}$	$F_{L(t-161)}, F_{L(t-162)},$ $F_{L(t-160)}, F_{L(t-163)},$ $F_{L(t-159)}, F_{L(t-29)},$ $F_{L(t-145)}, F_{L(t-158)},$ $F_{L(t-141)}, F_{L(t-65)}$	$F_W^0, F_W^1, F_D^7,$ $F_{L(t-24)}, F_{L(t-25)},$ $F_{L(t-26)}, F_D^1,$ $F_{L(t-28)}, F_{L(t-27)},$ $F_{L(t-29)}$	$F_{L(t-24)}, F_{L(t-162)},$ $F_{L(t-161)}, F_{L(t-160)},$ $F_{L(t-30)}, F_{L(t-29)},$ $F_{L(t-25)}, F_W^0,$ $F_{L(t-163)},$ $F_{L(mean,d-7)}$	$F_{L(mean,d-7)},$ $F_{L(t-159)},$ $F_{L(t-147)}, F_{L(t-146)},$ $F_{L(t-148)}, F_{L(max,d-7)},$ $F_{L(t-24)}, F_{L(t-25)},$ $F_{L(t-30)}, F_{L(t-26)}$
Hour 10	$F_{L(t-158)}, F_{L(t-159)},$ $F_{L(t-157)},$ $F_{L(mean,d-7)},$ $F_{L(t-160)}, F_{L(t-156)},$ $F_{L(t-24)}, F_{L(t-154)},$ $F_{L(t-147)}, F_{L(t-153)}$	$F_{L(t-161)}, F_{L(t-160)},$ $F_{L(t-162)}, F_W^0, F_W^1,$ $F_{L(t-159)}, F_{L(t-158)},$ $F_{L(t-157)}, F_{L(t-154)},$ $F_{L(t-155)}, F_{L(t-159)}$	$F_W^0, F_W^1, F_D^7, F_D^6,$ $F_{L(t-26)}, F_{L(t-25)},$ $F_{L(t-27)}, F_{L(t-24)},$ $F_{L(t-28)}, F_D^1$	$F_{WW}^1, F_W^0, F_{L(t-159)},$ $F_{L(t-25)}, F_{L(t-160)},$ $F_{L(t-24)}, F_{L(t-161)},$ $F_{L(t-26)}, F_{L(t-28)},$ $F_{L(t-27)}$	$F_{L(t-159)}, F_{L(t-158)},$ $F_{L(t-160)}, F_{L(t-157)},$ $F_{L(mean,d-7)},$ $F_{L(t-156)},$ $F_{L(t-25)}, F_{L(t-27)},$ $F_{L(t-28)}, F_{L(t-26)}$
Hour 11	$F_{L(t-159)}, F_{L(t-157)},$ $F_{L(t-158)}, F_{L(t-156)},$ $F_{L(mean,d-7)},$ $F_{L(t-153)}, F_{L(t-155)},$ $F_{L(t-152)},$ $F_{L(t-160)}, F_{L(t-154)}$	$F_{L(t-160)}, F_{L(t-162)},$ $F_{L(t-161)}, F_{L(t-159)},$ $F_{L(t-158)}, F_W^0, F_W^1,$ $F_{L(t-154)}, F_{L(t-156)},$ $F_{L(t-155)}$	$F_W^0, F_W^1,$ $F_D^7, F_{L(t-26)},$ $F_{L(t-27)}, F_{L(t-25)},$ $F_{L(t-33)}, F_{L(t-24)},$ $F_{L(t-34)}, F_{L(t-28)}$	$F_W^1, F_W^0, F_{L(t-26)},$ $F_{L(t-27)}, F_{L(t-25)},$ $F_{L(t-161)}, F_{L(t-157)},$ $F_{L(t-160)}, F_{L(t-24)},$ $F_{L(t-158)}$	$F_{L(t-157)}, F_{L(t-156)},$ $F_{L(t-155)}, F_{L(t-153)},$ $F_{L(t-154)}, F_{L(t-158)},$ $F_{L(t-26)}, F_{L(t-25)},$ $F_{L(t-27)}, F_{L(t-28)}$

Therefore, a feature analysis for each hour is required to choose the best features for improving the accuracy of STLF.

Optimal Feature Subset Selection Process

According to the trend of feature score curves of diverse feature selection methods, the first 36 to 50 features are chosen as the optimal features for modeling [30]. By analyzing the autocorrelation of the lag variables, 50 features were selected for very-short-term load forecasting [41]. When selecting a feature subset, most studies did not give a specific threshold for selecting the optimal feature subset. In this study, the performance of features which ranked in descending order based on feature score were estimated by the MAPE which was chosen as the threshold for selecting the optimal feature subset by adding features one-by-one to the feature subset.

Figure 4 shows the MAPE curves of different feature selection methods and predictor-based feature selection processes. As shown in each subplot in Figure 4, the MAPE was reduced and reached a minimum value with an increase in the number of features. For example, the MAPE of MI for hour 5 and the MAPEs of BPNN, CART, GPR, RF, and SVR when using the top feature were 4.587%, 4.743%, 4.618%, 5.196%, and 4.718%, respectively. When 20 features were used, the MAPEs were reduced to 3.901%, 4.555%, 4.008%, 4.160%, and 3.831%, respectively. The MAPEs of different predictors decreased in different levels, indicating that the 20 features made a positive contribution to a better prediction model build. A similar conclusion can be summarized by analyzing other curves. The dimension of each optimal feature subset and its MAPE is marked by different colored circles corresponding to different predictors.

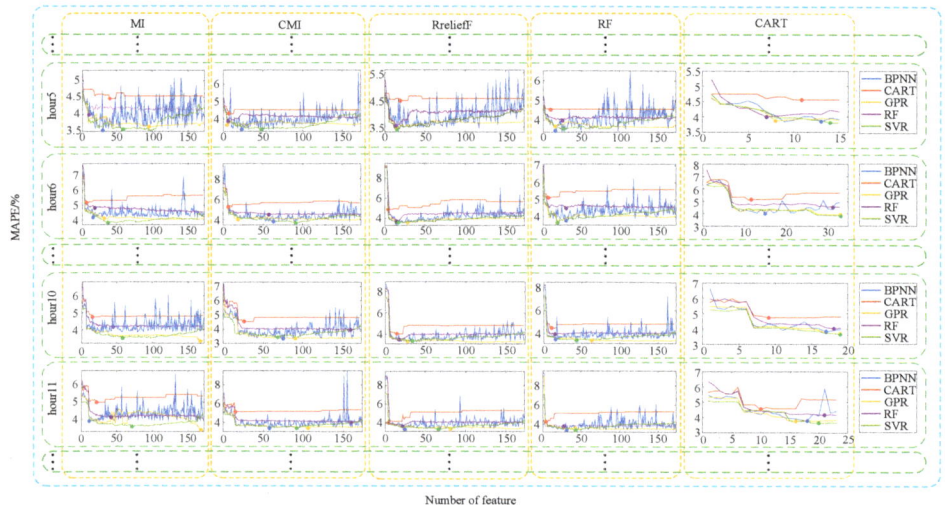

Figure 4. Mean absolute percentage error (MAPE) curves of combinations of feature selection methods and forecasting methods for selecting feature subset.

The following conclusions can be drawn from Table 3 and Figure 4:

(1) The feature permutation estimated by different feature selection methods varies.

(2) The dimension of the optimal feature subset and its MAPE depends on the predictor-based feature selection method.

(3) The optimal feature subset selected by the same predictor-based feature selection method for the predictive target of different hours is different.

Table 4 shows the MAPE and the dimension of the optimal feature subset corresponding to using MI as the feature selection method and RF as the prediction model. The Table shows that 1 to 6 am, the dimension of optimal feature subset is less than at 7 to 19 pm, as the same as the forecasting error. This is because people are less active at night and there are fewer factors affecting the load than during the day.

The MAPE and the dimension of optimal feature subset corresponding to 1:00 were carried out by different feature selection methods and forecasting methods shown in Table 5. The MAPEs are in 3% to 4% which means the performance of forecasters were similar after feature selection. By analysis of the feature dimension, we could find there is huge difference between the number of the feature of the optimal feature subset that selected by different feature selection methods, which caused by the different evaluation criterions.

Table 4. Optimal feature subset construction of different hours with mutual information (MI) + random forest (RF) for New England.

Time	MAPE	FD	Time	MAPE	FD
1	3.294	34	13	4.663	41
2	3.419	22	14	4.926	33
3	3.632	9	15	5.190	38
4	3.783	30	15	5.351	46
5	4.008	9	17	5.547	31
6	4.828	18	18	5.358	98
7	5.456	61	19	5.117	136
8	5.314	59	20	4.506	23
9	4.526	64	21	4.376	28
10	4.171	45	22	4.779	9
11	4.147	42	23	4.794	41
12	4.414	67	24	4.847	72

Remark: FD means the feature dimension.

Table 5. Optimal feature subset construction of 1:00 with different methods for New England.

Method	CART		RF		SVR		ANN		GPR	
	MAPE	FD	MAPE	FD	MAPE	FD	MAPE	FD	MAPE	FD
MI	3.741	7	3.294	34	3.064	10	3.226	8	3.087	119
CMI	3.729	2	3.447	20	3.043	13	3.062	47	3.052	134
CART	3.729	3	3.422	11	3.068	11	3.270	9	3.245	8
RF	3.741	7	3.533	51	3.140	41	3.069	18	3.099	81
RreliefF	3.741	10	3.310	26	3.043	18	3.269	9	3.019	134

The details of the dimension of the optimal feature subset and its MAPE are shown in Appendix A Table A1 to Table A2. Based on a longitudinal comparison, the dimension of optimal feature subsets selected by different feature selection methods with same-hour predictors were different. For instance, the horizon of the hour-2 MAPE calculated by various methods was from 3.107% to 4.050%. The combination RreliefF + SVR method had the smallest MAPE and lower feature subset dimension.

By the horizontal comparison, the dimension of optimal feature subsets selected by the same feature selection method with the same-hour predictor varied. For example, the horizon of dimension of the feature subset corresponding to different hours selected by the RreliefF + SVR method ranged from 13 to 109 and the MAPE range was 3.043% to 4.558%. In addition, the number of features for a night hour was less than the day hour, indicating that the day load components were more complex and more difficult to forecast.

In conclusion, the characteristic of the load to predict for different hours varies; therefore, the load needs a special feature set to build a predictor for special hours. The necessity of using one kind of structure of modular time-scale prediction and feature selection for the load of different hours was verified.

6.1.2. Forecasting Result of Method Combinations with Optimal Feature Subsets for New England Load Data

To test the performance of diverse method combinations with the optimal feature subset, we used a special week for our experiment.

The effect of temperature on the loads in summer and winter is large, and severe fluctuations make accurate forecasting difficult. Therefore, two weeks were chosen randomly from the summer and winter of 2013 for testing. The summer period was from 28 July to 3 August, and the winter period was from 22 to 28 December. As shown in Figure 5, the predictive load of each combined method was fit with the true summer load. The average error of the various methods are shown in Table 6. The top-three combined methods were CART + SVR, RreliefF + RF, and RreliefF + SVR, and the MAPEs were 3.634%, 3.710%, and 4.204%, respectively. The predictive load of each combined method in winter is shown in Figure 6, each of the predicted loads matched the actual load except for Tuesday and Wednesday which corresponded to Christmas day and the day before. As is shown in Table 7, the first three combined methods were RreliefF + SVR, CART + GPR, and CART + SVR, and the MAPEs were 4.207%, 4.754%, and 4.770%, respectively.

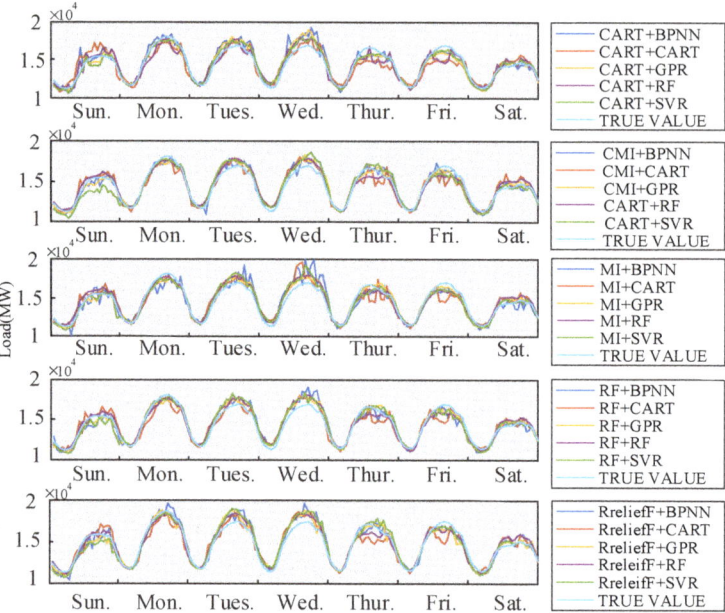

Figure 5. Prediction from 28 July to 3 August 2013.

Table 6. Comparison of different combined methods.

Method		CART	RF	GPR	BPNN	SVR
	MAPE	5.027	4.376	4.223	5.705	4.286
MI	MAE	849.194	732.926	709.848	962.649	720.361
	RMSE	1191.968	871.897	988.378	1323.916	921.862
	MAPE	4.672	4.423	4.299	4.457	4.880
CMI	MAE	784.550	719.337	699.910	566.609	809.988
	RMSE	1016.001	931.743	942.492	715.524	1027.936
	MAPE	6.179	4.936	4.449	4.910	3.634
CART	MAE	1034.009	833.712	752.653	823.088	599.284
	RMSE	1282.515	1077.501	961.304	1142.275	753.655
	MAPE	4.936	4.231	4.381	4.291	4.262
RF	MAE	833.712	711.268	815.776	711.438	705.789
	RMSE	1077.501	855.686	915.139	969.156	916.701
	MAPE	4.577	3.710	4.239	4.270	4.204
RreliefF	MAE	786.561	629.120	717.094	710.419	700.174
	RMSE	1072.662	781.775	1045.609	922.320	910.103

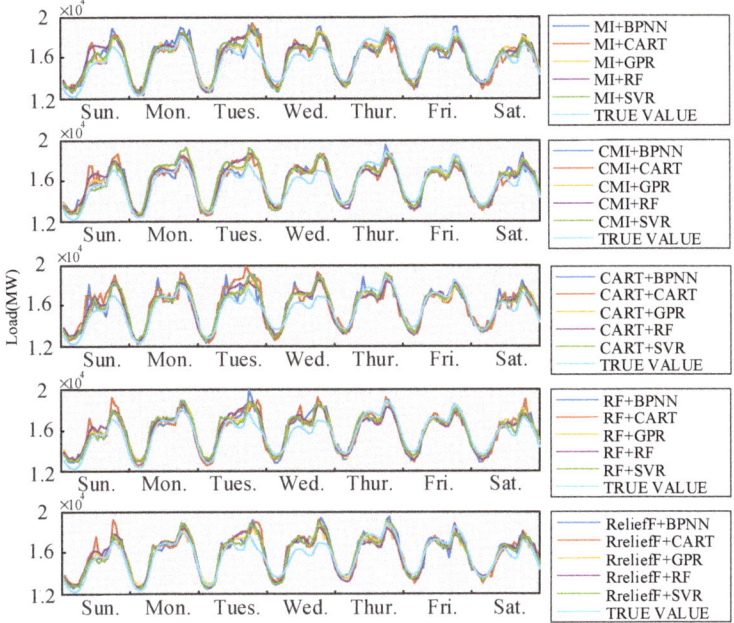

Figure 6. Prediction from 22 to 28 December 2013.

Table 7. Comparison of different combined methods.

Method		CART	RF	GPR	BPNN	SVR
	MAPE	5.420	5.783	4.862	5.823	4.977
MI	**MAE**	809.153	855.560	706.073	868.632	734.877
	RMSE	1052.017	1038.861	875.357	1059.056	897.331
	MAPE	5.479	5.515	4.862	5.072	5.262
CMI	**MAE**	814.890	821.482	710.464	733.701	788.030
	RMSE	1029.141	983.674	867.158	941.800	956.224
	MAPE	6.876	5.154	4.754	5.206	4.770
CART	**MAE**	1027.157	763.678	704.088	776.566	705.472
	RMSE	1307.768	1031.547	892.356	1055.224	911.921
	MAPE	5.154	5.421	4.817	5.190	4.540
RF	**MAE**	763.678	795.999	697.702	757.221	666.295
	RMSE	1031.547	955.704	858.767	961.667	849.553
	MAPE	4.985	4.830	5.026	4.689	4.207
RreliefF	**MAE**	741.379	713.534	749.809	702.243	628.159
	RMSE	1019.697	893.103	1034.086	931.176	810.417

For the full verification of different method combinations, the entire test set was used for the contrast experiment. The results of different evaluated criteria for the proposed forecasting approach applied by 25 method combinations are presented for day-ahead load forecasting in Table 8. The forecast errors of the different methods varied. For example, the error of MI-based SVR was close to that of the GPR. The MAPEs for the MI-based SVR and GPR were 4.872% and 4.785%, the RMSEs were 1196.775 MW and 1141.372 MW, and the MAEs were 773.447 MW and 755.325 MW, respectively. Based on these observations, the forecast errors of the SVR with any feature selection method was below 5% (marked in bold) except with RF. In addition, the MAPEs of GPR with CMI and RF were below 5% as well.

Table 8. Error of load forecasting of different methods with proposed forecasting approach for the whole test set.

Feature Selection Method	Forecaster	Evaluated Criterion		
		MAPE (%)	RMSE (MW)	MAE (MW)
MI	CART	6.021	1360.445	934.560
	RF	5.536	1260.281	864.385
	SVR	**4.872**	1196.775	773.447
	BPNN	5.491	1320.809	865.842
	GPR	**4.785**	1141.372	755.325
CMI	CART	6.088	1371.643	945.217
	RF	5.364	1235.216	841.376
	SVR	**4.870**	1225.231	776.654
	BPNN	5.054	1179.931	793.064
	GPR	**4.758**	1135.260	750.937
CART	CART	6.495	1493.344	1013.322
	RF	5.364	1228.542	837.765
	SVR	**4.794**	1158.022	758.601
	BPNN	5.414	1270.671	847.104
	GPR	5.018	1176.996	790.088
RF	CART	5.883	1322.730	911.334
	RF	5.385	1236.724	843.334
	SVR	5.534	1260.281	834.385
	BPNN	5.287	1248.014	827.752
	GPR	**4.839**	1244.614	761.119
RreliefF	CART	5.804	1898.190	1305.192
	RF	5.202	1220.145	816.788
	SVR	**4.746**	1229.229	759.143
	BPNN	5.175	1244.537	812.642
	GPR	5.543	1410.293	883.576

By comparison of the results, the RreliefF + SVR method showed the best performance with the least MAPE.

6.2. Load Forecasting for Singapore

To further verify the applicability of the proposed approach, the load data from Singapore was used to perform the load forecasting experiments.

6.2.1. Feature Selection for Hour Loads

First, using the same method used in Section 6.1.1, the score of the feature corresponding to the predictive target at different hours was computed by different feature selection methods. Then, the optimal feature subset was obtained based on the MAPE of different subsets forecast by a predictor.

Table 9 shows the MAPE and the dimension of the optimal feature subset corresponding to using MI as the feature selection method and RF as the prediction model. The Table shows that 1 to 7 am, the dimension of optimal feature subset is less than at 8 to 19 pm, as the same as the forecasting error. Similar to the analysis result of 4, this is because people are less active at night and there are fewer factors affecting the load than during the day.

Table 9. Optimal feature subset construction of different hours with MI + RF for Singapore.

Time	MAPE	FD	Time	MAPE	FD
1	1.349	72	13	2.353	49
2	1.138	64	14	2.376	42
3	1.112	61	15	2.387	48
4	1.137	66	15	2.486	44
5	1.201	79	17	2.534	57
6	1.453	75	18	2.258	62
7	1.836	57	19	2.049	49
8	2.229	55	20	1.793	43
9	2.389	55	21	1.632	64
10	2.359	52	22	1.526	59
11	2.379	59	23	1.485	45
12	2.332	58	24	1.529	55

The MAPE and the dimension of optimal feature subset corresponding to 1:00 were carried out by different feature selection methods and forecasting methods shown in Table 10. The MAPEs are in 1.0% to 1.6% which means the performance of forecasters were similar after feature selection. While by analysis the feature dimension, we could find there is huge difference between the number of the feature of the optimal feature subset and that selected by different feature selection methods, which is caused by the different evaluation criteria.

Table 10. Optimal feature subset construction of 1:00 with different methods for Singapore.

Method	CART		RF		SVR		ANN		GPR	
	MAPE	FD	MAPE	FD	MAPE	FD	MAPE	FD	MAPE	FD
MI	1.595	59	1.349	72	1.225	74	1.349	47	1.170	75
CMI	1.528	11	1.266	31	1.209	43	1.239	17	1.148	122
CART	1.559	14	1.303	26	1.103	56	1.210	16	1.169	60
RF	1.594	72	1.371	5	1.186	58	1.242	17	1.163	38
RreliefF	1.530	7	1.300	10	1.197	21	1.242	17	1.159	95

As is shown in Appendix A Table A3 to Table A4, considering both the MAPEs and the dimensions, the optimal feature subsets were used for the load forecasting of the Singapore data. Similar to the conclusion summarized in Table 4, the different optimal feature subsets employed various feature selection methods and forecasters.

6.2.2. Forecasting Results of Method Combinations with Optimal Feature Subsets for Singapore Load Data

To test the performance of diverse combined methods with the optimal feature subset, the data of special weeks were used for the experiment.

Two weeks were chosen randomly from the summer and winter of 2015 for testing as is shown in Figures 7 and 8. The summer week included the days from 21 to 27 June and the winter week included days from 8 to 14 November. The results are shown in Figure 7 and Table 11. It was found that the GPR, RF, and SVR methods showed a better ability to forecast the summer loads. The MAPEs of the combinations of MI + GPR, CMI + GPR, RF + GPR, RreliefF + GPR, CMI + SVR, and RreliefF + SVR were less than 1.5%. The outstanding combined method was RreliefF + GPR whose MAPE was 1.402%, MAE was 74.400 MW, and RMSE was 93.092 MW. By observing Figure 8 and Table 12, the RreliefF + GPR method showed the best performance with an MAPE of 3.567%, an MAE of 200.711 MW, and an RMSE of 224.017 MW. The predictive results of GPR and SVR with different feature selection methods were better than those of the CART, BPNN, and RF methods.

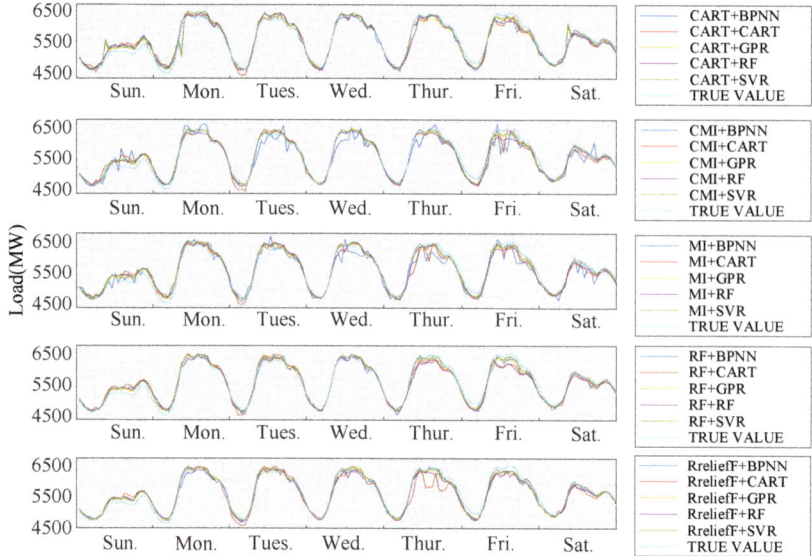

Figure 7. Prediction from 21 to 27 June 2015.

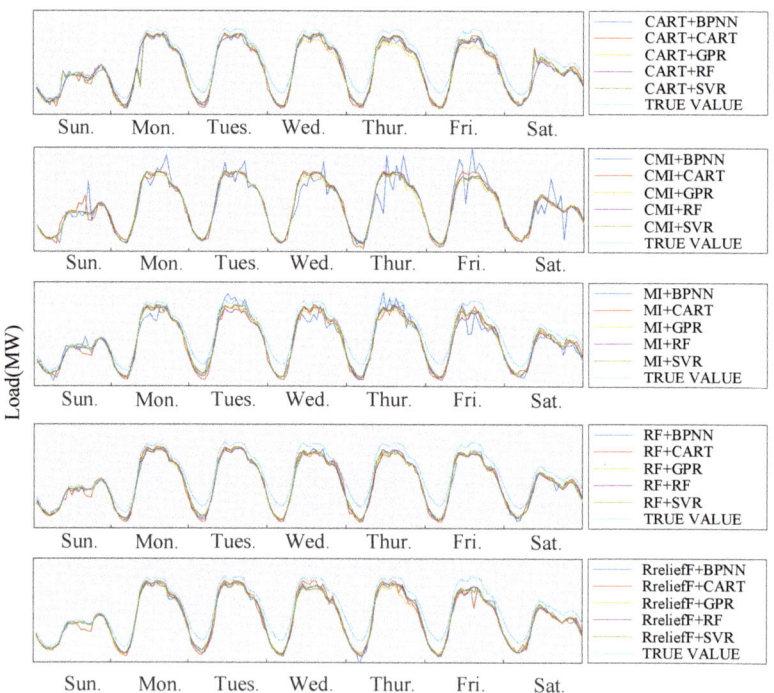

Figure 8. Prediction from 8 to 14 November 2015.

Table 11. Comparison of different combined methods.

Method		CART	RF	GPR	BPNN	SVR
	MAPE	2.321	2.145	1.439	2.719	1.662
MI	MAE	128.596	119.058	79.453	153.693	91.493
	RMSE	162.801	137.462	99.346	202.410	110.360
	MAPE	2.117	1.867	1.419	3.165	1.482
CMI	MAE	115.395	103.810	78.407	180.786	81.177
	RMSE	150.781	134.390	99.425	322.873	102.596
	MAPE	2.420	2.136	1.645	1.963	1.911
CART	MAE	132.823	118.571	91.358	108.851	106.369
	RMSE	175.615	143.584	139.408	160.930	152.349
	MAPE	2.213	2.000	1.435	1.702	1.404
RF	MAE	123.568	112.369	77.803	94.085	77.236
	RMSE	148.988	146.759	97.686	117.295	95.627
	MAPE	2.720	1.862	1.428	1.917	1.402
RreliefF	MAE	154.605	103.586	79.134	105.902	74.400
	RMSE	201.458	128.291	101.035	127.631	93.092

Table 12. Comparison of different combined methods.

Method		CART	RF	GPR	BPNN	SVR
	MAPE	3.895	3.854	3.806	4.273	3.637
MI	MAE	217.339	217.647	215.942	243.454	204.362
	RMSE	250.640	240.934	236.196	283.816	232.913
	MAPE	3.573	3.518	3.899	5.023	3.585
CMI	MAE	200.803	197.387	221.095	288.472	200.942
	RMSE	229.837	217.891	239.055	390.638	229.780
	MAPE	3.868	3.587	4.115	3.897	3.599
CART	MAE	215.523	200.915	234.630	219.650	201.124
	RMSE	260.178	225.501	272.193	254.684	235.158
	MAPE	3.799	3.711	3.851	3.871	3.599
RF	MAE	212.788	209.019	218.327	218.218	201.083
	RMSE	245.087	231.296	236.664	241.936	230.831
	MAPE	3.981	3.895	4.104	3.935	3.567
RreliefF	MAE	222.013	219.243	233.919	221.717	200.711
	RMSE	262.683	242.705	254.076	247.552	224.017

To further verify the superiority of the proposed method based on feature subsets of different hours, the entire test data from Singapore was used for validation. Detailed information about the test data is shown in Table 2 in Section 5.2. Table 13 shows the average predictive error of the different combined methods. It indicates that, based on MI, the CMI, RF, RreliefF, and SVR methods achieved the minimum errors with MAPEs of 1.471%, 1.440%, 1.387%, and 1.373%, respectively. Of all the combined methods, the RreliefF + SVR method worked best with an MAPE of 1.373%, an MAE of 75.118 MW, and an RMSE of 147.585 MW.

Table 13. Error of load forecasting of different methods with proposed forecasting strategy for the whole test set.

Feature Selection Method	Forecaster	Evaluated Criterion		
		MAPE (%)	RMSE (MW)	MAE (MW)
MI	CART	2.019	172.293	112.003
	RF	1.668	157.946	92.817
	SVR	**1.474**	**154.191**	**80.67**
	BPNN	2.551	218.916	145.116
	GPR	1.492	147.726	82.693
CMI	CART	2.174	189.964	121.050
	RF	1.623	156.450	90.309
	SVR	**1.440**	**151.230**	**78.764**
	ANN	3.072	332.424	177.185
	GPR	1.538	148.127	85.497
CART	CART	2.219	201.990	123.030
	RF	**1.733**	**164.604**	**96.589**
	SVR	1.748	188.225	96.562
	BPNN	1.954	192.515	109.282
	GPR	1.774	183.266	99.119
RF	CART	2.012	172.188	111.418
	RF	1.641	160.659	91.235
	SVR	**1.387**	**148.926**	**75.885**
	BPNN	1.663	158.088	92.355
	GPR	1.461	145.833	81.011
RreliefF	CART	2.075	177.441	116.199
	RF	1.608	155.962	89.551
	SVR	**1.373**	**147.585**	**75.118**
	BPNN	1.669	157.988	92.890
	GPR	1.446	144.170	80.283

By analyzing the load forecasting results for Singapore, the combination of RreliefF and SVR was the most accurate method.

6.3. Comparison and Discussion

6.3.1. Comparison of Forecasting Methods without Feature Selection for New England and Singapore

In this section, a comparison of the proposed method and the traditional method (which only builds a single predictor for forecasting without feature selection) based on the data of New England and Singapore was carried out to verify the necessity of forecasting by a modular predictor.

The histograms of the error and the training time duration of different forecasting methods using New England data are displayed in Figure 9. As shown in Figure 9a, the MAPE of the SVR that adopted the proposed method was almost half that of the SVR using the traditional method. The MAPE of other predictors employing the proposed method without the feature selection step decreased in different levels compared with the predictors employing the traditional method. By analyzing the MAE in Figure 9b and the RMSE in Figure 9c, a similar conclusion can be obtained. In addition, it is noted that the model training time of the proposed method decreased because of the smaller modeling training set. Therefore, the decreased error and training time reflect the advantages of the proposed method and confirms the necessity of employing a modular predictor.

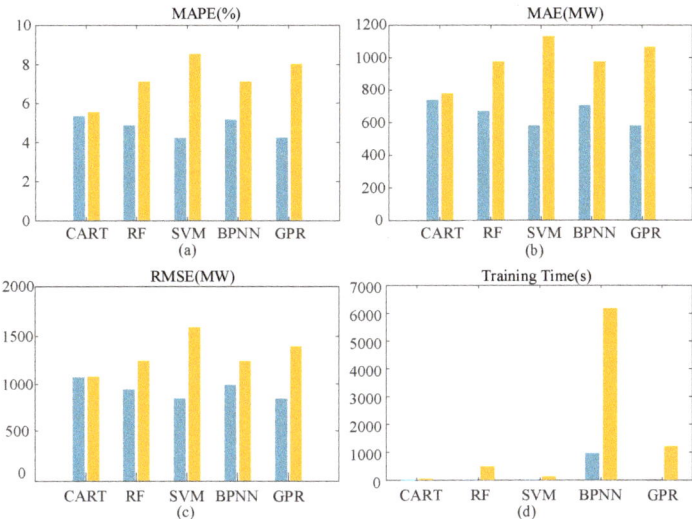

Figure 9. Comparison of error and time of training a model with traditional and proposed approaches.

The values of MAPE, MAE, and RMSE and the training time of each forecaster with different approaches based on the data of New England and Singapore are shown in Table 14. The results for New England indicate that the MAPE values of CART, RF, SVR, BPNN, and GPR with the proposed method were reduced by 0.182%, 2.253%, 4.294%, 1.953%, and 3.775% compared with the CART, RF, SVR, BPNN, and GPR with the traditional approach, respectively. Similarly, the results for Singapore also verified the superior performance of the proposed method.

Table 14. Comparison of the error of different forecasting approaches with original feature set.

Method	Forecaster	Test for New England				Test for Singapore			
		MAPE (%)	MAE (MW)	RMSE (MW)	Time (s)	MAPE (%)	MAE (MW)	RMSE (MW)	Time (s)
The proposed method	CART	5.348	738.641	1067.723	0.106	2.166	116.742	209.413	0.275
	RF	4.867	671.261	941.661	10.445	1.930	105.306	199.974	10.776
	SVR	**4.228**	**580.80**	**849.806**	**0.431**	**1.914**	**103.356**	**196.145**	**0.405**
	BPNN	5.167	705.324	986.974	962.457	3.133	174.104	285.083	844.257
	GPR	4.242	581.889	844.0766	2.102	1.523	82.573	170.478	1.569
The traditional method	CART	5.530	778.083	1076.316	7.976	3.597	196.391	273.112	2.601
	RF	7.120	975.272	1235.783	486.263	2.088	114.732	209.064	402.743
	SVR	8.522	1130.870	1556.371	123.394	5.067	267.048	361.547	91.623
	BPNN	7.120	975.272	1235.783	6170.835	4.864	267.416	408.305	4686.007
	GPR	8.017	1065.700	1387.252	1219.056	5.072	287.277	405.181	1054.359

6.3.2. Comparison of Forecasting Approaches with Feature Selection for New England and Singapore

A comparison between the proposed method and traditional method with feature selection was performed on the New England and Singapore datasets. The results of the proposed method with feature selection are shown in Table 8 (New England) and Table 13 (Singapore), and the results of the traditional method with feature selection are shown in Table 15. The results indicate that the error was reduced in different levels by adopting the proposed method. The largest reduction in

MAPE resulted from the CMI + SVR and CART + BPNN methods with MAPEs of 2.799% and 3.072%, respectively. The minimum error was achieved by the RreliefF + SVR combination with MAPEs of 4.746% (New England) and 1.373% (Singapore). In conclusion, the forecasted results obtained by the proposed method were better than those of the traditional method regardless of the predictor used. The most accurate combined method was RreliefF + SVR.

Table 15. Error of load forecasting of different methods with traditional forecasting approach for the whole test set.

Feature Selection Method	Forecaster	Test for New England			Test for Singapore		
		MAPE (%)	MAE (MW)	RMSE (MW)	MAPE (%)	MAE (MW)	RMSE (MW)
MI	CART	8.452	1269.711	1701.808	3.247	178.082	239.891
	RF	5.911	920.201	1339.227	1.855	103.612	168.744
	SVR	7.587	1116.529	1521.691	4.246	222.376	314.547
	BPNN	5.854	909.553	1390.574	2.103	115.674	176.764
	GPR	5.680	881.310	1296.119	2.161	118.833	180.018
CMI	CART	8.420	1267.213	1705.926	3.320	182.186	241.884
	RF	5.645	878.479	1281.361	1.838	102.269	164.790
	SVR	7.669	1134.308	1560.853	4.206	219.965	313.680
	BPNN	7.697	1173.160	1929.675	2.053	113.328	173.493
	GPR	6.562	1029.708	1558.377	2.104	115.482	175.308
CART	CART	8.420	1267.213	1705.926	3.212	175.834	238.396
	RF	5.970	921.976	1318.980	1.940	108.871	175.704
	SVR	7.635	1127.940	1506.504	4.170	217.423	312.793
	BPNN	6.044	922.497	1404.137	5.026	278.908	462.199
	GPR	5.904	920.084	1372.375	2.860	161.948	250.843
RF	CART	8.056	1212.114	1653.079	3.262	179.431	242.135
	RF	5.483	858.934	1306.136	1.833	102.516	167.013
	SVR	7.316	1081.404	1482.864	4.147	216.703	310.201
	BPNN	5.348	831.493	1196.481	1.790	99.181	160.264
	GPR	5.774	902.872	1321.057	1.951	108.686	169.148
RreliefF	CART	8.056	1212.114	1653.079	3.188	174.799	237.592
	RF	5.506	866.377	1333.577	2.003	111.688	176.002
	SVR	7.350	1081.259	1464.366	4.319	226.854	319.170
	BPNN	5.789	894.686	1320.901	1.958	107.762	168.592
	GPR	6.015	967.163	1682.298	2.130	117.884	188.138

7. Conclusions

Accurate day-ahead load forecasting enhances the stability of grid operations and improves the social benefits of power systems. To improve the accuracy of day-ahead load forecasting, a novel modular parallel forecasting model with feature selection was proposed. Load data from New England and Singapore were used to test the proposed method. The experimental results show the advantages of the proposed method as follows:

(1) A modular predictor consisting of 24 independent predictors can efficiently capture load characteristics with respect to different hours and thereby avoid the inaccurate analysis of a single predictor.

(2) The feature selection adopted for the load corresponding to different hours analyzes the relevance between the feature and a special load. Each optimal feature subset of different dimension benefits the building of a more-accurate predictor.

(3) To serve the demand of dispatch departments of different regions, the interval time $p = 24$ was chosen for structuring a general candidate feature set that met the requirements of the power system.

Future work will concentrate on predictor parameter optimization and improve the efficiency of forecasting in the modeling process and applying the proposed method to probabilistic load forecasting.

Author Contributions: L.L. put forward to the main idea and design the whole venation of this paper. L.X. and Z.H. did the experiments and prepared the manuscript. N.H. guided the experiments and paper writing. All authors have read and approved the final manuscript.

Acknowledgments: This work is supported by the National Nature Science Foundation of China (No. 51307020), the Science and Technology Development Project of Jilin Province (No. 20160411003XH), the Science and Technology Project of Jilin Province Education Department (No. JJKH20170219KJ), Major science and technology projects of Jilin Institute of Chemical Technology (No. 2018021), and Science and Technology Innovation Development Plan Project of Jilin City (No. 201750239).

Conflicts of Interest: The authors declare no conflict of interest.

Appendix A

Table A1. Optimal feature subset construction of different hours from 1:00 to 12:00 with different methods for New England.

	Time Point	1:00		2:00		3:00		4:00		5:00		6:00		7:00		8:00		9:00		10:00		11:00		12:00	
	Error	MAPE	FD	MAPE	FD	MAPE	FD	MAPE	FD	MAPE	FD	MAPE	FD	MAPE	FD	MAPE	FD	MAPE	FD	MAPE	FD	MAPE	FD	MAPE	FD
MI	CART	3.741	7	3.769	2	4.071	12	4.083	84	4.472	40	5.140	6	4.949	164	4.748	164	4.627	164	4.765	15	4.978	21	5.529	130
	RF	3.294	34	3.419	22	3.632	9	3.783	30	4.008	9	4.828	18	5.456	61	5.314	59	4.526	64	4.171	45	4.147	42	4.414	67
	SVR	3.064	10	3.167	28	3.189	27	3.314	23	3.553	59	3.852	38	4.353	57	4.327	47	3.682	61	3.510	58	3.598	72	3.789	23
	ANN	3.226	8	3.329	7	3.422	26	3.897	27	3.521	30	4.215	16	4.889	40	4.345	29	3.891	23	3.848	31	3.911	11	4.342	32
	GPR	3.087	119	3.226	115	3.359	9	3.381	99	3.629	96	4.087	36	4.476	54	4.432	102	3.805	65	3.645	49	3.781	36	3.852	31
CMI	CART	3.729	2	3.769	2	4.058	7	4.192	16	4.296	8	5.242	6	4.926	99	4.523	27	5.076	15	4.523	27	5.076	15	5.413	106
	RF	3.447	20	3.447	23	3.590	12	3.717	13	3.848	6	4.505	57	4.750	40	4.531	130	4.136	49	3.949	159	4.009	159	4.281	124
	SVR	3.043	13	3.126	12	3.238	12	3.341	4	3.375	48	3.722	91	4.008	60	3.972	73	3.469	88	3.351	68	3.448	93	3.667	88
	ANN	3.062	47	3.123	42	3.134	28	3.329	53	3.365	23	3.821	64	4.178	83	4.167	35	3.590	78	3.341	75	3.418	57	3.576	63
	GPR	3.052	134	3.189	23	3.288	18	3.366	16	3.593	21	4.017	18	4.128	150	3.911	158	3.517	168	3.352	91	3.455	106	3.612	88
CART	CART	3.729	3	4.050	6	4.071	4	4.134	6	4.558	6	5.596	13	4.958	9	4.751	4	4.634	7	4.725	10	4.524	19	5.512	10
	RF	3.422	11	3.511	5	3.589	6	3.615	12	3.963	7	4.511	32	4.512	20	4.367	11	4.062	22	3.989	18	4.151	21	4.546	11
	SVR	3.068	11	3.167	11	3.548	11	3.433	12	3.798	14	3.846	33	3.804	20	3.870	18	3.524	22	3.629	19	3.633	20	4.260	18
	ANN	3.270	9	3.301	9	3.670	5	3.483	5	3.836	13	4.012	15	3.974	19	4.081	16	3.921	17	3.775	17	3.806	18	4.397	11
	GPR	3.245	8	3.280	8	3.526	11	3.458	8	3.858	8	3.911	33	3.753	20	3.872	18	3.707	22	3.659	19	3.732	16	4.360	11
RF	CART	3.741	7	3.769	2	4.059	9	4.128	44	4.552	9	5.084	7	4.807	52	4.656	5	4.337	6	4.464	10	4.147	3	4.155	6
	RF	3.533	51	3.679	27	3.790	28	3.777	11	3.991	25	4.522	30	4.624	12	4.416	9	3.973	26	3.922	15	3.801	28	4.227	7
	SVR	3.140	41	3.512	14	3.312	42	3.469	11	3.5518	26	3.6654	19	3.9069	27	3.8894	46	3.3732	31	3.3966	44	3.376	43	3.7329	51
	ANN	3.069	18	3.255	20	3.097	21	3.469	17	3.486	16	3.816	27	4.262	11	4.278	13	3.682	11	3.542	19	3.424	31	3.923	14
	GPR	3.099	81	3.239	80	3.338	64	3.447	150	3.632	16	3.679	19	4.208	15	3.964	56	3.522	86	3.359	63	3.484	43	3.787	37
RnelielF	CART	3.741	10	3.769	2	4.059	8	4.156	42	4.475	20	4.917	4	4.729	38	4.646	15	4.443	7	4.030	15	4.417	4	4.448	20
	RF	3.310	26	3.390	20	3.466	17	3.560	19	3.528	14	3.764	14	4.534	30	4.294	23	3.643	10	3.514	17	3.680	22	4.006	18
	SVR	3.043	18	3.107	19	3.233	19	3.351	30	3.434	16	3.928	14	3.716	34	3.648	21	3.320	34	3.205	34	3.407	66	3.594	53
	ANN	3.269	9	3.306	14	3.338	13	3.368	17	3.445	16	3.555	15	4.193	35	3.760	23	3.399	19	3.329	20	3.427	25	3.820	24
	GPR	3.019	134	3.156	152	3.329	32	3.346	117	3.460	16	3.578	29	3.811	34	3.715	24	3.414	22	3.333	28	3.358	81	3.807	29

Table A2. Optimal feature subset construction of different hours from 13:00 to 24:00 with different methods for New England.

Time Point		13:00		14:00		15:00		16:00		17:00		18:00		19:00		20:00		21:00		22:00		23:00		24:00	
Error		MAPE	FD	MAPE	FD	MAPE	FD	MAPE	FD	MAPE	FD	MAPE	FD	MAPE	FD	MAPEE	FD	MAPEE	FD	MAPEE	FD	MAPE	FD	MAPEE	FD
MI	CART	4.654	164	5.143	164	5.764	164	5.406	164	6.170	41	6.313	6	6.066	12	5.388	9	5.166	14	5.105	12	5.198	6	5.086	8
	RF	4.663	41	4.926	33	5.190	38	5.351	46	5.547	31	5.358	98	5.117	136	4.506	23	4.376	28	4.779	9	4.794	41	4.847	72
	SVR	3.927	23	4.263	35	4.456	49	4.518	82	4.645	68	4.516	69	4.418	69	4.165	21	3.867	24	3.901	73	3.860	154	4.039	115
	ANN	4.042	19	4.362	14	4.626	35	4.799	29	4.746	37	5.262	46	4.849	33	4.371	15	3.919	27	4.599	42	4.261	27	4.203	35
	GPR	3.974	24	4.177	36	4.524	29	4.689	28	4.616	32	4.823	21	4.770	32	4.272	24	4.010	26	4.157	24	4.291	22	4.446	18
CMI	CART	5.459	14	5.045	53	5.519	24	5.406	91	5.985	18	6.382	5	5.973	5	5.299	19	5.134	20	5.178	19	5.164	8	4.994	14
	RF	4.445	146	4.608	151	4.843	150	5.064	162	5.326	164	5.352	157	5.136	126	4.496	136	4.394	126	4.176	145	4.731	153	4.804	133
	SVR	3.923	107	4.102	97	4.339	99	4.539	86	4.501	105	4.540	70	4.398	87	3.944	113	3.717	90	3.763	94	3.895	93	3.972	111
	ANN	3.846	64	3.827	53	4.304	54	4.452	49	4.698	86	4.500	72	4.416	94	4.273	156	3.780	32	3.943	76	3.971	78	3.902	117
	GPR	3.916	105	4.080	40	4.474	53	4.686	94	4.469	172	4.455	168	4.709	76	4.508	16	4.116	39	4.075	49	4.304	103	4.432	118
CART	CART	4.655	11	5.141	8	5.768	12	5.412	9	6.320	11	6.955	8	6.393	9	5.978	13	5.324	11	5.742	9	5.214	6	5.182	7
	RF	4.587	19	4.846	15	5.113	28	5.409	22	5.390	25	5.720	22	5.444	17	4.634	15	4.665	24	5.001	29	4.700	7	4.806	7
	SVR	4.105	15	4.371	21	4.393	28	4.882	22	4.804	24	5.515	22	4.804	26	4.236	30	4.099	24	4.256	21	4.228	18	4.168	20
	ANN	4.205	15	4.713	18	4.666	18	4.931	20	5.248	16	5.205	22	4.986	26	4.302	26	4.248	24	4.441	18	4.152	17	3.988	22
	GPR	4.174	13	4.507	21	4.573	28	5.109	22	5.054	23	5.124	21	4.864	28	4.424	32	4.304	39	4.545	18	4.405	17	4.321	21
RF	CART	4.448	12	5.022	23	5.476	6	5.350	7	5.678	5	6.129	5	6.040	22	5.273	11	4.820	26	5.175	98	4.980	5	5.080	45
	RF	4.445	7	4.617	55	4.845	109	5.045	52	5.226	73	5.296	70	5.101	85	4.484	68	4.450	76	4.780	75	4.711	64	4.810	50
	SVR	3.971	100	3.983	35	4.122	82	4.635	94	4.556	32	4.589	40	4.725	101	4.089	37	7.848	25	3.847	97	3.936	90	3.996	141
	ANN	4.101	17	4.638	16	4.585	32	4.882	13	5.205	17	5.275	11	5.227	11	4.413	17	3.890	10	4.411	10	4.328	23	4.336	19
	GPR	4.069	48	4.091	74	4.515	42	4.662	41	4.912	32	4.577	159	4.886	56	4.491	43	4.026	70	4.182	81	4.367	79	4.497	18
RrelierF	CART	4.574	20	4.949	21	5.529	29	5.405	108	5.817	22	6.238	23	5.860	24	5.321	24	4.796	37	4.721	23	5.005	9	4.986	20
	RF	4.365	17	4.716	17	4.908	74	5.065	136	5.293	15	5.166	16	5.029	81	4.403	14	4.368	140	4.613	14	4.612	20	4.619	16
	SVR	3.693	65	3.876	66	4.104	52	4.426	49	4.496	67	4.528	41	4.558	68	3.983	52	3.809	44	3.803	109	4.008	43	4.056	38
	ANN	4.089	19	4.246	26	4.644	20	4.902	39	5.000	25	5.216	16	5.161	31	4.633	30	4.379	23	4.218	14	4.566	24	4.291	21
	GPR	3.980	40	4.161	34	4.411	33	4.632	36	4.748	45	4.472	160	4.391	170	4.301	45	3.976	43	4.284	34	4.001	172	4.324	16

Remark. The FD in Table 4 means feature dimension.

Table A3. Optimal feature subset construction of different hours from 1:00 to 12:00 with different methods for Singapore.

Time Point		1:00		2:00		3:00		4:00		5:00		6:00		7:00		8:00		9:00		10:00		11:00		12:00	
Error		**MAPE**	**FD**	**MAPE**	**FD**	**MAPE**	**FD**	**MAPE**	**FD**	**MAPE**	**FD**	**MAPE**	**FD**	**MAPE**	**FD**	**MAPE**	**FD**	**MAPE**	**FD**	**MAPE**	**FD**	**MAPE**	**FD**	**MAPE**	**FD**
MI	CART	1.595	59	1.479	57	1.402	57	1.482	10	1.535	6	1.624	108	2.041	29	2.727	62	2.515	48	2.526	43	2.615	59	2.451	58
	RF	1.349	72	1.138	64	1.112	64	1.137	66	1.201	79	1.453	75	1.836	57	2.229	55	2.389	55	2.359	52	2.379	59	2.332	58
	SVR	1.225	74	1.057	61	1.056	61	1.025	57	1.114	59	1.377	90	1.401	57	1.565	43	1.749	44	1.723	55	1.796	58	1.738	57
	ANN	1.349	47	1.179	56	1.133	56	1.276	59	1.262	76	1.453	33	1.558	30	1.926	48	2.323	58	2.319	53	2.321	48	2.424	21
	GPR	1.170	75	0.955	109	0.904	109	0.963	86	1.025	110	1.281	101	1.452	94	1.859	110	2.021	113	2.039	41	1.952	98	1.876	99
CMI	CART	1.528	11	1.470	4	1.386	4	1.396	9	1.475	4	1.632	62	1.998	72	2.752	108	2.709	135	2.534	124	2.531	114	2.485	125
	RF	1.266	31	1.093	15	1.305	15	1.052	30	1.133	50	1.416	44	1.847	23	2.191	49	2.333	42	2.345	38	2.353	43	2.237	43
	SVR	1.209	43	1.027	28	0.950	33	1.082	35	1.123	55	1.316	65	1.387	39	1.508	34	1.651	52	1.732	33	1.692	44	1.631	44
	ANN	1.239	17	1.062	21	1.034	28	1.072	31	1.149	29	1.312	27	1.542	23	1.979	31	2.202	23	2.107	28	2.037	24	2.129	47
	GPR	1.148	122	0.930	138	0.882	163	0.900	167	1.037	73	1.090	122	1.470	27	1.879	27	2.054	43	2.012	48	1.925	135	1.892	50
CART	CART	1.559	14	1.528	14	1.403	3	1.482	12	1.608	2	1.675	143	2.323	50	2.718	7	2.950	18	2.875	19	2.593	102	2.478	102
	RF	1.303	26	1.113	12	1.100	21	1.105	32	1.214	25	1.431	38	1.917	44	2.289	19	2.417	45	2.458	45	2.544	56	2.477	19
	SVR	1.103	56	0.995	36	1.031	17	1.001	73	1.038	22	1.138	32	1.575	83	1.653	25	1.803	73	1.863	46	1.916	53	1.845	53
	ANN	1.210	16	1.061	21	1.043	15	1.037	24	1.122	32	1.252	31	1.876	22	1.981	27	2.082	23	2.158	19	2.254	28	2.274	13
	GPR	1.169	60	1.015	74	0.902	120	0.918	115	1.066	25	1.216	33	1.415	173	1.813	172	1.959	172	1.903	172	1.893	172	1.851	173
RF	CART	1.594	72	1.583	19	1.425	2	1.525	11	1.577	9	1.672	14	2.147	8	2.456	4	2.754	10	2.710	21	2.615	42	2.488	29
	RF	1.371	5	1.089	24	1.081	22	1.105	31	1.190	35	1.381	21	1.608	10	2.042	9	2.336	10	2.239	18	2.334	41	2.238	19
	SVR	1.186	58	0.993	13	0.904	30	0.972	38	1.035	27	1.080	39	1.370	20	1.483	27	1.617	30	1.588	29	1.682	30	1.649	23
	ANN	1.242	17	1.016	18	0.926	28	1.014	44	1.033	24	1.149	23	1.461	11	1.689	14	1.945	25	1.930	31	1.953	9	1.952	15
	GPR	1.163	38	0.949	76	0.897	76	0.897	60	0.946	105	1.115	27	1.427	27	1.835	30	1.982	31	1.966	26	1.952	35	1.882	31
RreliefF	CART	1.530	7	1.506	9	1.283	10	1.395	8	1.513	9	1.574	13	1.754	24	2.579	38	2.579	38	2.464	40	2.412	43	2.295	42
	RF	1.300	10	1.083	18	1.042	38	1.070	31	1.178	12	1.173	14	1.448	18	2.109	19	2.248	57	2.216	59	2.204	64	2.191	9
	SVR	1.197	21	1.019	13	1.003	14	1.047	14	1.098	59	1.080	26	1.343	38	1.464	24	1.620	42	1.671	43	1.679	42	1.678	34
	ANN	1.242	17	1.016	18	0.926	28	1.014	44	1.033	24	1.149	23	1.645	11	1.689	14	1.945	25	1.930	31	1.953	9	1.952	15
	GPR	1.159	95	0.950	94	0.910	95	0.925	96	0.989	88	1.128	16	1.442	22	1.780	18	1.886	18	1.901	37	1.876	41	1.778	42

Table A4. Optimal feature subset construction of different hours from 13:00 to 24:00 with different methods for Singapore.

Time Point		13:00		14:00		15:00		16:00		17:00		18:00		19:00		20:00		21:00		22:00		23:00		24:00	
Error		MAPE	FD	MAPE	FD	MAPE	FD	MAPE	FD	MAPE	FD	MAPE	FD	MAPE	FD	MAPE	FD	MAPE	FD	MAPE	FD	MAPE	FD	MAPE	FD
MI	CART	2.501	75	2.522	37	2.700	43	2.637	45	2.619	58	2.398	51	2.113	87	1.884	49	1.790	64	1.588	90	1.614	66	1.820	13
	RF	2.353	49	2.376	42	2.387	48	2.486	44	2.534	57	2.258	62	2.049	49	1.793	43	1.632	64	1.526	59	1.485	45	1.529	55
	SVR	1.795	49	1.850	42	1.878	43	1.898	43	2.010	54	1.883	54	1.629	111	1.469	39	1.317	94	1.372	40	1.337	41	1.301	75
	ANN	2.280	25	2.362	36	2.466	23	2.334	32	2.259	44	2.324	29	1.936	45	1.797	52	1.682	65	1.482	41	1.450	38	1.390	33
	GPR	1.912	95	2.036	40	2.032	43	2.098	43	2.095	102	1.821	170	1.752	101	1.554	44	1.357	94	1.304	91	1.283	102	1.253	119
CMI	CART	2.368	111	2.749	126	2.517	127	2.665	113	2.481	125	2.423	110	1.983	115	1.884	139	1.779	66	1.587	121	1.580	7	1.667	5
	RF	2.263	37	2.307	36	2.309	36	2.349	28	2.342	35	2.160	31	1.933	40	1.672	41	1.593	30	1.466	29	1.456	8	1.451	53
	SVR	1.702	49	1.768	34	1.792	45	1.914	42	1.937	41	1.806	33	1.646	32	1.442	27	1.359	33	1.301	41	1.324	54	1.352	37
	ANN	2.232	27	2.125	30	2.163	43	2.300	36	2.381	23	2.218	39	1.761	31	1.851	21	1.488	15	1.405	34	1.358	22	1.377	25
	GPR	1.913	48	1.978	45	1.994	44	2.055	40	2.054	102	1.884	128	1.667	171	1.448	172	1.296	172	1.244	171	1.239	125	1.282	92
CART	CART	2.486	103	2.557	4	2.630	4	2.734	4	2.626	151	2.398	104	2.132	15	1.940	20	1.911	86	1.591	58	1.638	156	1.734	7
	RF	2.414	22	2.463	18	2.493	46	2.622	7	2.653	8	2.352	17	1.909	20	1.903	27	1.760	24	1.448	25	1.499	23	1.503	45
	SVR	1.871	52	1.935	54	1.885	54	2.115	40	2.188	39	1.843	54	1.601	87	1.606	87	1.511	69	1.322	69	1.284	47	1.227	85
	ANN	2.218	19	2.213	18	2.202	19	2.387	19	2.404	31	2.179	19	1.832	32	1.773	19	1.754	12	1.438	33	1.453	22	1.351	17
	GPR	1.871	172	1.950	173	1.948	168	1.981	171	2.011	169	1.824	168	1.663	172	1.442	168	1.285	169	1.235	168	1.205	166	1.203	169
RF	CART	2.501	67	2.759	27	2.673	7	2.651	34	2.617	38	2.416	35	1.999	13	1.794	27	1.750	17	1.583	52	1.638	41	1.692	2
	RF	2.272	36	2.323	33	2.328	35	2.364	34	2.396	39	2.098	26	1.881	25	1.628	19	1.527	10	1.427	16	1.410	15	1.478	24
	SVR	1.670	28	1.751	23	1.857	22	1.853	39	1.944	36	1.789	16	1.577	57	1.394	22	1.299	56	1.240	15	1.195	12	1.244	38
	ANN	1.972	11	2.256	10	2.224	8	2.245	10	2.326	13	1.949	21	1.822	33	1.551	7	1.419	12	1.323	17	1.244	19	1.396	32
	GPR	1.876	44	1.981	30	2.125	34	2.023	54	2.108	42	1.893	91	1.767	35	1.502	32	1.377	13	1.269	16	1.244	18	1.265	47
RelieF	CART	2.323	41	2.532	41	2.662	42	2.533	24	2.366	58	2.220	42	1.949	66	1.766	7	1.674	15	1.605	21	1.624	12	1.642	10
	RF	2.201	84	2.298	9	2.243	17	2.262	14	2.271	35	1.998	14	1.713	18	1.468	21	1.366	17	1.308	17	1.337	17	1.400	24
	SVR	1.714	41	1.768	20	1.799	37	1.801	15	1.815	14	1.646	23	1.564	20	1.409	11	1.258	16	1.265	24	1.252	24	1.299	21
	ANN	1.972	11	2.256	10	2.224	8	2.244	10	2.326	13	1.949	21	1.822	33	1.551	7	1.419	12	1.322	12	1.244	19	1.396	21
	GPR	1.856	29	1.882	41	1.861	31	1.937	41	1.936	30	1.789	33	1.645	95	1.450	35	1.316	56	1.296	41	1.279	72	1.208	170

References

1. He, Y.; Xu, Q.; Wan, J.; Yang, S. Short-term power load probability density forecasting based on quantile regression neural network and triangle kernel function. *Energy* **2016**, *114*, 498–512. [CrossRef]
2. Nikmehr, N.; Najafi-Ravadanegh, S. Optimal operation of distributed generations in micro-grids under uncertainties in load and renewable power generation using heuristic algorithm. *IET Renew. Power Gener.* **2015**, *9*, 982–990. [CrossRef]
3. Duan, Z.Y.; Gutierrez, B.; Wang, L. Forecasting Plug-In Electric Vehicle Sales and the Diurnal Recharging Load Curve. *IEEE Trans. Smart Grid* **2014**, *5*, 527–535. [CrossRef]
4. Ferlito, S.; Adinolfi, G.; Graditi, G. Comparative analysis of data-driven methods online and offline trained to the forecasting of grid-connected photovoltaic plant production. *Appl. Energy* **2017**, *205*, 116–129. [CrossRef]
5. Ferruzzi, G.; Cervone, G.; Delle Monache, L.; Graditi, G.; Jacobone, F. Optimal bidding in a Day-Ahead energy market for Micro Grid under uncertainty in renewable energy production. *Energy* **2016**, *106*, 194–202. [CrossRef]
6. Feng, Y.H.; Ryan, S.M. Day-ahead hourly electricity load modeling by functional regression. *Appl. Energy* **2016**, *170*, 455–465. [CrossRef]
7. Bindiu, R.; Chindris, M.; Pop, G.V. Day-Ahead Load Forecasting Using Exponential Smoothing. *Sci. Bull. Petru Maior Univ. Tîrgu Mureş* **2009**, *6*, 89–93.
8. Al-Hamadi, H.M.; Soliman, S.A. Fuzzy short-term electric load forecasting using Kalman filter. *IEE Proc.-Gener. Transm. Distrib.* **2012**, *153*, 217–227. [CrossRef]
9. Lee, C.M.; Ko, C.N. Short-term load forecasting using lifting scheme and ARIMA models. *Expert Syst. Appl.* **2011**, *38*, 5902–5911. [CrossRef]
10. Luy, M.; Ates, V.; Barisci, N.; Polat, H.; Cam, E. Short-Term Fuzzy Load Forecasting Model Using Genetic–Fuzzy and Ant Colony–Fuzzy Knowledge Base Optimization. *Appl. Sci.* **2018**, *8*, 864. [CrossRef]
11. Xiao, L.Y.; Shao, W.; Liang, L.L.; Wang, C. A combined model based on multiple seasonal patterns and modified firefly algorithm for electrical load forecasting. *Appl. Energy* **2016**, *167*, 135–153. [CrossRef]
12. Khotanzad, A.; Zhou, E.; Elragal, H. A neuro-fuzzy approach to short-term load forecasting in a price-sensitive environment. *IEEE Trans. Power Syst.* **2002**, *17*, 1273–1282. [CrossRef]
13. Felice, M.D.; Yao, X. Short-Term Load Forecasting with Neural Network Ensembles: A Comparative Study Application Notes. *IEEE Comput. Intell. Mag.* **2012**, *6*, 47–56. [CrossRef]
14. Ahmad, A.; Javaid, N.; Alrajeh, N.; Khan, Z.A.; Qasim, U.; Khan, A. A Modified Feature Selection and Artificial Neural Network-Based Day-Ahead Load Forecasting Model for a Smart Grid. *Appl. Sci.* **2015**, *5*, 1756–1772. [CrossRef]
15. Che, J.X.; Wang, J.Z.; Tang, Y.J. Optimal training subset in a support vector regression electric load forecasting model. *Appl. Soft Comput.* **2012**, *12*, 1523–1531. [CrossRef]
16. Dudek, G. Short-Term Load Forecasting Using Random Forests. *Intell. Syst.* **2015**, *323*, 821–828.
17. Lloyd, R.J. GEFCom2012 hierarchical load forecasting: Gradient boosting machines and Gaussian processes. *Int. J. Forecast.* **2014**, *30*, 369–374. [CrossRef]
18. Che, J.X.; Wang, J.Z. Short-term load forecasting using a kernel-based support vector regression combination model. *Appl. Energy* **2014**, *132*, 602–609. [CrossRef]
19. Božić, M.; Stojanović, M.; Stajić, Z.; Stajić, N. Mutual Information-Based Inputs Selection for Electric Load Time Series Forecasting. *Entropy* **2013**, *15*, 926–942. [CrossRef]
20. Rong, G.; Liu, X. Support vector machine with PSO algorithm in short-term load forecasting. In Proceedings of the 2008 Chinese Control and Decision Conference, Yantai, China, 2–4 July 2008; pp. 1140–1142.
21. Ma, L.H.; Zhou, S.; Lin, M. Support Vector Machine Optimized with Genetic Algorithm for Short-Term Load Forecasting. In Proceedings of the International Symposium on Knowledge Acquisition and Modeling IEEE, Wuhan, China, 21–22 December 2008; pp. 654–657.
22. Zhang, Y.J.; Peng, X.Y.; Peng, Y.; Pang, J.Y.; Liu, D.T. Weighted bagging gaussion process regression to predict remaining useful life of electro-mechanical actuator. In Proceedings of the Prognostics and System Health Management Conference, Chengdu, China, 19–21 October 2016; pp. 1–6.
23. Lahouar, A.; Slama, J.B.H. Day-ahead load forecast using random forest and expert input selection. *Energy Convers. Manag.* **2015**, *103*, 1040–1051. [CrossRef]

24. Ghofrani, M.; West, K.; Ghayekhloo, M. Hybrid time series-bayesian neural network short-term load forecasting with a new input selection method. In Proceedings of the 2015 IEEE Power & Energy Society General Meeting, Denver, CO, USA, 26–30 July 2015; pp. 1–5.

25. Chandrashekar, G.; Sahin, F. *A Survey on Feature Selection Methods*; Pergamon Press, Inc.: New York, NY, USA, 2014.

26. Kohavi, R.; John, G.H. Wrappers for feature subset selection. *Artif. Intell.* **1996**, *97*, 273–324. [CrossRef]

27. Hu, Z.Y.; Bao, Y.K.; Chiong, R.; Xiong, T. Mid-term interval load forecasting using multi-output support vector regression with a memetic algorithm for feature selection. *Energy* **2015**, *84*, 419–431. [CrossRef]

28. Goldberg, D.E. *Genetic Algorithms in Search, Optimization and Machine Learning*; Addison-Wesley Longman Publishing Co., Inc.: Boston, MA, USA, 1990; pp. 2104–2116.

29. Hyojoo, S.; Kim, C. Forecasting Short-term Electricity Demand in Residential Sector Based on Support Vector Regression and Fuzzy-rough Feature Selection with Particle Swarm Optimization. *Procedia Eng.* **2015**, *118*, 1162–1168.

30. Isabelle, G.; Elisseeff, A. An introduction to variable and feature selection. *J. Mach. Learn. Res.* **2003**, *3*, 1157–1182.

31. Koprinska, I.; Rana, M.; Agelidis, V.G. Correlation and instance based feature selection for electricity load forecasting. *Knowl.-Based Syst.* **2015**, *82*, 29–40. [CrossRef]

32. Hu, Z.Y.; Bao, Y.K.; Xiong, T.; Chiong, R. Hybrid filter–wrapper feature selection for short-term load forecasting. *Eng. Appl. Artif. Intell.* **2015**, *40*, 17–27. [CrossRef]

33. Abedinia, O.; Amjady, N.; Zareipour, H. A New Feature Selection Technique for Load and Price Forecast of Electrical Power Systems. *IEEE Trans. Power Syst.* **2016**, *32*, 62–74. [CrossRef]

34. Raza, M.Q.; Khosravi, A. A review on artificial intelligence based load demand forecasting techniques for smart grid and buildings. *Renew. Sustain. Energy Rev.* **2015**, *50*, 1352–1372. [CrossRef]

35. Li, S.; Goel, L.; Wang, P. An ensemble approach for short-term load forecasting by extreme learning machine. *Appl. Energy* **2016**, *170*, 22–29. [CrossRef]

36. Kononenko, I. Theoretical and Empirical Analysis of ReliefF and RReliefF. *Mach. Learn. J.* **2003**, *53*, 23–69.

37. Breiman, L.I.; Friedman, J.H.; Olshen, R.A.; Stone, C.J. Classification and Regression Trees (CART). *Biometrics* **1984**, *40*, 17–23.

38. Breiman, L. Random Forest. *Mach. Learn.* **2001**, *45*, 5–32. [CrossRef]

39. He, Y.Y.; Liu, R.; Li, H.Y.; Wang, S.; Lu, X.F. Short-term power load probability density forecasting method using kernel-based support vector quantile regression and Copula theory. *Appl. Energy* **2107**, *185*, 254–266. [CrossRef]

40. Yu, F.; Xu, X.Z. A short-term load forecasting model of natural gas based on optimized genetic algorithm and improved BP neural network. *Appl. Energy* **2014**, *134*, 102–113. [CrossRef]

41. Seeger, M. Gaussian processes for machine learning. *J. Neural Syst.* **2011**, *14*, 69–106. [CrossRef] [PubMed]

42. ISO New England Load Data. Available online: https://www.iso-ne.com/isoexpress/web/reports/pricing/-/tree/zone-info (accessed on 11 November 2014).

43. Singapore Load Data. Available online: https://www.emcsg.com/PriceInformation#download (accessed on 19 December 2016).

44. Sheela, K.G.; Deepa, S.N. Review on Methods to Fix Number of Hidden Neurons in Neural Networks. *Math. Prob. Eng.* **2013**, *6*, 389–405. [CrossRef]

Article

Optimal Micro-PMU Placement Using Mutual Information Theory in Distribution Networks

Zhi Wu [1], Xiao Du [1], Wei Gu [1,*], Ping Ling [2], Jinsong Liu [2] and Chen Fang [2]

[1] School of Electrical Engineering, Southeast University, Nanjing 210096, China; zwu@seu.edu.cn (Z.W.);
    duxiaoseu@163.com (X.D.)
[2] State Grid Shanghai Electric Power Co, Electric Power Research Institute, Shanghai 200090, China;
    pingling_shh@163.com (P.L.); jinsongliu@163.com (J.L.); chenfang_shh@163.com (C.F.)
* Correspondence: wgu@seu.edu.cn

Received: 28 June 2018; Accepted: 19 July 2018; Published: 23 July 2018

Abstract: Micro-phasor measurement unit (μPMU) is under fast development and becoming more and more important for application in future distribution networks. It is unrealistic and unaffordable to place all buses with μPMUs because of the high costs, leading to the necessity of determining optimal placement with minimal numbers of μPMUs in the distribution system. An optimal μPMU placement (OPP) based on the information entropy evaluation and node selection strategy (IENS) using greedy algorithm is presented in this paper. The uncertainties of distributed generations (DGs) and pseudo measurements are taken into consideration, and the two-point estimation method (2PEM) is utilized for solving stochastic state estimation problems. The set of buses selected by improved IENS, which can minimize the uncertainties of network and obtain system observability is considered as the optimal deployment of μPMUs. The proposed method utilizes the measurements of smart meters and pseudo measurements of load powers in the distribution systems to reduce the number of μPMUs and enhance the observability of the network. The results of the simulations prove the effectiveness of the proposed algorithm with the comparison of traditional topological methods for the OPP problem. The improved IENS method can obtain the optimal complete and incomplete μPMU placement in the distribution systems.

Keywords: micro-phasor measurement unit; mutual information theory; stochastic state estimation; two-point estimation method

---

1. Introduction

Nowadays, more and more distributed generations (DGs) are integrated in distribution systems. One of the advantages of DG is that it can provide clean energy and diminish the emissions of CO_2. The integrations of DGs would also cause bidirectional power flow and great uncertainties, which makes the supervision and operation of distribution more complicated. It is necessary to use different strategies to improve the reliability, efficiency, and safety in planning and operation of distribution, such as fault analysis [1,2], dynamic operation and control strategies [3], and the improvement of transient stability [4]. Therefore, the distribution system needs powerful and accurate monitoring meting devices. Phasor measurement unit (PMU) is the current most advanced metering device of synchronized measurement technology which plays an important role in wide-area measurement system [5]. Phasor measurement unit can provide real-time and high-accurate magnitude and phase angle measurements of both voltage and current. Based on PMU measurements, many applications, such as state estimation, fault location, outage management, and event detection can be exploited [6,7]. For example, a hierarchical architecture for monitoring the distribution grid based on PMU data is proposed [8]. A linear model which considers PMU location for the observability assessment in different contingencies is presented [9]. Currently, PMU has been widely applied in the transmission network, but not in the distribution network. With the development of PMU technology and the integration of DGs, it is promising to deploy PMUs in distribution level. The number of nodes in the distribution network is far larger than that of the transmission network. So, it is important to study the optimal PMU placement (OPP) problem considering the characteristics of distribution network.

Though great merits PMU has, it is quite difficult for PMUs to replace the supervisory control and data acquisition (SCADA) system due to expensive cost of PMUs. Thus, the PMUs and SCADA are expected to coexist in the power system for a long time in future [10]. The PMU measurements and traditional measurements from SCADA can be collectively used for improvement of state estimation [11–14] and the accurate power system model parameter estimation [15,16]. An algorithm to use synchrophasor data conditioning in the prefix part of the existing linear state estimation formulation is presented [17]. Phasor measurement unit measurements can be integrated to refine the estimations according to the measurements from SCADA [18,19]. Traditional measurements such as power flow measurements (PFMs) and injection measurements (IMs) can be considered in the optimal PMU placement model with PMU measurements in [20,21].

Typically, there are two main categories in the problem of OPP models. The first one aims to calculate the minimal number and locations of PMUs to ensure the full observability of power system. Topological [22] and numerical [23] algorithms are two common methods for solving the OPP problem. The concept of spanning tree of full rank is constructed when the network is considered to be fully observable in the topological methodology. Based on such graph theory, great numbers of algorithms have been proposed to do observability analysis by considering different constraints. A methodology based on graph theoretical observability analysis for complete system observability is proposed [24]. An integer linear programming method is utilized for optimal PMU placement considering various arrangements of lines connections at complex buses [25]. Great numbers of heuristic algorithms, such as Tabu search [26] and immunity genetic [27], have been widely used to solve the OPP problem, which are classified and compared with different points of views [28]. A binary semi-definite programing (BSDP) model is utilized to make power system numerically observable in the presence of conventional measurements [29]. The channel limits of PMUs are taken into consideration in the formulation of OPP model in Reference [30]. The iterated local search (ILS) metaheuristic method is used to minimize the size of PMU configuration which makes the network observable [31]. An upgraded binary harmony search algorithm is presented to attain the minimum number of PMUs and their relevant locations considering the different installation cost of PMUs at different buses [32]. A modified greedy algorithm is proposed to solve the OPP problem under both normal operating and contingency conditions [33]. The second category is to realize some specific applications instead of full observability. For instance, a mixed-integer linear programming (MILP) framework is proposed for placing minimal PMUs to locate any fault in transmission system [34]. Different contingency conditions in power systems including measurement losses, line outages, and communication constraints are considered in the optimal PMU placement model [35]. An iterative linear program algorithm is applied to meet the prescribed synchrophasor availability profile in a smart grid in [36]. A fast greedy algorithm is used to strategically place secure PMUs at important buses to enhance the security of network and defend against data injection attacks [37]. The bad data detection and identification capability of the power system can be highly improved according to the optimal PMU placement [38]. An optimal placement for power system dynamic state estimation is presented by using empirical observability Gramian in [39]. A systematic framework is proposed for enhancing the situational awareness of the system operator using PMU placement [40]. A two-stage methodology for online identification of power system dynamic signature using PMU measurements and data mining is discussed in [41]. The multinomial logistic regression is utilized to place PMU optimally for identifying a single line outage in a power grid [42].

Besides the above methods, another PMU placement method using information-theoretic criterion called mutual information is presented in [43]. It is stated that typical methods are very likely to result in suboptimal placement and significant performance loss when only topological observability criterion is centered around. The information gain achieved by the PMU measurements is modeled as Shannon mutual information (MI) to obtain the full observability and incomplete observability. The PMU placement results based on information-theoretic criterion have been proved the effectiveness of the integration of mutual information to OPP model. In [25], the information-theoretic criterion could only be applied in the DC power flow model which cannot work in the AC power flow mode.

Phasor measurement units have not been widely applied at the distribution level due to great challenges in both technical and economic aspects [44]. To overcome these problems, several laboratories such as Power Standards Lab and Lawrence Berkeley National Lab have devoted to developing a novel powerful micro-phasor measurement unit (μPMU) and studied its practical and potential distribution system applications [45,46]. An advanced predictive analytics application for monitoring, protection, and control of distribution system assets using μPMU technology is presented in [47]. The diagnostic applications promising for future work are discussed for the presence of high penetrations of DGs. Despite the powerful functions μPMU has, it still requires a great number of μPMUs to obtain full observability which makes the cost of placement unaffordable. Therefore, the conventional measurements from smart meters such as feeder terminal units (FTUs) need to be considered in the placement model. Also, the data from historical database is necessarily utilized to generate pseudo measurements of load power as injection measurements by using load forecasting methods.

With the increasing DGs and the use of pseudo measurements in distribution level, the measurements errors need to be considered which results in stochastic state estimation. Few studies have been carried out about the stochastic state estimation in the literature. However, two-point estimate method (2PEM) which has been used to handle the uncertain variables based on the deterministic problem in mathematics field has been applicable for solving uncertainty problems in the field of electric system [48,49]. For instance, it has been used to account for uncertainties in the optimal power flow problem in electricity markets in [48] and to quantify the power transfer capability uncertainty in [49]. Thus, 2PEM is utilized to solve stochastic state estimation problem in this paper.

This paper proposes a novel optimal μPMU placement methodology by using the information entropy evaluation and node selection strategy (IENS) based on the mutual information theory. The results of stochastic state estimation which solved by 2PEM are used for the calculation of mutual information gain. The improved IENS method is also presented with two important rules. With the integration of pseudo measurements and FTU measurements, the proposed improved IENS can obtain the optimal μPMU placement for both complete and incomplete observability.

The contribution of this paper can be summarized as follows:

(1) The 2PEM is proposed to solve the stochastic state estimation considering the measurement errors of distribution network caused by DGs and pseudo injection measurements.
(2) The differential entropy of mutual information is proposed to evaluate the uncertainty of network which can be used in the AC power flow mode in distribution level.
(3) The improved IENS is proposed to obtain the optimal μPMU placement for both complete and incomplete observability under the improvement of initial IENS.

The rest of the paper is organized as follows. In Section 2, the formulation of mathematical model and IENS and improved IENS are illustrated. In Section 3, different case studies of revised IEEE 123-bus test system for complete and incomplete observability are presented. The conclusions are noted in Section 4.

2. Mathematical Formulation of Optimal μPMU Placement

In this section, the mathematical model of proposed method is elaborated in detail. The measurement errors of the distribution system are taken into consideration when using DGs and pseudo injection measurements obtained by load forecasting methods. The differential entropy of mutual information theory is firstly illustrated to assess the uncertainty of network under specific measurement configurations. Then 2PEM is proposed to solve the stochastic state estimation problem and standard deviation and mean of state variables can be calculated. Finally, IENS and improved INES are presented to obtain the optimal μPMU set.

2.1. Differential Entropy for Assessing Uncertainty of Network

As shown in [43], maximizing the mutual information is equivalent to minimizing the state estimation error covariance matrix. The concept of information gain is also used to assess the uncertainty of the distribution network.

Different from the mutual information only used in DC power flow model in [43], the differential entropy in this paper can be utilized to model the uncertainties of network using system states in AC power flow model:

$$I(x) = -\int_{-\infty}^{+\infty} f(x) \log f(x) dx = -\int_{-\infty}^{+\infty} \left(2\pi\sigma^2\right)^{-\frac{1}{2}} e^{-(x-\mu)^2/2\sigma^2} \ln\left[\left(2\pi\sigma^2\right)^{-\frac{1}{2}} e^{-(x-\mu)^2/2\sigma^2}\right] dx = \frac{1}{2}\left(\ln\left(2\pi\sigma^2\right) + 1\right) \quad (1)$$

where $I(x)$ is differential entropy for the continuous variable x, and x represents magnitude and phasor of voltage in this paper, σ and μ is the standard deviation and mean of x. The uncertainty of the network under specific measurement configurations can be assessed by the above equation according to the standard deviation of state variables.

The standard deviation σ used in Equation (1) can be calculated through 2PEM for stochastic state estimation, which will be introduced in the following part.

2.2. Stochastic State Estimation Using Two-Point Estimation Method

The deterministic state estimation model is firstly introduced, and the formulation of stochastic state estimation model and two-point estimation method comes next.

The formulation for deterministic state estimation including both μPMU and SCADA measurements is adopted here, just as the estimator with phasor measurements mixed with traditional measurements in Reference [11], given by:

$$\begin{bmatrix} z_1 \\ z_2 \end{bmatrix} = \begin{bmatrix} h_1(x) \\ h_2(x) \end{bmatrix} + \begin{bmatrix} \varepsilon_1 \\ \varepsilon_2 \end{bmatrix} \quad (2)$$

where x is the state variables of network, z_1 is the vector of traditional measurements from SCADA, and z_2 is the vector of measurements obtained from μPMUs, $h(x)$ is the nonlinear function of state vector, ε_1 and ε_2 is the measurement error vector of SCADA measurements and μPMU measurements, with the covariance matrix W_1 and W_2.

$$W_1 = \begin{bmatrix} \sigma_1^2 & 0 & 0 \\ 0 & \ddots & 0 \\ 0 & 0 & \sigma_{m_1}^2 \end{bmatrix} \quad W_2 = \begin{bmatrix} \sigma_1^2 & 0 & 0 \\ 0 & \ddots & 0 \\ 0 & 0 & \sigma_{m_2}^2 \end{bmatrix} \quad (3)$$

where σ_i^2 is the variance of ith measurements, m_1 and m_2 is the number of SCADA measurements and μPMU measurements, respectively.

The Jacobian matrix $H(x)$ is usually obtained by following derivation:

$$H(x) = \begin{bmatrix} H_1(x) \\ H_2(x) \end{bmatrix} = \begin{bmatrix} \dfrac{\partial h_1(x)}{\partial x} \\ \dfrac{\partial h_2(x)}{\partial x} \end{bmatrix} \quad (4)$$

It is considered to be a nonlinear problem and Newton iterative method is usually used to solve this kind of problem. Deterministic weighted least square (WLS) state estimation is solved by following iterative equation:

$$x_{q+1} = x_q + G(x_q)\left[H_1^T W_1^{-1}\right](z_1 - h_1(x_q))$$
$$+ G(x_q)\left[H_2^T W_2'^{-1}\right](z_2 - h_2(x_q)) \tag{5}$$

where q is the number of iteration, $G(x)$ is the gain matrix calculated by

$$G(x_q) = \left[H_1^T(x_q) W_1^{-1} H_1(x_q) + H_2^T(x_q) W_2'^{-1} H_2(x_q)\right]^{-1} \tag{6}$$

W is the block diagonal matrix given by

$$W = \begin{bmatrix} W_1 & 0 \\ 0 & W_2' \end{bmatrix} = \begin{bmatrix} W_1 & 0 \\ 0 & RW_2 R^T \end{bmatrix} \tag{7}$$

where R is the general rotation matrix [25]. According to the WLS iterative method, state variables of the network can be calculated when it reaches the required accuracy.

Various methods and techniques such as linear regression models, autoregressive and moving average models, and artificial neural networks have been applied in the field of load forecasting. The pseudo injection measurements of load power can be obtained according to the database of distribution management system by using certain load forecasting method which is not the key part in this paper. It is inevitable to have prediction errors in pseudo measurements which results in uncertainty in the state estimation of power system.

Taking the forecasting errors of loads and DGs into consideration, the deterministic state estimation then turns to be the stochastic one. As described in [50,51], two-point estimate method is a variation of point estimation estimate method, and it can be used to decompose Equation (2) into sub-problems by using two deterministic values of every uncertain variable on both sides of corresponding mean. The results of stochastic state estimation can be obtained by 2 runs of the deterministic state estimation for each uncertain variables in the measurement model, once for the value above the mean, once for the value below the mean, and other variables are set to be corresponding means. For example, if there are m uncertain measurements, then only $2m$ runs of deterministic state estimation are needed. Then the statistical results like mean, variance, and probability density function of state variables could be acquired after the calculation of stochastic state estimation. The uncertainty of the network could be assessed by calculation of mutual information gain using Equation (1).

In the state estimation, let Y denote the random variable with probability density function (PDF) $f_Y(y)$ where Y is the measurements vector in state estimation model. For nonlinear function $X = h'(Y)$ where X is the state variables vector of distribution network. The procedure for calculating stochastic state estimation using two-point estimation method can be summarized as follows:

$$Y = [y_1, \ldots, y_n, y_{n+1}, \ldots, y_{n+n_1}] \tag{8}$$

(1) Determine the number of uncertain variables of pseudo measurements as n, and the number of certain measurements obtained from PMU and SCADA as n_1.

(2) Set $E(X) = 0$ and $E(X^2) = 0$.

(3) Set $t = 1$, and carry out the following steps until $t = n$.

(4) Calculate concentrations $y_{t,1}$, $y_{t,2}$, locations of concentrations $\xi_{t,1}$, $\xi_{t,2}$ and its probabilities $P_{t,1}$, $P_{t,2}$

$$\xi_{t,1} = \sqrt{n}, \; \xi_{t,2} = -\sqrt{n} \tag{9}$$

$$P_{t,1} = P_{t,2} = \frac{1}{2n} \tag{10}$$

$$y_{t,1} = \mu_{Y,t} + \xi_{t,1}\sigma_{Y,t} \tag{11}$$

$$y_{t,2} = \mu_{Y,t} + \xi_{t,2}\sigma_{Y,t} \tag{12}$$

where $\mu_{Y,t}$ and $\sigma_{Y,t}$ is the mean and the standard deviation of Y_t according to the measurement information.

(1) Run the deterministic state estimation for $y_{t,i}$ by using $Y = [\mu_{Y,1}, \mu_{Y,2}, \ldots, y_{t,i}, \ldots, \mu_{Y,n}, y_{n+1}, \ldots, y_{n+n_1}]$.

(2) Update $E(X)$ and $E(X^2)$

$$E(X) \cong \sum_{t=1}^{n} \sum_{i=1}^{2} \left(P_{t,i} h'([\mu_{Y,1}, \mu_{Y,2}, \ldots, \mu_{t,i}, \ldots, \mu_{Y,n}, y_{n+1}, \ldots, y_{n+n_1}]) \right) \qquad (13)$$

$$E(X^2) \cong \sum_{t=1}^{n} \sum_{i=1}^{2} \left(P_{t,i} h'([\mu_{Y,1}, \mu_{Y,2}, \ldots, \mu_{t,i}, \ldots, \mu_{Y,n}, y_{n+1}, \ldots, y_{n+n_1}])^2 \right) \qquad (14)$$

Calculate the mean and the standard deviation of state variables and then $t = t + 1$.

$$\mu_X = E(X) \qquad (15)$$

$$\sigma_X = \sqrt{E(X^2) - \mu_X^2} \qquad (16)$$

According to the calculation of mean and the standard deviation of state variables for stochastic state estimation by 2PEM, the uncertainties of network can be evaluated by Equation (1) under certain configuration of μPMU placement.

2.3. Information Entropy Evaluation and Node Selection Strategy for μPMU Sets

After the illustration of differential entropy and two-point estimate method, the following part aims to illustrate the IENS and improved IENS for calculating the optimal μPMU placement to maximize the information gain of the distribution system and obtain the observability of the network.

2.3.1. Information Entropy Evaluation and Node Selection Strategy

It is assumed that pseudo measurements of injections powers of all buses can be acquired according to the historical database in the distribution energy management using load forecasting method. FTU measurements are also integrated with pseudo measurements to enhance the observability of distribution network.

In general, the greedy algorithm is used to obtain the set of optimal μPMU placement sequentially following an incremental expansion strategy in IENS.

The steps of IENS are introduced as follows:

Step One:

(1) Define the set of candidate buses from which to choose for the installation of new μPMU: $B_c = \{b_{1_c}, b_{2_c}, \ldots, b_{n_c}\}$. The location of new μPMU is selected from the buses in B_c. It is assumed to contain all the buses in the network if there is no mandatory μPMU allocated beforehand. The bus to be installed with new μPMU will be discarded from B_c after the selection of new μPMU.

(2) Define the set of buses for the installation of μPMU as $B_s = \{b_{1_a}, b_{2_a}, \ldots, b_{n_a}\}$. The buses in B_s would be installed with μPMUs. B_s is null if there is no μPMU allocated beforehand. The bus to be installed with new μPMU will be added into B_s after the selection of new μPMU.

(3) Set the number of μPMUs to be installed in the network as n_s.

Step Two:

Run stochastic state estimation using 2PEM and obtain the statistical results under initial measurement configuration which consists of pseudo measurements and FTU measurements. The initial differential entropy E_0 of network can be calculated by Equation (17):

$$E = \frac{1}{N}\sum_{i=1}^{N}\left(1 + \log\left(2\pi\sigma_{Vi}^2\right) + \log\left(2\pi\sigma_{\theta i}^2\right)\right) \tag{17}$$

where N is the number of all buses, $\sigma_{V_i}, \sigma_{\theta_i}$ is the standard deviation of the voltage amplitude and phase angle at bus i.

Step Three:

(1) Run the following part:

For $l = 1, 2, \ldots, n_c$:

(a) Build a new set: $\boldsymbol{B}_s^l = [b_1, b_2, \ldots, b_{n_a} | b_l]$ where first n_a columns are n_a buses already installed with μPMUs and last column means the lth bus candidate for the location of μPMU.

(b) Add μPMU measurements of $\boldsymbol{B}_s^l$ into initial measurement configuration as new measurement configuration. Then run stochastic state estimation by using 2PEM under lth measurement configuration and calculate its differential entropy E_l using Equation (17).

End

---

(2) Find bus k which maximizes the improvement in information gain of differential entropy.

$$b_k = \arg\left(\max_l(|E_0 - E_l|)\right) \tag{18}$$

Then $E_0 = E_k$, excludes bus k from $\boldsymbol{B}_c$, adds bus k into $\boldsymbol{B}_s$,

$$\boldsymbol{B}_c \leftarrow \boldsymbol{B}_c \setminus \{b_k\} \; and \; \boldsymbol{B}_s = \boldsymbol{B}_s \cup \{b_k\} \tag{19}$$

Step Four:

If the current number of installed μPMU satisfies the desired number n_s, then output the set $\boldsymbol{B}_s$ as the installation set of μPMUs; otherwise turn to Step Three.

The optimal μPMU set can be obtained according to IENS. Usually, n_s the number of μPMUs to be allocated in the network is decided by the project budget which is expected to be as much as possible. However, the upper limit of μPMU should not exceed n_{TM}, the number of optimal placement calculated by topological method for network full observability. An integrated model based on topological method is presented considering the effects of the zero injections buses (ZIBs) and conventional measurements (CMs) such as power flow measurements and injection measurements in [22]. The model of injection measurements is considered the same as that of ZIBs. This method is applied in this paper to determine the maximum number of μPMUs to be stalled in the network.

2.3.2. Selection Rules to Be Noticed

The IENS and topological method in Reference [22] is applied on a 11-bus test system where a FTU placed on line l_{1-2} as shown in Figure 1. The FTU can measure the voltage magnitude of the tail bus of installed line and the power flow of the line.

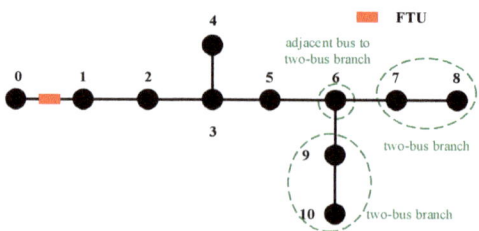

Figure 1. 11-bus test system.

According to the power flow measurements, the optimal μPMU placement obtained by topological method is shown in Figure 2. It needs only 4 μPMUs to make the network full observable.

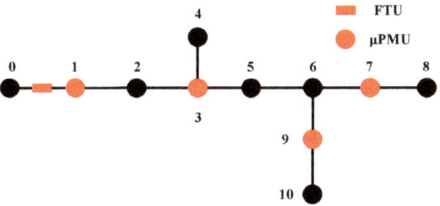

Figure 2. Optimal placement by topological method.

Take the number of results by topological method as the required number of μPMUs to be installed in IENS: $n_s = 4$, the placement by IENS is shown in Figure 3.

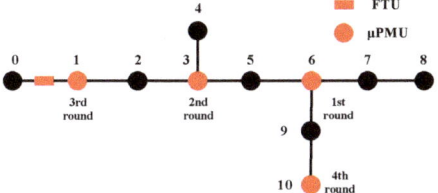

Figure 3. Optimal placement by information entropy evaluation and node selection strategy (IENS).

The sequence of the selected candidate bus in order is b_6, b_3, b_1, b_{10}. It is reasonable to install μPMU at b_6 since it can obtain the maximum information gain at the first round. Then it comes to b_3 and b_1. After the selection of b_6, b_3, b_1, the fourth bus to be installed with μPMU is b_{10}. However, the results calculated by IENS obviously cannot obtain the full observability for the 11-bus test system since b_8 is unobservable.

Compare the results of IENS with the results of topological method, the major reason for the unobservability of placement of IENS is the selection of b_6. Although the selection of b_6 can maximize the information gain of the network in the first round, it results in additional 2 μPMUs to make b_8, b_{10} observable, which means it needs 5 μPMUs to make 11-bus test system by IENS. When 2 μPMUs are located at b_7 and b_9 instead of b_6 and b_{10}, the placement can obtain the full observability just as shown in Figure 2. Considering the full observability of 11-bus network, b_6 may not be the ideal location for μPMU. To sum up, the bus which has one or more two-bus branches cannot be the selection of new μPMU. For instance, branch 7–8 and 9–10 is the two-bus branch of the bus b_6 as shown in Figure 1, and b_6 would not be the selection of μPMU considering the full observability of the network.

Thus the node selection part needs to be improved with the combination of characteristics of the placement of topological method for full observability. After the application of IENS on different networks for many times, two rules are summarized to be observed to improve the observability of IENS. The rules of the selection of candidate bus for μPMU should be proposed as follows:

Rule 1:

Find the bus k which maximizes the improvement in information gain of differential entropy by using Equation (9), if bus k has one or more two-bus branches, then add the buses adjacent to bus k on two-bus branches into new set B_{ak}, sort the buses in B_{ak} by information gain and find the bus q which maximizes the improvement of information gain, then $b_{add} = b_q$; if there is no two-bus branch collected to bus k, then $b_{add} = b_k$. (Bus b_{add} represents the selected bus to be installed with new μPMU in current round).

For example, b_6 is the bus which maximizes the improvement in information gain in the 11-bus test system, result of IENS proves that b_6 is not the ideal location for μPMU. Then according to Rule 1, b_6 has two-bus braches 7–8 and 9–10, adds b_7 and b_9 into set $\mathbf{B}_{ak}$, sort b_7 and b_9 by the information gain, and find the bus which maximizes the improvement of information gain as the selection bus for installation of μPMU.

The simulation test on 11-bus test system shows that the location of new μPMU cannot simply be the bus which maximizes the information gain of network. This kind of bus is not the optimal location for new μPMU when it has one or more two-bus branches. Taking the full observability into consideration, after finding the bus k which can maximize the information gain of differential entropy of network, the selection bus to be installed for μPMU should be determined by Rule 1.

Rule 2:

The terminal bus cannot be installed with μPMU in the distribution network.

Considering the radial structure of distribution system, since a μPMU can measure both the voltage magnitude and phasor angle of associated bus and current magnitude and phasor along all lines collected to this bus, the μPMU should not be placed at terminal bus.

2.3.3. Improved Information Entropy Evaluation and Node Selection Strategy

According to the rules above, the improved IENS can be modified based on the IENS with Rules 1 and 2 in node selection part.

In the simulation of 11-bus test system, the result of improved IENS is same as the result of topological method in Figure 2, which also needs four μPMUs to make network full observable. The order of the locations of four μPMUs is 7, 3, 9, and 1. b_6 should be the installation of new μPMU in the first selection since it obtains the maximal information gain. However, b_7 turns to be the location for μPMU according to Rule 1 since b_7 has larger improvement of information gain than b_9. Then b_3, b_9, and b_1 is selected to be installed with μPMU in the following round due to their maximization of improvement of information gain.

The process of improved IENS combined with Rules 1 and 2 for optimal μPMU set is shown in Figure 4.

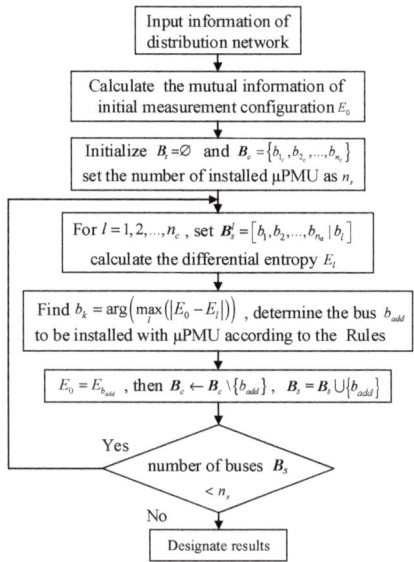

Figure 4. Block diagram of improved IENS.

3. Case Studies

The modified IEEE 123 test system is used to verify the effectiveness of proposed method. The layout of the test system is shown in Figure 5. The test system contains five distribution generations denoted by gray rectangles. Details of the test system can be referred to in Reference [52].

It is assumed that seven FTUs have been installed in the test distribution system. The locations of FTUs are on lines 1–2, 55–58, 36–120, 22–24, 68–98, 77–87, 88–90 which are denoted by red rectangles in Figure 5.

Three types of measurements with different accuracy values are considered in this paper. The settings of their maximum percentage errors are as follows:

Pseudo measurements: 50%. These measurements are obtained by load forecasting methods according to the historical data.

FTU measurements: 2%.

PMU measurements: it is assumed to be 1% total vector error in the worst case.

The simulation is performed using MATLAB 2017a, on Xeon E3-1230 3.30-GHz personal computer with 8 G memory.

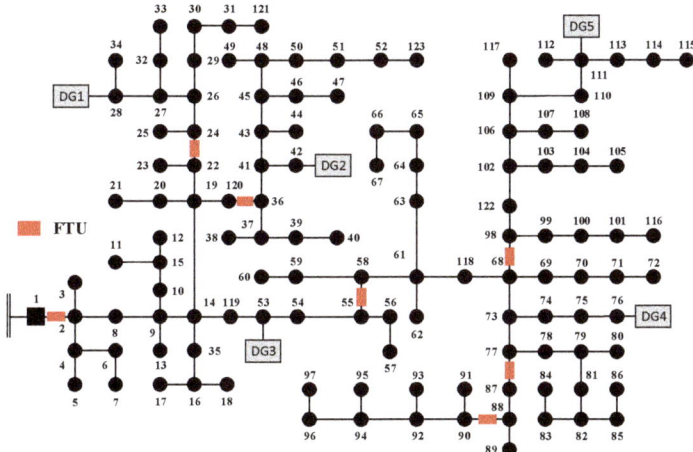

Figure 5. The modified Institute of Electrical and Electronics Engineers (IEEE) 123 test system.

3.1. Optimal Placement for Full Observability by Improved IENS

Considering the measurements of seven FTUs depicted in Figure 5, the minimal number of μPMUs to make modified 123-bus system full observable is calculated to be 45 by topological method. However, it needs 46 μPMUs to make system observable using genetic algorithm. The drawback of heuristic algorithms such as genetic algorithm is that it is difficult to get the global optimal solution while the topological one can. The optimal μPMUs placements for full observability with and without FTU measurements in modified 123-bus system are shown in Table 1, the results show that the integration of FTUs helps reduce the number of μPMUs.

According to the results by topological method, the required number of μPMUs is set to be 45 in improved IENS. In this case, the pseudo measurements of injection power of all buses in the network are assumed to be acquired in improved IENS for the selection of μPMU set. Under the initial measurement configuration, the mutual information gain E_0 is calculated with the pseudo injection measurements and FTU measurements. Based on the incremental expansion strategy of improved IENS, the locations of 45 μPMUs can be obtained in order as: 2, 9, 20, 61, 22, 68, 56, 79, 107, 109, 41, 88,

32, 24, 28, 59, 71, 92, 48, 75, 15, 101, 111, 43, 54, 106, 83, 85, 46, 94, 37, 64, 66, 4, 104, 31, 90, 52, 114, 16, 6, 96, 39, 99, 29, the optimal deployment of μPMUs is shown in Figure 6.

Table 1. Minimal micro-phasor measurement unit (μPMU) numbers for full observability with and without feeder terminal unit (FTU) measurements.

	With FTU Measurements	Without FTU Measurements
Topological method	45	47
Improved IENS	45	47
Genetic method	46	48

According to the mutual information theory, the first several buses are expected to be selected with the maximal information gain for the installations of μPMUs. For example, in the first six selection of μPMUs: 2, 9, 20, 61, 22, 68, b_2, b_{61}, and b_{68} are the buses adjacent to four buses which means more μPMUs measurements can be acquired. Thus, maximal information gain would be obtained when μPMUs are deployed at these buses.

Take the determination of second selection bus for μPMU as illustration, b_{14} should be the installation of new μPMU in the second selection since it obtains the maximal information gain after the first selection. However, b_{14} has a two-bus branch 9–13 and it could not be the selection bus for μPMU according to Rule 1. It is easily to be understood that another μPMU needs to be allocated at b_9 to make b_{13} observable if the second μPMU is located at b_{14}. Therefore, b_9 is determined to be second bus for the location of new μPMU according to Rule 1. So as the selection of b_{20} and b_{22}. According to Rule 2, there is no μPMU to be installed at the terminal bus in the network. The results calculated by improved IENS can obtain the full observability of the network which has the same effect of the placement of topological method with the identical number of μPMUs.

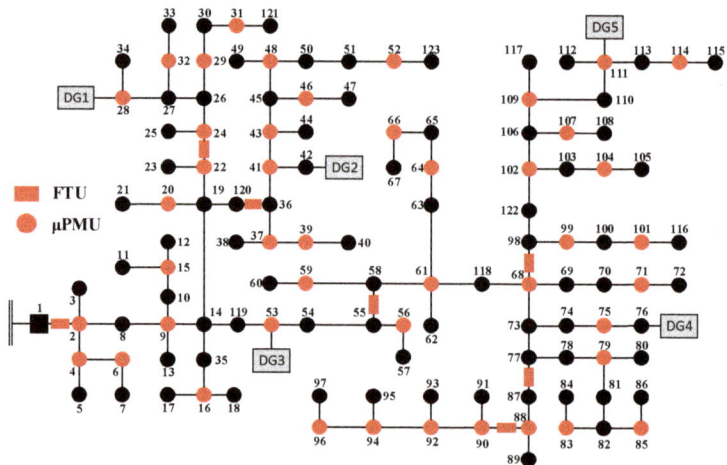

Figure 6. Optimal μPMU placement by improved IENS.

The pseudo measurements of DGs are usually considered to have more measurement errors than the pseudo measurements of loads. According to the proposed mutual information theory, the bus installed with DG has the priority to be placed with μPMU since that bus has more uncertainties. For example, when DG3 is located at b_{54}, the μPMU would also be located at b_{54} instead of b_{53}.

3.2. Incomplete Observability Analysis

The full observability of the distribution system can be obtained when enough µPMUs are deployed in the network. However, such µPMUs cannot be installed in one time due to the huge cost of placement, and only part of them can be placed. With the consideration of partial placement, the µPMU placement for maximal observability with limited number are studied in both the topological method and improved IENS method.

It is assumed that all pseudo measurements of injection power can be obtain in the ideal situation which rarely happens in reality. Only part of the pseudo injection measurements of the buses can be acquired for the state estimation according to the distribution management system. The different ratios of acquired pseudo injection measurements should be taken into account for the incomplete observability, the observable capability is used to assess the µPMU placements of different required numbers using numerical method. The observable capability is evaluated by the number of configurations which can make network observable with the µPMU placement divided by the number of all configurations in the set.

The case is still tested on the modified IEEE 123 test system. To evaluate the observable capability of the µPMU placement under different ratios of pseudo measurements, numerical simulation needs to be conducted. Two sets of pseudo measurements configurations with different ratios are considered, one is 90% and the other is 80%, which means only 80% or 90% pseudo injection measurements of buses can be obtained in a pseudo measurement configuration. Each set has 10,000 different configurations in which the pseudo injection measurements of different ratios are randomly generated first. For example, in the configuration of set with 90% pseudo measurement in modified IEEE 123 test system, about the pseudo injection powers of 111 buses can be used for observability analysis. Then the µPMU placement will be tested to be observable or not by numerical method under 10,000 different configurations. The percentage of observable placements under 10,000 configurations is considered to be the observable capability of the µPMU placement.

In improved IENS, the optimal µPMU set for full observability is calculated in the order of information gain. When it comes to the circumstance that the required µPMU number n_s is smaller than the number for full observability, the n_s buses can be easily selected from the optimal µPMU set which can make system full observable. However, it is hard for topological method to choose n_s µPMUs for incomplete observability since the topological method can only obtain the optimal placement for full observability. For simplicity, n_s buses are selected randomly from the results of topological method for full observability as 500 different placements. These placements are tested by numerical method with the integration of pseudo measurements configurations and the mean observability capability is compared with the one of improved IENS.

The observability capability of results of improved IENS and topological method under different circumstances are shown in Table 2.

Table 2. Observability capability of improved IENS and topological method under different circumstances.

Number of µPMUs	90% Pseudo Measurement Configurations		80% Pseudo Measurement Configurations	
	Topological Method (Mean)	Improved IENS	Topological Method (Mean)	Improved IENS
40	90.08%	97.00%	65.66%	88.40%
35	78.93%	82.50%	38.26%	42.40%
30	66.55%	73.20%	18.79%	26.70%

As shown in Table 2, the mean observable capability of results of topological method in 500 configurations is selected to be compared with the observable capability of results of Improved IENS. The observable capability of both topological method and improved IENS seem to be better when the numbers of µPMUs increased. Under both of 80% and 90% pseudo measurement configurations, the observable capabilities of improved IENS are better than the topological method. Due to the methodology of improved IENS, the incremental expansion strategy helps the mutual information of network nearly maximal at the incremental placement of μPMUs which obtains better observable capability than the topological method.

Take 40 μPMUs to be installed under 90% pseudo measurements configurations as an example; the observable capability of improved IENS is 97%, which is larger than the mean value of 500 placements of topological method. The observable capability of improved IENS is still larger than the mean value of topological method when the number of required μPMUs is 30 or 35. The observable capability of improved IENS outbalances the average level of the placements according to the results of topological method. The placement of both improved IENS and topological method seems to have better observable capability when the pseudo measurement configurations increased from 80 to 90%.

3.3. Effects of Two Rules

According to the results by topological method, the required number of μPMUs is set to be 45 in IENS and improved IENS. Also, the pseudo measurements of injection power of all buses in the network are used in improved IENS and IENS. Under $n_s = 45$, the results of both IENS and improved IENS are shown in Table 3 in the order of node selection. The results of three methods are tested for observability through numerical method. The observability of corresponding methods are shown in Table 3.

Table 3. Optimal μPMU placements of three methods.

Method	Optimal μPMU Placement	Tested by Numerical Method
Topological method	2, 4, 6, 9, 15, 16, 20, 22, 24, 28, 29, 31, 32, 37, 39, 41, 43, 46, 48, 52, 54, 56, 59, 61, 64, 66, 68, 71, 75, 79, 83, 85, 88, 90, 92, 94, 96, 98, 101, 104, 107, 109, 111, 114, 122	observable
IENS	2, 14, 68, 53, 61, 77, 106, 41, 27, 90, 9, 55, 15, 79, 24, 111, 48, 122, 82, 94, 65, 37, 16, 99, 70, 46, 75, 96, 20, 30, 101, 103, 51, 59, 114, 6, 124, 121, 19, 58, 123, 109, 88, 5, 28	unobservable
Improved IENS	2, 9, 20, 61, 22, 68, 56, 79, 107, 109, 41, 88, 32, 24, 28, 59, 71, 92, 48, 75, 15, 101, 111, 43, 54, 106, 83, 85, 46, 94, 37, 64, 66, 4, 104, 31, 90, 52, 114, 16, 6, 96, 39, 99, 29	observable

The result of topological method and improved IENS is tested to be observable by using a numerical method, while the result calculated by IENS is unobservable. As depicted in Figures 7 and 8, the buses in the green ellipses are the main differences between results of IENS and topological method. Note the area surrounded by green ellipses, μPMUs are mostly located at the buses adjacent to the terminal buses in Figure 7 while μPMUs are not in Figure 8. In the area 1, 3, 5, and 6 the buses in areas are all observable in Figure 7, while are not observable in Figure 8. These areas need more μPMUs for observable due to the suboptimal placement of μPMUs. The number of μPMUs will decrease effectively when μPMU is installed at the bus which is adjacent to the terminal bus in the green areas in Figure 8. Especially in the area 7, which contains b_{81}, b_{82}, b_{83}, b_{84}, b_{85}, b_{86}, the information gain would be larger when the μPMU is placed at b_{82}, but b_{84} and b_8 would be out of observability if there is no other μPMU in this area. It needs three μPMUs to make area 7 observable in Figure 8, while only two μPMUs are needed in Figure 7.

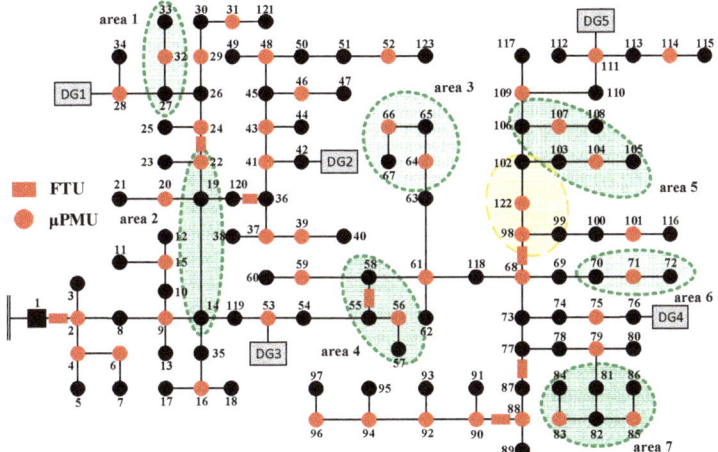

Figure 7. Optimal µPMU placement by topological method.

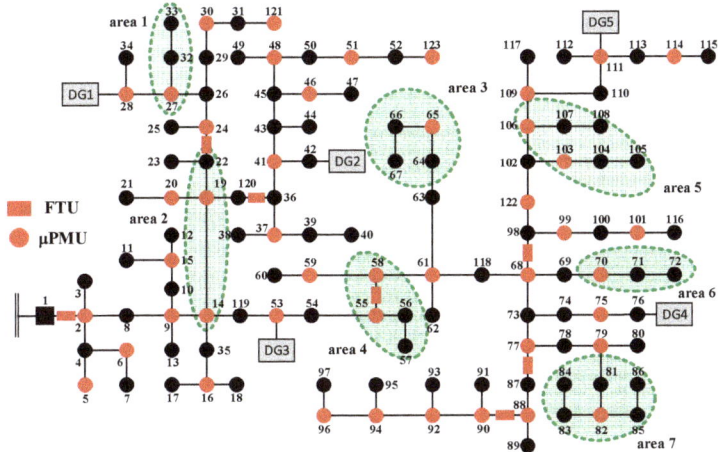

Figure 8. Optimal µPMU placement by IENS.

With the compliance of rules, the placement of µPMUs calculated by improved IENS is shown in Table 3 and Figure 9, the results are proved to be full observable under the test of numerical method with the same µPMU number of topological method. The placement of improved IENS is quite similar to the results in Figure 7 except the yellow area.

According to the Rules 1 and 2, the buses in the green areas in Figure 9 can be full observable under the optimal locations of µPMUs. The µPMUs are deployed at the buses adjacent to the terminal bus which cooperates with other µPMUs, making the network full observable. The results prove the effectiveness of improved IENS for full observability compared with the results of topological method with the same number of µPMUs.

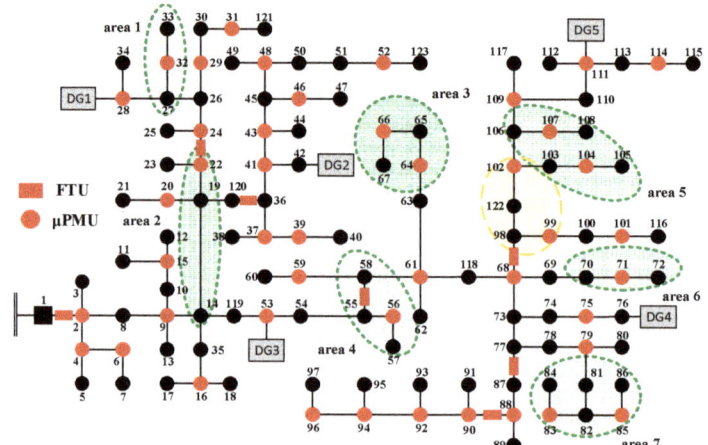

Figure 9. Optimal μPMU placement by improved IENS.

3.4. Limitations of the Improved IENS

Although the improved IENS has good performance in both complete and incomplete observability, it still has some limitations. The proposed method requires the integration of pseudo measurements in the stochastic state estimation, and the pseudo measurements are assumed to be obtained from historical data using load forecasting method. However, such historical data information is hard to be acquired in the actual distribution. Also, the proposed method only focuses on the observability of the network. The accuracy of state estimation, stability in fault and limitations in μPMU channels have not been taken into consideration.

4. Conclusions

This paper presents an optimal μPMU placement based on IENS using greedy algorithm. The differential entropy of mutual information theory is introduced and utilized to evaluate the uncertainty of distribution network in AC power flow mode using the results of 2PEM. By using mutual information theory, the IENS method is proposed first. However, the effectiveness of IENS is not satisfied enough and could not obtain full observability under the same number of placement of topological method. With the consideration of characteristic of the placement of topological method, improved IENS is presented with two rules based on the IENS strategy. The improved IENS proves to have the same effect as topological method in complete observability, using 45 μPMUs to make modified IEEE 123 test system full observable. As shown in Table 2, the improved IENS has better observable capability when the required μPMUs cannot make system full observable compared with topological method. The placement seems to have better observable capability when the pseudo measurement configurations increase. The results on the simulations prove the effectiveness of improved IENS both in full observability and incomplete observability. The proposed method only focuses on optimal placement under normal operation, and the reliability such as N-1 PMU loss will be considered in future work.

Author Contributions: The contribution of Z.W. is review and editing, the contribution of X.D. is writing he original draft, review and editing, the contribution of W.G. is review and editing, the contribution of P.L. is data curation and investigation, the contribution of J.L. is conducting formal analysis and supervision, the contribution of C.F. is providing methodology and resources.

Funding: This work was supported by National Key Research and Development Program of China (Grant No. 2017YFB0902900) and the State Grid Corporation of China.

Conflicts of Interest: The authors declare no conflicts of interest.

Nomenclature

Sets and Indices

B_c	The set of candidate buses where the installation of new micro-phasor measurement unit (μPMU) is selected from.
B_s	The set of buses for the installation of μPMU, the location of new μPMU will be added in this set.
B_s^l	The set of buses B_s at lth iteration.
B_{ak}	The set of buses which contains the buses adjacent to bus k on two-bus branches.
l_{i-j}	The line connected between bus i and bus j.
b_i	The ith bus.
b_k	The bus k which maximizes the improvement in information gain of differential entropy.
b_{add}	The selected bus to be installed with new μPMU in current round.

Parameters

σ	The standard deviation of variable x.
μ	The mean of variable x.
z	Vector of measurements.
ε	Error vector of measurements.
σ_i^2	Variance of ith measurements.
$H(x)$	The Jacobian matrix.
m_i	The number of measurements.
W_i	The covariance matrix of measurements.
W	The block diagonal matrix.
q	The number of iteration in weighted least square (WLS) state estimation.
R	The general rotation matrix.
Y	The measurements vector in state estimation.
y_i	The ith measurement in state estimation.
n	The number of uncertain variables of pseudo measurements.
n_1	The number of certain measurements obtained from phasor measurement unit (PMU) and supervisory control and data acquisition (SCADA) system.
$E(X)$	The expectation of state variables vector.
$E(X^2)$	The expectation of square of state variables vector.
$y_{t,i}$	The concentration of measurement at step t.
$\xi_{t,i}$	The location of concentration of measurement at step t.
$P_{t,1}$	The probability of concentration of measurement at step t.
$y_{t,i}$	The concentration of measurement at step t.
Y_t	The measurements vector at step t.
$\mu_{Y,t}$	The mean value of Y_t, obtained from measurement information.
$\sigma_{Y,t}$	The standard deviation of Y_t, obtained from measurement information.
μ_X	The mean value of state variables X.
σ_X	The standard deviation of state variables X.
E_0	The initial differential entropy of the network.
E	The differential entropy of the network.
N	The number of all buses in the network.
σ_{V_i}	The standard deviation of voltage amplitude at bus i.
σ_{θ_i}	The standard deviation of voltage phase angle at bus i.
n_c	The number of candidate buses which can be the location for new μPMU.
l	The number of round in the information entropy evaluation and node selection strategy (IENS).
E_l	The differential entropy of the network at lth iteration.
n_s	The number of μPMUs decided to be installed in the network according to the budget.
n_{TM}	The number of optimal placement calculated by topological method for network full observability.

Variables

x	State variables of network, including magnitude and phasor angle of voltage.
X	The state variables vector in state estimation.
$h(x)$	Nonlinear function of state variables.
$I(x)$	Differential entropy for the continuous variable x.

References

1. Ou, T.C. A novel unsymmetrical faults analysis for microgrid distribution systems. *Int. J. Electr. Power Energy Syst.* **2012**, *43*, 1017–1024. [CrossRef]
2. Ou, T.C. Ground fault current analysis with a direct building algorithm for microgrid distribution. *Int. J. Electr. Power Energy Syst.* **2013**, *53*, 867–875. [CrossRef]
3. Ou, T.C.; Hong, C.M. Dynamic operation and control of microgrid hybrid power systems. *Energy* **2014**, *66*, 314–323. [CrossRef]
4. Ou, T.C.; Lu, K.H.; Huang, C.J. Improvement of Transient Stability in a Hybrid Power Multi-System Using a Designed NIDC (Novel Intelligent Damping Controller). *Energies* **2017**, *10*, 488. [CrossRef]
5. Ree, J.D.L.; Centeno, V.; Thorp, J.S.; Phadke, A.G. Synchronized Phasor Measurement Applications in Power Systems. *IEEE Trans. Smart Grid* **2010**, *1*, 20–27.
6. Bertsch, J.; Carnal, C.; Karlson, D.; Mcdaniel, J.; Vu, K. Wide-Area Protection and Power System Utilization. *Proc. IEEE* **2005**, *93*, 997–1003. [CrossRef]
7. Dong, Z.Y.; Xu, Y.; Zhang, P.; Wong, K.P. Using IS to Assess an Electric Power System's Real-Time Stability. *IEEE Intell. Syst.* **2013**, *28*, 60–66. [CrossRef]
8. Jamei, M.; Scaglione, A.; Roberts, C.; Stewart, E.; Peisert, S.; Mcparland, C.; Mceachern, A. Anomaly Detection Using Optimally-Placed μPMU Sensors in Distribution Grids. *IEEE Trans. Power Syst.* **2017**, *33*. [CrossRef]
9. Teimourzadeh, S.; Aminifar, F.; Shahidehpour, M. Contingency-Constrained Optimal Placement of Micro-PMUs and Smart Meters in Microgrids. *IEEE Trans. Smart Grid* **2017**, *1*. [CrossRef]
10. Askounis, D.T.; Kalfaoglou, E. The Greek EMS-SCADA: From the contractor to the user. *IEEE Trans. Power Syst.* **2000**, *15*, 1423–1427. [CrossRef]
11. Zhou, M.; Centeno, V.A.; Thorp, J.S.; Phadke, A.G. An Alternative for Including Phasor Measurements in State Estimators. *IEEE Trans. Power Syst.* **2006**, *21*, 1930–1937. [CrossRef]
12. Costa, A.S.E.; Albuquerque, A.; Bez, D. An estimation fusion method for including phasor measurements into power system real-time modeling. *IEEE Trans. Power Syst.* **2013**, *28*, 1910–1920. [CrossRef]
13. Glavic, M.; Cutsem, T.V. Reconstructing and tracking network state from a limited number of synchrophasor measurements. *IEEE Trans. Power Syst.* **2013**, *28*, 1921–1929. [CrossRef]
14. Manousakis, N.M.; Korres, G.N.; Aliprantis, J.N.; Vavourakis, G.P.; Makrinas, G.C.J. A two-stage state estimator for power systems with PMU and SCADA measurements. In Proceedings of the 2013 IEEE Grenoble Conference, Grenoble, France, 16–20 June 2013; pp. 1–6.
15. Ritzmann, D.; Wright, P.S.; Holderbaum, W.; Potter, B. A Method for Accurate Transmission Line Impedance Parameter Estimation. *IEEE Trans. Instrum. Meas.* **2016**, *65*, 2204–2213. [CrossRef]
16. Wydra, M. Performance and Accuracy Investigation of the Two-Step Algorithm for Power System State and Line Temperature Estimation. *Energies* **2018**, *11*, 1005. [CrossRef]
17. Jones, K.D.; Pal, A.; Thorp, J.S. Methodology for Performing Synchrophasor Data Conditioning and Validation. *IEEE Trans. Power Syst.* **2015**, *30*, 1121–1130. [CrossRef]
18. Avila-Rosales, R.; Rice, M.J.; Giri, J.; Beard, L.; Galvan, F. Recent experience with a hybrid SCADA/PMU on-line state estimator. In Proceedings of the 2009 Power & Energy Society General Meeting, Calgary, AB, Canada, 26–30 July 2009; pp. 1–8.
19. Chakrabarti, S.; Kyriakides, E.; Ledwich, G.; Ghosh, A. A comparative study of the methods of inclusion of PMU current phasor measurements in a hybrid state estimator. In Proceedings of the Power and Energy Society General Meeting, Providence, RI, USA, 25–29 July 2010; pp. 1–7.
20. Gou, B. Optimal Placement of PMUs by Integer Linear Programming. *IEEE Trans. Power Syst.* **2008**, *23*, 1525–1526. [CrossRef]
21. Kavasseri, R.; Srinivasan, S.K. Joint Placement of Phasor and Power Flow Measurements for Observability of Power Systems. *IEEE Trans. Power Syst.* **2011**, *26*, 1929–1936. [CrossRef]
22. Khajeh, K.G.; Bashar, E.; Rad, A.M.; Gharehpetian, G.B. Integrated Model Considering Effects of Zero Injection Buses and Conventional Measurements on Optimal PMU Placement. *IEEE Trans. Smart Grid* **2017**, *8*, 1006–1013.
23. Wu, F.F.; Monticelli, A. Network Observability: Theory. *IEEE Trans. Power Appar. Syst.* **1985**, *PAS 104*, 1042–1048. [CrossRef]

24. Xie, N.; Torelli, F.; Bompard, E.; Vaccaro, A. A graph theory based methodology for optimal PMUs placement and multiarea power system state estimation. *Electr. Power Syst. Res.* **2015**, *119*, 25–33. [CrossRef]

25. Khorram, E.; Jelodar, M.T. PMU placement considering various arrangements of lines connections at complex buses. *Int. J. Electr. Power Energy Syst.* **2018**, *94*, 97–103. [CrossRef]

26. Peng, J.; Sun, Y.; Wang, H.F. Optimal PMU placement for full network observability using Tabu search algorithm. *Int. J. Electr. Power Energy Syst.* **2006**, *28*, 223–231. [CrossRef]

27. Aminifar, F.; Lucas, C.; Khodaei, A.; Fotuhi-Firuzabad, M. Optimal Placement of Phasor Measurement Units Using Immunity Genetic Algorithm. *IEEE Trans. Power Deliv.* **2009**, *24*, 1014–1020. [CrossRef]

28. Nazari-Heris, M.; Mohammadi-Ivatloo, B. Application of heuristic algorithms to optimal PMU placement in electric power systems: An updated review. *Renew. Sustain. Energy Rev.* **2015**, *50*, 214–228. [CrossRef]

29. Korres, G.N.; Manousakis, N.M.; Xygkis, T.C.; Löfberg, J. Optimal phasor measurement unit placement for numerical observability in the presence of conventional measurements using semi-definite programming. *IET Gener. Transm. Distrib.* **2015**, *9*, 2427–2436. [CrossRef]

30. Manousakis, N.M.; Korres, G.N. Optimal PMU Placement for Numerical Observability Considering Fixed Channel Capacity—A Semidefinite Programming Approach. *IEEE Trans. Power Syst.* **2016**, *31*, 3328–3329. [CrossRef]

31. Hurtgen, M.; Maun, J.C. Optimal PMU placement using Iterated Local Search. *Int. J. Electr. Power Energy Syst.* **2010**, *32*, 857–860. [CrossRef]

32. Nazari-Heris, M.; Mohammadi-Ivatloo, B. Optimal placement of phasor measurement units to attain power system observability utilizing an upgraded binary harmony search algorithm. *Energy Syst.* **2015**, *6*, 201–220. [CrossRef]

33. Tran, V.-K.; Zhang, H.-S. Optimal PMU Placement Using Modified Greedy Algorithm. *J. Control Autom. Electr. Syst.* **2018**, *29*, 99–109. [CrossRef]

34. Pokharel, S.P.; Brahma, S. Optimal PMU placement for fault location in a power system. In Proceedings of the North American Power Symposium, Starkville, MS, USA, 4–6 October 2009; pp. 1–5.

35. Aminifar, F.; Khodaei, A.; Fotuhi-Firuzabad, M.; Shahidehpour, M. Contingency-Constrained PMU Placement in Power Networks. *IEEE Trans. Power Syst.* **2010**, *25*, 516–523. [CrossRef]

36. Sarailoo, M.; Wu, N.E. A new PMU placement algorithm to meet a specified synchrophasor availability. In Proceedings of the Innovative Smart Grid Technologies Conference, Minneapolis, MN, USA, 6–9 September 2016; pp. 1–5.

37. Kim, T.T.; Poor, H.V. Strategic Protection Against Data Injection Attacks on Power Grids. *IEEE Trans. Smart Grid* **2011**, *2*, 326–333. [CrossRef]

38. Chen, J.; Abur, A. Placement of PMUs to Enable Bad Data Detection in State Estimation. *IEEE Trans. Power Syst.* **2006**, *21*, 1608–1615. [CrossRef]

39. Qi, J.; Sun, K.; Kang, W. Optimal PMU Placement for Power System Dynamic State Estimation by Using Empirical Observability Gramian. *IEEE Trans. Power Syst.* **2015**, *30*, 2041–2054. [CrossRef]

40. Sodhi, R.; Sharieff, M.I. Phasor measurement unit placement framework for enhanced wide-area situational awareness. *IET Gener. Transm. Distrib.* **2015**, *9*, 172–182. [CrossRef]

41. Guo, T.Y.; Milanovic, J.V. Online Identification of Power System Dynamic Signature Using PMU Measurements and Data Mining. *IEEE Trans. Power Syst.* **2016**, *31*, 1760–1768. [CrossRef]

42. Kim, T.; Wright, S.J. PMU Placement for Line Outage Identification via Multinomial Logistic Regression. *IEEE Trans. Smart Grid* **2018**, *9*, 122–131. [CrossRef]

43. Li, Q.; Cui, T.; Weng, Y.; Negi, R.; Franchetti, F.; Ilic, M.D. An Information-Theoretic Approach to PMU Placement in Electric Power Systems. *IEEE Trans. Smart Grid* **2013**, *4*, 446–456. [CrossRef]

44. Abdelsalam, H.A.; Abdelaziz, A.Y.; Mukherjee, V. Optimal PMU placement in a distribution network considering network reconfiguration. In Proceedings of the International Conference on Circuit, Power and Computing Technologies, Nagercoil, India, 20–21 March 2014; pp. 191–196.

45. Meier, A.V.; Culler, D.; Mceachern, A.; Arghandeh, R. Micro-synchrophasors for distribution systems. In Proceedings of the Innovative Smart Grid Technologies Conference, Washington, DC, USA, 19–22 February 2014; pp. 1–5.

46. Meier, A.V.; Stewart, E.; Mceachern, A.; Andersen, M.; Mehrmanesh, L. Precision Micro-Synchrophasors for Distribution Systems: A Summary of Applications. *IEEE Trans. Smart Grid* **2017**, *8*, 2926–2936. [CrossRef]

47. Stewart, E.; Stadler, M.; Roberts, C.; Reilly, J.; Dan, A.; Joo, J.Y. Data-driven approach for monitoring, protection, and control of distribution system assets using micro-PMU technology. *CIRED Open Access Proc. J.* **2017**, *2017*, 1011–1014. [CrossRef]
48. Verbic, G.; Canizares, C.A. Probabilistic Optimal Power Flow in Electricity Markets Based on a Two-Point Estimate Method. *IEEE Trans. Power Syst.* **2006**, *21*, 1883–1893. [CrossRef]
49. Su, C.L.; Lu, C.N. Two-point estimate method for quantifying transfer capability uncertainty. *IEEE Trans. Power Syst.* **2005**, *20*, 573–579. [CrossRef]
50. Rosenblueth, E. Two-point estimates in probabilities. *Appl. Math. Model.* **1981**, *5*, 329–335. [CrossRef]
51. Hong, H.P. An efficient point estimate method for probabilistic analysis. *Reliab. Eng. Syst. Saf.* **1998**, *59*, 261–267. [CrossRef]
52. Chen, X.; Wu, W.; Zhang, B. Robust Restoration Method for Active Distribution Networks. *IEEE Trans. Power Syst.* **2016**, *31*, 4005–4015. [CrossRef]

Article

A Novel Multi-Population Based Chaotic JAYA Algorithm with Application in Solving Economic Load Dispatch Problems

Jiangtao Yu [1,2], Chang-Hwan Kim [1], Abdul Wadood [1], Tahir Khurshiad [1] and Sang-Bong Rhee [1,*]

[1] Department of Electrical Engineering, Yeungnam University, Gyeongsan 38541, Korea; yujiangtao0221@gmail.com (J.Y.); kranz@ynu.ac.kr (C.-H.K.); wadood@ynu.ac.kr (A.W.); tahir@ynu.ac.kr (T.K.)
[2] Department of Electronic Information and Electrical Engineering, Anyang Institute of Technology, Anyang 455000, China
* Correspondence: rrsd@yu.ac.kr; Tel.: +82-10-3564-0970

Received: 1 July 2018; Accepted: 22 July 2018; Published: 26 July 2018

Abstract: The economic load dispatch (ELD) problem is an optimization problem of minimizing the total fuel cost of generators while satisfying power balance constraints, operating capacity limits, ramp-rate limits and prohibited operating zones. In this paper, a novel multi-population based chaotic JAYA algorithm (MP-CJAYA) is proposed to solve the ELD problem by applying the multi-population method (MP) and chaotic optimization algorithm (COA) on the original JAYA algorithm to guarantee the best solution of the problem. MP-CJAYA is a modified version where the total population is divided into a certain number of sub-populations to control the exploration and exploitation rates, at the same time a chaos perturbation is implemented on each sub-population during every iteration to keep on searching for the global optima. The proposed MP-CJAYA has been adopted to ELD cases and the results obtained have been compared with other well-known algorithms reported in the literature. The comparisons have indicated that MP-CJAYA outperforms all the other algorithms, achieving the best performance in all the cases, which indicates that MP-CJAYA is a promising alternative approach for solving ELD problems.

Keywords: JAYA algorithm; multi-population method (MP); chaos optimization algorithm (COA); economic load dispatch problem (ELD); optimization methods

1. Introduction

With the issues of global warming and depletion of classical fossil fuels, saving energy and reducing the operational cost have become the key topics in power systems nowadays. The economic load dispatch problem (ELD) is a crucial issue of power system operation that minimizes the operational cost while satisfying a set of physical and operational constraints imposed by generators and system limitations [1]. A large number of conventional optimization methods have been applied successfully for solving the ELD problem such as gradient method [2], lambda iteration method [3], semi-definite programming [4], quadratic programming [5], dynamic programming [6], Lagrangian relaxation method [7] and linear programming [8]. However, they suffer from difficulties when dealing with problems with nonconvex objective function and complex constraints, which tends to exhibit highly non-linear, non-convex and non-smooth characteristics with a number of local optima [9].

To overcome these drawbacks, meta-heuristic methods are proposed, such as genetic algorithm (GA) [10], particle swarm optimization (PSO) [11], tabu search (TS) [12], artificial bee colony algorithm (ABC) [13], firefly algorithm [14], harmony search (HS) [15] and teaching-learning-based optimization (TLBO) [16]. Additionally, hybrid meta-heuristic optimization approaches built by the combination between conventional methods and meta-heuristic methods or among the meta-heuristic methods have also been reported to deal with the ELD problem, such as DE-PSO method [17], HS-DE method [18], GA-PS-SQP algorithm [19] and Quantum-PSO method [20]. Even though hybrid methods offer much faster convergence rates, the combination may lead to increased numbers of parameters which causes more difficulties in selecting the proper value for each one. Hence, a new method with strong searching ability and less number of control parameters is needed.

The JAYA algorithm is a newly developed yet advanced heuristic algorithm for solving constrained and unconstrained optimization problems [21]. Different from other algorithms requiring for algorithm-specific parameters in addition to common parameters, the JAYA algorithm does not require any algorithm-specific parameters except for two common parameters named the population size (N_{pop}) and the number of iteration (N_{iter}). This significant benefit makes it popular in various real-world optimization problems such as optimum power flow [22], heat exchangers [23], photovoltaic models [24], thermal devices [25], MPPT of PV system [26], constrained mechanical design optimization [27], modern machining processes [28] and PV-DSTATCOM [29]. However, as a newly developed algorithm, the JAYA algorithm still has some disadvantages even though the number of parameters is less and the convergence rate is accelerated. Since there is only guidance as approach to get close to the best solution and get away from the worst solution, the population diversity may not be maintained efficiently, easily leading to local optimal solutions.

The multi-population based optimization method (MP) is applied for improving the search diversity by dividing the whole population into a certain number of sub-populations and distributing them throughout the search area so that the problem changes can be monitored more effectively. The MP method is aimed at maintaining population diversity during the search period by distributing different sub-populations to different search spaces. Each population is used to either intensify or diversifying the search process [30,31]. The interaction among the sub-populations occurs by dividing and merging process as long as a change in the solution is detected. Branke proposed a multi-population evolutionary algorithm in [32]. Turky and Abdullah proposed a multi-population electromagnetic algorithm and a multi-population harmony search algorithm in [33,34]. Nseef proposed a multi-population artificial bee colony algorithm in [35]. The published literature have demonstrated that employing MP method is useful for maintaining the population diversity when dealing with various problem changes.

Its worthy to be noted that the MP optimization method has superior behaviors because [36]:

(1) By dividing the whole population into sub-populations, population diversity can be maintained since the sub-populations are located in different regions of the problem landscape.
(2) With the ability to search various regions simultaneously, it is able to track the movement of optimum value more effectively.
(3) Population-based optimization algorithms can be easily integrated with MP method.

At the same time the chaotic optimization algorithm (COA) which adopts chaotic sequences instead of random sequences is also employed here. Due to the non-repetitive characteristics of chaotic sequences, the COA can execute with shorter execution time and more robust mechanisms than stochastic ergodic searches that depending on random probabilities. It also has the feature of easy implementation in meta-heuristic algorithms, such as chaotic evolutionary algorithms [37], chaotic ant swarm optimization [38], chaotic harmony search algorithm [39], chaotic particle swarm optimization [40], chaotic firefly algorithm [41]. The choice of chaotic sequences is justified theoretically by their unpredictability, i.e., by their spread-spectrum characteristic, non-periodic, complex temporal behavior and ergodic properties. Simulation results from the abovementioned literature have demonstrated that the application of deterministic chaotic signals to meta-heuristic algorithms is a promising strategy in engineering applications. In this paper, COA has been applied twice:

(1) During the initialization step, chaotic sequences generated by a chaotic map are used to initialize the initial solutions.

(2) During the iteration step, COA is conducted to search further around the solution obtained by former algorithm to enhance the global convergence and to prevent to be trapped on local optima.

Based on the descriptions above, a novel multi-population based chaotic JAYA algorithm (MP-CJAYA) is proposed in this paper. It is a modified version of JAYA algorithm where the total population is divided into sub-populations by the MP method to control the exploration and exploitation rates, meanwhile a chaos perturbation is implemented on each sub-population during every iteration to keep on searching for the global optima. The MP-CJAYA algorithm is applied for solving the ELD cases with constraints including valve-point effects, power balance constraints, operating capacity limits, ramp-rate limits and prohibited operating zones. In all the experimented ELD cases, the proposed MP-CJAYA has produced the most competitive results.

The rest of this paper is arranged as follows: In Section 2, the problem formulation of ELD problem is constructed. The basic JAYA, the compared CJAYA and the proposed MP-CJAYA algorithms are described in Section 3. The experimental results and comparisons of MP-CJAYA with other algorithms are presented and analyzed in Section 4. Finally, the conclusions and future work are given in Section 5.

2. Problem Formulation

The ELD problem is described as an objective function to minimize the total fuel cost while satisfying different constraints, we adopt the problem formulation described in [16,42].

2.1. Objective Function

The objective function is to sum up all the costs of committed generators as expressed below:

$$\min F = \sum_{i=1}^{n} F_i(P_i) \tag{1}$$

where n is the total generator number in power systems, $F_i(P_i)$ is the cost function of ith generator with output P_i.

Approximately, the cost function can be expressed as a quadratic polynomial by the following equation:

$$F_i(P_i) = a_i P_i^2 + b_i P_i + c_i \tag{2}$$

where a_i, b_i, c_i are the cost coefficients of ith generator, which are constants.

In reality, a higher-order non-linearity rectified sinusoid contribution is usually added to the cost function to model the valve-point effect, which is given below:

$$F_i(P_i) = a_i P_i^2 + b_i P_i + c_i + \left| e_i \times \sin(f_i \times (P_i^{\min} - P_i)) \right| \tag{3}$$

where e_i and f_i are cost coefficients of ith generator due to valve-point effect, while $P_i^{\min}$ is the minimum output for generator i.

According to the discussion above, the objective function of ELD problem considering the valve-point effect can be represented as:

$$\min F = \sum_{i=1}^{n} \left(a_i P_i^2 + b_i P_i + c_i + \left| e_i \times \sin(f_i \times (P_i^{\min} - P_i)) \right| \right) \tag{4}$$

2.2. Constrained Functions

2.2.1. Power Losses

The total power generated by available units must equal to the summation of the demanded power and the system power loss, which can be formulated as:

$$\sum_{i=1}^{n} P_i = P_{demand} + P_{loss} \tag{5}$$

where P_{demand} and P_{loss} is the value of the demanded power and the whole power loss in the system respectively. P_{loss} is calculated by Kron's formula:

$$P_{loss} = \sum_{i=1}^{n}\sum_{j=1}^{n} P_i B_{ij} P_j + \sum_{i=1}^{n} B_{i0} P_i + B_{00} \tag{6}$$

where B_{ij}, B_{i0}, B_{00} are the loss coefficients that generally can be assumed to be constants under a normal operating condition.

2.2.2. Generating Capacity

The real output P_i generated by a available unit must be ranged between its minimum limit and maximum limit:

$$P_i^{min} \leq P_i \leq P_i \leq P_i^{max} \tag{7}$$

where P_i^{min} and P_i^{max} are the minimum and maximum limits of ith generator.

2.2.3. Ramp Rate Limit

In practical circumstances, the output power P_i can not be adjusted immediately, the operating range is restricted by the ramp-rate limit constraint expressed below:

$$\max(P_i^{min}, P_i^0 - DR_i) \leq P_i \leq \min(P_i^{max}, P_i^0 + UR_i) \tag{8}$$

where P_i is the present power output, P_i^0 is the previous power output, UR_i and DR_i is the up-ramp and down-ramp limit of generator i respectively.

2.2.4. Prohibited Operating Zones

For generator with prohibited operating zones (POZs), which are the sets of output power ranges where the generator can not work, the feasible operating zones are as discontinuous as follows:

$$\begin{aligned} P_i^{min} &\leq P_i \leq P_{i,1}^{lower} \\ P_{i,j-1}^{upper} &\leq P_i \leq P_{i,j}^{lower} \\ P_{i,n_i}^{upper} &\leq P_i \leq P_i^{max} \end{aligned} \tag{9}$$

where j is the index of POZs, n_i is the total number of POZs where $j \in [1, n_i]$, $P_{i,j}^{lower}$ and $P_{i,j}^{upper}$ are the lower and upper bounds of the jth POZ of the ith unit, respectively.

3. The Proposed MP-CJAYA Algorithm

Since the proposed MP-CJAYA algorithm is a hybrid of the basic JAYA, COA and MP methods, it is quite necessary to observe the relative strength of each constituent when solving the ELD problem, so three different algorithms are studied:

(1) The basic JAYA algorithm: The classical JAYA algorithm with standard parameters; it is selected to compare its performance at solving different ELD cases with the other two algorithms.

(2) The compared CJAYA algorithm: The basic JAYA algorithm combined by COA but without the MP method.

(3) The proposed MP-CJAYA algorithm: The basic JAYA algorithm integrated with both the COA and MP methods.

3.1. The Basic JAYA Algorithm

The JAYA algorithm is a powerful heuristic algorithm proposed by Rao for solving optimization problems. It always attempts to get success to reach the best solution as well as move far away from the worst solution. Different from most of the other heuristic algorithms, JAYA is free from algorithm-specific parameters, only two common parameters named the population size N_{pop} and the number of iterations N_{iter} are required [21].

Suppose the objective function is $F(X)$ which is required to be minimized or maximized. Let $F(X)_{best}$ and $F(X)_{worst}$ represent the best value and the worst value of $F(X)$ among the entire candidate solutions during each iteration. Let $X_{j,k,i}$ be the value of the jth variable for the kth candidate during the ith iteration, then the new modified value $X'_{j,k,i}$ by JAYA algorithm is calculated by:

$$X'_{j,k,i} = X_{j,k,i} + r_{1,j,i} \times (X_{j,best,i} - |X_{j,k,i}|) - r_{2,j,i} \times (X_{j,worst,i} - |X_{j,k,i}|) \tag{10}$$

where $X'_{j,k,i}$ is the updated value of $X_{j,k,i}$. $X_{j,best,i}$ and $X_{j,worst,i}$ are the values of the jth variable for $F(X)_{best}$ and $F(X)_{worst}$ during the ith iteration respectively. $r_{1,j,i}$ and $r_{2,j,i}$ are two random numbers ranged in [0, 1]. The term '$r_{1,j,i} \times (X_{j,best,i} - |X_{j,k,i}|)$' indicates the tendency of the solution to move closer to the best solution and the term '$r_{2,j,i} \times (X_{j,worst,i} - |X_{j,k,i}|)$' indicates the tendency of the solution to avoid the worst solution. Suppose $F(X)'$ is the modified value of $F(X)$, if $F(X)'$ provides better value than $F(X)$, then $X_{j,k,i}$ is replaced by $X'_{j,k,i}$ and $F(X)$ is replaced by $F(X)'$; otherwise, keep the old value. All the values of new obtained $X_{j,k,i}$ and $F(X)$ at the end of every iteration are maintained and become the inputs to the next iteration [21].

The procedure for the basic JAYA algorithm to solve ELD problem is described as follows:

Step 1: Set parameters. Common parameters of JAYA are initialized in this step. The first one is the population size (N_{pop}) which represents how many solutions will be generated; the second one is the maximum iteration number (N_{JAYA_iter}) which indicates the stopping condition during the calculation; the last one is the total number of generators (N_{gen}) for N_{gen}-units system.

Set the iteration counter as *iter*.

Step 2: Initialize the solution. A set of initial solutions are randomly generated as follows:

$$X_{j,k,i} = X_j^{min} + (X_j^{max} - X_j^{min}). * rand(N_{pop}, N_{gen}) \tag{11}$$

where $j \in [1, N_{gen}], k \in [1, N_{pop}], i \in [1, N_{JAYA\_iter}]$, X_j^{min} and X_j^{max} are the lower and upper limits of jth generator given by generating capacity limits in Equation (7).

Step 3: Apply constraints. Apply the constraints in Section 2.2 by using Equations (5)–(9).

Step 4: Evaluate the solution. Calculate the objective function (cost function) by using Equation (3) with considering the valve-point effect or Equation (2) without considering the valve-point effect to obtain the initial value $F(X)$.

Set *iter* = 1.

Step 5: Determine the best and worst. Choose $X_{j,best,i}$ and $X_{j,worst,i}$ according to the value of $F(X)_{best}$ and $F(X)_{worst}$, which means the lowest and highest value among all the populations.

Step 6: Generate new solution. Generate new output $X'_{j,k,i}$ by Equation (10).

Step 7: Apply constraints. Apply the constraints in Section 2.2 by using Equations (5)–(9).

Step 8: Evaluate the new solution. Calculate the new objective function value $F(X)'$ by Equation (3) with considering the valve-point effect or Equation (2) without considering the valve-point effect.

Step 9: Compare. The new $F(X)'$ is compared with the old $F(X)$, the values are updated as follows:

If $F(X)' < F(X)$
then $F(X) = F(X)'$ and $X_{j,k,i} = X'_{j,k,i}$;
Otherwise, keep the old value.

Step 10: Check the stopping condition. If the current iteration number *iter* $< N_{JAYA\_iter}$, then *iter* $=$ *iter* $+ 1$ and return to Step 5. Otherwise, stop the procedure.

3.2. The Compared CJAYA Algorithm

In this chapter, the Chaos Optimization Algorithm (COA) is combined with the basic JAYA algorithm to form the compared CJAYA algorithm. COA has used chaotic map for new search surface during every iteration, which is a discrete-time dynamical system running in chaotic state:

$$Z(k+1) = f(Z(k)) \ (k = 0, 1, 2, 3, ...) \tag{12}$$

A widely used logistic map which appears in nonlinear dynamics of biological population evidencing chaotic behavior is shown below [43].

$$Z_i(k+1) = \alpha \times Z_i(k)(1 - Z_i(k)) \tag{13}$$

where i is the serial number of chaotic variables, k is the iteration number. The initial value of the ith chaotic variable is $Z_i(0)$ where $Z_i(0) \notin \{0.0, 0.25, 0.5, 0.75, 1.0\}$. $\alpha = 4$ is used in this paper. It is obvious that $Z_i(k+1) \in (0,1)$ under the conditions of $Z_i(0) \in (0,1)$.

The procedure for the CJAYA algorithm to solve ELD problem is provided here, the symbol $*$ denotes a new added step compared with the basic JAYA:

Step 1: Set parameters. Common parameters of CJAYA are initialized in this step. The population size (N_{pop}), the maximum iteration number (N_{JAYA_iter}) and the total number of generators (N_{gen}) are as the same as basic JAYA. However, one more parameter (N_{COA_iter}) is introduced which represents the maximum iteration number by COA.

Set the iteration counter as *iter*.

Step 2: *Generate chaotic sequence.* The chaotic sequence $Z_{j,k,q}$ is generated by Logistic map in this step, where j denoting the number of generators of the system, k denoting the population number and q denoting the number of iteration by COA, which is shown in the following equation:

$$Z_{j,k,q} = 4 \times Z_{j,k-1,q}(1 - Z_{j,k-1,q}) \tag{14}$$

Here $j \in [1, N_{gen}]$, $k \in [1, N_{pop}]$, $q \in [1, N_{COA\_iter}]$.

Step 3: Initialize the solution. By the carrier wave method, the set of initial variable $X_{j,k,i}$ can be transformed to chaos variables by:

$$X_{j,k,i} = X_j^{min} + (X_j^{max} - X_j^{min}). * Z_{j,k,q} \tag{15}$$

where X_j^{min} and X_j^{max} are the lower and upper limits of jth generator given by generating capacity limits in Equation (7).

Step 4: Apply constraints. As the same as Step 3 in Section 3.1.
Step 5: Evaluate the solution. As the same as Step 4 in Section 3.1.
Step 6: Determine the best and worst. As the same as Step 5 in Section 3.1.
Step 7: Generate new solution. As the same as Step 6 in Section 3.1.
Step 8: Apply constraints. As the same as Step 7 in Section 3.1.
Step 9: Evaluate the new solution. As the same as Step 8 in Section 3.1.
Step 10: Compare. As the same as Step 9 in Section 3.1.

Step 11∗: Apply COA. In the former step we have obtained the best set of solutions $X_{j,k,i}$ up to now, then the second carrier wave method can be performed by:

$$X'_{j,k,i} \;=\; X_{j,k,i} + R \times Z_{j,k,q} \tag{16}$$

where R is a constant, $R \times Z_{j,k,q}$ generates chaotic states with small ergodic ranges around current $X_{j,k,i}$ to seek further for improving the quality of current solutions. Then the generated neighborhood solutions will be compared with current solutions to check if they give better objective function values by the following steps:

(1)　Apply constraints. As the same as Step 7 in Section 3.1.
(2)　Evaluate the new solution. As the same as Step 8 in Section 3.1.
(3)　Compare. As the same as Step 9 in Section 3.1.

Step 12: Check the stopping condition. If the current iteration number *iter* $< N_{JAYA\_iter}$, then *iter* $=$ *iter* $+ 1$ and return to Step 6. Otherwise, stop the procedure.

3.3. The Proposed MP-CJAYA Algorithm

In this section, Multi-population based optimization method (MP) is combined with CJAYA algorithm to form the proposed MP-CJAYA algorithm. Figure 1 presents the flowchart of the proposed MP-CJAYA algorithm, the pseudo code of the proposed MP-CJAYA is described in Algorithm 1. The whole steps of MP-CJAYA to solve ELD problem is described as follows, the symbol ∗ denotes a newly added step compared with CJAYA:

Step 1: Set parameters. Common parameters of MP-CJAYA are initialized in this step. The population size (N_{pop}), the maximum iteration number (N_{JAYA_iter}), the total number of generators (N_{gen}) and the maximum COA iteration number (N_{COA_iter}) are as the same as basic JAYA and CJAYA. However, another important parameter (K) is introduced which represents the divided number of sub-populations, so the population size of the sub-populations (N_{sub_pop}) is:

$$N_{sub\_pop} \;=\; N_{pop}/K \tag{17}$$

Set the iteration counter as *iter*.
Step 2: Generate chaotic sequence. As the same as Step 2 in Section 3.2.
Step 3: Initialize the solution. As the same as Step 3 in Section 3.2.
Step 4: Apply constraints. As the same as Step 3 in Section 3.1.
Step 5: Evaluate the solution. As the same as Step 4 in Section 3.1.
Step 6∗: Divide the population. The entire population is divided into K sub-populations with population size of N_{sub_pop} by Equation (17). It is noted that the solutions in the whole population are randomly assigned to a sub-population, each sub-population is arranged to explore a different area of the whole search space.

The following steps are performed on each sub-population:
Step 7: Determine the best and worst. As the same as Step 5 in Section 3.1.
Step 8: Generate new solution. As the same as Step 6 in Section 3.1.
Step 9: Apply constraints. As the same as Step 7 in Section 3.1.
Step 10: Evaluate the new solution. As the same as Step 8 in Section 3.1.
Step 11: Compare. As the same as Step 9 in Section 3.1.
Step 12: Apply COA. As the same as Step 11 in Section 3.2.
Step 13: Check the stopping condition. If the current iteration number *iter* reaches N_{JAYA_iter}, stop the loop and report the best solution; otherwise follow the next step and set *iter* $=$ *iter* $+ 1$.
Step 14∗: Merge the sub-populations. All the sub-populations are merged together to form one population, then for re-divide the population go to Step 6.

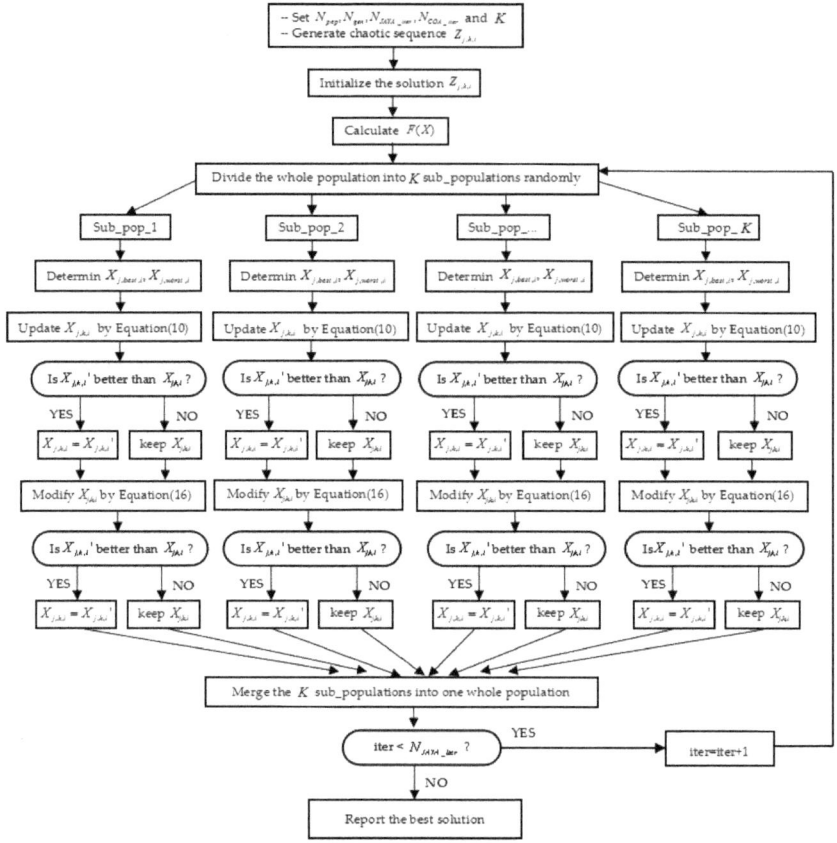

Figure 1. Flow chart of the MP-CJAYA Algorithm.

Algorithm 1 Pseudo code of the MP-CJAYA Algorithm

Begin
Initialize N_{pop}, N_{JAYA_iter}, N_{gen}, N_{COA_iter} and K;
Generate initial solution $X_{j,k,i}$ by chaotic sequence;
Calculate objective function value $F(X)$;
Set $iter = 1$
While $iter < N_{JAYA\_iter}$ **do**
Divide the whole population P into K sub-populations by Equation (17) randomly
$P_1, P_2, ..., P_{K-1}, P_K$
For $m = 1 \rightarrow K$ **do**
Confirm $X_{j,best,i}$ and $X_{j,worst,i}$ within P_m
For $k = 1 \rightarrow N_{sub\_pop}$ **do**
Generate new solution $X'_{j,k,i}$ by Equation (10)
If $F(X'_{j,k,i})$ is better than $F(X_{j,k,i})$ **then**
$X_{j,k,i} = X'_{j,k,i}$
$F(X_{j,k,i}) = F(X'_{j,k,i})$
Else
Keep the old value

End if
End for
For $k = 1 \rightarrow N_{sub\_pop}$ **do**
Generate new solution $X'_{j,k,i}$ by Equation (16)
If $F(X'_{j,k,i})$ is better than $F(X_{j,k,i})$ **then**
$X_{j,k,i} = X'_{j,k,i}$
$F(X_{j,k,i}) = F(X'_{j,k,i})$
Else
Keep the old value
End if
End for
End for
Merge the sub-populations $(P_1, P_2, ..., P_{K-1}, P_K)$ into P
$iter = iter + 1$
End while

---

4. Experimental Results and Analysis

In this section, the basic JAYA, the compared CJAYA and the proposed MP-CJAYA algorithms are applied on the following ELD cases to test their performances:

Case I.   3-units system for load demand of 850 MW.
Case II.   13-units system for load demand of 2520 MW.
Case III. 40-units system for load demand of 10500 MW.
Case IV.   6-units system for load demand of 1263 MW.
Case V.   15-units system for load demand of 2630 MW.

Since for meta-heuristic algorithms, parameter setting is critical for the quality of their performances, so the parameters used in the cases above are all listed below. All the cases are run in MATLAB 2016 under windows 7 on Intel(R) Core(TM) i5-6500 CPU 3.20 GHz, with 8 GB RAM.

4.1. Case I: 3-Units System for Load Demand of 850 MW

All detailed data are provided in [44]. The common parameters and constraint conditions are given in Table 1. The cost value of F_{mean} and F_{best} obtained by JAYA, CJAYA and MP-CJAYA are compared with GA [45], EP [45], EP-SQP [45], PSO [45], PSO-SQP [45], CPSO [46] and CPSO-SQP [46] in Table 2. The best cost are highlighted in bold font. Obviously, all the compared algorithms give the same best cost of 8234.07 $/h, except for GA who did not meet the load demand. However, JAYA, CJAYA and MP-CJAYA are able to give continuously decreasing values of F_{best} and MP-CJAYA achieves the minimum value of 8223.29 $/h, as well as the minimum value of F_{mean} which is 8232.06 $/h. To observe the cost convergence characteristics more visually, Figure 2 depicts one randomly chosen convergence curve from 20 times of independent runs (N_{runs}). We can see that JAYA has been trapped into local optimum at about 320 iterations and CJAYA has also settled down at around 230 iterations, but MP-CJAYA has showed extraordinary fast convergence ability at the beginning of 10 iterations and reached global optimum at approximately 200 iterations. It reveals that MP-CJAYA has faster convergence rate compared with JAYA and CJAYA due to its strong searching ability. Figure 3 shows the distribution outlines of F_{best} at each independent run time. In case of MP-CJAYA, the value of F_{best} after each run remains more or less steady, whereas in CJAYA the value of F_{best} varies much more than MP-CJAYA, while JAYA shows the worst stability of F_{best} with maximum cost as much as 8800 $/h. This indicates that MP-CJAYA is more consistent and robust than CJAYA and JAYA.

Table 1. Parameters and constraint conditions of the ELD cases.

	Case I			Case II			Case III			Case IV			Case V		
	JAYA	CJAYA	MP-CJAYA	JAYA	CJAYA	MP-CJAYA	JAYA	CJAYA	MP-CJAYA	JAYA	CJAYA	MP-CJAYA	JAYA	CJAYA	MP-CJAYA
N_{pop}	20	20	20	50	50	50	100	100	100	20	20	20	100	100	100
N_{JAYA_iter}	500	500	500	3000	3000	3000	5000	5000	5000	1000	1000	1000	5000	5000	5000
N_{COA_iter}	-	-	20	-	20	20	-	30	30	-	20	20	-	30	30
N_{sub_pop}	-	-	10	-	-	10	-	-	20	-	-	10	-	-	20
N_{runs}	20	20	20	30	30	30	50	50	50	20	20	20	50	50	50
Valve-point effect		●			●			●							
Ramp-rate limit											●			●	
POZ											●			●	
P_{loss}											●			●	

Table 2. Best outputs for 3-units system ($P_D = 850$ MW).

Unit	GA [45]	EP [45]	EP-SQP [45]	PSO [45]	PSO-SQP [45]	CPSO [46]	CPSO-SQP [46]	JAYA	CJAYA	MP-CJAYA
1	398.700	300.264	300.267	300.268	300.267	300.267	300.266	350.3314	350.0254	350.2464
2	399.600	400.000	400.000	400.000	400.000	400.000	400.000	400.0000	400.0000	400.0000
3	50.100	149.736	149.733	149.732	149.733	149.733	149.734	99.6453	99.9511	99.7576
P_{total}(MW)	848.400	850.000	850.000	850.000	850.000	850.000	850.000	849.977	849.977	850.004
F_{mean} ($/h)	8234.72	8234.16	8234.09	8234.72	8234.07	NA	NA	8382.10	8289.41	**8232.06**
F_{best} ($/h)	8222.07	8234.07	8234.07	8234.07	8234.07	8234.07	8234.07	8230.23	8226.18	**8223.29**

NA indicates the cost value is not found.

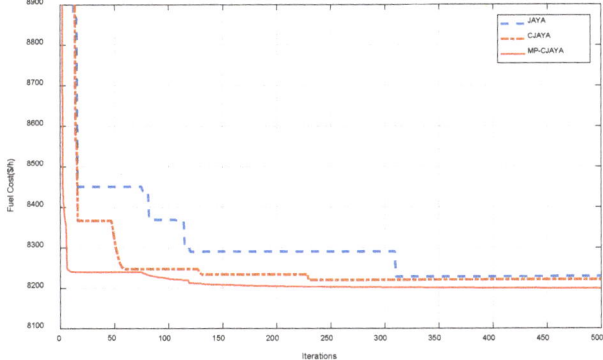

Figure 2. Fuel cost convergence characteristic of 3-units system (P_D = 850 MW).

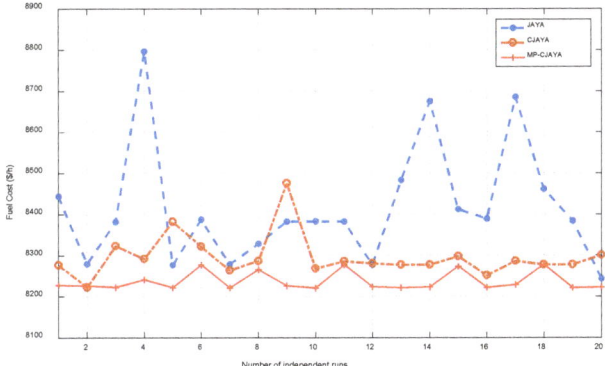

Figure 3. Fuel cost for 20 independent runs of 3-units system (P_D = 850 MW).

4.2. Case II: 13-Units System for Load Demand of 2520 MW

As the same as case I, all detailed data are provided in [44]. Since the increasing number of generators causes more non-linearity and complexity, N_{pop}, N_{JAYA_iter} and N_{runs} have all increased in this case, which are given in Table 1. The best individual of dispatched outputs obtained by different methods including GA [47], SA [47], HSS [47], EP-SQP [45], PSO-SQP [45], CPSO [46], CPSO-SQP [46], JAYA, CJAYA and MP-CJAYA are reported in Table 3. The best cost are highlighted in bold font. It is observed that the minimum value of F_{mean} and F_{best} are both achieved by MP-CJAYA, which is 24,228.1331 \$/h and 24,175.5444 \$/h respectively. In Figure 4 the convergence curve of MP-CJAYA is compared with JAYA and CJAYA, it can be observed that JAYA has been trapped into a local optimum in about 1300 iterations, while CJAYA has the same problem at around 1500 iterations. However, the proposed MP-CJAYA has greatly accelerated the convergence rate and reached the best value within only 750 iterations. Figure 5 is the distribution outlines of F_{best} at each run time. Once again, it can be easily compared that MP-CJAYA shows the most robust characteristic among the three versions of JAYA due to most of its independent runs have achieved getting close to the best individual. All the comparisons above real that MP-CJAYA has greatly improved the best cost, the mean cost, the convergence rate and the consistency of the solution.

Table 3. Best outputs for 13-units system (P_D = 2520 MW).

Unit	GA [47]	SA [47]	HSS [47]	EP-SQP [45]	PSO-SQP [45]	CPSO [46]	CPSO-SQP [46]	JAYA	CJAYA	MP-CJAYA
1	628.32	668.40	628.23	628.3136	628.3205	628.32	628.31	628.3185	628.3185	628.3183
2	356.49	359.78	299.22	299.1715	299.0524	299.83	299.83	299.2009	299.1992	299.0170
3	359.43	358.20	299.17	299.0474	298.9681	299.17	299.16	306.9105	299.1993	299.1428
4	159.73	104.28	159.12	159.6399	159.4680	159.70	159.73	159.7339	159.7330	159.5714
5	109.86	60.36	159.95	159.6560	159.1429	159.64	159.73	159.7337	159.7331	159.6930
6	159.73	110.64	158.85	158.4831	159.2724	159.67	159.73	159.7338	159.7331	159.6801
7	159.63	162.12	157.26	159.6749	159.5371	159.64	159.73	109.8673	159.7330	159.7270
8	159.73	163.03	159.93	159.7265	158.8522	159.65	159.73	159.7342	159.7330	159.7328
9	159.73	161.52	159.86	159.6653	159.7845	159.78	159.73	159.7340	159.7331	159.5119
10	77.31	117.09	110.78	114.0334	110.9618	112.46	109.07	114.8012	110.0403	111.0288
11	75.00	75.00	75.00	75.00	75.00	74.00	77.40	114.8001	114.7994	77.1661
12	60.00	60.00	60.00	60.00	60.00	56.50	55.00	92.4018	55.0000	55.0014
13	55.00	119.58	92.62	87.5884	91.6401	91.64	92.85	55.0027	55.0000	92.3862
P_{total}(MW)	2520	2520	2520	2520	2520	2520	2520	2519.97	2519.96	2519.98
F_{mean}($/h)	NA	NA	NA	NA	NA	NA	NA	24,476.5247	24,385.7604	24,228.1331
F_{best}($/h)	24,398.23	24,970.91	24,275.71	24,266.44	24,261.05	24,211.56	24,190.97	24,220.7529	24,178.8040	24,175.5444

NA indicates the cost value is not found.

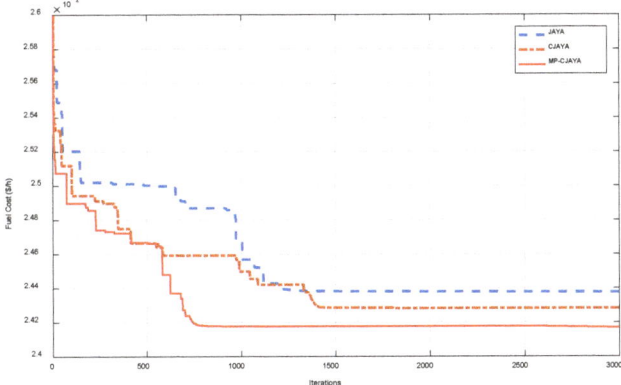

Figure 4. Fuel cost convergence characteristic of 13-units system (P_D = 2520 MW).

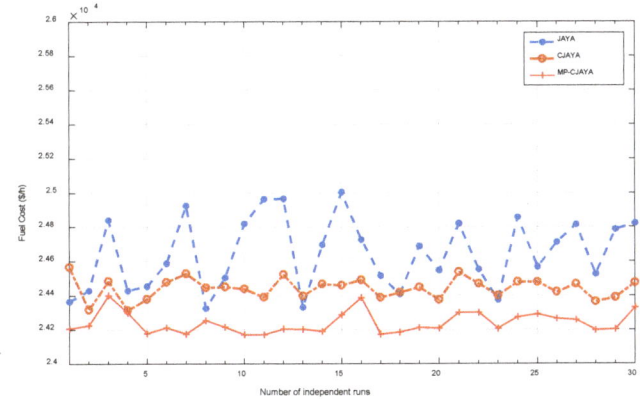

Figure 5. Fuel cost for 30 independent runs of 13-units system (P_D = 2520 MW).

4.3. Case III: 40-Units System for Load Demand of 10,500 MW

In order to investigate the effectiveness of MP-CJAYA for larger scale power system, it is further evaluated by 40 generating units with load demand of 10,500 MW, which is the largest system of ELD problem considering the valve-point effect in the available literature. Considering the increased number of generators and the much more complex solution space, N_{pop}, N_{JAYA_iter}, N_{COA_iter}, N_{sub_pop} and N_{runs} have all increased, as shown in Table 1. The results comparison from methods PSO-LRS [48], NPSO [48], NPSO-LRS [48], SPSO [49], PC-PSO [49], SOH-PSO [49], JAYA, CJAYA and MP-CJAYA are shown in Table 4. The minimum value of F_{mean} and F_{best} are highlighted in bold font. It is observed that MP-CJAYA has achieved the minimum value of F_{best} among all the values by above-mentioned methods, which is 121,480.10 $/h. What's more, the minimum value of F_{mean} is also achieved by MP-CJAYA, which is 121,861.08 $/h. In Figure 6 the convergence curve of MP-CJAYA is compared with JAYA and CJAYA, it can easily be observed that CJAYA performs better than JAYA due to the local searching ability provided by COA, while MP-CJAYA shows superiority over CJAYA due to the extra searching diversification provided by MP method.

Table 4. Best outputs for 40-units system (P_D = 10,500 MW).

Unit	PSO-LRS [48]	NPSO [48]	NPSO-LRS [48]	SPSO [49]	PC-PSO [49]	SOH-PSO [49]	JAYA	CJAYA	MP-CJAYA
1	111.9858	113.9891	113.9761	113.97	113.98	110.80	114.0000	113.5264	114.0000
2	110.5273	113.6334	113.9986	114.00	114.00	110.80	111.6651	110.7998	110.7998
3	98.5560	97.5500	97.4141	109.19	97.26	97.40	119.9876	120.0000	97.3999
4	182.9266	180.0059	179.7327	179.77	179.51	179.73	188.2606	179.7331	179.7331
5	87.7254	97.0000	89.6511	97.00	89.38	87.80	96.9763	97.0000	93.1276
6	139.9933	140.0000	105.4044	91.01	105.20	140.00	139.9488	140.0000	140.0000
7	259.6628	300.0000	259.7502	259.87	259.55	259.60	264.0949	300.0000	300.0000
8	297.7912	300.0000	288.4534	286.99	286.90	284.60	299.9814	284.5997	284.5997
9	284.8459	284.5797	284.6460	284.09	284.71	284.60	284.9042	284.5997	284.5997
10	130.0000	130.0517	204.8120	204.05	206.24	130.00	130.0908	130.0000	130.0000
11	94.6741	243.7131	168.8311	168.40	166.52	94.00	94.0011	94.0000	94.0000
12	94.3734	169.0104	94.00	94.00	94.00	94.00	94.0000	94.0000	94.0000
13	214.7369	125.0000	214.7663	212.30	214.56	304.52	125.1028	125.0000	125.0000
14	394.1370	393.9662	394.2852	393.76	392.76	304.52	394.2529	394.2794	394.2794
15	483.1816	304.7586	304.5187	303.62	306.24	394.28	484.1262	394.2794	394.2794
16	304.5381	304.5120	394.2811	392.05	394.88	394.28	304.5950	394.2794	394.2794
17	489.2139	489.6024	489.2807	489.49	489.26	489.28	490.8265	489.2794	489.2794
18	489.6154	489.6087	489.2832	489.35	489.82	489.28	489.3438	489.2794	489.2794
19	511.1782	511.7903	511.2845	512.39	510.62	511.28	511.3775	511.1294	511.1294
20	511.7336	511.2624	511.3049	511.21	511.68	511.27	512.1395	511.1294	511.1294
21	523.4072	523.3274	523.2853	522.61	523.52	523.28	523.6621	523.2794	523.2794
22	523.4599	523.2196	523.2853	523.65	523.26	523.28	523.3534	523.2794	523.2794
23	523.4756	523.4707	523.2797	523.06	523.98	523.28	524.9677	523.2794	523.2794
24	523.7032	523.0661	523.2994	520.72	523.21	523.28	524.2850	523.2794	523.2794
25	523.7854	523.3978	523.2865	524.86	523.54	523.28	522.9279	523.2794	523.2794
26	523.2757	523.2897	523.2936	525.22	523.10	523.28	523.2298	523.2794	523.2794
27	10.0000	10.0208	10.0000	10.00	10.00	10.00	10.0000	10.0000	10.0000
28	10.6251	10.0927	10.0000	10.00	10.00	10.00	10.0047	10.0000	10.0000
29	10.0727	10.0621	10.0000	10.00	10.00	10.00	10.0000	10.0000	10.0000
30	51.3321	88.9456	89.0139	87.64	89.05	97.00	97.0000	97.0000	87.7999
31	189.8048	189.9951	190.0000	190.00	190.00	190.00	190.0000	190.0000	190.0000
32	189.7386	190.0000	190.0000	190.00	190.00	190.00	189.9503	190.0000	190.0000
33	189.9122	190.0000	190.0000	190.00	190.00	190.00	190.0000	190.0000	190.0000
34	199.3258	165.9825	199.9998	200.00	200.00	185.20	169.8860	164.7998	200.0000

Table 4. *Cont.*

Unit	PSO-LRS [48]	NPSO [48]	NPSO-LRS [48]	SPSO [49]	PC-PSO [49]	SOH-PSO [49]	JAYA	CJAYA	MP-CJAYA
35	199.3065	172.4153	165.1397	167.18	164.78	164.80	199.8549	200.0000	200.0000
36	192.8977	191.2978	172.0275	172.12	172.89	200.00	199.9896	200.0000	200.0000
37	110.0000	109.9893	110.0000	110.00	110.00	110.00	109.9712	110.0000	110.0000
38	109.8628	109.9521	110.0000	110.00	110.00	110.00	109.9977	110.0000	110.0000
39	92.8751	109.8733	93.0962	95.58	94.24	110.00	109.9871	110.0000	110.0000
40	511.6883	511.5671	511.2996	510.85	511.36	511.28	511.2250	511.2794	511.2794
P_{total}(MW)	10,499.9452	10,499.9989	10,499.9871	10,500	10,500	10,500	10,499.97	10,499.97	10,499.97
F_{mean}($/h)	122,558.4565	122,221.3697	122,209.3185	NA	NA	121,853.57	122,581.85	121,926.77	**121,861.08**
F_{best}($/h)	122,035.7946	121,704.7391	121,664.43	122,049.66	121,767.89	121,501.14	121,799.88	121,516.97	**121,480.10**

NA indicates the cost value is not found.

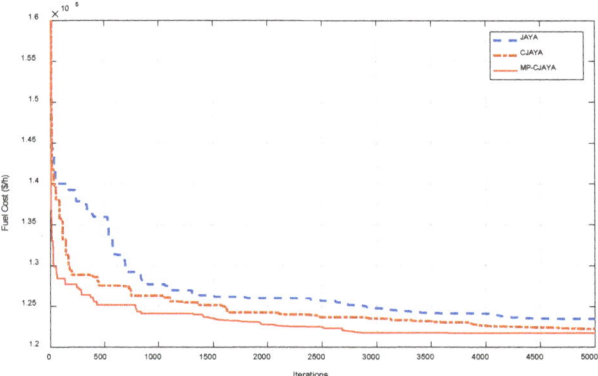

Figure 6. Fuel cost convergence characteristic of 40-units system (P_D = 10500 MW).

Figure 7 is the distribution outlines of F_{best} within 50 times of independent runs. Once again, it can be observed that MP-CJAYA shows the most robust characteristic among the three versions of JAYA because most of the F_{best} value keeps steady and very close to the best individual. The comparisons have verified that MP-CJAYA get better results than all of the other algorithms in best cost, mean cost, convergence rate and consistency when dealing with larger scale power system.

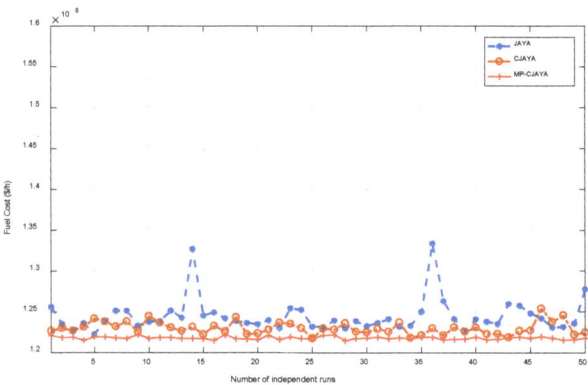

Figure 7. Fuel cost for 50 independent runs of 40-units system (P_D = 10,500 MW).

4.4. Case IV: 6-Units System for Load Demand of 1263 MW

In this case, the three versions of JAYA are applied to 6-units system with constraints of ramp rate limit, prohibited operating zones (*POZs*) and transmission loss (P_{loss}), as shown in Table 1. The generator data and B-coefficients have been taken from [50]. For every generator it has two *POZs*, this problem causes challenging complexity to find the global optima because of increasing number of non-convex decision spaces.

The best individual achieved by MP-CJAYA, as well the other algorithms such as SA [51], GA [51], TS [51], PSO [51], MTS [51], PSO-LRS [48], NPSO [48], NPSO-LRS [48], JAYA and CJAYA have been recorded in Table 5. It can be observed that MP-CJAYA provides the lowest F_{best} among all the methods as 15,446.17 \$/h, while CJAYA and JAYA provide the second and third lowest F_{best} as 15,446.71 \$/h and 15,447.09 \$/h. Furthermore, the best cost F_{best}, the worst cost F_{worst} and the mean cost F_{mean} of the three version of JAYA algorithms are also compared with those above-mentioned methods and summarized in Table 6. It can be found that MP-CJAYA is superior to all the other compared methods and achieves the minimum value of F_{best}, F_{worst} and F_{mean} at the same time, which are highlighted in bold font. Figure 8 is the distribution outlines of F_{bes}, it can be noticed that MP-CJAYA shows the most robust characteristic and the value keeps almost steady within 20 independent runs, which has greatly surpassed JAYA and a little surpassed CJAYA. One randomly chosen convergence curve of fuel cost is shown in Figure 9, from which we can see that MP-CJAYA is extraordinary fast in convergence rate and approaches global optimum within only about 60 iterations. It all demonstrates that MP-CJAYA has the strongest capabilities of handling ELD problems with different constraint conditions.

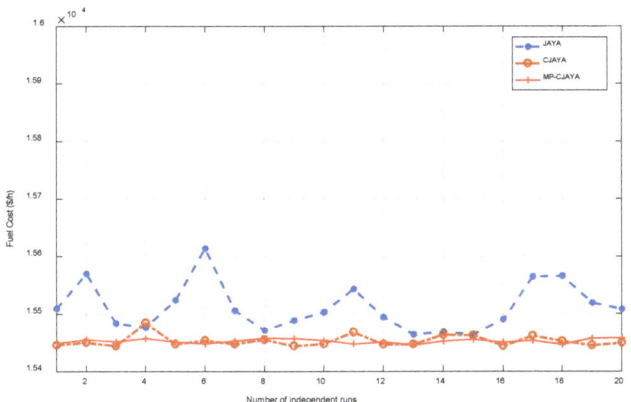

Figure 8. Fuel cost for 20 independent runs of 6-units system (P_D = 1263 MW).

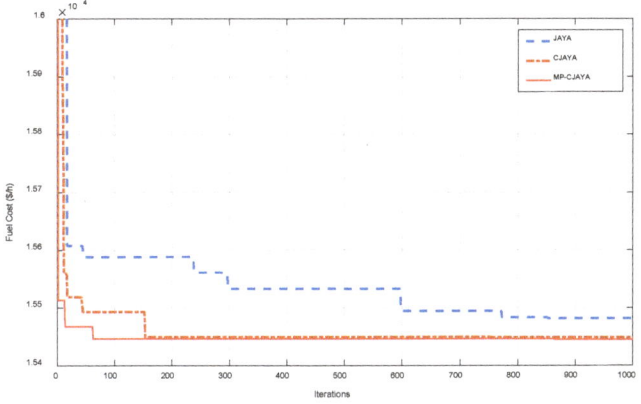

Figure 9. Fuel cost convergence characteristic of 6-units system (P_D = 1263 MW).

Table 5. Best outputs for 6-units system (P_D = 1263 MW).

Generator	SA [51]	GA [51]	TS [51]	PSO [51]	MTS [51]	PSO-LRS [48]	NPSO [48]	NPSO-LRS [48]	JAYA	CJAYA	MP-CJAYA
1	478.1258	462.0444	459.0753	447.5823	448.1277	447.4440	447.4734	446.96	457.9858	452.3884	444.7000
2	163.0249	189.4456	185.0675	172.8387	172.8082	173.3430	173.1012	173.3944	176.8785	162.1065	171.1458
3	261.7146	254.8535	264.2094	261.3300	262.5932	263.3646	262.6804	262.3436	250.0717	256.4885	253.8111
4	125.7665	127.4296	138.1222	138.6812	136.9605	139.1279	139.4156	139.5120	129.3748	142.1863	134.8118
5	153.7056	151.5388	154.4716	169.6781	168.2031	165.5076	165.3002	164.7089	172.8886	170.7924	175.4557
6	93.7965	90.7150	74.9900	85.8963	87.3304	87.1698	87.9761	89.0162	88.4618	91.5015	95.6913
P_{total} (MW)	1276.1339	1276.0270	1275.94	1276.0066	1276.0232	1275.95	1275.96	1275.94	1275.6611	1275.4637	1275.6158
P_{loss} (MW)	13.1317	13.0268	12.9422	13.0066	13.0205	12.9571	12.9470	12.9361	12.6665	12.4444	12.6030
F_{best} ($/h)	15,461.10	15,457.96	15,454.89	15,450.14	15,450.06	15,450.00	15,450.00	15,450.00	15,447.09	15,446.71	**15,446.17**

Table 6. Results comparison of 6-units system (P_D = 1263 MW).

	$F_{best}(\$/h)$	$F_{worst}(\$/h)$	$F_{mean}(\$/h)$
SA [51]	15,461.10	15,545.50	15,488.98
GA [51]	15,457.96	15,524.69	15,477.71
TS [51]	15,454.89	15,498.05	15,472.56
PSO [51]	15,450.14	15,491.71	15,465.83
MTS [51]	15,450.06	15,453.64	15,451.17
PSO-LRS [48]	15,450.00	15,455.00	15,454.00
NPSO [48]	15,450.00	15,454.00	15,452.00
NPSO-LRS [48]	15,450.00	15,452.00	15,450.50
JAYA	15,447.09	15,622.16	15,500.11
CJAYA	15,446.71	15,484.34	15,461.62
MP-CJAYA	**15,446.17**	**15,451.68**	**15,449.23**

4.5. Case V: 15-Units System for Load Demand of 2630 MW

In the last case, the three versions of JAYA are applied to a larger 15-units system with the same constraints as in case 4, the system data and B-coefficients have been taken from [50]. There are 4 generators having *POZs*. Generators 2, 5 and 6 have three *POZs* and generator 12 has two *POZs*. Considering that these *POZs* result in non-convex decision spaces consisting of 192 convex sub-spaces, the value of N_{pop}, N_{JAYA_iter}, N_{COA_iter}, N_{sub_pop} and N_{runs} are all increased compared to Case IV to cope with the challenges.

The best outputs from JAYA, CJAYA, MP-CJAYA and other algorithms including SA [51], GA [51], TS [51], PSO [51], MTS [51], TSA [52], DSPSO-TSA [52] and AIS [53] are summarized in Table 7. From the table we can observe that DSPSO-TSA has provided lower F_{best} than JAYA, but it is not as lowest as CJAYA and MP-CJAYA, which obtains 32,710.0768 \$/h and 32,706.5158 \$/h respectively and ranks the second and first best value among all the algorithms. Furthermore, in addition to the best cost F_{best}, the worst cost F_{worst} and the mean cost F_{mean} of the three version of JAYA algorithms are also compared with those above-mentioned methods in Table 8. It can be found that MP-CJAYA achieves the minimum value of F_{best}, F_{worst} and F_{mean} at the same time, which are highlighted in bold font. Figure 10 is the distribution outlines of F_{best}, we can notice that MP-CJAYA exhibits the best consistency in achieving minimum F_{best} within 50 independent runs. One randomly chosen convergence curve is shown in Figure 11, from which we can see that CJAYA has improved the convergence rate and accuracy of basic JAYA, while MP-CJAYA has made further improvements of CJAYA in the rate of approaching the lowest cost. From the analysis above, it can be concluded that MP-CJAYA has the strongest capabilities of handling larger size of ELD problems with different constraint conditions.

Table 7. Best outputs for 15-units system (P_D = 2630 MW).

Unit	SA [51]	GA [51]	TS [51]	PSO [51]	MTS [51]	TSA [52]	DSPSO-TSA [52]	AIS [53]	JAYA	CJAYA	MP-CJAYA
1	453.6646	445.5619	453.5374	454.7167	453.9922	440.500	453.627	441.159	455.0000	455.0000	455.0000
2	377.6091	380.0000	371.9761	376.2002	379.7434	346.800	379.895	409.587	379.9848	380.0000	380.0000
3	120.3744	129.0605	129.7823	129.5547	130.0000	110.880	129.482	117.298	130.0000	130.0000	130.0000
4	126.2668	129.5250	129.3411	129.7083	129.9232	122.460	129.923	131.258	129.9821	130.0000	130.0000
5	165.3048	169.9659	169.5950	169.4407	168.0877	177.740	168.956	151.011	169.6535	170.0000	170.0000
6	459.2455	458.7544	457.9928	458.8153	460.0000	459.110	459.907	466.258	460.0000	460.0000	460.0000
7	422.8619	417.9041	426.8879	427.5733	429.2253	406.410	429.971	423.368	429.0688	430.0000	430.0000
8	126.4025	97.8230	95.1680	67.2834	104.3097	107.550	103.673	99.948	81.7235	106.1556	71.8662
9	54.4742	54.2933	76.8439	75.2673	35.0358	107.270	34.909	110.684	51.3258	25.0000	58.9683
10	149.0879	144.2214	133.5044	155.5899	155.8829	140.560	154.593	100.229	146.6714	160.0000	160.0000
11	77.9594	77.3002	68.3087	79.9522	79.8994	78.470	79.559	32.057	79.1805	80.0000	80.0000
12	93.9489	77.0371	79.6815	79.8947	79.9037	74.170	79.388	78.815	80.0000	80.0000	80.0000
13	25.0022	31.1537	28.3082	25.2744	25.0220	31.950	25.487	23.568	25.0000	25.0000	25.0000
14	16.0636	15.0233	17.7661	16.7318	15.2586	37.380	15.952	40.258	27.7503	15.0000	15.0000
15	15.0196	33.6125	22.8446	15.1967	15.0796	22.470	15.640	36.906	15.0000	15.0000	15.0000
P_{total} (MW)	2663.29	2661.23	2661.53	2661.19	2661.36	2663.70	2660.96	2662.04	2660.3408	2661.1556	2660.8346
P_{loss} (MW)	33.2737	31.2363	31.4100	31.1697	31.3523	33.8110	30.9520	32.4075	30.3442	31.1643	30.8346
F_{best} ($/h)	32,786.40	32,779.81	32,762.12	32,724.17	32,716.87	32,918.00	32,715.06	32,854.00	32,716.8706	32,710.0768	**32,706.5158**

Table 8. Results comparison of 15-units system (P_D = 2630 MW).

	$F_{best}(\$/h)$	$F_{worst}(\$/h)$	$F_{mean}(\$/h)$
SA [51]	32,786.40	33,028.95	32,869.51
GA [51]	32,779.81	33,041.64	32,841.21
TS [51]	32,762.12	32,942.71	32,822.84
PSO [51]	32,724.17	32,841.38	32,807.45
MTS [51]	32,716.87	32,796.15	32,767.21
TSA [52]	32,917.87	33,245.54	33,066.76
DSPSO-TSA [52]	32,715.06	32,730.39	32,724.63
AIS [53]	32,854.00	32,892.00	32,873.25
JAYA	32,716.8706	32,967.8314	32,789.1472
CJAYA	32,710.0768	32,828.6554	32,740.0719
MP-CJAYA	**32,706.5158**	**32,708.8736**	**32,706.7150**

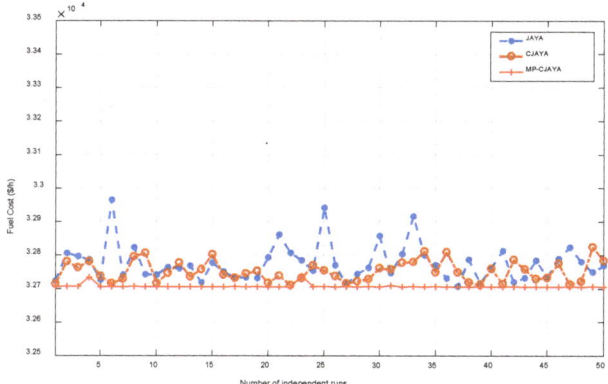

Figure 10. Fuel cost for 50 independent runs of 15-units system (P_D = 2630 MW).

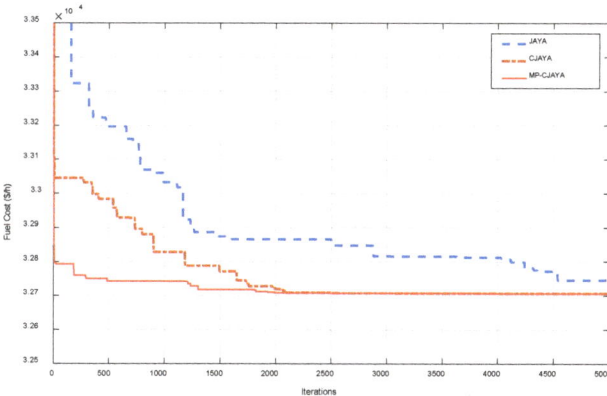

Figure 11. Fuel cost convergence characteristic of 15-units system (P_D = 2630 MW).

5. Discussion and Conclusions

A novel multi-population based chaotic JAYA algorithm (MP-CJAYA) is proposed in this paper. By introducing the MP method and chaotic map to the basic JAYA algorithm, both the global exploration capability and the local searching capability have been greatly improved. MP-CJAYA is employed in five typical ELD cases to compare the performances with other well-established algorithms in terms of best solutions, convergence rate and robustness. The results have proved that MP-CJAYA algorithm has outstanding superiority to all the other compared algorithms in all cases.

It is noteworthy that for most of the meta-heuristic algorithms, parameter setting is critical for the quality of their results. But for MP-CJAYA, it does not require for specific algorithm parameters except for common parameters. What's more, it is observed that the common parameter population size (N_{pop}) does not affect the performance of its final optimal solution significantly, as shown in Figure 12. With increased N_{pop} of 30, 50, 100 and 200 under the same circumstances, a slightly steady improvement of the convergence rate can be observed at initial part of the iteration. However, after about 5000 iterations, the differences among those curves become difficult to be observed and they all have reached the same best solution, which has proved that MP-CJAYA algorithm is not highly dependent on the common parameter N_{pop}.

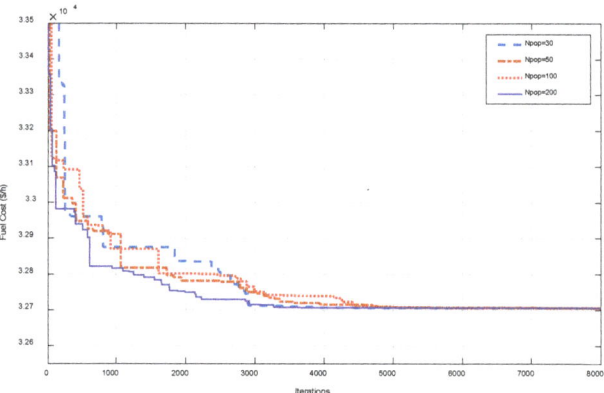

Figure 12. Convergence characteristics of MP-CJAYA with varying population sizes for case V.

As a newly proposed meta-heuristic algorithm, even though MP-CJAYA has gained the most outstanding superiority in this paper, it still has not been used for solving other optimization issues, except for the ELD problem. Hence, authors are planning to apply it to different kinds of optimization issues in the future to broaden its applications, such as multiple fuel options, micro grid power dispatch problems and multi-objective scheduling optimization problems.

Author Contributions: Conceptualization, J.Y., C.-H.K. and S.-B.R.; Data curation, J.Y.; Formal analysis, J.Y., A.W. and T.K.; Investigation, J.Y., C.-H.K., A.W. and T.K.; Methodology, J.Y. and S.-B.R.; Software, J.Y. and C.-H.K.; Supervision, S.-B.R.; Writing—Original Draft, J.Y.; Writing—Review & Editing, A.W. and T.K.

Funding: This research received no external funding.

Acknowledgments: Authors would like to thank Yeungnam University for all the supports in terms of fellowship to Jiangtao Yu.

Conflicts of Interest: The authors declare no conflict of interest.

References

1. Narimani, MR.; Joo, J.-Y.; Crow, M.L. Dynamic Economic Dispatch with Demand Side Management of Individual Residential Loads. In Proceedings of the 2015 North American Power Symposium (NAPS), Charlotte, NC, USA, 4–6 October 2015.

2. Dodu, J.C.; Martin, P.; Merlin, A.; Pouget, J. An optimal formulation and solution of short-range operating problems for a power system with flow constraints. *Proc. IEEE* **1972**, *60*, 54–63. [CrossRef]

3. Chen, C.L.; Wang, S.C. Branch-and-bound scheduling for thermal generating units. *IEEE Trans. Energy Convers.* **1993**, *8*, 184–189. [CrossRef]

4. Jubril, A.; Olaniyan, O.; Komolafe, O.; Ogunbona, P.O. Economic-emission dispatch problem: A semi-definite programming approach. *Appl. Energy* **2014**, *134*, 446–455. [CrossRef]

5. Papageorgiou, L.G.; Fraga, E.S. A mixed integer quadratic programming formulation for the economic dispatch of generators with prohibited operating zones. *Electr. Power Syst. Res.* **2007**, *77*, 1292–1296. [CrossRef]

6. Liang, Z.X.; Glover, J.D. A zoom feature for a dynamic programming solution to economic dispatch including transmission losses. *IEEE Trans. Power Syst.* **1992**, *7*, 544–550. [CrossRef]

7. El-Keib, A.A.; Ma, H.; Hart, J.L. Environmentally constrained economic dispatch using the Lagrangian relaxation method. *IEEE Trans. Power Syst.* **1994**, *9*, 1723–1729. [CrossRef]

8. Farag, A.; Al-Baiyat, S.; Cheng, T. Economic load dispatch multiobjective optimization procedures using linear programming techniques. *IEEE Trans. Power Syst.* **1995**, *10*, 731–738. [CrossRef]

9. Niknam, T. A new fuzzy adaptive hybrid particle swarm optimization algorithm for non-linear, non-smooth and non-convex economic dispatch problem. *Appl. Energy* **2010**, *87*, 327–339. [CrossRef]

10. Chiang, C.L. Improved genetic algorithm for power economic dispatch of units with valve-point effects and multiple fuels. *IEEE Trans. Power Syst.* **2005**, *20*, 1690–1699. [CrossRef]

11. Park, J.B.; Lee, K.S.; Shin, J.R.; Lee, K.Y. A particle swarm optimization for economic dispatch with non-smooth cost functions. *IEEE Trans. Power Syst.* **2005**, *20*, 34–42. [CrossRef]

12. Lin, W.M.; Cheng, F.S.; Tsay, M.T. An improved tabu search for economic dispatch with multiple minima. *IEEE Trans. Power Syst.* **2002**, *17*, 108–112. [CrossRef]

13. Secui, D.C. A new modified artificial bee colony algorithm for the economic dispatch problem. *Energy Convers. Manag.* **2015**, *89*, 43–62. [CrossRef]

14. Yang, X.S.; Sadat Hosseini, S.S.; Gandomi, A.H. Firefly algorithm for solving non-convex economic dispatch problems with valve loading effect. *Appl. Soft Comput.* **2012**, *12*, 1180–1186. [CrossRef]

15. Dos Santos Coelho, L.; Mariani, V.C. An improved harmony search algorithm for power economic load dispatch. *Energy Convers. Manag.* **2009**, *50*, 2522–2526. [CrossRef]

16. He, X.Z.; Rao, Y.Q.; Huang, J.D. A novel algorithm for economic load dispatch of power systems. *Neurocomputing* **2016**, *171*, 1454–1461. [CrossRef]

17. Niknam, T.; Mojarrad, H.D.; Meymand, H.Z. A novel hybrid particle swarm optimization for economic dispatch with valve-point loading effects. *Energy Convers. Manag.* **2011**, *52*, 1800–1809. [CrossRef]

18. Wang, L.; Li, L.P. An effective differential harmony search algorithm for the solving non-convex economic load dispatch problems. *Int. J. Electr. Power Energy Syst.* **2013**, *44*, 832–843. [CrossRef]

19. Alsumait, J.S.; Sykulski, J.; Al-Othman, A. A hybrid GA-PS-SQP method to solve power system valve-point economic dispatch problems. *Appl. Energy* **2010**, *87*, 1773–1781. [CrossRef]

20. Mahdi, F.P.; Vasant, P. Quantum particle swarm optimization for economic dispatch problem using cubic function considering power loss constraint. *IEEE Trans. Power Syst.* **2002**, *17*, 108–112.

21. Rao, R.V. Jaya: A simple and new optimization algorithm for solving constrained and unconstrained optimization problems. *Int. J. Ind. Eng. Comput.* **2016**, *7*, 19–34.

22. Warid, W.; Hizam, H.; Mariun, N.; Abdul-Wahab, N.I. Optimal power flow using the JAYA algorithm. *Energies* **2016**, *9*, 678. [CrossRef]

23. Rao, R.V.; Saroj, A. Multi-objective design optimization of heat exchangers using elitist-JAYA algorithm. *Energy Syst.* **2018**, *9*, 305–341. [CrossRef]

24. Yu, K.J.; Liang, J.J.; Qu, B.Y. Parameters identification of photovoltaic models using an improved JAYA optimization algorithm. *Energy Convers. Manag.* **2017**, *150*, 742–753. [CrossRef]

25. Rao, R.V.; More, K. Design optimization and analysis of selected thermal devices using self-adaptive JAYA algorithm. *Energy Convers. Manag.* **2017**, *140*, 24–35. [CrossRef]

26. Huang, C.; Wang, L.; Yeung, R.S. A prediction model guided JAYA algorithm for the PV system maximum power point tracking. *IEEE Trans. Sustain. Energy* **2018**, *9*, 45–55. [CrossRef]

27. Rao, R.V.; Waghmare, G. A new optimization algorithm for solving complex constrained design optimization problems. *Eng. Opt.* **2017**, *49*, 60–83. [CrossRef]

28. Rao, R.V.; Rai, D.P.; Balic, J. A multi-objective algorithm for optimization of modern machining processes. *Eng. Appl. Artif. Intell.* **2017**, *61*, 103–125. [CrossRef]

29. Mishra, S.; Ray, P.K. Power quality improvement using photovoltaic fed DSTATCOM based on JAYA optimization. *IEEE Trans. Sustain. Energy* **2016**, *7*, 1672–1680. [CrossRef]

30. Nguyen, T.T.; Yang, S.; Branke, J. Evolutionary dynamic optimization: a survey of the state of the art. *Swarm Evol. Comput.* **2012**, *6*, 1–24. [CrossRef]

31. Cruz, C.; González, J.R.; Pelta, D.A. Optimization in dynamic environments: a survey on problems methods and measures. *Soft Comput.* **2011**, *15*, 1427–1448. [CrossRef]

32. Branke, J.; Kaußler, T.; Schmidt, C.; Schmeck, H. A multi-population approach to dynamic optimization problems. *Evol. Des. Manuf.* **2000**, 299–309. [CrossRef]

33. Turky, A.M.; Abdullah, S. A multi-population electromagnetic algorithm for dynamic optimization problems. *Appl. Soft Comput.* **2014**, *22*, 474–482. [CrossRef]

34. Turky, A.M.; Abdullah, S. A multi-population harmony search algorithm with external archive for dynamic optimization problems. *Inf. Sci.* **2014**, *272*, 84–95. [CrossRef]

35. Nseef, S.K.; Abdullah, S.; Turky, A.; Kendall, G. An adaptive multi-population artificial bee colony algorithm for dynamic optimization problems. *Knowl. Based Syst.* **2016**, *104*, 14–23. [CrossRef]

36. Li, C.; Nguyen, T.T.; Yang, M.; Yang, S.; Zeng, S. Multi-population methods in un-constrained continuous dynamic environments: the challenges. *Inf. Sci.* **2015**, *296*, 95–118. [CrossRef]

37. Caponetto, R.; Fortuna, L.; Fazzino, S.; Gabriella, M. Chaotic sequences to improve the performance of evolutionary algorithms. *IEEE Trans. Evol. Comput.* **2003**, *7*, 289–304. [CrossRef]

38. Li, Y.; Wen, Q.; Li, L.; Peng, H. Hybrid chaotic ant swarm optimization. *Chaos Solitons Fractals* **2009**, *42*, 880–889. [CrossRef]

39. Alatas, B. Chaotic harmony search algorithms. *Appl. Math. Comput.* **2010**, *216*, 2687–2699. [CrossRef]

40. Chuang, L.-Y.; Tsai, S.-W.; Yang, C.-H. Chaotic catfish particle swarm optimization for solving global numerical optimization problems. *Appl. Math. Comput.* **2011**, *217*, 6900–6916. [CrossRef]

41. Dos Santos Coelho, L.; Mariani, V.C. Firefly algorithm approach based on chaotic Tinkerbell map applied to multivariable PID controller tuning. *Comput. Math. Appl.* **2012**, *64*, 2371–2382. [CrossRef]

42. Narimani, M.R.; Joo, J.-Y.; Crow, M. Multi-objective dynamic economic dispatch with demand side management of residential loads and electric vehicles. *Energies* **2017**, *10*, 624. [CrossRef]

43. Heidari-Bateni, G.; McGillem, C.D. A chaotic direct-sequence spread-spectrum communication system. *IEEE Trans. Commun.* **1994**, *42*, 1524–1527. [CrossRef]

44. Sinha, N.; Chakrabarti, R.; Chattopadhyay, P. Evolutionary programming techniques for economic load dispatch. *IEEE Trans. Evol. Comput.* **2003**, *7*, 83–94. [CrossRef]

45. Victoire, T.A.A.; Jeyakumar, A.E. Hybrid PSO–SQP for economic dispatch with valve-point effect. *Electr. Power Syst. Res.* **2004**, *71*, 51–59. [CrossRef]

46. Cai, J.; Li, Q. A hybrid CPSO–SQP method for economic dispatch considering the valve-point effects. *Energy Convers. Manag.* **2012**, *53*, 175–181. [CrossRef]

47. Bhagwan Das, D.; Patvardhan, C. Solution of Economic Load Dispatch using real coded Hybrid Stochastic Search. *Int. J. Electr. Power Energy Syst.* **1999**, *21*, 165–170. [CrossRef]

48. Selvakumar, I.; Thanushkodi, K. A new particle swarm optimization solution to nonconvex economic dispatch problems. *Electr. Power Syst. Res.* **2007**, *22*, 42–51. [CrossRef]

49. Chaturvedi, K.T.; Pandit, M. Self-Organizing Hierarchical Particle Swarm Optimization for Nonconvex Economic Dispatch. *IEEE Trans. Power Syst.* **2008**, *23*, 1079–1087. [CrossRef]

50. Gaing, Z.L. Particle swarm optimization to solving the economic dispatch considering the generator constraints. *IEEE Trans. Power Syst.* **2003**, *18*, 1187–1195. [CrossRef]

51. Pothiya, S.; Ngamroo, I.; Kongprawechnon, W. Application of multiple tabu search algorithm to solve dynamic economic dispatch considering generator constraints. *Energy Convers. Manag.* **2008**, *49*, 506–516. [CrossRef]

52. Khamsawang, S.; Jiriwibhakorn, S. DSPSO–TSA for economic dispatch problem with nonsmooth and noncontinuous cost functions. *Energy Convers. Manag.* **2010**, *51*, 365–375. [CrossRef]

53. Panigrahi, B.K.; Yadav, S.R.; Agrawal, S.; Tiwari, M.K. A clonal algorithm to solve economic load dispatch. *Electr. Power Syst. Res.* **2007**, *77*, 1381–1389. [CrossRef]

Article

Parameter Estimation of Electromechanical Oscillation Based on a Constrained EKF with C&I-PSO

Yonghui Sun [1,*], Yi Wang [1], Linquan Bai [2], Yinlong Hu [1], Denis Sidorov [3] and Daniil Panasetsky [3]

[1]   College of Energy and Electrical Engineering, Hohai University, Nanjing 210098, China;
      wyhhu@hhu.edu.cn (Y.W.); ylhu@hhu.edu.cn (Y.H.)
[2]   ABB Inc., Raleigh, NC 27606, USA; linquan.bai@ieee.org
[3]   Melentiev Energy Systems Institute, Russian Academy of Sciences, Irkutsk 664033, Russia;
      dsidorov@isem.irk.ru (D.S.); panasetsky@gmail.com (D.P.)
*   Correspondence: yhsun@hhu.edu.cn; Tel.: +86-130-0516-9126

Received: 18 July 2018; Accepted: 6 August 2018; Published: 8 August 2018

Abstract: By combining together the extended Kalman filter with a newly developed C&I particle swarm optimization algorithm (C&I-PSO), a novel estimation method is proposed for parameter estimation of electromechanical oscillation, in which critical physical constraints on the parameters are taken into account. Based on the extended Kalman filtering algorithm, the constrained parameter estimation problem is formulated via the projection method. Then, by utilizing the penalty function method, the obtained constrained optimization problem could be converted into an equivalent unconstrained optimization problem; finally, the C&I-PSO algorithm is developed to address the unconstrained optimization problem. Therefore, the parameters of electromechanical oscillation with physical constraints can be successfully estimated and better performed. Finally, the effectiveness of the obtained results has been illustrated by several test systems.

Keywords: constrained parameter estimation; extended Kalman filter; power systems; C&I particle swarm optimization; ringdown detection

1. Introduction

Over the past few decades, there have been a lot of concerns about estimation of electromechanical modes since it can offer a substantial amount of important information about power system stability [1–5]. Therefore, it is necessary to track these parameters in real time to monitor the system stability and prevent blackouts [6–8]. In general, electromechanical oscillations are divided into two groups: local and inter-area, which refer to the oscillations between nearby generators and distant generators, respectively [3].

Generally speaking, there are mainly two different methods for the parameters estimation of power system oscillation: one of them is the model-based method, and the other is the measurement-based method. In the model-based method, the governing equations of the studied model are linearized near the present operating point. In contrast, a linear model can be estimated directly from measurement data in the measurement-based method. Therefore, in recent years, for complex modern power systems, the measurement-based method is generally considered easier than the model-based method and is widely used [3,9–17]. Normally, two types of measurement data can be obtained by using phasor measurement units (PMUs). One of them is the ambient data, which can be sampled from a power system at a stable operating point [10]; the other one is designed as the ringdown data, which is generated from a power system with a major disturbance. The ringdown data contains key oscillatory information and is the main focus of this paper. In recent years, for ringdown data detection, several methods have been developed. In [11,12], some methods are investigated that are mainly based on frequency estimation and harmonic detection. Furthermore, the matrix pencil method [13], the Kung's algorithm [14], the rotational invariance technique [15], and the linear prediction methods such as the Prony method [16,17] were proposed and widely utilized for ringdown detection. However, most of the aforementioned approaches cannot be used for real-time application since they are sliding window-based methods, in which the calculations are performed only when a window of data is received [3]. Recently, a novel method for ringdown detection was proposed based on the traditional EKF in [3], where the damping ratio and frequency of ringdown data were modeled in the state vector; then these parameters could be easily estimated by using EKF. However, it should be pointed out that, in the above mentioned results, some critical physical constraints on parameters were ignored; thus, the practical value of the proposed method could be limited.

The Kalman filter is known as an optimal estimator and has been widely used in aerospace, the state estimation of power systems and other fields [18–23]. EKF is a famous nonlinear estimator, which is developed by combining the conventional Kalman filter and the linearized technology-based Taylor series expansion. Due to its advantages and convenience, it has been used in many nonlinear state estimation problems. However, in the application of EKF, some known information of the signal is sometimes either ignored or addressed heuristically. For example, the damping ratio and the frequency of oscillation signal are usually meet with the positive constraint condition. In general, for a constrained estimation problem, it is radically difficult within the recursive framework for the traditional EKF to incorporate the constraints into system states. To deal with this problem, many approaches have been developed, such as the projection method, the mean square method [24,25], etc., by which the constrained estimation problem can be successfully converted to the constrained optimization problem after each iteration of EKF [26,27].

The constrained optimization problems are common in many practical applications, such as optimal power flow calculations, allocation problems, and structural optimizations [28]. Until now, various solutions have been developed to address these optimization problems. The most popular method of them is particle swarm optimization (PSO) [29–32], which has been widely used in many optimization problems. It has the ability to solve difficult optimization problems and converge quickly to a solution [33–36]. Although the PSO was developed primarily to address unconstrained optimization problems; nevertheless, it performs well when applied to constrained optimization problems. In this case, by adding a penalty term into the objective function, an equivalent unconstrained optimization problem can be obtained; then PSO can be used to address the unconstrained optimization problem. However, in practice, although the traditional PSO algorithm can converge quickly, it is always easy to fall into local optimum [37–41], so the global optimization solutions cannot be obtained easily. Therefore, in this paper, a modified PSO algorithm (namely, C&I-PSO) is developed to address the equivalent unconstrained optimization problem, in which a constrictive factor and a linear decreasing inertia weight (LDIW) are introduced to improve the performance of global searches and local searches [42]. By this method, in the early search stage, a large inertia weight is utilized to extend the search region and avoid the occurrence of premature problems. In the later stage, a small inertia weight is used to obtain a more accurate local search such that a more accurate global optimization solution can be achieved. To the best of the authors' knowledge, in existing work, there have been little literature that considers parameter estimations of electromechanical oscillation with physical constraints. Based on the above discussion, with the consideration of the physical constraints on parameters, a constrained EKF in combination with C&I-PSO algorithm is proposed in this paper for parameter estimation of the ringdown signal.

The rest of this paper is organized as follows: In Section 2, the state space model for the parameters estimation of ringdown signal is provided. In Section 3, by incorporating the projection method, the traditional EKF with inequality constraints on system states is presented. In Section 4, the C&I-PSO algorithm for the equivalent unconstrained optimization problem is proposed, and the new parameter estimation algorithm is presented. In Section 5, some simulation results are provided to verify the usefulness and effectiveness of the proposed results. Finally, the conclusions are presented in Section 6.

2. State-Space Model of the Ringdown Signal

As stated in [3,7], the ringdown signal can be expressed by the sequence of exponentially damped sinusoids (EDS):

$$y(t) = \sum_{i=1}^{N} A_i e^{-\delta_i t} \cos(\omega_i t + \phi_i) + n(t), \tag{1}$$

where N represents the number of EDSs, A_i is a known constant coefficient, δ_i is the damping ratio, ω_i represents the frequency, ϕ_i is the initial phase value, the subscript i indicates the ith component of signal, and $n(t)$ is considered a white Gaussian noise sequence with zero mean.

Assume that the signal sampling frequency is f_s. Then the ringdown signal containing multiple EDS signals can be deduced:

$$S_k = \sum_{i=1}^{N} A_i e^{-\delta_i \frac{k}{f_s}} \cos(\omega_i \frac{k}{f_s} + \phi_i) + n_k. \tag{2}$$

Define $4N$ state variables as [3]:

$$\begin{cases} x_{4i-3,k} = A_i e^{-\delta_i \frac{k}{f_s}} \cos(\omega_i \frac{k}{f_s}) \\ x_{4i-2,k} = A_i e^{-\delta_i \frac{k}{f_s}} \sin(\omega_i \frac{k}{f_s}) \\ x_{4i-1,k} = \omega_i \\ x_{4i,k} = \delta_i \end{cases}, \tag{3}$$

where the subscript i indicates the ith component of the signal. Based on Equation (3), the $k+1$ instant state equation of the ringdown signal can be formulated as:

$$\begin{cases} x_{4i-3,k+1} = e^{-\frac{x_{4i,k}}{f_s}} \left(x_{4i-3,k} \cos(\frac{x_{4i-1,k}}{f_s}) - x_{4i-2,k} \sin(\frac{x_{4i-1,k}}{f_s}) \right) + \omega_{4i-3,k} \\ x_{4i-2,k} = e^{-\frac{x_{4i,k}}{f_s}} \left(x_{4i-3,k} \sin(\frac{x_{4i-1,k}}{f_s}) + x_{4i-2,k} \cos(\frac{x_{4i-1,k}}{f_s}) \right) + \omega_{4i-2,k} \\ x_{4i-1,k} = x_{4i-1,k} + \omega_{4i-1,k} \\ x_{4i,k} = x_{4i,k} + \omega_{4i,k} \end{cases}, \tag{4}$$

where $\omega_{4i-l,k}$ ($l = 0, 1, 2, 3$) denotes the system noise and modeling variations, which are usually taken as the zero-mean white noises. The in-phase signal is given by the state variables $x_{4i-3,k}$, $x_{4i-2,k}$ is the quadrature signal, $x_{4i-1,k}$ represent the frequency ω_i to be identified, $x_{4i,k}$ is the damping factor to be estimated, and the subscript i indicates the ith component of ringdown signal.

The observation can be obtained by:

$$y_k = \sum_{i=1}^{N} k_{2i-1} x_{4i-3,k} + k_{2i} x_{4i-2,k} + n_k, \tag{5}$$

where N indicates the total number of EDSs, the subscript i indicates the ith component of the ringdown signal, $k_{2i-1} = \cos(\phi_i)$, $k_{2i} = -\sin(\phi_i)$, and n_k is a white Gaussian noise sequence with zero mean.

Remark 1. *It should be pointed out that most of the monitoring data of electromechanical oscillation in real power systems are stable or positively damped, which means the damping factor is positive, even in some poorly damped cases. Therefore, it is reasonable to assume the damping ratio and frequency are positive under this case. Based on the above discussion, following the idea proposed in [3], the next step of this research is to explore new methods to estimate parameters of electromechanical oscillation by taking into account the physical constraints.*

3. Extended Kalman Filter with Inequality Constraints

In this section, the traditional EKF for parameter identification will be introduced in advance. Then, the projection method is utilized to turn the constrained estimation problem into an equivalent constrained optimization problem.

3.1. Traditional Extended Kalman Filter

EKF is usually used for states or the parameters estimation of nonlinear systems, which is developed by combining the conventional Kalman filter and the linearized technology based on Taylor series expansion.

Consider the following nonlinear stochastic system:

$$\begin{cases} x_{k+1} = f(x_k, u_k) + w_k \\ y_k = h(x_k) + v_k \end{cases}, \tag{6}$$

where $f(\cdot)$ is the system function, $h(\cdot)$ is the output function, both of them are nonlinear functions that could be linearized by Taylor series expansion, x_k is the n-dimensional system state, u_k is the input vector, w_k is the state noise process vector denoting disturbances and modeling errors, y_k is the m-dimensional observation vector, and v_k is the measurement noise. Here w_k and v_k are usually considered as the zero-mean Gaussian white noises.

For convenience, the following formulas are defined in advance:

$$\hat{x}_k = E[(x_k | y_k^*)], \ \hat{P}_k = E[(x_k - \hat{x}_k)(x_k - \hat{x}_k)^T] \tag{7}$$

$$\tilde{x}_k = E[(x_k | y_{k-1}^*)], \ \tilde{P}_k = E[(x_k - \tilde{x}_k)(x_k - \tilde{x}_k)^T] \tag{8}$$

where $\hat{x}_k$ is the state estimation, $\hat{P}_k$ is the estimation error covariance matrix, $\tilde{x}_k$ is the state prediction, and $\tilde{P}_k$ is the prediction error covariance matrix, the subscript k indicating time instant.

The prediction step:

$$\begin{cases} \tilde{x}_{k+1} = f(\hat{x}_k, u_k) \\ \tilde{P}_{k+1} = F_k \hat{P}_k F_k^T + Q_k \end{cases}, \tag{9}$$

The filtering step:

$$\begin{cases} \hat{x}_{k+1} = \tilde{x}_{k+1} + G_{k+1}[y_{k+1} - h(\tilde{x}_{k+1})] \\ G_{k+1} = \tilde{P}_{k+1} H_{k+1}^T (H_{k+1} \tilde{P}_{k+1} H_{k+1}^T + R_{k+1})^{-1} \\ \hat{P}_{k+1} = (I - G_{k+1} H_{k+1}) \tilde{P}_{k+1} \end{cases}, \tag{10}$$

where G_{k+1} is the Kalman filter gain vector at $k+1$ time instant, and the Jacobian matrices F_k and H_{k+1} can be derived by:

$$F_k = \frac{\partial f(x_k, u_k)}{\partial x_k}|_{x_k = \hat{x}_k}, \ H_{k+1} = \frac{\partial h(x_{k+1}, u_{k+1})}{\partial x_{k+1}}|_{x_{k+1} = \tilde{x}_{k+1}} \tag{11}$$

Q_k and R_k are the covariance matrices of state noise and measurement noise, respectively, and they are usually defined as:

$$Q_k = E[w_k w_k^T], \ R_k = E[v_k v_k^T] \tag{12}$$

Based on the model of ringdown signal (4) and (5), by using the EKF algorithm, some related equations for estimating parameters of the ringdown signal are obtained:

$$f(x_k) = \begin{bmatrix} M_1 \\ M_2 \\ \vdots \\ M_i \\ \vdots \\ M_N \end{bmatrix}, \quad M_i = \begin{bmatrix} x_{4i-3,k} \\ x_{4i-2,k} \\ x_{4i-1,k} \\ x_{4i,k} \end{bmatrix} \tag{13}$$

The function of $h(x_k)$, and the covariance matrix Q_k, R_k can be also obtained:

$$H = (\ k_1 k_2 0 0 \quad \cdots \quad k_{2i-1} k_{2i} 0 0 \quad \cdots \quad k_{2N-1} k_{2N} 0 0\ ), \quad h(x_k) = H x_k \tag{14}$$

$$Q_k = \begin{pmatrix} q_1 & 0 & \cdots & 0 \\ 0 & q_2 & \cdots & 0 \\ \vdots & \vdots & \ddots & \vdots \\ 0 & 0 & \cdots & q_{4N} \end{pmatrix}, \quad R_k = E[v_k v_k^T] = [r] \tag{15}$$

Remark 2. *In [3], a new method was proposed for ringdown detection, where the traditional EKF method was used to estimate the parameters. However, it should be pointed out that both the frequency and damping ratio of the ringdown signal are usually positive. Thus, the practical value of the method will be inevitably reduced if such state constraints are ignored. In order to obtain a better and more practical estimation algorithm, a new constrained method is proposed to estimate the parameters of the ringdown signal by accounting for the physical constraints.*

3.2. The Projection Method

In most cases, it is necessary to specify constraints on parameter values. Therefore, in order to solve the constrained parameter identification problem, the constrained estimation problem can be successfully converted by using the projection method to the equivalent minimum constrained optimization problem at each iteration.

Consider the nonlinear dynamical system (6) with the constraint as follows:

$$U_c x \leq u_c, \tag{16}$$

where U_c indicates a known matrix, which has s lines and n columns. s represents the number of constraints, n indicates the number of state variables, usually the rank of U_c is assumed to be s.

By using the projection method, the unconstrained state estimate $\tilde{x}$ can be projected onto the constraint surface directly, then the constrained EKF problem is expressed by:

$$\min_{\tilde{x}} (\tilde{x} - \hat{x})^T W_p (\tilde{x} - \hat{x}), \quad s.t. \quad U_c \tilde{x} \leq u_c \tag{17}$$

where $W_p > 0$ is a weighting matrix. For simplicity, W_p is chosen as the identify matrix in this paper.

The constrained EKF could be inferred by seeking out an estimation $\tilde{x}$, which satisfies the constraint condition (16) and can be obtained by solving the minimum constrained optimization problem expressed in (17). Therefore, with the projection method, the constrained state estimation problem can be addressed successfully by solving the constrained optimization after each iteration.

4. The C&I Particle Swarm Optimization

In this section, the penalty function method is utilized to change the minimum constrained optimization problem into an equivalent unconstrained optimization problem. Then, the C&I-PSO algorithm is developed to solve the equivalent unconstrained optimization problem.

4.1. The Penalty Function Approach

Due to its simple principle, the penalty function method has been used in many constrained optimization problems; after adding a penalty term, it can be applied to the constrained optimization problem, and an equivalent unconstrained optimization problem can be obtained.

Motivated by the idea proposed in [34], a non-stationary multistage assignment penalty function will be utilized to address the minimum constrained optimization, which can be expressed by:

$$f_a(x) = f(x) + h_d(k)H_p(x), \ x \in R^n \tag{18}$$

where $f(x)$ represents the minimum constrained objective function expressed in (17). $h_d(k)$ indicates a dynamical modified penalty value, k represents the iteration number of the algorithm, and $H_p(x)$ denotes a penalty factor given by:

$$H_p(x) = \sum_{i=1}^{m} \theta_f(\beta_i(x))\beta_i(x)^{\alpha(\beta_i(x))}, \tag{19}$$

where $\beta_i(x) = \max\{0, g_i(x)\}, (i = 1, \ldots s)$, which is a relative violated function of the constraints, and $g(x) = U_c x - u_c$ are the constraints expressed in (17); $\theta_f(\beta_i(x))$ represents a multi-stage assignment function; $\alpha(\beta_i(x))$ is the power of the penalty function.

Based on the above method, the minimum constrained optimization problem expressed in (17) can be turned into an equivalent minimum unconstrained optimization problem, and the unconstrained objective function can be expressed by (18). In order to address this minimum unconstrained optimization problem, the C&I-PSO approach will be introduced in the next part of this paper.

4.2. The C&I-PSO Algorithm

The PSO is one of the most popular swarm intelligence optimization algorithms [29]; its main idea is based simulating simplified social models, such as bird flocking and fish schooling. It begins with an initial population, which is generated in random positions in the search space. Without loss of generality, the D-dimensional search space is assumed to be $S \subseteq R^D$; the swarm has N particles that each having neither weight nor volume and each holding its own position and velocity. These particles update their positions over finite iterations. For the ith particle at iteration L, the position and velocity vectors are defined as $X_i^L = (x_{i1}^L, x_{i2}^L, \cdots x_{iD}^L)$ and $V_i^L = (v_{i1}^L, v_{i2}^L, \cdots v_{iD}^L)$, respectively. The best historical position visited by particle i is taken as a point in S, which is denoted as $Pbest_i^L = (pbest_{i1}^L, pbest_{i2}^L, \cdots, pbest_{iD}^L)$. The best $Gbest^L = (gbest_1^L, gbest_2^L, \cdots gbest_D^L)$. In the traditional PSO, each particle moves toward the *Particle Best* and *Global Best* positions by the specified velocity:

$$V_i^{L+1} = \psi(L) \times V_i^L + c_1 r_{1,i}\left(Pbest_i^L - X_i^L\right) + c_2 r_{2,i}\left(Gbest^L - X_i^L\right), \tag{20}$$

$$X_i^{L+1} = X_i^L + \gamma \times V_i^{L+1}, \tag{21}$$

where γ is a *constrictive factor* introduced to constrict and control velocities, and $\psi(L)$ is a *linear decreasing inertia weight*, which can be obtained by:

$$\psi(L) = \psi_{start} \times (\psi_{start} - \psi_{end}) \times (T_{max} - L)/T_{max}, \tag{22}$$

where ψ_{start} represents the inertia weight of initial iteration point, wend indicates the inertia weight of final iteration point, usually chosen as $\psi_{start} = 0.9$ and $\psi_{end} = 0.4$ respectively, L is the number of iteration and $T_{\max}$ represents the maximum iteration number.

Remark 3. *It is worth pointing out that, for solving the optimization problem in parameter estimation of electromechanical oscillation, a modified PSO named the C&I-PSO is proposed, in which a constrictive factor and a linearly decreasing inertia weight are introduced to improve the performance of local and global search. By using the proposed C&I-PSO, the premature problem of the traditional PSO can be successfully addressed, and much better solutions to the global optimization problem can be achieved if compared with the traditional PSO. Now we can present the proposed approach for parameter estimation, which is formulated as the following Algorithm 1:*

Algorithm 1: Constrained EKF with C&I-PSO

1: **Initialization:** Set appropriate values for $\hat{P}_0, \hat{x}_0, Q_0, R_0, S_t, M_t$;

2: **for** $k = 0$ to S_t **do**

3: Calculate the value of state prediction $\tilde{x}_{k+1}$ and prediction error covariance $\tilde{P}_{k+1}$:

4: $\qquad\qquad\qquad \tilde{x}_{k+1} \leftarrow f(\hat{x}_k, u_k), \ \tilde{P}_{k+1} \leftarrow F_k \hat{P}_k F_k^T + Q_k$;

5: Compute the Kalman gain matrix at time instant $k + 1$:

6: $\qquad\qquad\qquad G_{k+1} \leftarrow \tilde{P}_{k+1} H_{k+1}^T \left(H_{k+1} \tilde{P}_{k+1} H_{k+1}^T + R_{k+1} \right)^{-1}$;

7: Update the state estimation $\hat{x}_{k+1}$ and estimation error covariance:

8: $\qquad\qquad \hat{x}_{k+1} \leftarrow \tilde{x}_{k+1} + G_{k+1}[y_{k+1} - h(\tilde{x}_{k+1})], \ \hat{P}_{k+1} \leftarrow (I - G_{k+1} H_{k+1}) \tilde{P}_{k+1}$;

9: **if** $\min(\omega_1, \delta_1, \omega_2, \delta_2 \cdots \omega_N, \delta_N) < 0$ **then**

10: $\qquad\qquad\qquad \min_{\tilde{x}} (\tilde{x} - \hat{x})^T W_p (\tilde{x} - \hat{x}), s.t. \ U_c \tilde{x} \leq u_c$;

11: $\quad$ **while** $L \leq M_t$ **do**

12: $\qquad\qquad\qquad f_a(\tilde{x}) = f(\tilde{x}) + h_d(L) H_p(\tilde{x}), \tilde{x} \in R^n$;

13: $\qquad$ **for** $i = 1$ to D **do**

14: $\qquad\qquad$ Update $Gbest$, $Pbest_i$ and calculate particle new velocity and position:

15: $\qquad\qquad\qquad \psi(L) \leftarrow \psi_{start} \times (\psi_{start} - \psi_{end}) \times (M_t - L)/M_t$,

16: $\qquad\qquad V_i^{L+1} \leftarrow \psi(L) \times V_i^L + c_1 r_{1,i} \left(Pbest_i^{iter} - X_i^L \right) + c_2 r_{2,i} \left(Gbest^L - X_i^L \right)$,

17: $\qquad\qquad\qquad X_i^{L+1} = X_i^L + \gamma \times V_i^{L+1}$;

18: $\qquad$ **end for**

19: $\qquad\qquad\qquad L = L + 1$;

20: $\quad$ **end while**

21: $\qquad\qquad\qquad \hat{x}_{k+1} = Gbest$;

22: **end if**

23: $\qquad\qquad\qquad k = k + 1$;

24: **end for**

Remark 4. *By using the proposed algorithm, the problem of parameter estimation for the ringdown signal with physical constraints can be solved successfully. It should be noted that not only can the parameters of the ringdown signal be truly identified with physical constraints, but a shorter convergence time can also be expected if compared with the existing results in [3]. Therefore, the proposed algorithm is expected to be much more realistic than the previous work.*

5. Simulation Results and Discussions

In this section, three different test systems are provided to demonstrate the performance of the proposed method. The parameters in the penalty function are chosen using the same values as those in [34]. Specifically, the function $h_d(k)$ is set as $h_d(k) = k\sqrt{k}$, other related function values are given by:

$$\alpha(\beta_i(x)) = \begin{cases} 1 \text{ when } \beta_i(x) < 1 \\ 2 \text{ when } \beta_i(x) \geq 1 \end{cases}, \tag{23}$$

$$\theta_f(\beta_i(x)) = \begin{cases} 10 & \text{when } \beta_i(x) < 0.001 \\ 20 & \text{when } 0.001 \leq \beta_i(x) < 0.1 \\ 100 & \text{when } 0.1 \leq \beta_i(x) \leq 1 \\ 300 & \text{when } \beta_i(x) > 1 \end{cases}. \tag{24}$$

In addition, in order to evaluate the whole performance of the proposed approach and conventional EKF method effectively, the root-mean-square deviation (RMSD) used in [43] is adopted:

$$RMSD = \sqrt{\frac{\sum\limits_{k=1}^{k=n}\sum\limits_{i=1}^{i=m}(\hat{x}_{k,i} - x_{k,i})^2}{n}}, \tag{25}$$

where $\hat{x}_{k,i}$ is the estimation result of the *i*th corresponding parameter, $x_{k,i}$ is the true value of the *i*th corresponding parameter, *m* is the number of parameters, and *n* is the number of time steps.

5.1. Test System 1: Ringdown Signal Composed of One EDS Signal

At first, the ringdown signal in [7] is utilized to evaluate performance of the proposed algorithm. The damping ratio and frequency of this ringdown signal are 0.01 and 1 rad/s, respectively, which is provided by:

$$y_1(t) = e^{-0.01t}\cos(t) + n(t). \tag{26}$$

The input ringdown signal is shown in Figure 1. By utilizing the Matrix Pencil [13], Prony [17], EKF [3] methods and the proposed algorithm, the damping factor and frequency could be estimated. Figure 2 shows the estimation results, and the RMSD comparison of different approaches is also provided in Figure 3. It can be found that the proposed approach is able to estimate parameters correctly in a short time, while EKF and other methods need a long time to converge to the real values. It follows from these simulation results that, a better convergence and steady state performance could be obtained if compared with those by using EKF, Matrix Pencil and Prony methods.

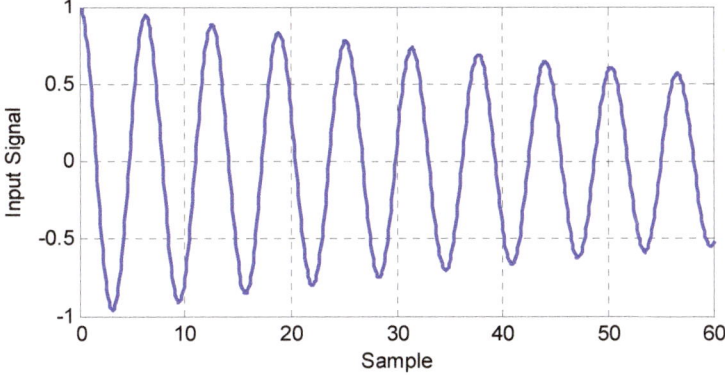

Figure 1. The input signal of test system 1.

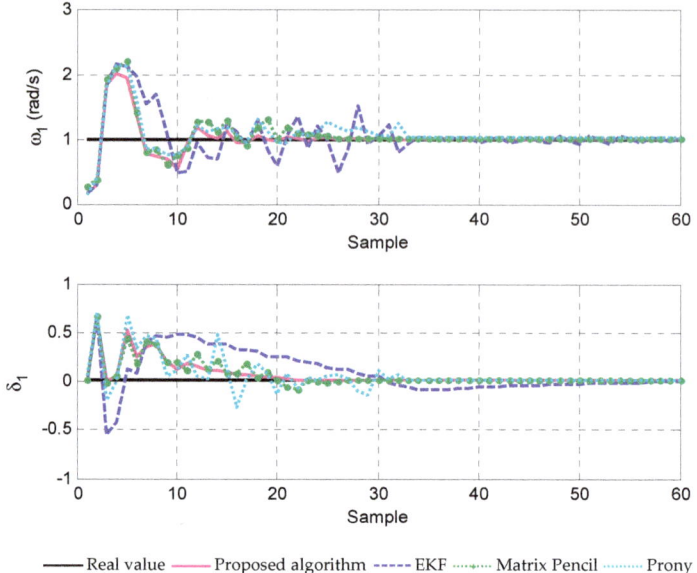

Figure 2. Estimation of ω_1, δ_1 by using different methods.

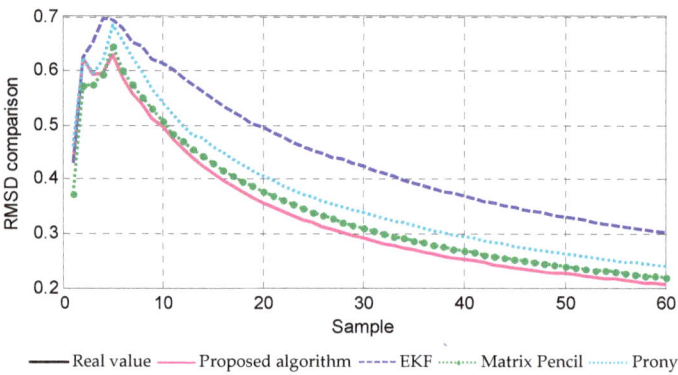

Figure 3. Comparisons of average estimation errors.

In order further to demonstrate the performance of the proposed algorithm, the final identification results of different methods are also shown in Table 1. It is observed from the results that the proposed method with smallest estimation errors than other approaches in [3,13,17]; which proves that the proposed method can estimate the parameters of ringdown signal accurately.

Table 1. Identified results of different methods.

Method	Damping Factor δ_1	Frequency ω_1 (rad/s)
Matrix Pencil [13]	0.0093	1.0104
Prony [17]	0.0087	1.0223
EKF [3]	0.0091	1.0109
Proposed Method	0.0099	1.0001

5.2. Test System 2: Ringdown Signal Composed of Two EDS Signals

In this case, a ringdown signal consisting of two EDS signals is considered:

$$y(t) = y_2(t) + y_3(t) + n(t), \tag{27}$$

where:

$$y_2(t) = e^{-0.005t} \cos(0.2t), \ y_3(t) = e^{-0.001t} \cos(0.6t). \tag{28}$$

and $n(t)$ denotes white noise. The input signal is presented in Figure 4. The estimation results of frequency and damping factor by using different methods are shown in Figures 5 and 6. Figure 7 shows the RMSD comparison of the methods. It can be seen from these simulation results that a better convergence and steady state performance could be obtained if compared with those when using other methods in [3,13,17].

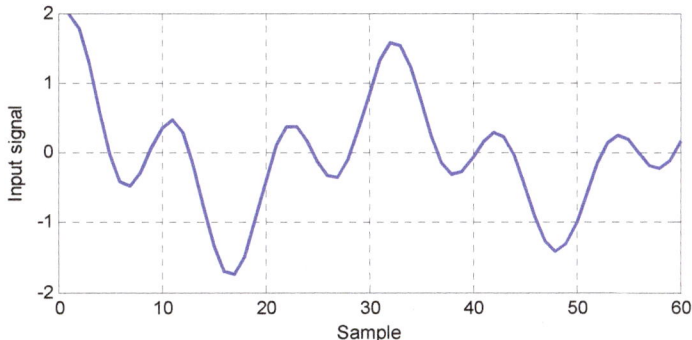

Figure 4. The input signal of test system 2.

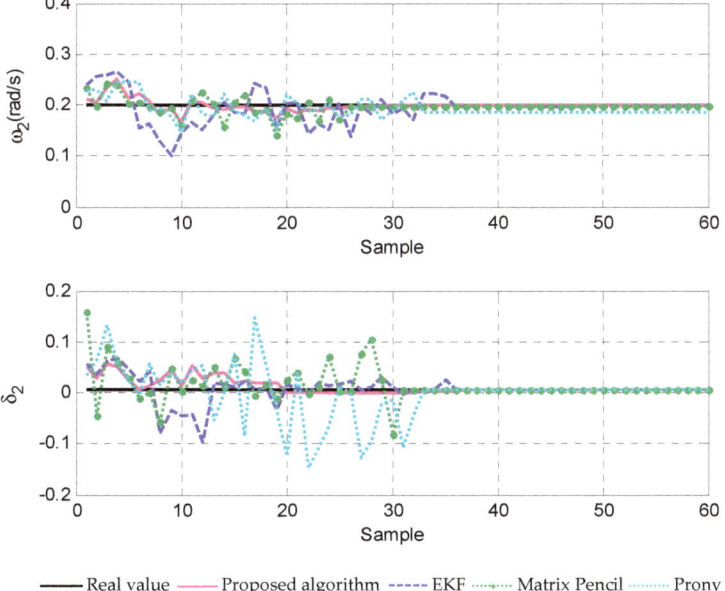

Figure 5. Estimation of ω_2, δ_2 by using different methods.

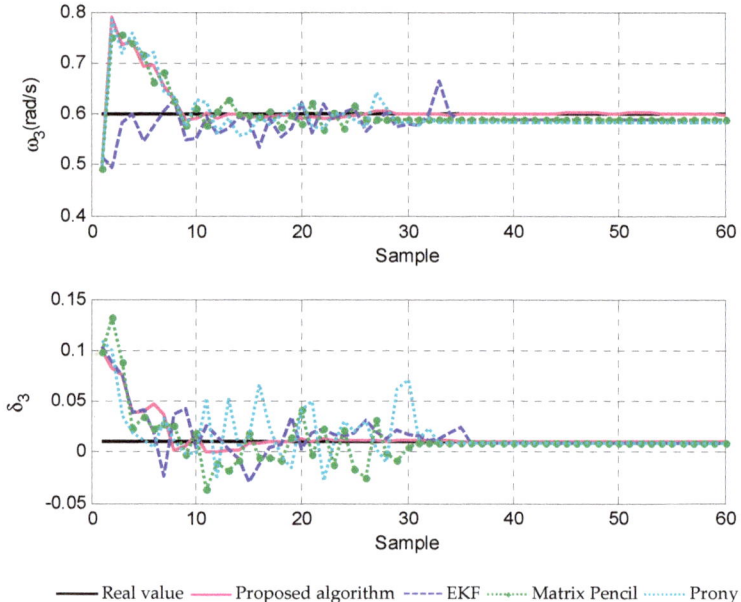

Figure 6. Estimation of ω_3, δ_3 by using different methods.

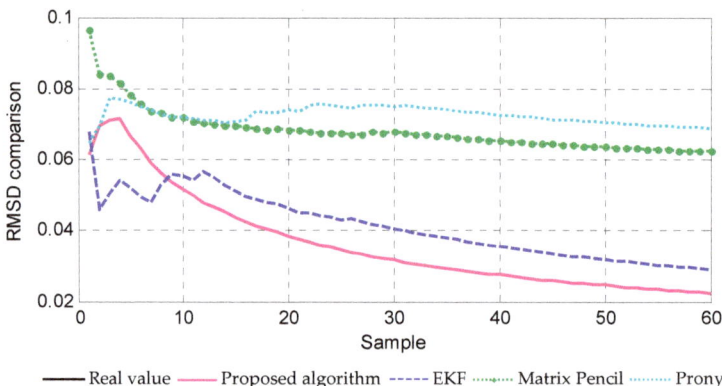

Figure 7. Comparisons of average estimation errors.

In addition, the final identification results of different methods are also provided in Table 2, which further proves superior performance of the proposed method.

Table 2. Identified results of different methods.

Method	Damping factor δ_2/δ_3	Frequency ω_2/ω_3 (rad/s)
Matrix Pencil [13]	0.0049/0.0094	0.1983/0.5887
Prony [17]	0.0045/0.0089	0.1864/0.5843
EKF [3]	0.0048/0.0092	0.1967/0.5876
Proposed Method	0.0050/0.0101	0.2001/0.6001

5.3. Test System 3: WSCC Model

In this case, the WSCC model is taken as the test system [42], which has three generators and nine buses. The network is depicted in Figure 8a. The normal frequency of G2 is 60 HZ, at $t = 5.1$ s, the load at bus 5 is removed. This disturbance causes an inter-area sustained oscillation in G2 with the frequency of $\omega_4 = 2.4$ rad/s, and the damping factor is $\delta_4 = 0.3$. Figure 8b presents the event recorded in the bus frequency signal.

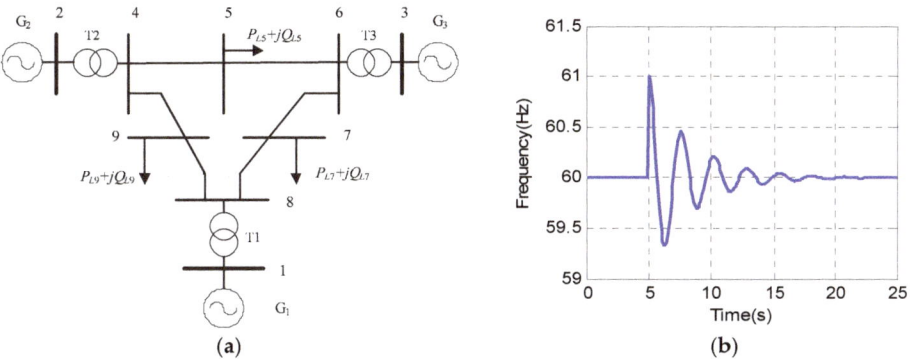

Figure 8. (**a**) The model of WSCC power system; (**b**) The input signal of test system 3.

Figure 9 shows the estimated frequency and damping factor when using the proposed algorithm and other methods. Figure 10 illustrates the RMSD comparison of different methods. It could be found from these simulation results that, in this case, the proposed algorithm can estimate the parameters effectively with better convergence and steady state performance than EKF and other methods. Finally, the final identification results of other methods are also provided in Table 3, which demonstrate the superior performance of the proposed mehtod. Therefore, it can be concluded that the proposed algorithm possesses much more practical value than those shown in previous work.

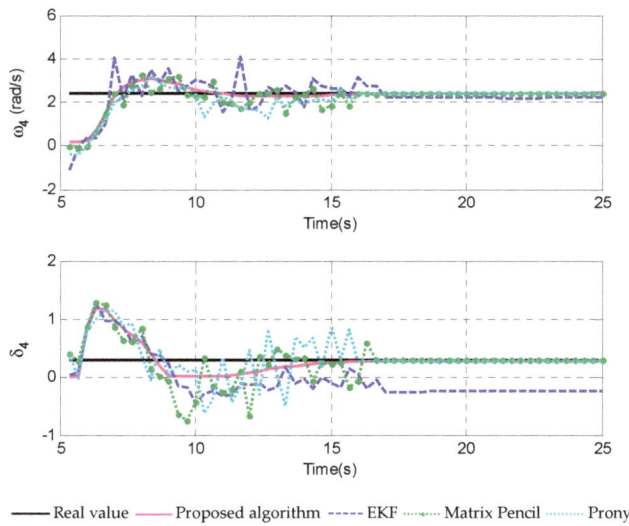

Figure 9. Estimation of ω_4, δ_4 by using different methods.

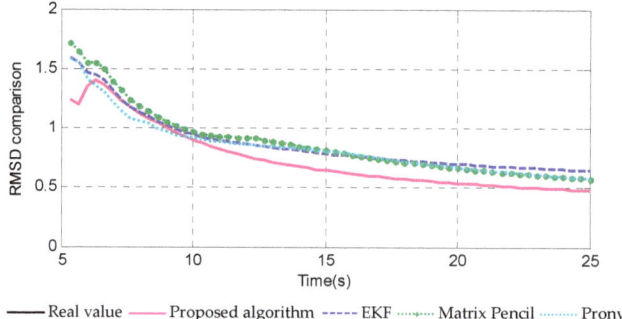

Figure 10. Comparisons of average estimation errors.

Table 3. Identified results of different methods.

Method	Damping Factor δ_4	Frequency ω_4 (rad/s)
Matrix Pencil [13]	0.2808	2.3875
Prony [17]	0.2732	2.3873
EKF [3]	−0.0488	2.3666
Proposed Method	0.2998	2.4000

6. Conclusions

In this paper, with the consideration of the physical constraints on parameters, the constrained parameter estimation problem of ringdown signals has been discussed in detail. Based on the EKF algorithm and the projection method, the constrained estimation problem was converted into a minimum constrained optimization issue. Thus, the constrained optimization issue can be solved successfully by the proposed C&I-PSO with the penalty function method. Finally, the proposed algorithm was successfully applied to estimate the constrained parameters of the ringdown signal. Simulation results have demonstrated in most cases that the proposed algorithm can achieve much better performance and makes more practical sense than using the Matrix Pencil, Prony, and traditional EKF approaches.

Author Contributions: Y.S. performed the experiments and evaluated the data. Y.W. developed the concept, conceived the experiments, and wrote the original manuscript. L.B., Y.H., D.S. and D.P. reviewed and edited the manuscript.

Funding: This work was supported in part by the National Natural Science Foundation of China under Grants 61673161 and 61603122, in part by the Natural Science Foundation of Jiangsu Province of China under GrantBK20161510, in part by the Fundamental Research Funds for the Central Universities of China under Grants 2017B13914 and 2017B655X14, in part by Six Talent Peaks High Level Project of Jiangsu Province under Grant XNY-004, in part by Postgraduate Research & Practice Innovation Program of Jiangsu Province under Grant KYCX17_0484, and in part by the research projects III.17.3.1, III.17.4 of the program of fundamental research of the Siberian Branch of the Russian Academy of Sciences, reg No. AAAA-A17-117030310442-8, No. AAAA-A17-117030310438-1.

Conflicts of Interest: The authors declare no conflict of interest.

References

1. Alizadeh, M.; Kojori, S.; Ganjefar, S. A modular neural block to enhance power system stability. *IEEE Trans. Power Syst.* **2013**, *28*, 4849–4856. [CrossRef]
2. Sidorov, D.; Panasetsky, D.; Šmádl, V. Non-Stationary Autoregressive Model for On-Line Detection of Inter-Area Oscillations in Power Systems. In Proceedings of the IEEE PES Innovative Smart Grid Technologies Conference Europe (ISGT Europe), Gothenberg, Sweden, 11–13 October 2010.

3. Yazdanian, M.; Mehrizi-Sani, A.; Mojiri, M. Estimation of electromechanical oscillation parameters using an extended Kalman filter. *IEEE Trans. Power Syst.* **2015**, *30*, 2994–3002. [CrossRef]

4. Jiang, T.; Yuan, H.; Jia, H.; Zhou, N.; Li, F. Stochastic subspace identification-based approach for tracking inter-area oscillatory modes in bulk power system utilizing synchrophasor measurements. *IET Gener. Transm. Distrib.* **2015**, *9*, 2409–2418. [CrossRef]

5. Wu, F.; Ju, P.; Zhang, X.-P.; Qin, C.; Peng, G.J.; Huang, H.; Fang, J. Modeling, control strategy, and power conditioning for direct-drive wave energy conversion to operate with power grid. *Proc. IEEE* **2013**, *101*, 925–941. [CrossRef]

6. Jiang, T.; Bai, L.; Li, F.; Jia, H.; Hu, Q.; Jin, X. Synchrophasor measurement-based correlation approach for dominant mode identification in bulk power systems. *IET Gener. Transm. Distrib.* **2016**, *10*, 2710–2719. [CrossRef]

7. Yazdanian, M.; Mehrizi-Sani, A.; Mojiri, M. A Novel Approach for Ringdown Detection Using Extended Kalman Filter. In Proceedings of the IECON 2013—39th Annual Conference of the IEEE Industrial Electronics Society, Vienna, Austria, 10–13 November 2013.

8. Ju, P.; Handschin, E.; Karlsson, D. Nonlinear dynamic load modelling: Model and parameter estimation. *IEEE Trans. Power Syst.* **1996**, *11*, 1689–1697. [CrossRef]

9. Peng, J.; Nair, N. Enhancing Kalman filter for tracking ringdown electromechanical oscillation. *IEEE Trans. Power Syst.* **2012**, *27*, 1042–1050. [CrossRef]

10. Pierre, J.; Trudnowski, D.; Donnely, M. Initial results in electromechanical mode identification from ambient data. *IEEE Trans. Power Syst.* **1997**, *12*, 1245–1251. [CrossRef]

11. Mojiri, M.; Bakhshai, A. An adaptive notch filter for frequency estimation of a periodic signal. *IEEE Trans. Auto. Cont.* **2004**, *49*, 314–318. [CrossRef]

12. Mojiri, M.; Karimi-Ghartemani, M.; Bakhshai, A. Processing of harmonics and interharmonics using an adaptive notch filter. *IEEE Trans. Power Deliv.* **2010**, *25*, 534–542. [CrossRef]

13. Jain, V. Filter analysis by use of pencil of functions: Part I. *IEEE Trans. Circuits Syst.* **1974**, *21*, 574–579. [CrossRef]

14. Kung, S.Y.; Arun, K.S.; Rao, B. State-space and singular-value decomposition-based approximation methods for the harmonic retrieval problem. *J. Opt. Soc. Amer.* **1983**, *73*, 1799–1811. [CrossRef]

15. Roy, R.; Paulraj, A.; Kailath, T. ESPRIT—A subspace rotation approach to estimation of parameters of cisoids in noise. *IEEE Trans. Acoust. Speech Signal Process.* **1986**, *34*, 1340–1342. [CrossRef]

16. Kumaresan, R.; Tufts, D. Estimating the parameters of exponentially damped sinusoids and pole-zero modeling in noise. *IEEE Trans. Acoust. Speech Signal Process.* **1982**, *30*, 833–840. [CrossRef]

17. Hildebrand, F.B. *Introduction to Numerical Analysis, ser. Dover Books on Advanced Mathematics*, 2nd ed.; McGraw-Hill: New York, NY, USA, 1956.

18. Wang, D.; Bao, Y.; Shi, J. Online lithium-ion battery internal resistance measurement application in state-of-charge estimation using the extended Kalman filter. *Energies* **2017**, *10*, 1284. [CrossRef]

19. Xia, B.; Wang, H.; Wang, M.; Sun, W.; Xu, Z.; Lai, Y. A new method for state of charge estimation of lithium-ion battery based on strong tracking cubature Kalman filter. *Energies* **2015**, *8*, 13458–13472. [CrossRef]

20. He, H.; Qin, H.; Sun, X.; Shui, Y. Comparison study on the battery SoC estimation with EKF and UKF algorithms. *Energies* **2013**, *6*, 5088–5100. [CrossRef]

21. Zhang, C.; Jiang, J.; Zhang, W.; Sharkh, S.M. Estimation of state of charge of lithium-ion batteries used in HEV using robust extended Kalman filtering. *Energies* **2012**, *5*, 1098–1115. [CrossRef]

22. Xia, B.; Wang, H.; Tian, Y.; Wang, M.; Sun, W.; Xu, Z. State of charge estimation of lithium-ion batteries using an adaptive cubature Kalman filter. *Energies* **2015**, *8*, 5916–5936. [CrossRef]

23. Li, W.; Wei, G.; Liu, Y. Weighted average consensus-based unscented Kalman filtering. *IEEE Trans. Cybern.* **2016**, *46*, 558–567. [CrossRef] [PubMed]

24. Simon, D.; Chia, T. Kalman filtering with state constraints: A survey of linear and nonlinear algorithms. *IET Control Theory Appl.* **2010**, *4*, 1303–1318. [CrossRef]

25. Simon, D.; Chia, T. Kalman filtering with state equality constraints. *IEEE Trans. Aerosp. Electron. Syst.* **2002**, *38*, 128–136. [CrossRef]

26. Julier, S.; Alveolar, J. On Kalman filtering with nonlinear equality constraints. *IEEE Trans. Signal Process.* **2007**, *55*, 2774–2784. [CrossRef]

27. Massicotte, D.; Morawski, R.; Barwicz, A. Incorporation of a positivity constraint into a Kalman filter based algorithm for correction of spectrometric data. *IEEE Trans. Instrum. Meas.* **1995**, *44*, 2–7. [CrossRef]

28. Li, W.; Jia, Y.; Du, J. Variance-constrained state estimation for nonlinearity coupled complex networks. *IEEE Trans. Cybern.* **2018**, *48*, 818–824. [CrossRef] [PubMed]

29. Mendes, R.; Kennedy, J.; Neves, J. The fully informed particle swarm: Simpler, maybe better. *IEEE Trans. Evol. Comput.* **2004**, *8*, 204–210. [CrossRef]

30. Zhang, Y.; Chiang, H. A novel consensus-based particle swarm optimization-assisted trust-tech methodology for large-scale global optimization. *IEEE Trans. Cybern.* **2017**, *47*, 2717–2729. [CrossRef] [PubMed]

31. Chou, S.; Huang, S. Stochastic set-based particle swarm optimization based on local exploration for solving the carpool service problem. *IEEE Trans. Cybern.* **2016**, *46*, 1771–1783. [CrossRef] [PubMed]

32. Chen, S.; Hsin, W. Weighted fuzzy interpolative reasoning based on the slopes of fuzzy sets and particle swarm optimization techniques. *IEEE Trans. Cybern.* **2015**, *45*, 1250–1261. [CrossRef] [PubMed]

33. Lee, J.; Kim, J.; Song, J.; Kim, Y.; Jung, S. A novel memetic algorithm using modified particle swarm optimization and mesh adaptive direct search for PMSM design. *IEEE Trans. Magn.* **2016**, *52*, 1–4. [CrossRef]

34. Abdelhafiz, A.; Behjat, L.; Ghannouchi, F. Generalized memory polynomial model dimension selection using particle swarm optimization. *IEEE Microw. Compon. Lett.* **2018**, *52*, 96–98. [CrossRef]

35. Ratnaweera, A.; Halgamuge, S.; Watson, H. Self-organizing hierarchical particle swarm optimizer with time-varying acceleration coefficients. *IEEE Trans. Evol. Comput.* **2004**, *8*, 240–255. [CrossRef]

36. Tang, Y.; Wang, Z.; Fang, J. Parameters identification of unknown delayed genetic regulatory networks by a switching particle swarm optimization algorithm. *Expert Syst. Appl.* **2011**, *38*, 2523–2535. [CrossRef]

37. Singh, S.; Singh, B. Optimized passive filter design using modified particle swarm optimization algorithm for a 12-pulse converter-fed LCI synchronous motor drive. *IEEE Trans. Ind. Appl.* **2014**, *50*, 2681–2689. [CrossRef]

38. Ting, T.; Rao, M.; Loo, C. A novel approach for unit commitment problem via an effective hybrid particle swarm optimization. *IEEE Trans. Power Syst.* **2006**, *21*, 411–418. [CrossRef]

39. Pehlivanoglu, Y. A new particle swarm optimization method enhanced with a periodic mutation strategy and neural networks. *IEEE Trans. Evol. Comput.* **2013**, *17*, 436–452. [CrossRef]

40. Selvakumar, A.; Thanushkodi, K. A new particle swarm optimization solution to nonconvex economic dispatch problems. *IEEE Trans. Power Syst.* **2007**, *22*, 42–51. [CrossRef]

41. Park, J.B.; Jeong, Y.W.; Shin, J.R.; Lee, K.Y. An improved particle swarm optimization for nonconvex economic dispatch problems. *IEEE Trans. Power Syst.* **2010**, *25*, 156–166. [CrossRef]

42. Anderson, P.M.; Fouad, A.A. *Power System Control and Stability*, 2nd ed.; IEEE Press: New York, NY, USA, 2003.

43. Wang, Y.; Sun, Y.; Wei, Z.; Sun, G. Parameters estimation of electromechanical oscillation with incomplete measurement information. *IEEE Trans. Power Syst.* **2018**. [CrossRef]

Article

Reactive Power Dispatch Optimization with Voltage Profile Improvement Using an Efficient Hybrid Algorithm [†]

Zahir Sahli [1,*]**, Abdellatif Hamouda** [2]**, Abdelghani Bekrar** [3] **and Damien Trentesaux** [3]

[1]   QUERE Laboratory, Electrical Engineering Department, University Ferhat Abbas Setif 1, Setif 19000, Algeria
[2]   QUERE Laboratory, Optics and fine mechanics Institute, University Ferhat Abbas Setif 1, Setif 19000, Algeria; a_hamouda1@yahoo.fr
[3]   LAMIH-UMR CNRS, University of Valenciennes, 59313 Valenciennes, France; abdelghani.bekrar@univ-valenciennes.fr (A.B.); damien.trentesaux@univ-valenciennes.fr (D.T.)
*   Correspondence: sah_zah@yahoo.fr
†   The present work is an extension of the paper "Hybrid PSO-tabu Search for the Optimal Reactive Power Dispatch Problem" presented at the IECON 2014-40th Annual Conference of the IEEE Industrial Electronics Society, 29 October–1 November 2014, Dallas, TX, USA.

Received: 12 July 2018; Accepted: 13 August 2018; Published: 16 August 2018

Abstract: This paper presents an efficient approach for solving the optimal reactive power dispatch problem. It is a non-linear constrained optimization problem where two distinct objective functions are considered. The proposed approach is based on the hybridization of the particle swarm optimization method and the tabu-search technique. This hybrid approach is used to find control variable settings (i.e., generation bus voltages, transformer taps and shunt capacitor sizes) which minimize transmission active power losses and load bus voltage deviations. To validate the proposed hybrid method, the IEEE 30-bus system is considered for 12 and 19 control variables. The obtained results are compared with those obtained by particle swarm optimization and a tabu-search without hybridization and with other evolutionary algorithms reported in the literature.

Keywords: optimal reactive power dispatch; loss minimization; voltage deviation; hybrid method; tabu search; particle swarm optimization

---

1. Introduction

Power systems are complex networks (Figure 1) used for generating and transmitting electric power, which is expected to consume minimal resources while providing maximum security and reliability. Optimal reactive power dispatch (ORPD) is a specific optimal power flow (OPF) problem that has a significant influence on the secure and economic operation of power systems [1,2]. The objectives of ORPD in power systems are to minimize active power losses and to improve the voltage profile by minimizing the load bus voltage deviations while satisfying a given set of operating and physical constraints. The ORPD then provides optimal control variable settings such as (generator bus voltages, output of static reactive power compensators, transformer tap-settings, shunt capacitors, etc.) [3,4]. Due to its significant influence on the secure and economic operation of power systems, ORPD has attracted increasing interest from electric power suppliers. Many approaches for solving the ORPD problem have been described in the literature: initially, several classical optimization methods such as the gradient-based approach [5,6], linear programming [7], non-linear programming [8,9], quadratic programming [10], and interior point [11], were used to solve this problem. However, these methods have some disadvantages in solving complex ORPD problems, namely, premature convergence properties, algorithmic complexity and the local minima trap [12].

In order to overcome these drawbacks, researchers have applied evolutionary and meta-heuristic algorithms such as the genetic algorithm (GA) [13], differential evolution (DE) [14–16], evolutionary programming (EP) [17], stud krill herd algorithm (SKHA) [18], whale optimization algorithm (WOA) [19], backtracking search algorithm (BSA) [20], Jaya Algorithm [21], moth-flame optimization (MFO) [22], symbiotic organism search (SOS) [23] and particle swarm optimization (PSO) [24,25]. PSO, in particular, has received increased attention from researchers because of its novelty and searching capability. It was developed through simulation of a simplified social system and has been found to be robust in solving continuous non-linear optimization problems [1]. Generally, PSO has a more global searching ability at the beginning of the run and a local search near the end of the run [1]. The PSO technique can generate high-quality solutions and has a more stable convergence characteristic than other stochastic methods. However, when solving complex multimodal problems, PSO can be trapped in local optima [26]. To overcome this drawback, PSO performance can be enhanced with few adjustments. Hybridization is one of these modifications or techniques which, nowadays, is a popular idea being applied to evolutionary algorithms in order to increase their efficiency and robustness [27].

Recently, hybrid PSO has provided promising results for problems such as the power loss minimization problem [28]. The novelty of this paper is that an efficient hybrid PSO with the tabu search (PSO-TS) method is implemented to solve the ORPD problem with two distinct objective functions, namely, active power loss minimization and the sum of the load bus voltage deviations. The proposed optimization approach was tested on an IEEE 30-bus system considering two case studies. To demonstrate the effectiveness of the proposed PSO-TS algorithm, the obtained results were compared with TS, PSO and with several methods published in the literature, namely:

- Biogeography Based Optimization (BBO) technique: This method has been developed based on the theory of biogeography which is nature's way of distributing species. It is mainly based on migration and mutation [29].
- Differential Evolution (DE) algorithm: Similar to the genetic algorithm, the DE algorithm is a population-based algorithm that uses crossover, mutation and selection operators [14].
- General passive congregation PSO (GPAC), local passive congregation PSO (LPAC) and coordinated aggregation (CA) are a development of the PSO algorithm using recent advances in swarm intelligence. GPAC and LPAC algorithms are based on the global and local-neighborhood variant PSOs, respectively, and the CA technique is based on the coordinated aggregation observed in swarms [28].
- CLPSO method introduces learning strategy in PSO. In this method, for each particle, besides its own best particle (*pbest*), other particles' *pbests* are used as exemplars. Each particle learns from all potential particles' *pbests* in the swarm [9].
- Interior point (IP) method is a conventional technique based on the primal-dual algorithm [11].

This paper is organized as follows: In Section 2, a brief description and mathematical formulation of optimal reactive power dispatch (ORPD) problem are provided. The hybrid PSO-tabu search approach is described in Section 3 along with a short description of the PSO algorithm and tabu search method. Simulation results and comparison with other methods are given in Section 4. Finally, a conclusion with future works is outlined in Section 5.

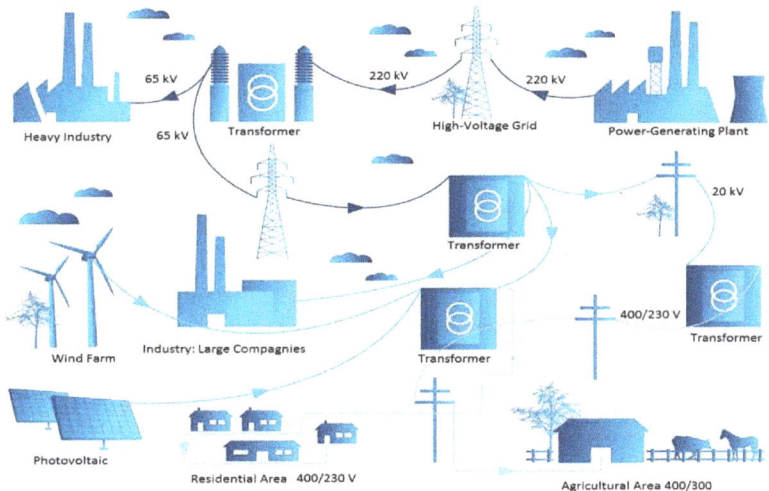

Figure 1. Schematic diagram of an electric power system.

2. ORPD Problem Formulation

ORPD is a highly constrained non-linear optimization problem in which a specific objective function is to be minimized while satisfying a number of nonlinear equality and inequality constraints. The objectives of the reactive power dispatch problem considered here are to minimize separately the whole system active power losses (P_{loss}) and the sum of the load bus voltage deviations (SVD) with the intention of improving the voltage profile of the power system. These objectives are achieved by proper adjustment of the control variables like generator voltage magnitudes, shunt capacitor sizes and transformer tap settings. The ORPD problem can be then stated as follows [14]:

For the power loss minimization:

$$\left\{ \begin{array}{l} \min J_1(x, u) \ subject\ to \\ \quad g(x, u) = 0 \\ \quad h(x, u) \le 0 \end{array} \right. \tag{1}$$

For the voltage deviation minimization:

$$\left\{ \begin{array}{l} \min J_2(x, u) \ subject\ to \\ \quad g(x, u) = 0 \\ \quad h(x, u) \le 0 \end{array} \right. \tag{2}$$

where:

- $J_1(x,u)$ and $J_2(x,u)$ are the transmission active power losses and SVD objective functions, respectively.
- g and h are the set of equality and inequality constraints, respectively.
- x is the state or dependent variables vector.
- u is the control or independent variables vector.

In this study, all control variables have been considered as continuous variables. The following sections outline this problem by detailing the objective functions.

2.1. Objective Functions

2.1.1. Power Losses Minimization

The first objective to be minimized is the system transmission active power losses. This objective function is expressed as follows [1]

$$J_1(x, u) = \sum_{k=1}^{N_L} g_k (V_i^2 + V_j^2 - 2 V_i V_j cos\theta_{ij}) \tag{3}$$

where:

- N_L is the number of transmission lines.
- V_i and V_j are the voltage magnitude at buses i and j, respectively.
- g_k is the conductance of branch k between buses i and j.
- θ_{ij} is the voltage angle difference between bus i and bus j.

The elements of the state variables vector "x" are load buses voltage (V_L), generators reactive power output (Q_G) and lines apparent power flow (S_L). The control variables vector "u" includes the generation buses voltage (V_G), the transformer tap settings (T) and the shunt VAR compensators (Q_C). Accordingly, the x vector can be written as follows:

$$x^T = [V_{L_1} \ldots V_{L_{N_{PQ}}}, Q_{G_1} \ldots Q_{G_{N_G}}, S_{L_1} \ldots S_{L_{N_L}}] \tag{4}$$

where N_G is the number of generators; N_{PQ} is the number of PQ buses (load buses);
u can be expressed as:

$$u^T = [V_{G_1} \ldots V_{G_{N_G}}, T_1 \ldots T_{N_T}, Q_{C_1} \ldots Q_{C_{N_C}}] \tag{5}$$

where:

- N_T is the number of tap regulating transformers.
- N_C is the number of shunt VAR compensations.

2.1.2. Minimization of Voltage Deviation

The bus voltage is one of the most important security and service quality indices. Improving the voltage profile can be achieved by minimizing the load buses voltage deviation, which is modeled as follows [29]:

$$J_2(x, u) = \sum_{1}^{N_{PQ}} \left| V_{L_i} - V_{ref} \right| \tag{6}$$

where:

- V_{Li} is the voltage magnitude at load bus i.
- V_{ref} is the voltage reference value which is equal to 1 p.u.

2.2. Problem Constraints

The considered objective functions for the ORPD problem are subject to several equality and inequality constraints [1] which will be detailed below.

2.2.1. Equality Constrains

These constraints reflect the physical laws governing the electrical system known as power flow equations. They are the expression of the balance between load demand (power loss included) and generated power. The power flow equations are given by:

$$P_{Gi} - P_{Di} - V_i \sum_{j=1}^{N_B} V_j (G_{ij} \cos \theta_{ij} + B_{ij} \sin \theta_{ij}) = 0 \tag{7}$$

$$Q_{Gi} - Q_{Di} - V_i \sum_{j=1}^{N_B} V_j (G_{ij} \sin \theta_{ij} - B_{ij} \cos \theta_{ij}) = 0 \tag{8}$$

where:

- P_{Gi}, Q_{Gi} are the respective active and reactive power of the ith generator.
- P_{Di}, Q_{Di} are the respective active and reactive power demand at bus i.
- N_B is the total number of buses; B_{ij}, G_{ij} are real and imaginary parts of $(i,j)^{th}$ element of the bus admittance matrix.

2.2.2. Inequality Constraints.

Inequality Constraints on Security Limits

Some limits are imposed for security purposes:

- Active power generated at slack bus

$$P_{G,slack}^{min} \leq P_{G,slack} \leq P_{G,slack}^{max} \tag{9}$$

- Load bus voltage

$$V_{L_i}^{min} \leq V_{L_i} \leq V_{L_i}^{max} \quad i \in N_{PQ} \tag{10}$$

- Generated reactive power

$$Q_{Gi}^{min} \leq Q_{Gi} \leq Q_{Gi}^{max} \quad i \in N_G \tag{11}$$

- Thermal limits: the apparent power flowing in line "L" must not exceed the maximum allowable apparent power flow value (S_L^{max})

$$S_L \leq S_L^{max} \quad L \in N_L \tag{12}$$

Inequality Constraints on Control Variable Limits

The different control variables are bounded as follows:

- Generator voltage limits

$$V_{G_i}^{min} \leq V_{G_i} \leq V_{G_i}^{max} \quad i \in N_{PV} \tag{13}$$

- Transformer tap limits

$$T_i^{min,} \leq T_i \leq T_i^{max} \quad i \in N_T \tag{14}$$

- Shunt capacitor limits

$$Q_{C_i}^{min} \leq Q_{C_i} \leq Q_{C_i}^{max} \quad i \in N_C \tag{15}$$

where:

- $P_{G,slack}$ is the real power generation at slack bus.

- V_{Gi} is the voltage magnitude at generator bus i.
- T_i is the tap ratio of transformer i.
- Q_{ci} is the reactive power compensation source at bus i.
- N_{PQ} is the number of PQ bus.
- $(.)^{max}$ and $(.)^{min}$ are the upper and lower the limits of the considered variables, respectively.

The objective functions, equality and inequality constraints are non-linear functions and they depend upon control variables. Therefore, ORPD is a constrained non-linear optimization problem with multiple local minima [30]. The equality constraints given by Equations (7) and (8) are met by solving the load-flow problem. The inequality constraints given by Equations (13)–(15) should be maintained during the solution evolution, while the inequality Equations (9)–(12) should be handled by additional techniques.

3. Proposed Hybrid Algorithm

Hybridization is a way of combining two techniques in a judicious manner, so that the resulting algorithm contains positive features of both algorithms [27]. The success of the meta-heuristics optimization algorithms depends to a large extent on the careful balance between two conflicting goals: exploration (diversification) and exploitation (intensification) [27]. In order to achieve these two goals, the algorithms use either local search techniques, global search approaches, or an integration of both, commonly known as hybrid methods [27]. For the ORPD problem, different hybridizations with PSO have been used to improve the algorithm's performance by avoiding premature convergence. For instance, PSO has been hybridized with the linear interior point method [31], fuzzy logic [32,33], Pareto optimal set [34], direct search method [35], differential evolution [36], a multi-agent systems [1], imperialist competitive algorithm [37], genetic algorithm [38] and eagle strategy [39]. Tabu search was used to solve OPF [40] and optimal reactive power planning [41] problems, but to the best of our knowledge, the hybridization of TS with PSO has never been used even though it was effective in solving other optimization-constrained problems [42]. Both algorithms (PSO, TS) and their hybridization (PSO-TS) for solving the ORPD problem are discussed in the following sections.

3.1. Particle Swarm Optimization

The concept of PSO was first suggested by Kennedy and Eberhart in 1995 [43]. PSO is a population-based evolutionary computation technique. The main idea is to evolve the population (particles) of initial solutions in a search space in order to find the best solution. This evolution is an analogy of the behavior of some species as they look for food, like a flock of birds or a school of fish [44]. These particles move through the search domain with a specified velocity in search of optimal solution. Each particle maintains a memory which helps it to keep track of its previous best position. The positions of the particles are distinguished as personal best and global best.

The swarm of particles evolves in the search space by modifying their velocities according to the following equations [27]:

$$v_i^{k+1} = w_i v_i^k + c_1 rand \times \left(pbest_i - x_i^k \right) + c_2 rand \times \left(gbest - x_i^k \right) \tag{16}$$

where:

- v_i^k is the current velocity of particle i at iteration k.
- w_i is the inertia weight.
- $rand$ is a random number between 0 and 1.
- c_1 and c_2 are the acceleration coefficients.
- $pbest_i$ is the best position of the current particle achieved so far.
- $gbest$ is the global best position achieved by all informants.
- x_i^k is the current position of particle i at iteration k.

The new position of each particle is given by the following equation:

$$x_i^{k+1} = x_i^k + v_i^{k+1} \tag{17}$$

The inertia weighting factor for the velocity of particle i is defined by the inertial weight approach [28].

$$w_i = w_{max} - \frac{w_{max} - w_{min}}{iter_{max}} \times k \tag{18}$$

where:

- $iter_{max}$ is the maximum number of iterations.
- k is the current number of iteration.
- w_{max} and w_{min} are the upper and lower limits of the inertia weighting factor, respectively.

The efficiency of PSO has been proved for a wide range of optimization problems. However, constrained non-linear optimization problems have not been widely studied with this method. Hu and Eberhart were the first to try to adapt PSO to constrained non-linear problems [45]. The difficulty in adapting meta-heuristics mainly involves the question of how to preserve the feasibility of solutions during different iterations.

A variety of approaches can be used to deal with feasibility in constrained non-linear optimization problems, which largely fall into two classes:

- Penalty function approaches, and
- Approaches preserving feasibility throughout evolutionary computation,

Each method has its advantages and disadvantages. A penalty function approach is used in this paper due to its simplicity of implementation and its proven efficiency for many constrained non-linear optimization problems [46]. Conversely, feasibility preserving methods are highly time-consuming. To use a penalty function method, a penalty factor associated with each violated constraint is added to the objective function in order to penalize infeasible solutions [47]. Therefore, the optimum is found when all the constraints are respected and the objective function is minimized. The ORPD objective function is then modified as follows [48]:

$$F_T = F + K_P(P_{G,slack} - P_{G,slack}^{lim})^2 + K_V \sum_{i=1}^{N_{PQ}} (V_{L_i} - V_{L_i}^{lim})^2$$
$$+ K_Q \sum_{i=1}^{N_G} (Q_{Gi} - Q_{Gi}^{lim})^2 + K_S \sum_{i=1}^{N_L} (S_{L_i} - S_{L_i}^{lim})^2 \tag{19}$$

where F is equal to J_1 given by Equation (3) in the case of the power losses minimization or equal to J_2 given by Equation (6) in the case of the voltage deviations minimization; K_P, K_V, K_Q and K_S are the penalty factors of the slack bus generator, bus voltage limit violation, generator reactive power limit violation, and line flow violation, respectively.

$P_{G,slack}^{lim}$, $V_{L_i}^{lim}$, Q_{Gi}^{lim} and $S_{L_i}^{lim}$ are defined as follows:

$$P_{G,slack}^{lim} = \begin{cases} P_{G,slack}^{min} & if \quad P_{G,slack} < P_{G,slack}^{min} \\ P_{G,slack}^{max} & if \quad P_{G,slack} > P_{G,slack}^{max} \end{cases} \tag{20}$$

$$V_{L_i}^{lim} = \begin{cases} V_{L_i}^{min} & if \quad V_{L_i} < V_{L_i}^{min} \\ V_{L_i}^{max} & if \quad V_{L_i} > V_{L_i}^{max} \end{cases} \tag{21}$$

$$Q_{Gi}^{lim} = \begin{cases} Q_{Gi}^{min} & if \quad Q_{Gi} < Q_{Gi}^{lim} \\ Q_{Gi}^{max} & if \quad Q_{Gi} > Q_{Gi}^{max} \end{cases} \tag{22}$$

$$S_{L_i}^{lim} = \begin{cases} S_{L_i}^{max} & if \quad S_{L_i} > S_{L_i}^{max} \\ 0 & if \quad S_{L_i} \leq S_{L_i}^{max} \end{cases} \tag{23}$$

3.2. Tabu Search Method

In 1986, Fred Glover proposed a new approach, called "tabu search" (TS). TS is a meta-heuristic that guides a local heuristic search procedure to explore the solution space beyond local optimality. This technique uses an operation called "move" to define the neighborhood of any given solution. One of the main components of TS is its use of adaptive memory, which creates a more flexible search behavior [49,50]. The simplest of these processes consists of recording in a tabu list the features of the visited regions in the space search, which provides a means to avoid revisiting already inspected solutions and thus avoid becoming trapped in local optima. Generally, the advantages of the TS optimization technique can be summarized as follows [40]:

- TS is characterized by its ability to avoid entrapment in a local optimal solution and to prevent the same solution being found by using the flexible memory of the search history.
- TS uses probabilistic transition rules to make decisions, rather than deterministic ones. Hence, TS is a kind of stochastic optimization algorithm that can search a complicated and uncertain area to find the global optimum. This makes TS more flexible and robust than conventional methods.
- TS uses adaptive memory processes for guiding the seeking in the problem search space. Therefore, it can easily deal with non-smooth, non-continuous and non-differentiable objective functions.

3.3. Hybrid PSO-Tabu Search Approach Applied to ORPD

Several arguments support the hybridization of PSO with TS. Firstly, PSO is a global population-based algorithm while TS proposes fast local search mechanism. Secondly, the incorporation of TS into PSO enables the algorithm to maintain population diversity. Finally, TS is integrated to prevent PSO from falling into local optima. To this end, TS is proposed to serve a local optimizer of the best local solutions (*pbest*). The *pbest* solutions of PSO are the inputs of the TS diversification procedure. For each solution "*s*", a list of neighborhoods is defined. Candidate solutions from these neighborhoods are examined and the best one becomes the new current solution that replaces "*s*". The move leading to the solution "*s*" is saved in the tabu list, called *best_list*. This process is repeated to produce successive new solutions until a defined stopping criterion is satisfied.

The neighborhoods of a solution "*s*" are defined by hyper-rectangles introduced in [51]. A hyper-rectangle of "*s*" with a radius "*r*" is the space containing solutions (*s′*) such that the distance between *s* and (*s′*) is less than "*r*". To generate m neighbors for the solution "*s*", *m* hyper-rectangles centered on "*s*" are created, and a point is randomly chosen from each of them. The best of the m chosen points then replaces "*s*". The search procedure of PSO-TS algorithm will terminate whenever the predetermined maximum number of generations is reached, or whenever the global best solution does not improve over a predetermined number of iterations. The diversification procedure is outlined in Algorithm 1 while, the general and detailed flowcharts of the proposed PSO-tabu search are given in Figures 2 and 3, respectively.

Algorithm 1 Tabu search procedure (Diversification)
Inputs
pbest; // best historical solution of particles
pbestval; solutions values
m; //neighborhood size
r; //radius of hyper-rectangles
eps; //threshold for accepting new solution
best_list = (*pbest*, *r*); // Initializing the tabu list *best_list*
Repeat
For each solution $s(V_{Gi}, T_i, Q_{ci})$ in *pbest*
//generation of *m* neighbors
i = 1
While *i* <= *m*
Generate the hyper-rectangle of radius *r*i* around *s*,
choose randomly a solution *NS* in the hyper- rectangle
If *NS* ∉ *best_list* then
add the move to *best_list*;
if *eval(NS)-pbestval(s)* ≤ *eps* then update *pbestval* and *pbest*
s = *NS*,
End if
i = *i* + 1;
End While
Until (stoping criteria)

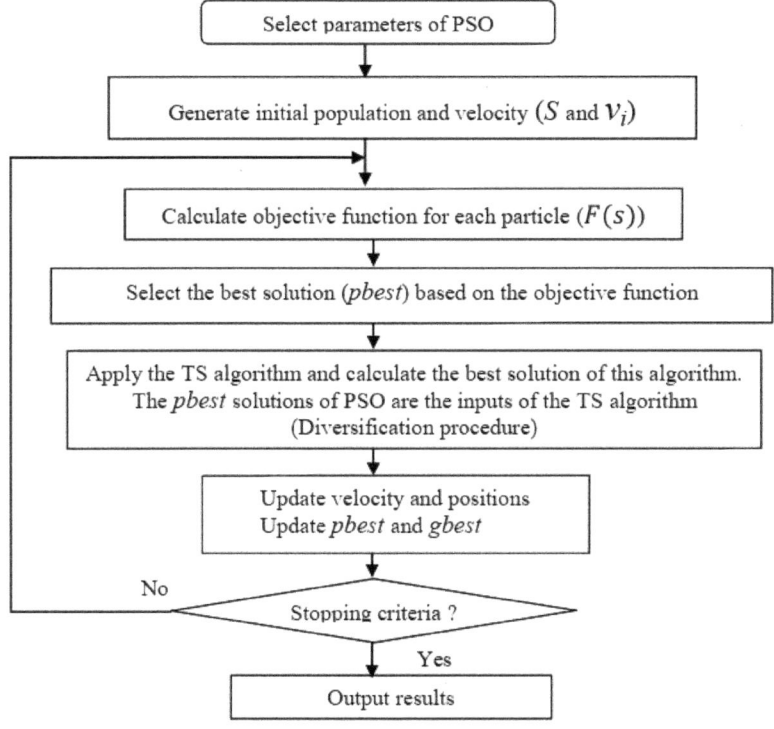

Figure 2. General flowchart of PSO-TS method.

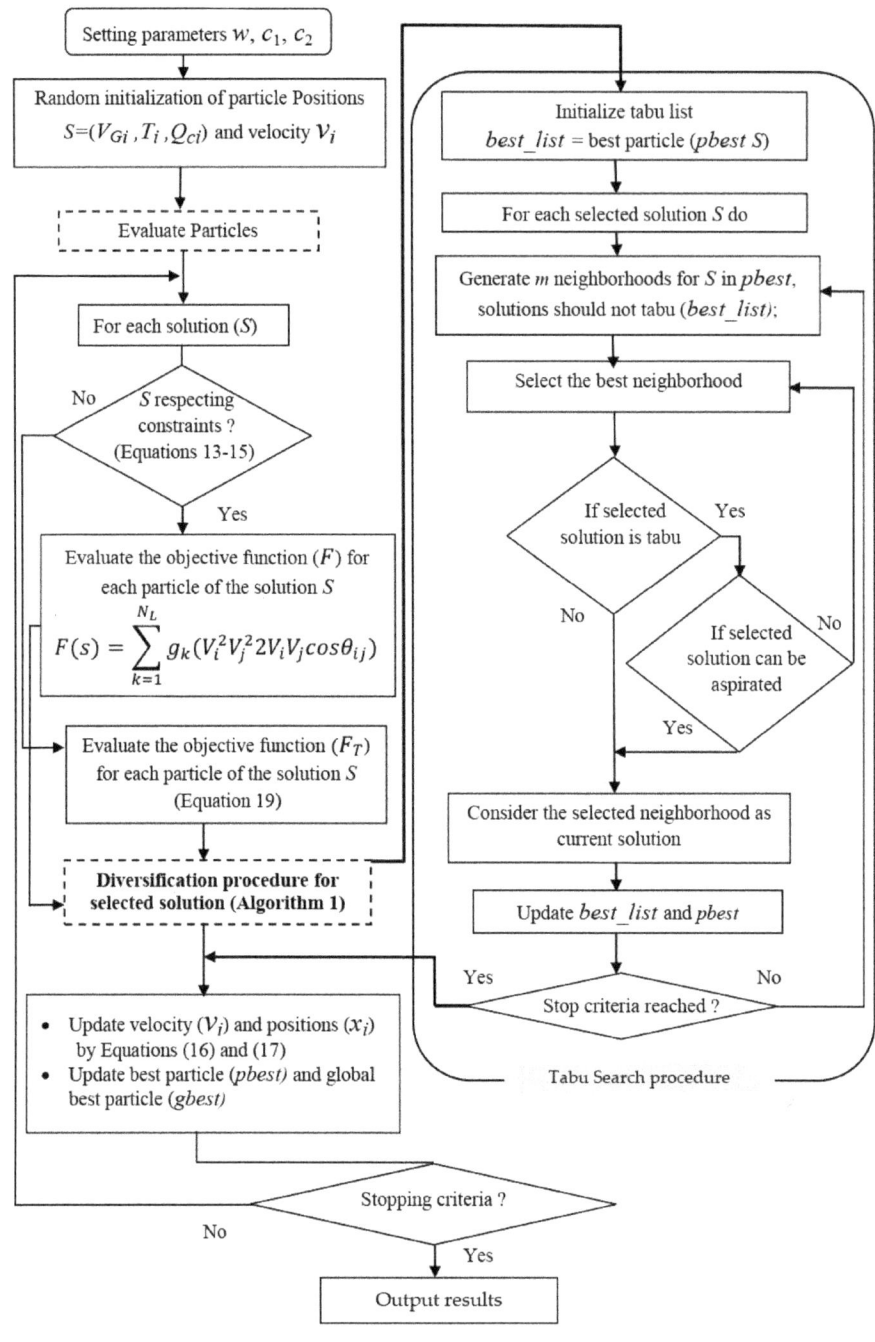

Figure 3. Detailed flowchart of the PSO-TS method.

4. Application and Results

In this study, the proposed PSO-TS based reactive power optimization approach was applied to an IEEE 30-bus power system (Figure 4). For the purpose of comparison, two reactive power injection schemes have been considered:

- Case 1: IEEE 30 bus system with 12 control variables [52].
- Case 2: IEEE 30 bus system with 19 control variables [6].

For both cases, two objective functions are considered: active power loss (Equation (3)) and bus voltage deviation (Equation (6)). In the study, all inequality constraints (Equations (9)–(15)) were taken into consideration. This is significantly different from related studies where only part of the inequality constraints is considered. The simulations were carried out using Matlab 7.3 on a Pentium® 3.4 GHz computer with 1 GB total memory. The PSO-TS parameter selection is a challenging task not only for this algorithm but also for other meta-heuristic algorithms. The parameter settings used in the proposed PSO-TS algorithm are determined through extensive experiments, including initial inertia weight, acceleration factors, number of generations, swarm size, tabu list length, total number of neighborhood and radius of neighborhood. Based on these results, the control parameter settings shown in Table 1 have been used in the proposed PSO-TS algorithm and for all simulation studies in both objective functions.

Table 1. Control parameter settings.

Parameters	Value
Initial inertia weight w	0.9 and decreased to 0.4
Acceleration factor c_1	2
Acceleration factor c_2	2
Maximum number of generations (PSO)	200
Swarm size	20
Tabu list length	7
Number of neighborhood	3
Radius of neighborhood	0.1
Maximum number of generations (TS)	1000

4.1. Case 1: IEEE 30 Bus with 12 Control Variables

This system contains six generator units connected to buses 1, 2, 5, 8, 11 and 13; four regulating transformers connected between the line numbers 6–9, 6–10, 4–12 and 27–28; and two shunt compensators connected to bus numbers 10 and 24. The transmission feeder numbers is of 41. The transmission line data and loads were taken from [52] and are shown in the Appendix A (Tables A1 and A2). The generator voltages, transformer tap settings and VAR injection of the shunt capacitors were considered as control variables. The voltage magnitudes of all the buses were between 0.95 and 1.1 p.u, the transformer tap settings were within the range of 0.9–1.1 p.u and the shunt capacitor sizes were within the interval of 0 to 30 MVAR [28]. There are 12 control variables in this case, namely, 6 generator voltages, 4 transformer taps and 2 capacitor banks. Two objective functions are considered in order to demonstrate the effectiveness of the proposed algorithm. The proposed PSO-TS algorithm is used to minimize separately the system active power losses and the voltage deviation of all load buses.

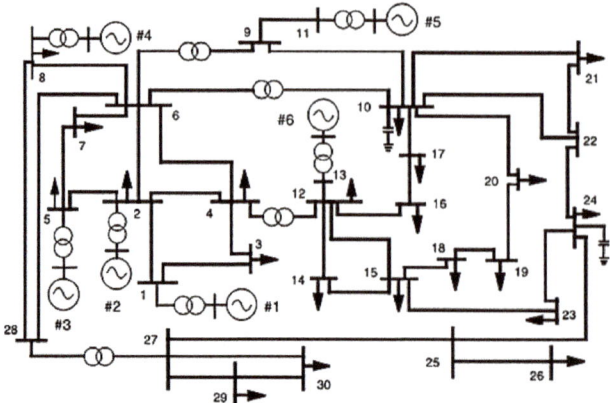

Figure 4. Single-line diagram of IEEE 30 bus test system.

4.1.1. Power Loss Minimization

The objective in this case is to minimize the total active power losses. Before minimization, the total power losses were 5.2783 MW. Table 2 summarize the results of the optimal settings and the system power losses obtained by the proposed PSO-TS approach, each of the two our techniques PSO and TS considered alone and different methods reported in [28,29], namely, CA, IP-OPF, LPAC, GPAC and BBO. These results show that the dispatch optimal solutions determined by the PSO-TS led to better results. Active power losses are lower than those found by TS, PSO and considered reference. Using PSO-TS algorithm, power losses range from 5.2783 MW to 4.6304 MW, indicating a reduction of 12.27%, while PSO and TS taken alone reduce power losses by only 1.03% and 5.61%, respectively. For the other optimization algorithms, the best result is given by BBO algorithm [29] which reduces the losses by 5.93%. It can be concluded that the proposed PSO-TS method is able to determine the near-global optimal solution. At the same time, the proposed method succeeded in keeping the dependent variables within their limits. Figure 5 shows the supremacy of PSO-TS algorithm over the other methods. The convergence characteristics of power loss objective function for this case are plotted in Figure 6. As the hardware and the software environments significantly affect the computational time, it is not possible to compare the computational time requirements of the different methods unless all the methods are run on the same hardware and programmed using the same environment. As a rough guide, however, the average time taken by PSO-TS in this case is 19 s.

Table 2. Experimental results of TS, PSO and PSO-TS algorithms (Case 1).

Control Variables	CA	IP-OPF	LPAC	GPAC	BBO	TS	PSO	PSO-TS
V_1	1.02282	1.10000	1.02342	1.02942	1.1000	1.0684	1.1000	1.0992
V_2	1.09093	1.05414	0.99893	1.00645	1.0943	1.0933	1.0943	1.0948
V_5	1.03008	1.10000	0.99469	1.01692	1.0804	1.0893	1.1000	1.0766
V_8	0.95000	1.03348	1.01364	1.03952	1.0939	1.0853	1.1000	1.0977
V_{11}	1.04289	1.10000	1.01647	1.03952	1.1000	1.0017	0.9505	1.0837
V_{13}	1.03921	1.01497	1.01101	1.04870	1.1000	1.0780	1.1000	1.0754
T_{6-9}	1.07894	0.99334	1.04247	1.04225	1.1000	0.9979	1.0547	0.9257
T_{6-10}	0.94276	1.05938	0.99432	0.99417	0.9058	0.9008	1.1000	1.0291
T_{4-12}	1.00064	1.00879	1.00061	1.00218	0.9521	1.0337	0.9000	0.9265
T_{27-28}	1.00693	0.99712	1.00694	1.00751	0.9638	0.9441	0.9468	0.9422
Q_{Sh10}	0.15232	0.15253	0.17737	0.17267	0.2891	0.1395	0.3000	0.2864
Q_{Sh24}	0.06249	0.08926	0.06172	0.06539	0.1007	0.1838	0.0000	0.1363
P_{loss} (MW)	5.09209	5.10091	5.09212	5.09226	4.9650	5.2240	4.9819	4.6304

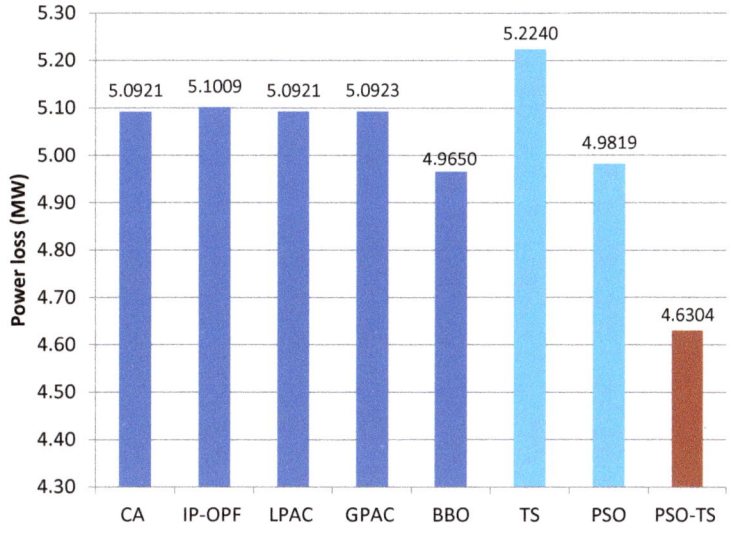

Figure 5. Comparative graph of the power losses (Case 1).

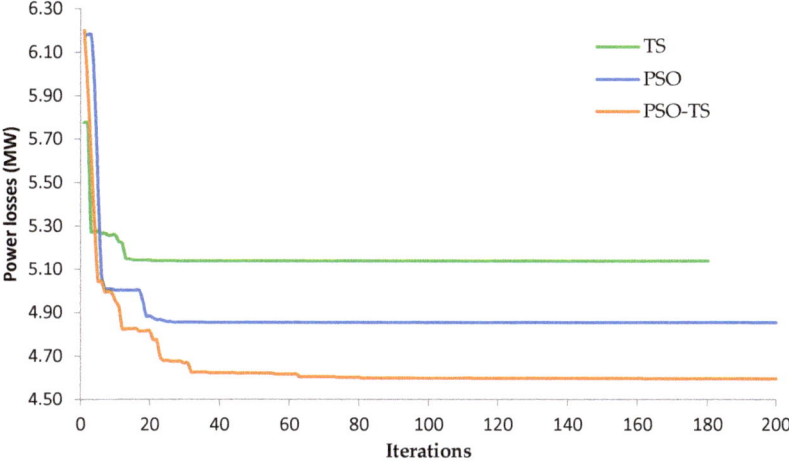

Figure 6. Convergence characteristic of the power losses (Case 1).

4.1.2. Voltage Deviation Minimization

The objective in this case is the minimization of the voltage deviations in order to improve the system voltage profile. The SVD and the optimal setting of control variables obtained by our PSO, TS, PSO-TS and different considered methods (CA, IP-OPF LPAC, GPAC and BBO) are listed in Table 3. The convergence characteristics of the objective BBOe function obtained by TS, PSO and PSO-TS are illustrated in Figure 7. Before minimization, the SVD was 0.619 p.u. As shown in Table 3, the obtained SVD using the proposed PSO-TS hybrid approach is 0.1113 p.u which means a reduction of 82.02% while the ones given by the mentioned methods are, respectively, 80.21%, 79.97%, 79.42%, 80.71%, 69.73% and 79.40%. These results clearly indicate that PSO-TS outperforms other methods in term of solution quality (see Figure 8).

Table 3. Experimental results of TS, PSO and PSO-TS algorithms (Case 1).

Control Variables	CA	IP-OPF	LPAC	GPAC	BBO	TS	PSO	PSO-TS
V_1	1.0890	1.10000	1.03879	1.00963	1.0033	1.0760	0.9875	1.0014
V_2	0.9500	0.99100	1.01776	1.00984	1.0071	1.0494	0.9513	1.0592
V_5	1.0860	0.96145	1.04863	1.01000	1.0189	1.0056	1.0641	1.0542
V_8	1.1000	0.95986	1.04993	1.03516	1.0148	1.0238	1.0596	1.0133
V_{11}	1.0021	1.10000	0.98373	1.03000	0.9908	1.0085	1.0972	0.9905
V_{13}	1.0279	0.95000	1.00524	1.00274	1.0697	0.9641	1.1000	1.0291
T_{6-9}	1.0287	0.99734	1.03054	1.02139	1.0039	0.9486	1.0344	0.9762
T_{6-10}	0.9000	1.08595	0.91429	0.93327	0.9000	0.9840	1.1000	1.0163
T_{4-12}	0.9929	1.00087	0.99469	0.99338	1.0490	0.9647	0.9000	0.9537
T_{27-28}	1.0248	1.00482	1.02078	1.02729	0.9546	1.0287	0.9516	0.9481
Q_{Sh10}	0.0000	0.11072	0.00000	0.04348	0.0924	0.0917	03000	0.2890
Q_{Sh24}	0.0000	0.15928	0.03586	0.00000	0.1244	0.2278	0.0440	0.0697
SVD (p.u)	0.12252	0.17328	0.12401	0.12737	0.1194	0.1874	0.1275	0.1113

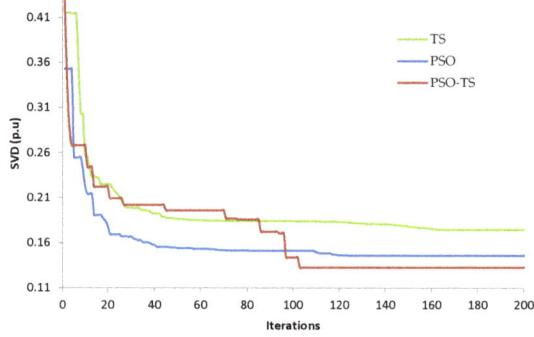

Figure 7. Convergence characteristic of the voltage deviation objective (SVD) (Case 1).

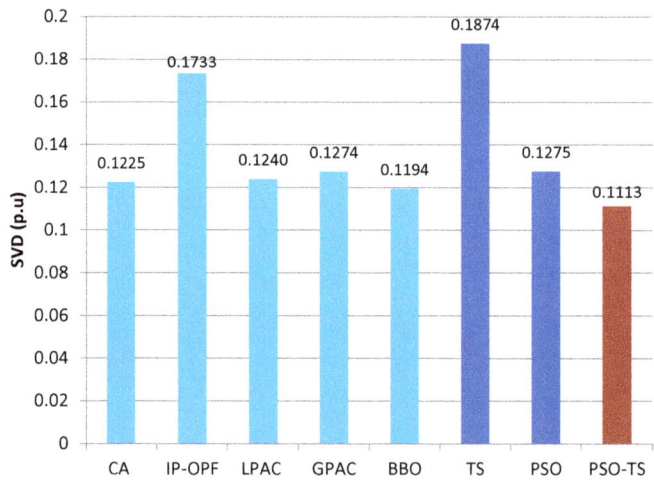

Figure 8. Comparative graph of the voltage deviation objective (SVD) (Case 1).

4.2. Case 2: IEEE 30 Bus with 19 Control Variables

In this case, the IEEE 30-bus system includes six generation buses, 24 load buses and 41 branches; 4 of them have tap-changing transformer as in the first case. In addition, buses 10, 12, 15, 17, 20, 21, 23, 24 and 29 were selected for receiving shunt capacitors. This IEEE 30-bus test system included

19 control variables. The constraint limits of the generator voltage magnitude and the tap settings of the regulating transformers are the same as those used in the first case. The capacitor sizes are considered as continuous variables and they must take their values from the interval of 0–5 MVAR. The transmission line data and the loads were taken from [6]. The active and reactive total loads are P_{load} = 2.834 p.u and Q_{load} = 1.262 p.u.

4.2.1. Power Losses Minimization

To demonstrate the superiority of the proposed algorithm in the minimization of transmission power losses (J_1), Table 4 shows the PSO-TS simulation results compared with those reported in the literature such as DE [14], BBO [48], and comprehensive learning PSO (CLPSO) [9]. The initial conditions for all these methods were the same and were taken from [6]. The total active power loss was initially 5.8322 MW, reduced to 4.5213 MW by the proposed method, i.e., a reduction in power losses by 22.48%. As shown in Figure 9, the proposed PSO-TS algorithm outperforms the cited meta-heuristic methods. Figure 10 shows the convergence characteristics of TS, PSO and PSO-TS approaches.

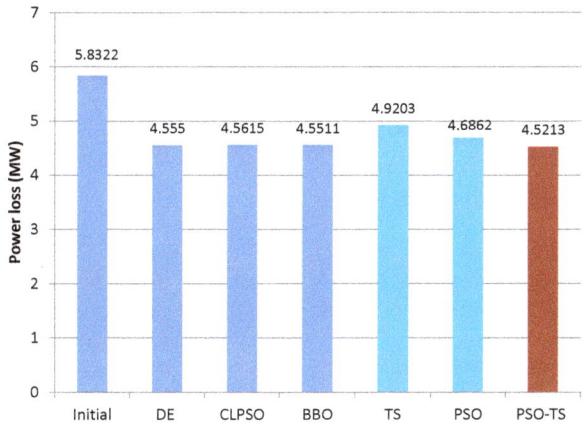

Figure 9. Comparative graph of the power losses (Case 2).

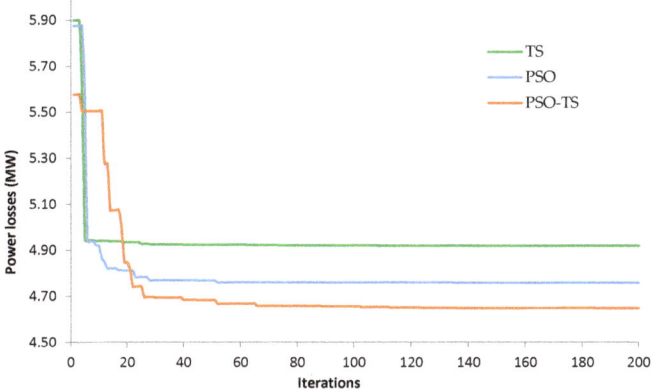

Figure 10. Convergence characteristic of the power losses (Case 2).

Table 4. Experimental results of TS, PSO and PSO-TS algorithms (Case 2).

Control Variables	Initial	DE	CLPSO	BBO	TS	PSO	PSO-TS
V_1	1.0500	1.1000	1.1000	1.1000	1.0835	1.1000	1.1000
V_2	1.0400	1.0931	1.1000	1.0944	1.0567	1.1000	1.0943
V_5	1.0100	1.0736	1.0749	1.0749	1.0671	1.0832	1.0749
V_8	1.0100	1.0756	1.1000	1.0768	1.0944	1.1000	1.0766
V_{11}	1.0500	1.1000	1.1000	1.0999	0.9873	0.9500	1.1000
V_{13}	1.0500	1.1000	1.1000	1.0999	1.0863	1.1000	1.1000
T_{6-9}	1.0780	1.0465	0.9154	1.0435	1.0745	1.1000	0.9744
T_{6-10}	1.0690	0.9097	0.9000	0.9011	0.9960	1.0953	1.0510
T_{4-12}	1.0320	0.9867	0.9000	0.9824	0.9678	0.9000	0.9000
T_{27-28}	1.0680	0.9689	0.9397	0.9691	1.0267	1.0137	0.9635
Q_{Sh10}	0.0000	0.0500	0.0492	0.0499	0.0146	0.0500	0.0500
Q_{Sh12}	0.0000	0.0500	0.0500	0.0498	0.0376	0.0500	0.0500
Q_{Sh15}	0.0000	0.0500	0.0500	0.0499	0.0000	0.0000	0.0500
Q_{Sh17}	0.0000	0.0500	0.0500	0.0499	0.0335	0.0500	0.0500
Q_{Sh20}	0.0000	0.0440	0.0500	0.0499	0.0019	0.0500	0.0386
Q_{Sh21}	0.0000	0.0500	0.0500	0.0499	0.0242	0.0500	0.0500
Q_{Sh23}	0.0000	0.0280	0.0500	0.0387	0.0307	0.0500	0.0500
Q_{Sh24}	0.0000	0.0500	0.0500	0.0498	0.0294	0.0500	0.0500
Q_{Sh29}	0.0000	0.0259	0.0500	0.0290	0.0399	0.0260	0.0213
P_{loss} (MW)	5.8322	4.5550	4.5615	4.5511	4.9203	4.6862	4.5213

4.2.2. Voltage Deviation Minimization

The SVD minimization has also been tested using the PSO-TS proposed method on the IEEE 30 bus with 19 control variables. The optimal control variables settings and the SVD obtained by the different methods are shown in Table 5. These results show that the optimal solutions determined by PSO-TS lead to lower SVD than those found by TS, PSO and DE (Figure 11). The PSO-TS algorithm has reduced the SVD from the initial state at 1.1521 p.u to 0.0866 p.u, representing a reduction of 92.48% compared with TS, PSO and DE, which reduced SVD by 86.63%, 91.27%, and 92.09%, respectively. This shows that the PSO-TS is well capable of determining the global or near-global optimum solution. The proposed method succeeded also in keeping the dependent variables within their limits. Figure 12 gives the SVD evolution over iterations of TS, PSO and PSO-TS methods.

Table 5. Experimental results of TS, PSO and PSO-TS algorithms (Case 2).

Control Variables	Initial State	DE	TS	PSO	PSO-TS
V_1	1.0500	1.0100	0.9518	0.9898	0.9867
V_2	1.0400	0.9918	1.0888	0.9529	0.9910
V_5	1.0100	1.0179	1.0502	1.0493	1.0244
V_8	1.0100	1.0183	1.0052	0.9988	1.0042
V_{11}	1.0500	1.0114	1.0730	1.0749	1.0106
V_{13}	1.0500	1.0282	1.0637	1.0404	1.0734
T_{6-9}	1.0780	1.0265	1.0137	1.0548	1.0725
T_{6-10}	1.0690	0.9038	1.0342	1.1000	0.9797
T_{4-12}	1.0320	1.0114	0.9993	0.9115	0.9273
T_{27-28}	1.0680	0.9635	0.9652	0.9458	0.9607
Q_{Sh10}	0.0000	0.0494	0.0355	0.0500	0.0095
Q_{Sh12}	0.0000	0.0108	0.0419	0.0500	0.0215
Q_{Sh15}	0.0000	0.0499	0.0032	0.0486	0.0226
Q_{Sh17}	0.0000	0.0023	0.0008	0.0500	0.0005
Q_{Sh20}	0.0000	0.0499	0.0491	0.0500	0.0359
Q_{Sh21}	0.0000	0.0490	0.0134	0.0500	0.0401
Q_{Sh23}	0.0000	0.0498	0.0382	0.0500	0.0427
Q_{Sh24}	0.0000	0.0496	0.0426	0.0500	0.0374
Q_{Sh29}	0.0000	0.0223	0.0306	0.0000	0.0210
SVD (p.u)	1.1521	0.0911	0.1540	0.1006	0.0866

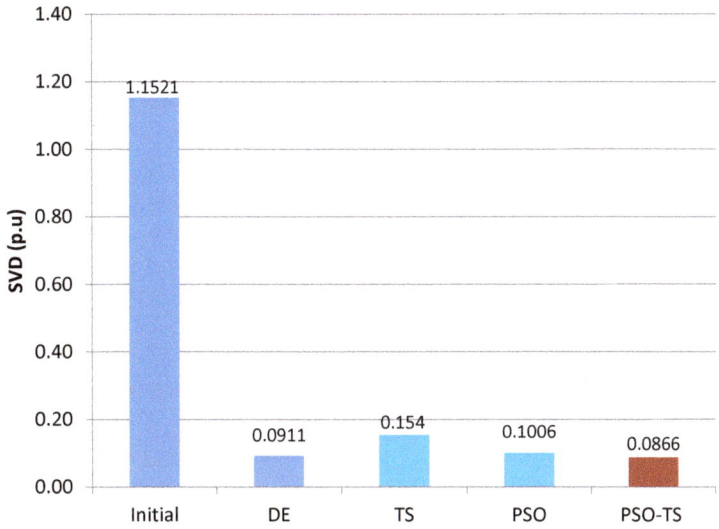

Figure 11. Comparative graph of the voltage deviation objective (SVD) (Case 2).

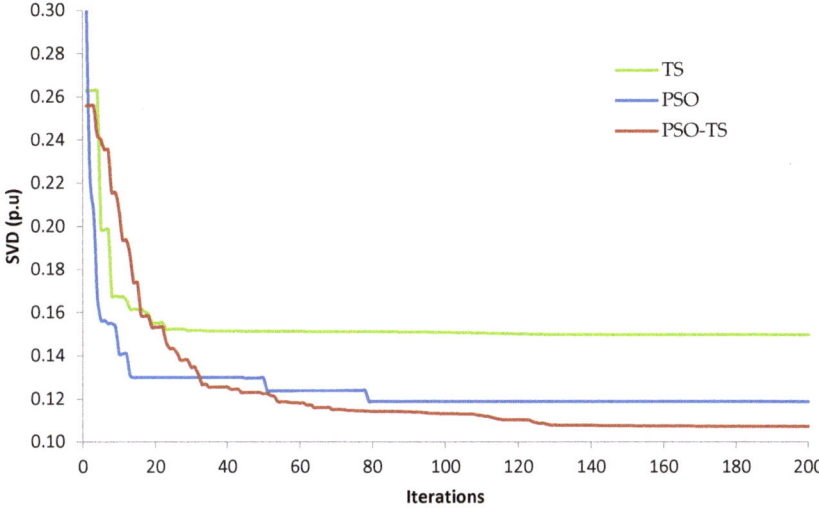

Figure 12. Convergence characteristic of the voltage deviation objective (SVD) (Case 2).

5. Conclusions

In this paper, a hybrid PSO-tabu search algorithm was proposed and successfully applied as a solution to the optimal reactive power dispatch problem. This problem was formulated as a highly constrained non-linear optimization problem where all realistic constraints were taken into consideration, including security inequalities, such as thermal constraints and real power generation constraint at the slack bus. To demonstrate the superiority of the proposed PSO-TS approach, simulation results were compared with TS, PSO and with various techniques available in the literature, such as CA, IP-OPF, LPAC, GPAC, BBO, DE and CLPSO. These simulation results show that the proposed PSO-TS algorithm gives superior solutions compared with these optimization techniques implemented using the same case studies. The presented results are very encouraging and indicate that the implementation of PSO-TS could be effective for solving the optimal power dispatch problem.

Future research should focus on analyzing other hybridization techniques in order to integrate discrete variables. Moreover, power network systems require optimizing more objectives at the same time, thus multi-objective optimization will be considered a future target.

Author Contributions: All the authors have contributed to this work including the production of the idea, the development of the algorithm, the analysis and interpretation of the results and finally the editing of the paper.

Funding: This research received no external funding.

Conflicts of Interest: The authors declare no conflict of interest.

Appendix A

Table A1. Transmission line date of IEEE 30-bus system.

Bus No.	Bus No.	R (p.u)	X (p.u)	B/2 (p.u)	Bus No.	Bus No.	R (p.u)	X (p.u)	B/2 (p.u)
1	2	0.0192	0.0575	0.0264	15	18	0.1073	0.2185	0
1	3	0.0452	0.1852	0.0204	18	19	0.0639	0.1292	0
2	4	0.0570	0.1737	0.0184	19	20	0.0340	0.0680	0
3	4	0.0132	0.0379	0.0042	10	20	0.0936	0.2090	0
2	5	0.0472	0.1983	0.0209	10	17	0.0324	0.0845	0
2	6	0.0581	0.1763	0.0187	10	21	0.0348	0.0749	0
4	6	0.0119	0.0414	0.0045	10	22	0.0727	0.1499	0
5	7	0.0460	0.1160	0.0102	21	22	0.0116	0.0236	0
6	7	0.0267	0.0820	0.0085	15	23	0.1000	0.2020	0
6	8	0.0120	0.0420	0.0045	22	24	0.1150	0.1790	0
6	9	0	0.2080	0	23	24	0.1320	0.2700	0
6	10	0	0.5560	0	24	25	0.1885	0.3292	0
9	11	0	0.2080	0	25	26	0.2544	0.3800	0
9	10	0	0.1100	0	25	27	0.1093	0.2087	0
4	12	0	0.2560	0	28	27	0	0.3960	0
12	13	0	0.1400	0	27	29	0.2198	0.4153	0
12	14	0.1231	0.2559	0	27	30	0.3202	0.6027	0
12	15	0.0662	0.1304	0	29	30	0.2399	0.4533	0
12	16	0.0945	0.1987	0	8	28	0.0636	0.2000	0.0214
14	15	0.2210	0.1997	0	6	28	0.0169	0.0599	0.0650
16	17	0.0824	0.1923	0					

Table A2. Load data of IEEE 30-bus system.

Bus No.	Active Load (p.u)	Reactive Load (p.u)	Bus No.	Active Load (p.u)	Reactive Load (p.u)
1	0.0000	0.0000	16	0.0350	0.0180
2	0.2170	0.1270	17	0.0900	0.0580
3	0.0240	0.0120	18	0.0320	0.0090
4	0.0760	0.0160	19	0.0950	0.0340
5	0.9420	0.1900	20	0.0220	0.0070
6	0.0000	0.0000	21	0.1750	0.1120
7	0.2280	0.1090	22	0.0000	0.0000
8	0.3000	0.3000	23	0.0320	0.0160
9	0.0000	0.0000	24	0.0870	0.0670
10	0.0580	0.0200	25	0.0000	0.0000
11	0.0000	0.0000	26	0.0350	0.0230
12	0.1120	0.0750	27	0.0000	0.0000
13	0.0000	0.0000	28	0.0000	0.0000
14	0.0620	0.0160	29	0.0240	0.0090
15	0.0820	0.0250	30	0.1060	0.0190

References

1. Zhao, B.; Guo, C.X.; Cao, Y.J. A multiagent-based particle swarm optimization approach for optimal reactive power dispatch. *IEEE Trans. Power Syst.* **2005**, *20*, 1070–1078. [CrossRef]
2. Agamah, S.; Ekonomou, L. A methodology for web-based power systems simulation and analysis using PHP programming. In *Electricity Distribution-Intelligent Solutions for Electricity Transmission and Distribution Networks*; Karampelas, P., Ekonomou, L., Eds.; Springer: Berlin, Germany, 2016.

3.   Lakshmi, M.; Ramesh, K.A. Optimal reactive power dispatch using crow search algorithm. *Int. J. Electr. Comput. Eng.* **2018**, *8*, 1423.

4.   Sulaiman, M.H.; Mustaffa, Z.; Mohamed, M.R.; Aliman, O. Using the gray wolf optimizer for solving optimal reactive power dispatch problem. *Appl. Soft Comput.* **2015**, *32*, 286–292. [CrossRef]

5.   Deeb, N.; Shahidehpour, S.M. Linear reactive power optimization in a large power network using the decomposition approach. *IEEE Trans. Power Syst.* **1990**, *5*, 428–438. [CrossRef]

6.   Lee, K.Y.; Park, Y.M.; Ortiz, J.L. A united approach to optimal real and reactive power dispatch. *Power Eng. Soc. Gen. Meet.* **1985**, *104*, 1147–1153.

7.   Horton, J.S.; Grigsby, L. Voltage optimization using combined linear programming & gradient techniques. *IEEE Trans. Power Syst.* **1984**, *103*, 1637–1643.

8.   Saachdeva, S.; Billington, R. Optimum network VAR planning by non linear programming. *IEEE Trans. Power Syst.* **1973**, *92*, 1217–1973. [CrossRef]

9.   Mahadevan, K.; Kannan, P.S. Comprehensive learning particle swarm optimization for reactive power dispatch. *Appl. Soft Comput.* **2010**, *10*, 641–652. [CrossRef]

10.   Quintana, V.H.; Santos-Nieto, M. Reactive power dispatch by successive quadratic programming. *IEEE Trans. Energy Convers.* **1989**, *4*, 425–435. [CrossRef]

11.   Granville, S. Optimal reactive power dispatch through interior point methods. *IEEE Trans. Power Syst.* **1994**, *4*, 136–146. [CrossRef]

12.   Polprasert, J.; Ongsakul, W.; Dieu, V.N. Optimal reactive power dispatch using improved pseudo-gradient search particle swarm optimization. *Electr. Power Compon. Syst.* **2016**, *44*, 518–532. [CrossRef]

13.   Abdullah, W.N.W.; Saibon, H.; Zain, A.A.M.; Lo, K.L. Genetic algorithm for optimal reactive power dispatch. In Proceedings of the International Conference on Energy Management and Power Delivery (EMPD), Singapore, 5 March 1998.

14.   Abou El Ela, A.A.; Abido, M.A.; Spea, S.R. Differential evolution algorithm for optimal reactive power dispatch. *Electr. Power Syst. Res.* **2011**, *81*, 458–464. [CrossRef]

15.   Biswas, P.P.; Suganthan, P.N.; Mallipeddi, R.; Amaratunga, G.A.J. Optimal power flow solutions using differential evolution algorithm integrated with effective constraint handling techniques. *Eng. Appl. Artif. Intell.* **2018**, *68*, 81–100. [CrossRef]

16.   Basu, M. Quasi-oppositional differential evolution for optimal reactive power dispatch. *Electr. Power Energy Syst.* **2016**, *78*, 29–40. [CrossRef]

17.   Karthikaikannan, D.; Sundarabalan, C.K. Optimal reactive power dispatch with static VAR compensator using harmony search algorithms. *Electron. J.* **2017**. [CrossRef]

18.   Pulluri, H.; Naresh, R.; Sharma, V. Application of stud krill herd algorithm for solution of optimal power flow problems. *Int. Trans. Electr. Energy Syst.* **2017**, *6*, 27. [CrossRef]

19.   Medani, K.B.O.; Sayah, S.; Bekrar, A. Whale optimization algorithm based optimal reactive power dispatch: A case study of the Algerian power system. *Electr. Power Syst. Res.* **2018**, *163*, 696–705. [CrossRef]

20.   Shaheen, A.M.; El-Sehiemy, R.A.; Farrag, S.M. Optimal reactive power dispatch using backtracking search algorithm. *Aust. J. Electr. Electron. Eng.* **2016**, *13*, 200–210. [CrossRef]

21.   Warid, W.; Hizam, H.; Mariun, N.; Abdul-Wahab, N. Optimal power flow using the jaya algorithm. *Energies* **2016**, *678*, 9. [CrossRef]

22.   Mei, R.N.S.; Sulaiman, M.H.; Mustaffa, Z.; Daniyal, H. Optimal reactive power dispatch solution by loss minimization using moth-flame optimization technique. *Appl. Soft Comput.* **2017**, *59*, 210–222.

23.   Anbarasan, P.; Jayabarathi, T. Optimal reactive power dispatch problem solved by symbiotic organism search algorithm. In Proceedings of the International Conference on Innovations in Power and Advanced Computing Technologies IEEE, Vellore, India, 21–22 April 2017.

24.   Bhattacharyya, B.; Saurav, R. PSO based bio inspired algorithms for reactive power planning. *Int. J. Electr. Power Energy Syst.* **2016**, *74*, 396–402. [CrossRef]

25.   Khaled, U.; Eltamaly, A.M.; Beroual, A. Optimal power flow using particle swarm optimization of renewable hybrid distributed generation. *Energies* **2017**, *10*, 1013. [CrossRef]

26.   Shaw, B.; Mukherjee, V.; Ghoshal, S.P. Solution of reactive power dispatch of power systems by an opposition-based gravitational search algorithm. *Int. J. Electr. Power Energy Syst.* **2014**, *55*, 29–40. [CrossRef]

27.   Thangaraj, R.; Pant, M.; Abraham, A.; Bouvry, P. Particle swarm optimization: Hybridization perspectives and experimental illustrations. *Appl. Math. Comput.* **2011**, *217*, 5208–5226. [CrossRef]

28. Vlachogiannis, J.G.; Lee, K.Y. A Comparative study on particle swarm optimization for optimal steady-state performance of power systems. *IEEE Trans. Power Syst.* **2006**, *21*, 1718–1728. [CrossRef]

29. Roy, P.K.; Ghoshal, S.P.; Thakur, S.S. Optimal VAR control for improvements in voltage profiles and for real power loss minimization using biogeography based optimization. *Int. J. Electr. Power Energy Syst.* **2012**, *43*, 830–838. [CrossRef]

30. Mallipeddi, R.; Jeyadevi, S.; Suganthan, P.N.; Baskar, S. Efficient constraint handling for optimal reactive power dispatch problems. *Swarm Evol. Comput.* **2012**, *5*, 28–36. [CrossRef]

31. Chuanwen, J.; Bompard, E. A hybrid method of chaotic particle swarm optimization and linear interior for reactive power optimization. *Math. Comput. Simul.* **2005**, *68*, 57–65. [CrossRef]

32. Bhattacharyya, B.; Goswami, S.K.; Bansal, R.C. Hybrid fuzzy particle swarm optimization approach for reactive power optimization. *J. Electr. Syst.* **2009**, *5*, 1–15.

33. Naderi, E.; Narimani, H.; Fathi, M.; Narimani, M.R. A novel fuzzy adaptive configuration of particle swarm optimization to solve large-scale optimal reactive power dispatch. *Appl. Soft Comput.* **2017**, *53*, 441–456. [CrossRef]

34. Li, Y.; Jing, P.; Hu, D.; Zhang, B.; Mao, C.; Ruan, X.; Miao, X.; Chang, D. Optimal reactive power dispatch using particle swarms optimization algorithm based Pareto optimal set. In *Advances in Neural Networks-ISNN 2009, Proceedings of the International Symposium on Neural Networks, Wuhan, China, 26–29 May 2009*; Springer: Berlin/Heidelberg, Germany, 2009.

35. Subbaraj, P.; Rajnarayanan, P.N. Hybrid particle swarm optimization based optimal reactive power dispatch. *Int. J. Comput. Appl.* **2010**, *1*, 65–70. [CrossRef]

36. Sayah, S.; Hamouda, A. A hybrid differential evolution algorithm based on particle swarm optimization for nonconvex economic dispatch problems. *Appl. Soft Comput.* **2013**, *13*, 1608–1619. [CrossRef]

37. Mehdinejad, M.; Mohammadi-Ivatloo, B.; Dadashzadeh-Bonab, R.; Zare, K. Solution of optimal reactive power dispatch of power systems using hybrid particle swarm optimization and imperialist competitive algorithms. *Int. J. Electr. Power Energy Syst.* **2016**, *83*, 104–116. [CrossRef]

38. Lenin, K.; Reddy, B.R.; Kalavathi, M.S. Hybrid genetic algorithm and particle swarm optimization (HGAPSO) algorithm for solving optimal reactive power dispatch problem. *Int. J. Electron. Electr. Eng.* **2013**, *1*, 262–268. [CrossRef]

39. Yapıcı, H.; Çetinkaya, N. An improved particle swarm optimization algorithm using eagle strategy for power loss minimization. *Math. Probl. Eng.* **2017**, *2017*, 1–11. [CrossRef]

40. Abido, M.A. Optimal power flow using tabu search algorithm. *Electr. Power Compon. Syst.* **2002**, *30*, 469–483. [CrossRef]

41. Zou, Y. Optimal reactive power planning based on improved tabu search algorithm. In Proceedings of the 2010 International Conference on Electrical and Control Engineering (ICECE), Wuhan, China, 25–27 June 2010.

42. Shen, Q.; Shi, W.M.; Wei, K. Hybrid particle swarm optimization and tabu search approach for selecting genes for tumor classification using gene expression data. *Comput. Biol. Chem.* **2008**, *32*, 53–60. [CrossRef] [PubMed]

43. Kennedy, J.; Eberhart, R. Particle swarm optimization. In Proceedings of the IEEE International Conference on Neural Networks, Perth, Australia, 27 November–1 December 1995.

44. Eberhart, R.C.; Kennedy, J. A new optimizer using particle swarm theory. In Proceedings of the Sixth International Symposium on Micro Machine and Human Science, Nagoya, Japan, 4–6 October 1995.

45. Hu, X.; Eberhart, R.C. Solving constrained nonlinear optimization problems with particle swarm optimization. In Proceedings of the Sixth World Multi-conference on Systemics, Cybernetics and Informatics, Orlando, FL, USA, 14–18 July 2002.

46. Özgür, Y. Penalty function methods for constrained optimization with genetic algorithms. *Math. Comput. Appl.* **2005**, *10*, 45–56.

47. Bouchekara, H.R.E.H.; Abido, M.A.; Boucherma, M. Optimal power flow using Teaching-Learning-Based Optimization technique. *Electr. Power Syst. Res.* **2014**, *114*, 49–59. [CrossRef]

48. Bhattacharya, A.; Chattopadhyay, P.K. Solution of optimal reactive power flow using biogeography-based optimization. *Electr. Electron. Sci. Eng.* **2010**, *3*, 269–277.

49. Glover, F.; Laguna, M. *Tabu Search*; Kluwer Academic Publishers: Norwell, MA, USA, 1997.

50. Pothiya, S.; Ngamroo, I.; Kongprawechnon, W. Application of multiple tabu search algorithm to solve dynamic economic dispatch considering generator constraints. *Energy Convers. Manag.* **2008**, *49*, 506–516. [CrossRef]
51. Chelouah, R.; Siarry, P. Enhanced continuous tabu search: An algorithm for the global optimization of multiminima functions. In *Metaheuristics Advances and Trends in Local Search Paradigms for Optimization*; Kluwer Academic Publishers: Norwell, MA, USA, 1999.
52. Power Systems Test Case Archive. Available online: www.ee.washington.edu/research/pstca/pf30/pg_tca30bus.htm (accessed on 20 September 2017).

Article

Maintenance Factor Identification in Outdoor Lighting Installations Using Simulation and Optimization Techniques

Ana Ogando-Martínez [1],*, Javier López-Gómez [1] and Lara Febrero-Garrido [2]

[1]   School of Industrial Engineering, University of Vigo, Campus Universitario, 36310 Vigo, Spain;
     javilopez@uvigo.es
[2]   Defense University Center, Spanish Naval Academy, Plaza de España, s/n, 36920 Marín, Spain;
     lfebrero@cud.uvigo.es
*   Correspondence: aogando@uvigo.es; Tel.: + 34-986-818-624

Received: 20 July 2018; Accepted: 14 August 2018; Published: 20 August 2018

Abstract: This document addresses the development of a novel methodology to identify the actual maintenance factor of the luminaires of an outdoor lighting installation in order to assess their lighting performance. The method is based on the combined use of Radiance, a free and open-source tool, for the modeling and simulation of lighting scenes, and GenOpt, a generic optimization program, for the calibration of the model. The application of this methodology allows the quantification of the deterioration of the road lighting system and the identification of luminaires that show irregularities in their operation. Values lower than 9% for the error confirm that this research can contribute to the management of street lighting by assessing real road conditions.

Keywords: artificial lighting; simulation; calibration; radiance; GenOpt; street light points

---

1. Introduction

Artificial lighting accounts for 19% of the worldwide electricity consumption [1]. There is substantial potential for savings in electricity-related costs because inefficient light sources continue to be used all over the world [1,2]. The International Energy Agency (IEA) [3] and the European Union (EU) have developed ambitious and wide-ranging energy policies and strategies for the future. The 2030 climate and energy framework of the EU establishes three key targets aimed at achieving a safer, more competitive, and more sustainable energy system. They are looking for: (i) at least 40% reduction in greenhouse gas (GHG) emissions since 1990, (ii) at least 27% of total energy consumed to be renewable energy and (iii) at least 27% enhancement in energy efficiency [4,5]. With regard to these energy reduction strategies, lighting can play a fundamental role because of it offers considerable potential for improvement. The Lighting Europe's Strategy Roadmap 2025 focuses efforts on profitable approaches such as LEDification, intelligent lighting systems, human-centric lighting and circular economy [6]. Different types of lamps such as incandescent lamps, halogen lamps, fluorescent lamps and Light Emitting Diodes (LEDs) are used in artificial lighting. Montoya et al. [7] analyzed the indoor lighting techniques used throughout history and its impact on energy saving and sustainability coming across that the recent advances are significant to improve the people's living conditions. LED technology is highlighted because of its huge potential to reduce energy consumption and its fast evolution to high performance energy transformation [2,5,8,9].

Indoor building lighting has been deeply studied [10–13], but this work is focused on outdoor lighting. Currently, street and road lighting facilities are one of the highest energy consumers owing to their inefficiency. They account for up to 2.3% of the global electricity consumption [14,15]. Moreover, high-quality outdoor lighting is an essential service for a city; it has economic benefits for municipalities, engenders a feeling of safety for inhabitants, aids the vision of drivers and pedestrians and contributes to energy efficiency [14,16,17]. Therefore, lighting optimization is an important field of study in modern society and several authors have carried out extensive research on this issue [2,14,16–19]. Nardelli et al. [2] evaluated the potential of LED technology and their results showed that LED light sources have an extensive lifecycle, good luminous efficiency and superior color rendering index, although the acquisition costs are still high. Yoomak et al. [16] assessed in-depth the performance of LED technology for roadway lighting applications. These authors simulated roadway lighting quality and then set an experimental scenario to evaluate the energy saving, making some conclusions regarding the benefits of LED luminaires. Other authors have evaluated the influence of lighting in obstacle detection, demonstrating that positioning has a strong influence at low illuminances, e.g., a hub-mounted lamp improved detection compared to a handlebar-mounted lamp [18]. Gago-Calderón et al. [17] studied LED photometric properties such as illuminance, uniformity, disability and discomfort glares and enhanced these properties by using an improved optic cover. Corte-Valiente et al. [14] presented a new algorithm to optimize street lighting installations by obtaining the overall illuminance uniformity. They found that the Levenberg–Marquardt back-propagation algorithm was the best at minimizing the error. Di Mascio et al. [19] studied the influence of the type of road pavement on maintenance costs of the lighting system in a tunnel, concluding that the required levels of illuminance are lower for concrete pavements to guarantee the same luminance values to offer the same luminance.

All these studies revealed the importance of lighting optimization in reducing energy consumption, to operate and manage costs of street lighting as well as to improve citizens' safety and visual comfort. In this regard, simulation is one of the most helpful and promising techniques to implement lighting optimization. Simulation allows the optimization of the use of lighting by revealing the maximum savings, i.e., minimizing electricity consumption per lux while complying with mandatory regulations and comfort indices. International Commission on Illumination (CIE) establishes the lighting requirements for outdoor lighting installations on roads with motor and pedestrian traffic according to lighting classes [20]. However, lighting simulation is still challenging because of the complexity of reflecting the real conditions of an environment and the users' needs within the environment. Baloch et al. [12] conducted a detailed investigation of the most popular lighting simulation tools, their applications and associated parameters, concluding that MATLAB is the most popular simulation tool for general purposes. Shaikh et al. [21] developed a multiagent control system with stochastic intelligent optimization in MATLAB, achieving an energy efficiency of 31.6%. Other authors have used tools such as EnergyPlus [22], DOE [23], Daysim [24], BuildOpt [25] or Radiance [26]. Bustamante et al. [27] used EnergyPlus and Radiance to evaluate two complex fenestration systems (CFS) in four different cities using mkSchedule as a tool to determine the maximum allowable irradiance that minimizes the energy consumption while meeting the visual comfort criteria. Vera et al. [28] combined EnergyPlus, Radiance and GenOpt [29] with the hybrid particle swarm optimization/Hooke–Jeeves (PSO/HJ) algorithm to perform the lighting simulation concluding that the optimization process is efficient and robust.

This article presents a novel methodology to reliably reflect the real outdoor lighting conditions in order to perform lighting optimization. Lighting simulation is accomplished using Radiance, and optimization is implemented by developing a deterministic calibration method using the PSO/HJ algorithm. Next, the complete procedure is assessed and validated by means of two different case studies: (i) a single street lamp facility in a dark room recreating real conditions, (ii) and a real road lighting facility applied to a specific real scene in Málaga, Spain. Although this study is validated in outdoor lighting conditions, its findings are also relevant to indoor lighting. The methodology described in this paper can be useful for street lighting maintenance, assessment of roadway conditions or street lighting regulation because it accurately reflects real conditions.

2. Methodology

The methodology proposed in this work can be observed in Figure 1, and it has two stages: (i) Modeling and simulation of artificial lighting using Radiance, developed by the Lawrence Berkeley National Laboratory, as a physically based lighting simulation engine to recreate the environmental conditions of a road; (ii) calibration of the lighting model by adjusting the maintenance factor of each luminaire according to the illuminance measured experimentally on the road using GenOpt in combination with Radiance.

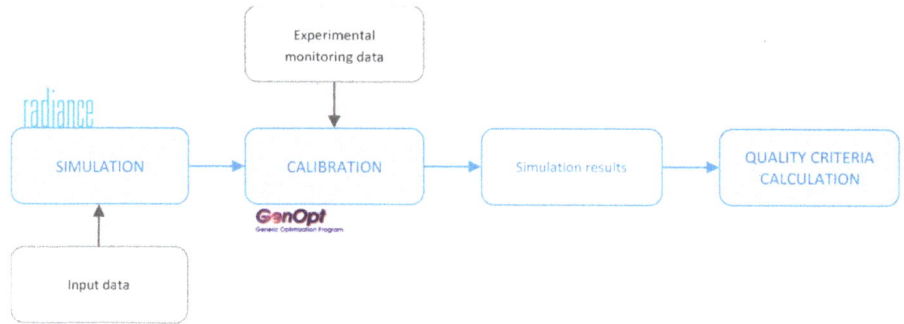

Figure 1. Simulation and calibration methodology diagram.

2.1. Modeling and Simulation of Artificial Lighting

First, a lighting scene is modeled using Radiance and following the procedure recommended by Ward et al. [26] for roadway lighting, including information about geometry and materials of the objects, the arrangement of artificial lights that make up the lighting system provided by lighting manufacturers and the sky radiance distribution. Radiance is a suite of tools that must be combined properly in order to carry out a lighting simulation. In this work tools like gensky, oconv or ies2rad are used to model the scene while rtrace or rcalc allow to perform the calculations related to the simulation.

The geometric model is elaborated in the simulation software, taking into account the characteristics of the objects. To simplify the model, the objects represented are the road, the lampposts and the luminaires in the analyzed area.

Following the classification applied by Radiance, the lampposts' material is modeled as metal to account for the specular component in the way light is reflected. In the case of road pavement, a more complex definition of material is used considering that there is a strong specular component in some source and viewing angles. The amount of light reflected on the surface of a road depends on the attributes of the pavement, the position of the observer and the point that the observer is visualizing, as well as the direction of the incident light at that point. For this work, the pavement description system established by the CIE in CIE 30-1976 [30] is adopted. A specular factor (S) and an average luminance coefficient (Q0) are used in the definition of the road material, whose values are standardized by the CIE (CIE 140-2000) [31] according to the type of pavement. To take into account all the parameters that influence the reflection of the light in the asphalt, the r-tables are used. The r-tables contain the reduced luminance coefficient in the angular intervals and in the directions given by the angles β (angle between plane of light incidence and plane of observation) and γ (angle of light incidence), considering α (angle of observation) at a constant value equal to $1°$. The standard values for the reduced luminance coefficients are defined in the CIE 144: 2001 [32] standard for each of the road classes.

Light sources are modeled based on their geometry and light properties, which is provided by manufacturers. The light information is presented in photometric files in the format of the lighting standard of Illuminating Engineering Society (IES). This type of file has the information that defines the luminance behavior of the luminaire. From the IES photometric file, the luminous distribution of the luminaire is extracted by the Radiance program ies2rad, which is represented by intensity tables for different planes and angles of light incidence.

The description of the scene is completed using gensky to generate the radiance distribution of the sky. The nighttime sky is represented by a glowing dome without sun, thus trying to approximate the radiation values that are reached at ground level. The installation location is taken into account by specifying factors such as latitude and longitude.

The complete model is generated through the oconv tool that creates an octree from the scene description files. The process continues with the simulation of the scene. Radiance is based on the backward ray tracing tracking algorithm to simulate light propagation [33]. The rtrace program is used to evaluate irradiance at the indicated coordinates from a point of view and gives the result in a RGB numerical value. Illuminance is calculated from irradiance by executing the rcalc tool.

2.2. Calibration of the Lighting Scene

With the aim of minimizing the errors made in the representation of the scene because of the simplifications and approximations applied, the model is subjected to a calibration process. A deterministic calibration methodology (shown in Figure 2) has been developed by implementing the Hooke–Jeeves algorithm, based on generalized pattern search (GPS). It is a hybrid algorithm for global optimization. This algorithm is implemented through GenOpt software, a tool that minimizes cost functions by working together with other simulation programs such as Radiance [29].

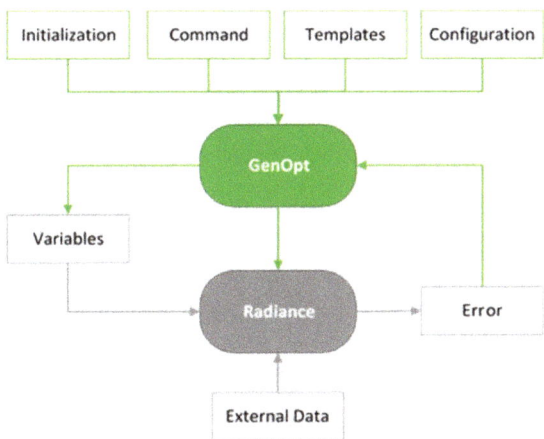

Figure 2. Calibration methodology applied to the simulation of outdoor lighting installations.

The Hooke–Jeeves search pattern method [34] creates a set of search directions iteratively combining exploratory moves with pattern moves or accelerations, both regulated by some heuristic rules. Therefore, an iterative process is applied. The process starts with the variables adopting a specified initial value and iteratively GenOpt applies the algorithm to vary the selected parameters, whether discrete or continuous, among the range of possible values considering the established limits. For each iteration, GenOpt executes the combination of Radiance programs with the allocated values and then calculates the cost of the objective function analyzing if it reaches optimal results. Consequently, a comparison can be established between the model predictions and the measurements obtained experimentally.

The coefficient of variation of the root mean squared error CV(RMSE) is the statistical index used to verify the results [35], which can be expressed as Equation (1), where $\hat{y}_i$ is the variable measure,

y_i represents the predicted value, $\bar{y}$ indicates the arithmetic mean of the measured samples and n is the number of measures.

$$\text{CV(RMSE)} = \frac{\sqrt{\frac{1}{n}\sum_{n=1}^{n}(\hat{y}_i - y_i)^2}}{\bar{y}} \tag{1}$$

In this work, the error is calculated considering the illuminance measured during the experimental data acquisition stage and the expected illuminance as a result of the simulation methods. The choice of factors that are varied depends on the conditions in which the experiment is performed but the calibration process mainly focuses on the modification of parameters that represent the individual maintenance factor of each luminaire.

The maintenance factor is calculated as indicated in CIE 154:2003 [36] depending on the lumen depreciation of the luminaire, the probability of lamps continuing in operation after a specific time and the accumulation of dirt on lamps and luminaires. The maintenance factor reaches a value of 0.67 for a typical low-maintenance installation, similar to the experimental systems studied in this work that are typically comprised of widely used high pressure sodium (HPS) lights and IP6 luminaires installed in a high-pollution environment. It should be noted that a value close to 1 for this variable represents a recently installed lamp, while a low value indicates that the lamp's properties have been depleted.

The calibrated model can be used to audit the state of the installation by comparing the different parameters calculated as indicated by CIE 140:2000 [31], using Radiance or any other simulation software, and the specifications for each type of road indicated in CIE 115:2010 [20]. The parameters of interest include the average luminance (L_{av}), overall uniformity of luminance (U_0), longitudinal uniformity of luminance (U_l), threshold increment (TI) and surround ratio (SR).

3. Experimental System

The method was tested to verify its effectiveness through the comparison of experimental measurements and results of several simulations performed in a simple installation in a controlled environment and in a more complex installation of a real road with actual traffic.

3.1. Single Street Lamp Facility

First, the experiments were conducted in a dark room with black walls and no natural light. The facility consists of a single street lamp, model VL-250 luminaire from the manufacturer Carandini (Barcelona, Spain), located on a column 3 m high and with an overhang of 1 m and one 150 W HPS light source.

Four experiments were carried out (E1, E2, E3 and E4) considering two similar light bulbs, from the manufacturers Philips (Amsterdam, Netherlands) and GE Lighting (OH, USA), and two different lamp configurations. The complete luminaire was installed (model VL-250/M) in two of the experiments while for the others, the luminaire was set up without the polycarbonate bowl (model VL-250/A), such as two different luminaires with two different photometric distributions. Table 1 lists information about the conducted experiments.

Table 1. Experiments combining different types of luminaires and lamps.

Experiment ID	Luminaire Model	Lamp Model	Power [W]	Luminous Flux [lm]
E1	VL-250/M	PHILIPS MASTER SON-T PIA Plus 150 W	150	17700
E2	VL-250/A	PHILIPS MASTER SON-T PIA Plus 150 W	150	17700
E3	VL-250/M	GE Lucalox LU150/100/XO/T/40	150	17600
E4	VL-250/A	GE Lucalox LU150/100/XO/T/40	150	17600

Luminaires with a high number of hours of operation were used in order to easily determine the losses that occur owing to the aging of the lamp and other factors such as dirtying, deterioration of the reflector or the erosion of the luminaire materials.

The data acquisition system used was an illuminance meter. The measurements were taken in a grid of 9×5 points on the ground plane separated from each other 1 m apart as indicated in Figure 3. This information was used for the calibration process which was focused to modify the maintenance factor of the installed luminaire.

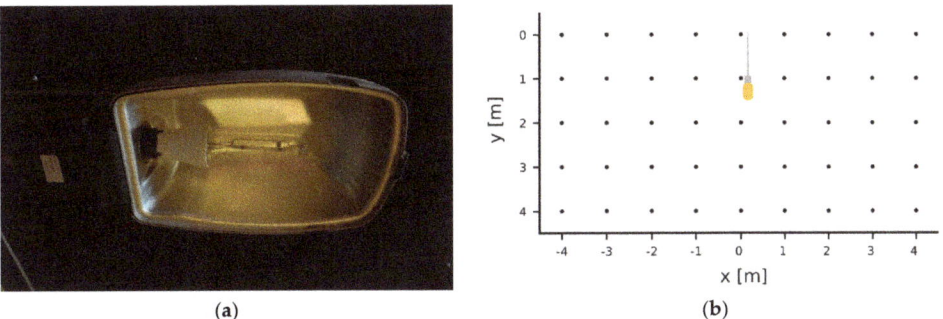

(a) (b)

Figure 3. Experimental system details: (**a**) luminaire; and (**b**) grid of 9×5 on the ground plane where the luminance was measured.

3.2. Road Street Lighting Facility

In the second set of experiments, the methodology was applied to a specific scene, located in the city of Málaga, Spain. The two-lane road considered in the experiment is approximately 330 m long and 7 m wide and includes 12 street lights. They are separated from each other at varying distances and are installed on one side of the road. Each luminaire is mounted on the top of a 13 m pole with an overhang of 1 m and involves one 250 W HPS light source. According to the CIE road surface classification CIE 30-1976 [30], the road pavement is considered as R3 whereas the light class can be considered M4 as per CIE 115:2010 [20].

Geometrical information about the road involved in the scene was acquired through the automatic processing of mobile laser scanning (MLS) point clouds. MLS represents the latest in Light Detection and Ranging (LiDAR) technology. The luminaire detection methodology developed by Puente et al. [37] for the detection of luminaires in tunnels was applied too, which provides precise coordinates to position the luminaires on the road. Data were collected using the Optech Lynx Mobile Mapper showed in Figure 4. The mobile system is composed of two LiDAR sensors, an illuminance meter, four digital cameras and a navigation system with global navigation and inertial measurement units (IMU) with a two-antenna heading measurement system, called global navigation satellite system (GNSS). A complete accuracy study and system review can be found in Puente et al. [38].

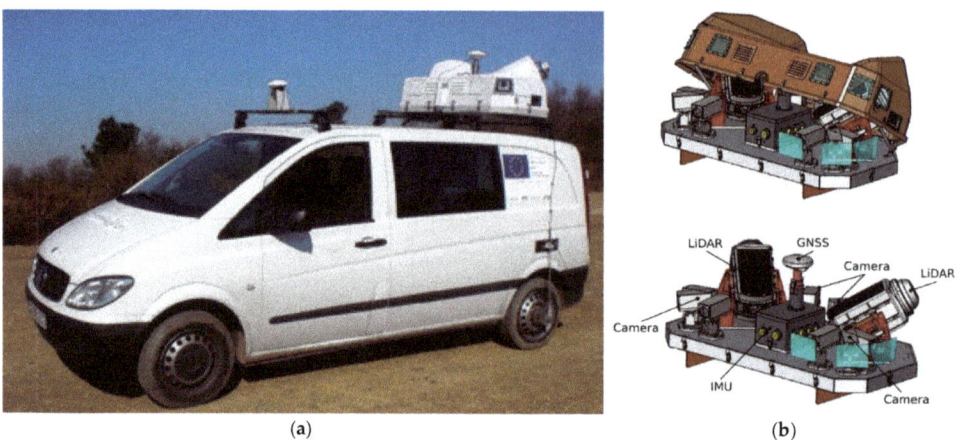

(a)   (b)

Figure 4. Optech Lynx mobile mapper used for the data collection: (**a**) vehicle; and (**b**) depiction of the system and the equipment installed on the vehicle [39].

Figure 5 shows a rendering of the scene, in which the arrangement of luminaires and horizontal road signs are presented. Horizontal signs were obtained using automatic detection algorithms developed by Riveiro [39].

The information generated by LiDAR techniques can be used for the modeling of scenes in Radiance thanks to the versatility of the tool that can be adapted and work with complex geometries of objects such as point clouds. The combination of data collection techniques with open source simulation software allows to automate the process and reduce the time invested in the development of the task.

(a)   (b)

Figure 5. Lighting scene used on the simulation and the calibration process: (**a**) 3D rendering and; (**b**) real scene.

The experimental measures of the illuminance were taken in a direction longitudinal to the road as the vehicle crosses the route. A sample of 127 points separated from each other by 2–3 m was collected. For mobile experiments, a discrepancy between the illuminance data and the coordinates collected is produced owing to the sensor response time. Considering that the vehicle that incorporates the measuring equipment moves at a constant speed, in addition to the depreciation of the luminaires, a new calibration parameter is taken into account, representing the delay that the sensor introduces in the measurements.

4. Results and Discussion

Results concerning the two proposed systems are presented and analyzed in this section. First, the methodology is applied in a controlled environment with a single luminaire; later, the procedure is repeated in a real scene with 12 luminaires.

4.1. Single Street Lamp Results

For each longitudinal section of the calculation grid described in Figure 3 and for each experimental system considered, the illuminance data collected for point (E_{exp}), the illuminance results of the initial simulation (E_{sim}) and the results of the simulation with the calibrated model (E_{simCal}) are shown in Figure 6. The graphs show the accuracy of the methodology. The results obtained from the calibrated simulation approximate the illuminance measured experimentally for the four study cases analyzed, especially in the areas closest to the luminaire, in which the value of the illuminance is higher, and therefore, so is the error produced. Figure 6 also includes the results of the simulation carried out under the assumption that the lamp has been installed recently and, therefore, has not suffered any depreciation of its properties.

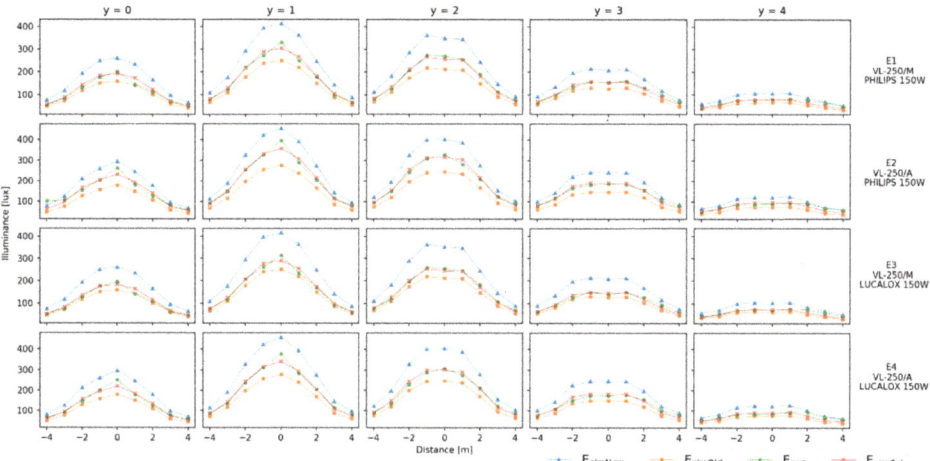

Figure 6. Illuminance in the longitudinal sections of the calculation grid considered (Figure 2) (E_{exp}: experimental illuminance; E_{simCal} and E_{simOld}: simulated illuminance after and before calibration; and E_{simNew}: simulated illuminance supposing that the lamps are newly installed).

Through the calibration process, it has been estimated that the installed luminaires for the experimental tests have suffered a reduction of 20–30% in their maintenance factor, as given in Table 2. Comparison with experiments in which the same model of light bulb has been included evinces that not only has the light source suffered losses, the complete luminaire has also aged. Tests in which the polycarbonate bowl has been installed (E1 and E3) provided a lower value for the calibrated variables, indicating that dirt and deterioration of the luminaires' materials have a negative impact on the lighting levels offered.

Table 2. Value of the Maintenance Factor and the CV(RMSE) for the different experiment cases, before and after calibration.

Experiment ID	Design Value	Calibrated Value	CV(RMSE) Initial Simulation	CV(RMSE) Calibrated Simulation	Reduction
E1	0.67	0.816	21.75%	7.23%	66.76%
E2	0.67	0.871	27.50%	8.27%	69.93%
E3	0.67	0.775	17.11%	7.14%	58.27%
E4	0.67	0.824	22.61%	7.59%	66.43%

In Table 2, the CV(RMSE) calculated for each experimental system is presented, matching the uncalibrated simulation with the calibrated one. After the calibration process, the metrics decrease by 58–70% depending on the experiment, achieving values lower than 8.27% for the CV(RMSE).

4.2. Road Street Lighting Results

Figure 7 shows the illuminance in several points of the analyzed sample, comparing the experimental data collected and the simulation results before and after the application of the calibration methodology. The calibration approximates the results of the initial simulation to the measurements collected. The similarity of the shape between calibrated simulation and experimental curves is high. The top values of illuminance correspond to areas very close to luminaires while the lower ones are more distant areas. This allows identification in the graph of the positions in which the light spots and the distances between them are located. The comparison of the curves resulting from the calibrated and uncalibrated simulation provides evidence that the deterioration cannot be considered uniform for all luminaires, being independent for each of them. The inclusion of the sensor delay in the calibration process prevents gaps between the results of the simulation and the real data, improving the method without increasing the computational cost.

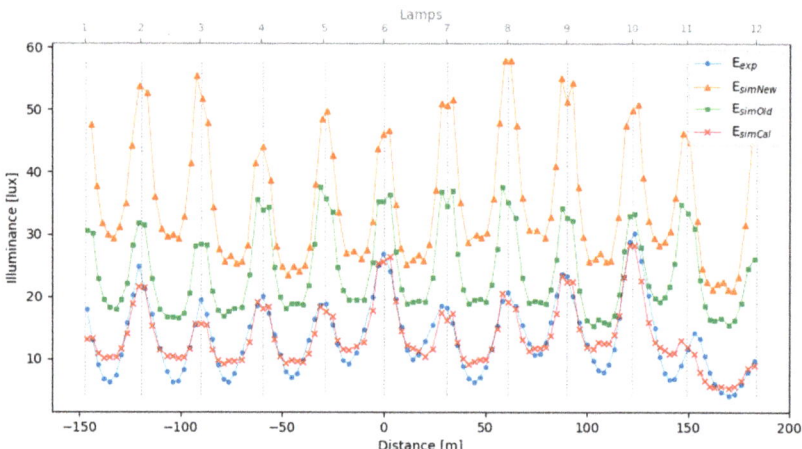

Figure 7. Illuminance at the points of the sample analyzed. (E_{exp}: experimental illuminance; E_{simCal} and E_{simOld}: simulated illuminance after and before calibration; and E_{simNew}: simulated illuminance supposing that the lamps are newly installed)

Table 3 includes the variables considered in the calibration and the values obtained at the end of the process. The results evince that the lamps in the case study show a behavior far removed from the behavior that is theoretically expected, which might be because of the high number of operational hours, the dirt or other technical defects. The analysis of the maintenance factor can facilitate decision-making in order to improve the efficiency of the installation, by identifying luminaires with inferior performance.

Table 3. Variables of the model for calibration process.

ID	Variable	Design Value	Calibrated Value
X1	Maintenance Factor street Lamp 1	1	0.269
X2	Maintenance Factor street Lamp 2	1	0.485
X3	Maintenance Factor street Lamp 3	1	0.360
X4	Maintenance Factor street Lamp 4	1	0.362
X5	Maintenance Factor street Lamp 5	1	0.317
X6	Maintenance Factor street Lamp 6	1	0.509
X7	Maintenance Factor street Lamp 7	1	0.230
X8	Maintenance Factor street Lamp 8	1	0.364
X9	Maintenance Factor street Lamp 9	1	0.458
X10	Maintenance Factor street Lamp 10	1	0.615
X11	Maintenance Factor street Lamp 11	1	0.205
X12	Maintenance Factor street Lamp 12	1	0.235
X13	Delay illuminance sensor [m]	0	−51.00

The CV(RMSE) that quantify the error in the simulations performed are calculated. The values listed in Table 4 show that, through calibration, a reduction of error by 80% is achieved by considering the variables that represent the aging of each street lamp and the delay of the light sensor. The CV(RMSE) reaches a value of 16.57% for the calibrated simulation.

Table 4. Values of the CV(RMSE) for the initial simulation and after the calibration of the simulation.

Statistical Error	Initial Simulation	Calibrated Simulation	Reduction
CV(RMSE)	84.19%	16.57%	80.31%

Taking into account the lighting class and ensuring that the requirements of the CIE 115:2010 standard are met, it is verified that the installation status is not adequate to provide the required services, as detailed in the Table 5. The parameters that quantify the disability glare (threshold increment) and the lighting of the surrounding areas achieve acceptable values as per the regulations, as well as according to some of the road surface luminance criteria of the carriageway (overall and longitudinal uniformity). However, the average luminance requirement is not fulfilled.

Table 5. Average luminance (L_{av}), overall uniformity of luminance (U_0), longitudinal uniformity of luminance (U_l), threshold increment (TI) and surround ratio (SR) of the road between Lamp 6 and Lamp 7 of the system.

	Simulation results		CIE 115:2010
L_{av}	0.82	$\geq$	1
U_0	0.56	$\geq$	0.4
U_l	0.76	$\geq$	0.6
TI	6	$\leq$	15
SR	0.73	$\geq$	0.5

Figure 8 displays false color renderings of the road surface as seen from the observer position, showing the light distribution in terms of luminance and illuminance. The results agree with the aging factors established for the simulation after the calibration process (Table 3). The higher illumination values are positioned around Lamp 6, which has suffered less deterioration.

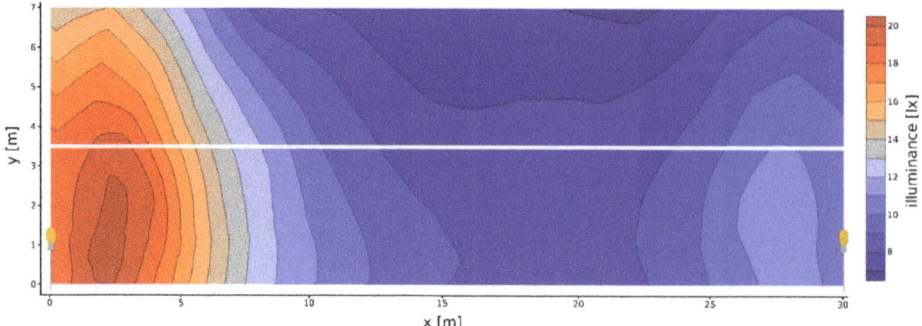

Figure 8. False color view for the illuminance [lux] of the two-lane road between Lamp 6 and Lamp 7 of the system.

5. Conclusions

In this paper, a method for modeling and calibrating outdoor lighting installations is presented and applied to two different cases: a first case study, composed of a single luminaire (studied under different configurations), was designed to test the methodology in a controlled environment, and a second case study, simulating a real scene composed of 12 luminaires distributed along a road and located in the south of Spain, allowed the validation of the process in more complex conditions and with the added complication of using measurements collected in motion for the calibration.

The methodology provides simulated illuminance measurements at points of a mesh positioned at different distances from the luminaire, taking into account the deterioration suffered over time by the group formed by lamp and luminaire. The first experimental system proves that the method is feasible for various configurations and ages of luminaires and light sources, reducing the CV(RMSE) below 9%.

The application of the process to a more complex scene shows that by adding the delay of the light sensor in the measurement as a calibration parameter, errors produced during the data collection of experimental illuminance can be corrected and the CV(RMSE) decreased by 80%. The identification of the real maintenance factor of each luminaire facilitates decision-making to improve the efficiency of the installation by recognizing the street lamps with poor performance. In addition to allowing the reproduction of the actual lighting conditions of a scene, the method can potentially be used in the identification of the lighting technology and model of the luminaire among a range of possibilities when these characteristics are unknown or uncertain.

The results of this work showed that, used together, Radiance and GenOpt form a useful tool to calibrate and simulate lighting models and to evaluate the features offered by outdoor lighting installations.

Author Contributions: Conceptualization, A.O.-M.; Data curation, J.L.-G.; Investigation, A.O.-M. and J.L.-G.; Methodology, A.O.-M.; Resources, J.L.-G. and L.F.-G.; Software, A.O.-M.; Supervision, L.F.-G.; Validation, J.L.-G. and L.F.-G.; Visualization, L.F.-G.; Writing—original draft, A.O.-M., J.L.-G. and L.F.-G.

Acknowledgments: Authors want to give thanks to the Xunta de Galicia (Grant ED481A). This investigation article was partially supported by CANDELA project, through the Xunta de Galicia CONECTA PEME 2016 (IN852A/81).

Conflicts of Interest: The authors declare no conflict of interest.

Energies **2018**, *11*, 2169

Abbreviations

The following abbreviations are used in this article:

IEA	International Energy Agency
EU	European Union
GHG	Greenhouse Gases
LED	Light Emitting Diode
CFS	Complex Fenestration Systems
PSO/HJ	Particle Swarm Optimization / Hooke-Jeeves
IES	Illuminating Engineering Society
CIE	International Commission on Illumination
MLS	Mobile Laser Scanning
LiDAR	Laser Imaging Detection and Ranging
IMU	Inertial Measurement Units
GNSS	Global Navigation Satellite System
CV(RMSE)	coefficient of variation of the root mean squared error

References

1. Waide, P.; Tanishima, S.; International Energy Agency. *Light's Labour's Lost. Policies for Energy-Efficient Lighting*; OECD Publications: Paris, France, 2006.
2. Nardelli, A.; Deuschle, E.; de Azevedo, L.D.; Pessoa, J.L.; Ghisi, E. Assessment of Light Emitting Diodes technology for general lighting: A critical review. *Renew. Sustain. Energy Rev.* **2017**, *75*, 368–379. [CrossRef]
3. IEA. Available online: https://www.iea.org/about/ (accessed on 18 August 2018).
4. EU 2030 Climate & Energy Framework. Available online: https://ec.europa.eu/clima/policies/strategies/2030_en (accessed on 18 August 2018).
5. Thielemans, S.; Di Zenobio, D.; Touhafi, A.; Lataire, P.; Steenhaut, K. DC Grids for Smart LED-Based Lighting: The EDISON Solution. *Energies* **2017**, *10*, 1454. [CrossRef]
6. Lighting Europe. Available online: https://www.lightingeurope.org/ (accessed on 18 August 2018).
7. Montoya, F.G.; Peña-García, A.; Juaidi, A.; Manzano-Agugliaro, F. Indoor lighting techniques: An overview of evolution and new trends for energy saving. *Energy Build.* **2017**, *140*, 50–60. [CrossRef]
8. Chang, M.H.; Das, D.; Varde, P.V.; Pecht, M. Light emitting diodes reliability review. *Microelectron. Reliab.* **2012**, *52*, 762–782. [CrossRef]
9. Liu, H.; Zhou, Q.; Yang, J.; Jiang, T.; Liu, Z.; Li, J. Intelligent Luminance Control of Lighting Systems Based on Imaging Sensor Feedback. *Sensors* **2017**, *17*, 321. [CrossRef] [PubMed]
10. Troncoso-Pastoriza, F.; Eguía-Oller, P.; Díaz-Redondo, R.P.; Granada-Álvarez, E. Generation of BIM data based on the automatic detection, identification and localization of lamps in buildings. *Sustain. Cities Soc.* **2018**, *36*, 59–70. [CrossRef]
11. Soori, P.K.; Vishwas, M. Lighting control strategy for energy efficient office lighting system design. *Energy Build.* **2013**, *66*, 329–337. [CrossRef]
12. Baloch, A.A.; Shaikh, P.H.; Shaikh, F.; Leghari, Z.H.; Mirjat, N.H.; Uqaili, M.A. Simulation tools application for artificial lighting in buildings. *Renew. Sustain. Energy Rev.* **2018**, *82*, 3007–3026. [CrossRef]
13. Haq, M.A.u.; Hassan, M.Y.; Abdullah, H.; Rahman, H.A.; Abdullah, M.P.; Hussin, F.; Said, D.M. A review on lighting control technologies in commercial buildings, their performance and affecting factors. *Renew. Sustain. Energy Rev.* **2014**, *33*, 268–279. [CrossRef]
14. Corte-Valiente, A.; Castillo-Sequera, J.; Castillo-Martinez, A.; Gómez-Pulido, J.; Gutierrez-Martinez, J. An Artificial Neural Network for Analyzing Overall Uniformity in Outdoor Lighting Systems. *Energies* **2017**, *10*, 175. [CrossRef]
15. Castillo-Martinez, A.; Ramon Almagro, J.; Gutierrez-Escolar, A.; del Corte, A.; Castillo-Sequera, J.; Gómez-Pulido, J.; Gutiérrez-Martínez, J. Particle Swarm Optimization for Outdoor Lighting Design. *Energies* **2017**, *10*, 141. [CrossRef]
16. Yoomak, S.; Jettanasen, C.; Ngaopitakkul, A.; Bunjongjit, S.; Leelajindakrairerk, M. Comparative study of lighting quality and power quality for LED and HPS luminaires in a roadway lighting system. *Energy Build.* **2018**, *159*, 542–557. [CrossRef]

17. Gago-Calderón, A.; Hermoso-Orzáez, M.; De Andres-Diaz, J.; Redrado-Salvatierra, G. Evaluation of Uniformity and Glare Improvement with Low Energy Efficiency Losses in Street Lighting LED Luminaires Using Laser-Sintered Polyamide-Based Diffuse Covers. *Energies* **2018**, *11*, 816. [CrossRef]

18. Fotios, S.; Qasem, H.; Cheal, C.; Uttley, J. A pilot study of road lighting, cycle lighting and obstacle detection. *Light. Res. Technol.* **2017**, *49*, 586–602. [CrossRef]

19. Moretti, L.; Cantisani, G.; Di Mascio, P. Management of road tunnels: Construction, maintenance and lighting costs. *Tunn. Undergr. Space Technol.* **2016**, *51*, 84–89. [CrossRef]

20. *CIE 115:2010 Lighting of Roads for Motor and Pedestrian Traffic*; International Commission on Illumination: Vienna, Austria, 2010.

21. Shaikh, P.H.; Nor, N.B.M.; Nallagownden, P.; Elamvazuthi, I.; Ibrahim, T. Intelligent multi-objective control and management for smart energy efficient buildings. *Int. J. Electr. Power Energy Syst.* **2016**, *74*, 403–409. [CrossRef]

22. Energyplus. *EnergyPlus Engineering Reference*; US Department of Energy: Washington, DC, USA, 2010.

23. York, D.A.; Cappiello, C.C. *DOE-2 Engineers Manual*; Lawrence Berkeley Lab: Berkeley, CA, USA; Los Alamos National Lab: Los Alamos, NM, USA, 1981.

24. Reinhart, C.F.; Walkenhorst, O. Validation of dynamic RADIANCE-based daylight simulations for a test office with external blinds. *Energy Build.* **2001**, *33*, 683–697. [CrossRef]

25. Wetter, M. BuildOpt—A new building energy simulation program that is built on smooth models. *Build. Environ.* **2005**, *40*, 1085–1092. [CrossRef]

26. Larson, G.W.; Shakespeare, R. *Rendering with Radiance: The Art and Science of Lighting Visualization*; Booksurge Llc: San Francisco, CA, USA, 2004.

27. Bustamante, W.; Uribe, D.; Vera, S.; Molina, G. An integrated thermal and lighting simulation tool to support the design process of complex fenestration systems for office buildings. *Appl. Energy* **2017**, *198*, 36–48. [CrossRef]

28. Vera, S.; Uribe, D.; Bustamante, W.; Molina, G. Optimization of a fixed exterior complex fenestration system considering visual comfort and energy performance criteria. *Build. Environ.* **2017**, *113*, 163–174. [CrossRef]

29. Wetter, M. GenOpt®—A Generic Optimization Program. In Proceedings of the Seventh International IBPSA Conference, Rio de Janeiro, Brazil, 13–15 August 2001; pp. 601–608.

30. *CIE 30-1976 Calculation and Measurement of Luminance and Illuminance in Road Lighting*; International Commission on Illumination: Vienna, Austria, 1976.

31. *CIE 140-2000 Road Lighting Calculations*; International Commission on Illumination: Vienna, Austria, 2000.

32. *CIE 144:2001 Road Surface and Road Marking Reflection Characteristics*; International Commission on Illumination: Vienna, Austria, 2001.

33. Ward, G.J. The RADIANCE Lighting Simulation and Rendering System. In Proceedings of the 21st Annual Conference on Computer Graphics and Interactive Techniques (SIGGRAPH), Orlando, FL, USA, 24–29 July 1994; pp. 459–472.

34. Wetter, M. *GenOpt Generic Optimization Program, User Manual*; Version 3.1.1; Simulation Research Group, Lawrence Berkeley National Laboratory: Berkeley, CA, USA, 2016.

35. Ruiz, R.G.; Bandera, F.C. Validation of Calibrated Energy Models: Common Errors. *Energies* **2017**, *10*, 1587. [CrossRef]

36. *CIE 154:2003 The Maintenance of Outdoor Lighting Systems*; International Commission on Illumination: Vienna, Austria, 2003.

37. Puente, I.; González-Jorge, H.; Martínez-Sánchez, J.; Arias, P. Automatic detection of road tunnel luminaires using a mobile LiDAR system. *Measurement* **2014**, *47*, 569–575. [CrossRef]

38. Puente, I.; González-Jorge, H.; Riveiro, B.; Arias, P. Accuracy verification of the Lynx Mobile Mapper system. *Opt. Laser Technol.* **2013**, *45*, 578–586. [CrossRef]

39. Riveiro, B.; González-Jorge, H.; Martínez-Sánchez, J.; Díaz-Vilariño, L.; Arias, P. Automatic detection of zebra crossings from mobile LiDAR data. *Opt. Laser Technol.* **2015**, *70*, 63–70. [CrossRef]

Article

Using Generalized Generation Distribution Factors in a MILP Model to Solve the Transmission-Constrained Unit Commitment Problem

Guillermo Gutierrez-Alcaraz [1],* and **Victor H. Hinojosa [2],***

[1] Department of Electrical Engineering, Tecnológico Nacional de México/I.T. Morelia, Morelia 58020, Michoacán, México
[2] Department of Electrical Engineering, Universidad Técnica Federico Santa María, Valparaíso 2390123, Chile
* Correspondence: ggutier@itmorelia.edu.mx (G.G.-A); victor.hinojosa@usm.cl (V.H.H.)

Received: 9 August 2018; Accepted: 24 August 2018; Published: 26 August 2018

Abstract: This study proposes a mixed-integer linear programming (MILP) model to figure out the transmission-constrained direct current (DC)-based unit commitment (UC) problem using the generalized generation distribution factors (GGDF) for modeling the transmission network constraints. The UC problem has been reformulated using these linear distribution factors without sacrificing optimality. Several test power systems (PJM 5-bus, IEEE-24, and 118-bus) have been used to validate the introduced formulation. Results demonstrate that the proposed approach is more compact and less computationally burdensome than the classical DC-based formulation, which is commonly employed in the technical literature to carry out the transmission network constraints. Therefore, there is a potential applicability of the accomplished methodology to carry out the UC problem applied to medium and large-scale electrical power systems.

Keywords: DC optimal power flow; power transfer distribution factors; generalized generation distribution factors; unit commitment

---

1. Introduction

The unit commitment (UC) optimization problem is the conventional formulation used by regulated companies and power pools to schedule the power generation units for supplying the load demand over a multi-hour to multi-day timeframe [1]. The UC problem consists of deciding which thermoelectric power units need to operate at each time period (1 h) in order to minimize the generation costs (fuel cost, startup, and shutdown costs), and to satisfy the operational technical constraints for the entire power system (spinning reserve and load), as well as for each power generation unit (minimum up/down times, minimum and maximum power, and load ramps) [2].

1.1. Literature Review

It is critical that transmission power flow constraints will be incorporated in the UC formulation, because most power grids are operating close to their security electrical margins [3]. Different linear transmission network formulations have been apply to model the transmission capacity limits in the UC optimization problem. However, most researchers use the classical DC-based power flow formulation [4–13], where the active power unit generation and the voltage phase angles are the decision variables used to carry out the operational problem. This problem consists of two analyses: (1) the nodal power balance equality constraints; and (2) the maximum transmission power flow inequality constraints. Based on the classical DC-based formulation and incorporating the transmission power flow constraints in the optimization problem, it is significantly increased the problem size becoming computationally more complex when it is applied to large-scale electrical power systems.

Determining the transmission power flow relationships, constraints and variables that have no influence in the mathematical formulation could be eliminated from the optimization problem [6]. Alternatively, linear sensitivity factors (LSF) have also been used in the technical literature [14–21] to determine the active power network constraints in the UC problem. The LSF formulation has the advantage of requiring fewer decision variables, as well as equality constraints, by excluding the phase voltage angles without sacrificing optimality. Nevertheless, the power flow sensitivity matrix is not sparse, and it could be precomputed offline and updated when the network topology changes due to an outage, maintenance, or a switching event [17].

The LSF are also known as partial transmission distribution factors (PTDF), and these linear factors are used to carry out the transmission-constrained UC problem by several researchers in the technical literature. In addition, this approach does not sacrifice the optimality in the mathematical formulation; i.e., the equivalence for both models has been demonstrated in several studies [12,20].

An algorithm for solving the UC problem by means of the Lagrangian relaxation approach is reported in [13], where the transmission power flow constraints are formulated as linear inequality constraints based on the LSF and the net power injected in each electrical bus. A similar approach is implemented in a three-phase algorithmic scheme to determine the UC problem reported in [16]. Benders decomposition is proposed in [15] for solving the UC problem. The transmission-constrained UC problem is decomposed into two problems: a master problem and a subproblem. The master problem solves the UC without transmission network limits using the augmented Lagrangian relaxation, and the subproblem must accomplish the transmission inequality constraints. In this study, the transmission power flow constraints are also formulated as linear constraints using the PTDF. On the other hand, a method for treating transmission network bottlenecks in a stochastic market model, where generators and loads are allocated into regional sub-systems or price areas, is reported in [18]. The market model is designed for long-term and medium-term scheduling of hydrothermal power system operation. When any of the interconnections are overloaded, power flow constraints are added to the area optimization problem using the PTDF. An effective approach for obtaining robust solutions to the security-constrained (SC) UC problem with load and wind uncertainty correlation is proposed in [19]. The SCUC model is solved by Benders decomposition. Transmission network constraints are modeled using the PTDF. A power-based network constrained UC model to deal with wind generation uncertainty is reported in [20]. The model schedules power-trajectories instead of the traditional energy-blocks, and it takes into account the inherent startup and shutdown power trajectories of thermal units. The PTDF are used to model the active power flow constraints. Additionally, an N–1 security-constrained UC approach is reported in [21]. The transmission constraints are formulated as linear constraints based on the classical DC power flow approach. The transmission-constrained UC problem is determined using the injection shift factors for modeling the pre-contingency constraints and the line outage distribution factors (LODF) for modeling the post-contingency power flow constraints (N–1 criterion).

1.2. Contributions

This study proposes to apply the GGDF, which are another LSF used to formulate the network transmission inequality constraints in the unit commitment problem. The GGDF relates the active power flow in the transmission lines or transformers with the generation power unit for a given electrical system [22]. In comparison with the PTDF-based formulation, the main advantage of the GGDF-based formulation is that the definition of the active power flow inequality constraints is enhanced. Notice that these linear distribution factors represent the portion of generation supplied by each power unit that flows on a specific transmission line. Another advantage is that, unlike PTDF, which uses a slack bus to calculate these values, the obtained GGDF is the same matrix using any slack bus to compute these factors. It is worth highlighting that, in the GGDF-based formulation: (1) the nodal power balance equality constraints are transformed using only one equality constraint, which is also used in the economic dispatch problem to supply the load of the customers; and (2) the transmission power flow inequality constraints are carried out using the GGDF and the active power generation of each unit. It should be also pointed out that the active power generation is the only decision variable in the operational optimization problem.

In the technical literature review, there is no evidence about the performance of the GGDF-based formulation applied to the transmission-constrained UC problem. In this study, the accuracy and performance of UC GGDF-based formulation are gauged and compared to both the classical DC- and PTDF-based formulations. It has accomplished several analyses by means of the proposed methodology and a commercial solver in order to evaluate the performance of the formulation applied to PJM 5-bus, IEEE-24, and 118-bus power systems. The results demonstrate that a superior performance is achieved for modeling medium-scale power systems and, mainly, it has improved the mathematical complexity of the optimization problem given in [12], as well as bringing great practical advantages for modeling the stochastic scheduling problems without sacrificing the UC optimality.

This paper is structured as follows: Section 2 presents the UC optimization problem, and Section 3 describes the simulation results. Finally, Section 4 concludes and suggests directions for future work.

2. Transmission-Constrained Unit Commitment (TCUC) Model

2.1. Linearized Generator Cost Modeling

The total generation cost is typically expressed as a quadratic cost curves (QCCs). To facilitate the UC optimization process with efficient mixed-integer linear programing (MILP) solvers, QCCs are piecewise linearized.

Unit *i*'s production cost function is given by the following equation:

$$C(p_i) = \alpha_i + \beta_i p_i + \gamma_i p_i^2 \tag{1}$$

where $C(p_i)$ is the total generation cost, p_i is the output power of the generator i, and α_i, β_i, and γ_i are the production cost factors. The generator total cost curve can be represented by a series of linear sections [23]. The linearized generator cost model could be mathematically formulated as follows:

$$p_i = \sum_{l=1}^{Lg} \Delta p_{i,l} \tag{2}$$

$$0 \leq \Delta p_{i,l} \leq \frac{P_i^{Max}}{Lg} \tag{3}$$

$$C(p_i) = \alpha_i + \sum_{l=1}^{Lg} k_{i,l} \Delta p_{i,l} \tag{4}$$

$$k_{i,l} = \beta_i + (2l - 1)\gamma_i \left(\frac{P_i^{Max}}{Lg} \right) \tag{5}$$

where $\Delta p_{i,l}$ represent segments of the *i*-th output power unit, P_i^{Max} is the maximum power generation of unit i, $k_{i,l}$ is a linear section in the total cost curve, and Lg is the number of segments.

2.2. Transmission Network Modeling

Most popular UC implementations have adopted the classical DC-based formulation to carry out: (1) the nodal power balance constraints; and (2) the transmission network constraints.

(1)   In the classical DC-based formulation, the power balance equality constraints are formulated using the following matrix equation:

$$B\theta = P - Pd \tag{6}$$

where B is the bus admittance matrix, θ is the vector of node phase angles, P is a power generation vector, and Pd is a power demand vector.

However, when the voltage phase angles are replaced in Equation (6) using the inverse of the admittance matrix B, only one equation is obtained to supply the load of the customers for both the PTDF- and GGDF-based formulations.

(2) Using the classical DC approach, the power flow, P_{mq}, through the transmission line between bus m and q, is defined using Equation (7):

$$P_{mq} = \frac{\theta_m - \theta_q}{X_{mq}} = B_{mq}\left(\theta_m - \theta_q\right) \tag{7}$$

where θ_m is the complex voltage angle at bus m, B_{mq} is the line susceptance between buses m–q, and X_{mq} is the line reactance between buses m–q.

Based on the technical literature review, the transmission power flows can also be expressed using the PTDF matrix as [12,24,25]:

$$P_{mq} = PTDF\left(A_g^t P - Pd\right) \tag{8}$$

where A_g is the generator-bus incidence matrix.

The PTDF matrix is a function of the transmission lines impedances. In addition, the PTDF matrix depends on a slack bus, which means that for any choice of slack bus, there will be a PTDF matrix that completely describes how the injections at each bus in the network affect the power flows throughout the transmission system. Note that every power injection is compensated by the slack bus.

It is worth mentioning that the PTDF factors must be precomputed and stored prior to the mathematical formulation and simulation.

On the other hand, the power flow through the transmission line between bus m and n, could be expressed using the GGDF [26]:

$$P_{mq} = GGDF_{mq,i}p_i \tag{9}$$

where the GGDF represents the portion of generation supplied by each generator that flows on a specific transmission line.

For any mathematical formulation, the maximum transmission power flows must be constrained considering the transmission thermal limits:

$$- P_{mq}^{Max} \leq P_{mq} \leq P_{mq}^{Max} \quad \forall m, q \in N \tag{10}$$

where N is the total number of buses, and P_{mq}^{Max} is the maximum power flow through the transmission line m–q.

Using the PTDF matrix, the transmission limit constraint is expressed as follows:

$$- P_{mq}^{Max} \leq PTDF_{mq}\left(A_g^t P - Pd\right) \leq P_{mq}^{Max} \quad \forall m, q \in N \tag{11}$$

Since the load of the customers is constant values for each time period; i.e., they are not affected by the UC problem so that they can be accordingly moved from Equation (11) to the left-hand (*lhs*) and right-hand (*rhs*) limits.

Using the GGDF matrix, the transmission limit constraints are expressed using the following equation:

$$- P_{mq}^{Max} \leq GGDF_{mq,i}p_i \leq P_{mq}^{Max} \quad \forall m, q \in N \tag{12}$$

Notice that previous task is avoided using the GGDF formulation, because the transmission power flows are only a function of GGDF and decision variables (active power generation of each unit).

See the Appendix A for a detailed derivation of the GGDF matrix.

In order to compare the three formulations, we show how to model, for a given period t, the nodal power balance equality constraints and the transmission power flow inequality constraints. Every formulation has been applied to the PJM 5-bus system, which is included in Matpower [27]. For the transmission network modeling, bus 1 has been used as the reference (slack) bus and a base power value of 100 MVA.

(1) Classical DC-based formulation using p.u. values:

$$
B_{bus}
\begin{bmatrix}
\theta_1 \\
\theta_2 \\
\theta_3 \\
\theta_4 \\
\theta_5
\end{bmatrix}
=
\begin{bmatrix}
P_1 - Pd_1 \\
P_2 - Pd_2 \\
P_3 - Pd_3 \\
P_4 - Pd_4 \\
P_5 - Pd_5
\end{bmatrix}
$$

$$
\begin{aligned}
-4 &\leq \tfrac{\theta_1 - \theta_2}{0.0281} \leq 4 \\
-10 &\leq \tfrac{\theta_1 - \theta_4}{0.0304} \leq 10 \\
-10 &\leq \tfrac{\theta_1 - \theta_5}{0.0064} \leq 10 \\
-10 &\leq \tfrac{\theta_2 - \theta_3}{0.0108} \leq 10 \\
-10 &\leq \tfrac{\theta_3 - \theta_4}{0.0297} \leq 10 \\
-2.4 &\leq \tfrac{\theta_4 - \theta_5}{0.0297} \leq 2.4
\end{aligned}
$$

(2) The PTDF-based formulation using MW values:

$$
P_1 + P_2 + P_3 + P_4 + P_5 = P_D
$$

$$
-
\begin{bmatrix}
400 \\
1000 \\
1000 \\
1000 \\
1000 \\
240
\end{bmatrix}
\leq PTDF
\begin{bmatrix}
P_1 - Pd_1 \\
P_2 - Pd_2 \\
P_3 - Pd_3 \\
P_4 - Pd_4 \\
P_5 - Pd_5
\end{bmatrix}
\leq
\begin{bmatrix}
400 \\
1000 \\
1000 \\
1000 \\
1000 \\
240
\end{bmatrix}
$$

(3) The GGDF-based formulation using MW values:

$$
P_1 + P_2 + P_3 + P_4 + P_5 = P_D
$$

$$
-
\begin{bmatrix}
400 \\
1000 \\
1000 \\
1000 \\
1000 \\
240
\end{bmatrix}
\leq GGDF
\begin{bmatrix}
P_1 \\
P_2 \\
P_3 \\
P_4 \\
P_5
\end{bmatrix}
\leq
\begin{bmatrix}
400 \\
1000 \\
1000 \\
1000 \\
1000 \\
240
\end{bmatrix}
$$

where:

$$
B_{bus} =
\begin{bmatrix}
224.7319 & -35.5872 & 0 & -32.8947 & -156.2500 \\
-35.5872 & 128.1798 & -92.5926 & 0 & 0 \\
0 & -92.5926 & 126.2626 & -33.6700 & 0 \\
-32.8947 & 0 & -33.6700 & 100.2348 & -33.6700 \\
-156.2500 & 0 & 0 & -33.6700 & 189.9200
\end{bmatrix}
$$

$$PTDF = \begin{bmatrix} 0 & -0.6698 & -0.5429 & -0.1939 & -0.0344 \\ 0 & -0.1792 & -0.2481 & -0.4376 & -0.0776 \\ 0 & -0.1509 & -0.2090 & -0.3685 & -0.8880 \\ 0 & 0.3302 & -0.5429 & -0.1939 & -0.0344 \\ 0 & 0.3302 & 0.4571 & -0.1939 & -0.0344 \\ 0 & 0.1509 & 0.2090 & 0.3685 & -0.1120 \end{bmatrix}$$

$$GGDF = \begin{bmatrix} 0.4414 & -0.2284 & -0.1015 & 0.2475 & 0.4070 \\ 0.3032 & 0.1240 & 0.0551 & -0.1343 & 0.2257 \\ 0.2554 & 0.1044 & 0.0464 & -0.1131 & -0.6327 \\ 0.1414 & 0.4716 & -0.4015 & -0.0525 & 0.1070 \\ -0.1586 & 0.1716 & 0.2985 & -0.3525 & -0.1930 \\ -0.2554 & -0.1044 & -0.0464 & 0.1131 & -0.3673 \end{bmatrix}$$

It should be pointed out that the equality and inequality constraints are equivalent for each formulation without sacrificing the UC optimality.

2.3. TCUC Mathematical Formulation

This paper uses the MILP formulation introduced in [28] as the reference formulation for the UC problem. The proposed TCUC model using the GGDF is mathematically formulated as follows:

$$Min \sum_{t=1}^{T} \sum_{i=1}^{Ng} \left[\alpha_i I_{i,t} + \sum_{j=1}^{Lg} k_{ij} \Delta p_{ij,t} + c_{i,t}^U + c_{i,t}^D \right] + \sum_{t=1}^{T} \sum_{j=1}^{N} c_{j,t}^{ENS} \tag{13}$$

$$\text{subject to:} \sum_{i=1}^{Ng} p_{i,t} + \sum_{j=1}^{N} ENS_{i,t} = \sum_{i=1}^{N} Pd_{i,t} \quad \forall t \in T \tag{14}$$

$$\sum_{i=1}^{Ng} p_{i,t} \geq \sum_{i=1}^{N} Pd_{i\ t} + R_t \quad \forall t \in T \tag{15}$$

$$p_{i,t} = \sum_{ij=1}^{Lg} \Delta p_{ij,t} \quad \forall i \in Ng \tag{16}$$

$$0 \leq \Delta p_{ij} \leq \frac{P_i^{Max}}{Lg} \quad \forall j \in Lg \tag{17}$$

$$\overline{p}_{i,t} \leq P_i^{Max} \left[I_{i,t} - z_{i,(t+1)} \right] + z_{i,(t+1)} SD_i \quad \forall t \in T \tag{18}$$

$$\overline{p}_{i,t} \leq p_{i,(t-1)} + RU_i I_{i,(t-1)} + SU_i y_{i,t} \quad \forall t \in T \tag{19}$$

$$\overline{p}_{i,t} \geq 0 \quad \forall t \in T \tag{20}$$

$$p_{i,t} \leq \overline{p}_{i,t} \quad \forall t \in T \tag{21}$$

$$P_i^{Min} I_{i,t} \leq p_{i,t} \quad \forall t \in T \tag{22}$$

$$p_{i,(t-1)} - p_{i,t} \leq RD_i I_{i,t} + SD_i z_{i,t} \quad \forall t \in T \tag{23}$$

$$\sum_{t=1}^{F_i} I_{i,t} = 0 \text{ where } F_i = Min\{[T, (DT_i - s_{i,0})][1 - I_{i,0}]\} \tag{24}$$

$$\sum_{n=t}^{t+DT_i-1} [1 - I_{i,t}] \geq DT_i z_{i,t} \quad \forall t = F_{i+1} \ldots T - DT_{t+1} \tag{25}$$

$$\sum_{n=t}^{T} [1 - I_{i,n} - z_{i,t}] \geq 0 \quad \forall t = T - DT_{i+2} \dots T \tag{26}$$

$$\sum_{t=1}^{L_i} [1 - I_{i,t}] = 0 \text{ where } L_i = Min[T, (UT_i - U_{i,0})I_{i,0}] \tag{27}$$

$$\sum_{n=t}^{t+UT_i-1} I_{i,n} \geq UT_i y_{i,t} \quad \forall t = L_{i+1} \dots T - UT_{t+1} \tag{28}$$

$$\sum_{n=t}^{T} [I_{i,n} - y_{i,n}] \geq 0 \quad \forall t = T - UT_{i+2} \dots T \tag{29}$$

$$-P_{mq,t}^{Max} \leq GGDFp_{i,t} \leq P_{mq,t}^{Max} \quad \forall t \in T \tag{30}$$

$$y_{i,t} - z_{i,t} = I_{i,t} - I_{i,t-1} \quad \forall t \in T \tag{31}$$

$$y_{i,t} + z_{i,t} \leq 1 \quad \forall t \in T \tag{32}$$

where Ng is the total number of generators, T is the total number of periods; $c_{i,t}^U, c_{i,t}^D$ are the startup and shutdown costs of unit i at period t ($), respectively; $c_{j,t}^{ENS}$ is the non-served energy cost of bus j at period t ($/MWh); $Pd_{i\ t}$ is the load demand of node i at period t (MW); R_t is the system spinning reserve at period t (MW); P_i^{Min} is the minimum output power of unit i (MW); SU_i and SD_i are the startup and shutdown ramp limits of unit i (MW/h), respectively; RU_i and RD_i are the ramp-up and ramp-down rate limits of unit i (MW/h), respectively; UT_i and DT_i are the minimum up and down time of unit i (h), respectively; $I_{i,t}$ is a binary variable equal to 1 for period t whether unit i is on and 0 otherwise (off); $y_{i,t}$ is a binary variable equal to 1 whether unit i is started up at the beginning of period t and 0 otherwise; and $z_{i,t}$ is a binary variable equal to 1 whether unit i is shutdown at the beginning of period t and 0 otherwise.

The objective function in Equation (13) minimizes the variable production costs, the startup and shutdown costs and the non-served energy cost. The UC problem is subject to the following equality and inequality constraints: load balance equality constraint (Equation (14)), system spinning reserve (Equation (15)), linearized production cost (Equations (16) and (17)), limits on power output (Equations (21) and (22)), generators' ramp rate (Equations (18)–(20) and (23)), generators' minimum downtime (Equations (24)–(26)), generators' minimum uptime (Equations (27)–(29)), transmission network (Equation (30)), and commitment, as well as startup and shutdown logic, of generating units (Equations (31)–(32)).

3. Results

In this section, the introduced mathematical formulation is applied to three electrical test systems: the PJM 5-bus system, the IEEE 24-bus reliability test system, as well as the IEEE 118-bus system. The proposed transmission-constrained UC approach is compared in terms of unit commitment costs and computational aspects using results obtained by other methodologies [12].

All simulations are performed on a personal computer (PC) running Windows® 10 with an Intel® Core i7, 2.7-GHz, 12 GB RAM, and 64-bit, using CPLEX® (12.7.1) under MATLAB® Code (Version 2014b, ITM, Mexico). Computing the simulation time is based on a set of 100 simulations.

3.1. PJM 5-Bus System

The PJM 5-bus system has five buses, six transmission lines, and five generators. Operational costs and system data are taken from [29]. Figure 1 depicts the one-line diagram for this system. In this test system, the spinning capacity requirements are set to $R_t = 0.03 \sum_{i=1}^{N} Pd_{i\ t}$ for all time periods (t).

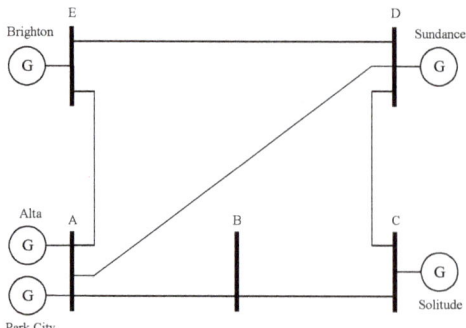

Figure 1. The PJM 5-bus system.

Table 1 lists the minimum up and down times and initial conditions for each power unit.

Table 1. Minimum up (*UT*) and down (*DT*) times.

Unit #	UT_i (hours)	DT_i (hours)	Initial Condition (hours)
Alta	5	3	0
Park City	5	3	0
Solitude	4	2	8
Sundace	3	2	8
Brighton	5	4	0

The hourly load percentage levels are taken from [30] (Table 4—RTS 96 system). The system's peak demand occurs at hour 21, and the minimum demand occurs at hour 4.

To fully illustrate the TCUC problem applied to this power system, the following mathematical equations have been presented in order to explain the optimization problem given in Equations (13)–(32):

$$Min \sum_{t=1}^{24} \begin{bmatrix} \alpha_{\text{Brighton}} I_{i,t} + \sum_{j=1}^{Lg} k_{\text{Brighton } j} \Delta p_{\text{Brighton } j,t} + c^U_{\text{Brighton},t} + c^D_{\text{Brighton},t} \\ \alpha_{\text{Sundance}} I_{\text{Sundance},t} + \sum_{j=1}^{Lg} k_{\text{Sundance } j} \Delta p_{\text{Sundance } j,t} + c^U_{\text{Sundance},t} + c^D_{\text{Sundance},t} \\ \alpha_{\text{Alta}} I_{\text{Alta},t} + \sum_{j=1}^{Lg} k_{\text{Alta } j} \Delta p_{\text{Alta } j,t} + c^U_{\text{Alta},t} + c^D_{\text{Alta},t} \\ \alpha_{\text{Park City}} I_{\text{Park City},t} + \sum_{j=1}^{Lg} k_{\text{Park City } j} \Delta p_{\text{Park City } j,t} + c^U_{\text{Park City},t} + c^D_{\text{Park City},t} \\ \alpha_{\text{Solitude}} I_{\text{Solitude},t} + \sum_{j=1}^{Lg} k_{\text{Solitude } j} \Delta p_{\text{Solitude } j,t} + c^U_{\text{Solitude},t} + c^D_{\text{Solitude},t} \end{bmatrix}$$

$$p_{\text{Brighton},t} + p_{\text{Sundance},t} + p_{\text{Alta},t} + p_{\text{Park City},t} + p_{\text{Solitude},t} = \sum_{i=1}^{N} Pd_{i,t} \quad \forall t = 1,\dots,24$$

$$p_{\text{Brighton},t} + p_{\text{Sundance},t} + p_{\text{Alta},t} + p_{\text{Park City},t} + p_{\text{Solitude},t} \geq \sum_{i=1}^{N} Pd_{i\ t} + R_t \quad \forall t = 1,\dots,24$$

$$p_{\text{Brighton},t} = \sum_{j=1}^{Lg} \Delta p_{\text{Brighton } j,t} \quad \forall t = 1,\dots,24$$

$$p_{\text{Sundance},t} = \sum_{j=1}^{Lg} \Delta p_{\text{Sundance } j,t} \quad \forall t = 1, \dots, 24$$

$$p_{\text{Alta},t} = \sum_{j=1}^{Lg} \Delta p_{\text{Alta } j,t} \quad \forall t = 1, \dots, 24$$

$$p_{\text{Park City},t} = \sum_{j=1}^{Lg} \Delta p_{\text{Park City } j,t} \quad \forall t = 1, \dots, 24$$

$$p_{\text{Solitude},t} = \sum_{j=1}^{Lg} \Delta p_{\text{Solitude } j,t} \quad \forall t = 1, \dots, 24$$

$$0 \le \Delta p_{\text{Brighton } j,t} \le \frac{P_{\text{Brighton}}^{Max}}{Lg} \quad \forall j = 1, \dots, Lg$$

$$0 \le \Delta p_{\text{Sundance } j,t} \le \frac{P_{\text{Sundance}}^{Max}}{Lg} \quad \forall j = 1, \dots, Lg$$

$$0 \le \Delta p_{\text{Alta } j,t} \le \frac{P_{\text{Alta}}^{Max}}{Lg} \quad \forall j = 1, \dots, Lg$$

$$0 \le \Delta p_{\text{Park City } j,t} \le \frac{P_{\text{Park City}}^{Max}}{Lg} \quad \forall j = 1, \dots, Lg$$

$$0 \le \Delta p_{\text{Solitude } j,t} \le \frac{P_{\text{Solitude}}^{Max}}{Lg} \quad \forall j = 1, \dots, Lg$$

$$\overline{p}_{\text{Brighton},t} \le P_{\text{Brighton}}^{Max}\left[I_{\text{Brighton},t} - z_{\text{Brighton},(t+1)}\right] + z_{\text{Brighton},(t+1)}SD_{\text{Brighton}} \quad \forall t = 1, \dots, 24$$

$$\overline{p}_{\text{Sundance},t} \le P_{\text{Sundance}}^{Max}\left[I_{\text{Sundance},t} - z_{\text{Sundance},(t+1)}\right] + z_{\text{Sundance},(t+1)}SD_{\text{Sundance}} \quad \forall t = 1, \dots, 24$$

$$\overline{p}_{\text{Alta},t} \le P_{\text{Alta}}^{Max}\left[I_{\text{Alta},t} - z_{\text{Alta},(t+1)}\right] + z_{\text{Alta},(t+1)}SD_{\text{Alta}} \quad \forall t = 1, \dots, 24$$

$$\overline{p}_{\text{Park City},t} \le P_{\text{Park City}}^{Max}\left[I_{\text{Park City},t} - z_{\text{Park City},(t+1)}\right] + z_{\text{Park City},(t+1)}SD_{\text{Park City}} \quad \forall t = 1, \dots, 24$$

$$\overline{p}_{\text{Solitude},t} \le P_{\text{Solitude}}^{Max}\left[I_{\text{Solitude},t} - z_{\text{Solitude},(t+1)}\right] + z_{\text{Solitude},(t+1)}SD_{\text{Solitude}} \quad \forall t = 1, \dots, 24$$

$$\overline{p}_{\text{Brighton},t} \le p_{\text{Brighton},(t-1)} + RU_{\text{Brighton}}I_{\text{Brighton},(t-1)} + SU_{\text{Brighton}}y_{\text{Brighton},t} \quad \forall t = 1, \dots, 24$$

$$\overline{p}_{\text{Sundance},t} \le p_{\text{Sundance},(t-1)} + RU_{\text{Sundance}}I_{\text{Sundance},(t-1)} + SU_{\text{Sundance}}y_{\text{Sundance},t} \quad \forall t = 1, \dots, 24$$

$$\overline{p}_{\text{Alta},t} \le p_{\text{Alta},(t-1)} + RU_{\text{Alta}}I_{\text{Alta},(t-1)} + SU_{\text{Alta}}y_{\text{Alta},t} \quad \forall t = 1, \dots, 24$$

$$\overline{p}_{\text{Park City},t} \le p_{\text{Park City},(t-1)} + RU_{\text{Park City}}I_{\text{Park City},(t-1)} + SU_{\text{Park City}}y_{\text{Park City},t} \quad \forall t = 1, \dots, 24$$

$$\overline{p}_{\text{Solitude},t} \le p_{\text{Solitude},(t-1)} + RU_{\text{Solitude}}I_{\text{Solitude},(t-1)} + SU_{\text{Solitude}}y_{\text{Solitude},t} \quad \forall t = 1, \dots, 24$$

$$\overline{p}_{\text{Brighton},t} \ge 0 \quad \forall t = 1, \dots, 24$$

$$\overline{p}_{\text{Sundance},t} \ge 0 \quad \forall t = 1, \dots, 24$$

$$\overline{p}_{\text{Alta},t} \ge 0 \quad \forall t = 1, \dots, 24$$

$$\overline{p}_{\text{Park City},t} \ge 0 \quad \forall t = 1, \dots, 24$$

$$\overline{p}_{\text{Solitude},t} \ge 0 \quad \forall t = 1, \dots, 24$$

$$p_{\text{Brighton},t} \le \overline{p}_{\text{Brighton},t} \quad \forall t = 1, \dots, 24$$

$$p_{\text{Sundance},t} \leq \overline{p}_{\text{Sundance},t} \quad \forall t = 1, \dots, 24$$

$$p_{\text{Alta},t} \leq \overline{p}_{\text{Alta},t} \quad \forall t = 1, \dots, 24$$

$$p_{\text{Park City},t} \leq \overline{p}_{\text{Park City},t} \quad \forall t = 1, \dots, 24$$

$$p_{\text{Solitude},t} \leq \overline{p}_{\text{Solitude},t} \quad \forall t = 1, \dots, 24$$

$$P_{\text{Brighton}}^{Min} I_{\text{Brighton},t} \leq p_{\text{Brighton},t} \quad \forall t = 1, \dots, 24$$

$$P_{\text{Sundance}}^{Min} I_{\text{Sundance},t} \leq p_{\text{Sundance},t} \quad \forall t = 1, \dots, 24$$

$$P_{\text{Alta}}^{Min} I_{\text{Alta},t} \leq p_{\text{Alta},t} \quad \forall t = 1, \dots, 24$$

$$P_{\text{Park City}}^{Min} I_{\text{Park City},t} \leq p_{\text{Park City},t} \quad \forall t = 1, \dots, 24$$

$$P_{\text{Solitude}}^{Min} I_{\text{Solitude},t} \leq p_{\text{Solitude},t} \quad \forall t = 1, \dots, 24$$

$$\sum_{j=k}^{k+DT_{\text{Brighton}}-1} \left[1 - I_{\text{Brighton},j}\right] \geq DT_{\text{Brighton}} z_{\text{Brighton},k} \quad \forall k = F_{i+1} \dots 24 - DT_{j+1}$$

$$\sum_{j=k}^{k+DT_{\text{Sundance}}-1} \left[1 - I_{\text{Sundance},j}\right] \geq DT_{\text{Sundance}} z_{\text{Sundance},k} \quad \forall k = F_{i+1} \dots 24 - DT_{j+1}$$

$$\sum_{j=k}^{k+DT_{\text{Alta}}-1} \left[1 - I_{\text{Alta},j}\right] \geq DT_{\text{Alta}} z_{\text{Alta},k} \quad \forall k = F_{i+1} \dots 24 - DT_{j+1}$$

$$\sum_{j=k}^{k+DT_{\text{Park City}}-1} \left[1 - I_{\text{Park City},j}\right] \geq DT_{\text{Park City}} z_{\text{Park City},k} \quad \forall k = F_{i+1} \dots 24 - DT_{j+1}$$

$$\sum_{j=k}^{k+DT_{\text{Solitude}}-1} \left[1 - I_{\text{Solitude},j}\right] \geq DT_{\text{Solitude}} z_{\text{Solitude},k} \quad \forall k = F_{i+1} \dots 24 - DT_{j+1}$$

$$\sum_{j=k}^{T} \left[1 - I_{\text{Brighton},j} - z_{\text{Brighton},k}\right] \geq 0 \quad \forall k = 24 - DT_{i+2} \dots 24$$

$$\sum_{j=k}^{T} \left[1 - I_{\text{Sundance},j} - z_{\text{Sundance},k}\right] \geq 0 \quad \forall k = 24 - DT_{i+2} \dots 24$$

$$\sum_{j=k}^{T} \left[1 - I_{\text{Alta},j} - z_{\text{Alta},k}\right] \geq 0 \quad \forall k = 24 - DT_{i+2} \dots 24$$

$$\sum_{j=k}^{T} \left[1 - I_{\text{Park City},j} - z_{\text{Park City},k}\right] \geq 0 \quad \forall k = 24 - DT_{i+2} \dots 24$$

$$\sum_{j=k}^{T} \left[1 - I_{\text{Solitude},j} - z_{\text{Solitude},k}\right] \geq 0 \quad \forall k = 24 - DT_{i+2} \dots 24$$

$$\sum_{j=k}^{k+UT_{\text{Brighton}}-1} I_{\text{Brighton},j} \geq UT_{\text{Brighton}} y_{\text{Brighton},k} \quad \forall k = L_{i+1} \dots 24 - UT_{i+1}$$

$$\sum_{j=k}^{k+UT_{\text{Sundance}}-1} I_{\text{Sundance},j} \geq UT_{\text{Sundance}} y_{\text{Sundance},k} \quad \forall k = L_{i+1} \dots 24 - UT_{i+1}$$

$$\sum_{j=k}^{k+UT_{\text{Alta}}-1} I_{\text{Alta},j} \geq UT_{\text{Alta}} y_{\text{Alta},k} \quad \forall k = L_{i+1} \dots 24 - UT_{i+1}$$

$$\sum_{j=k}^{k+UT_{\text{Park City}}-1} I_{\text{Park City},j} \geq UT_{\text{Park City}} y_{\text{Park City},k} \quad \forall k = L_{i+1} \dots 24 - UT_{i+1}$$

$$\sum_{j=k}^{k+UT_{\text{Solitude}}-1} I_{\text{Solitude},j} \geq UT_{\text{Solitude}} y_{\text{Solitude},k} \quad \forall k = L_{i+1} \dots 24 - UT_{i+1}$$

$$\sum_{j=k}^{T} \left[I_{\text{Brighton},j} - y_{\text{Brighton},k} \right] \geq 0 \quad \forall k = 24 - UT_{i+2} \dots 24$$

$$\sum_{j=k}^{T} \left[I_{\text{Sundance},j} - y_{\text{Sundance},k} \right] \geq 0 \quad \forall k = 24 - UT_{i+2} \dots 24$$

$$\sum_{j=k}^{T} \left[I_{\text{Alta},j} - y_{\text{Alta},k} \right] \geq 0 \quad \forall k = 24 - UT_{i+2} \dots 24$$

$$\sum_{j=k}^{T} \left[I_{\text{Park City},j} - y_{\text{Park City},k} \right] \geq 0 \quad \forall k = 24 - UT_{i+2} \dots 24$$

$$\sum_{j=k}^{T} \left[I_{\text{Solitude},j} - y_{\text{Solitude},k} \right] \geq 0 \quad \forall k = 24 - UT_{i+2} \dots 24$$

$$-\begin{bmatrix} 400 \\ 1000 \\ 1000 \\ 1000 \\ 1000 \\ 240 \end{bmatrix} \leq GGDF \begin{bmatrix} p_{\text{Brighton},t} \\ p_{\text{Sundance},t} \\ p_{\text{Alta},t} \\ p_{\text{Park City},t} \\ p_{\text{Solitude},t} \end{bmatrix} \leq \begin{bmatrix} 400 \\ 1000 \\ 1000 \\ 1000 \\ 1000 \\ 240 \end{bmatrix} \quad \forall t = 1, \dots, 24$$

$$y_{\text{Brighton},t} - z_{\text{Brighton},t} = I_{\text{Brighton},t} - I_{\text{Brighton},t-1} \quad \forall t = 1, \dots, 24$$

$$y_{\text{Sundance},t} - z_{\text{Sundance},t} = I_{\text{Sundance},t} - I_{\text{Sundance},t-1} \quad \forall t = 1, \dots, 24$$

$$y_{\text{Alta},t} - z_{\text{Alta},t} = I_{\text{Alta},t} - I_{\text{Alta},t-1} \quad \forall t = 1, \dots, 24$$

$$y_{\text{Park City},t} - z_{\text{Park City},t} = I_{\text{Park City},t} - I_{\text{Park City},t-1} \quad \forall t = 1, \dots, 24$$

$$y_{\text{Solitude},t} - z_{\text{Solitude},t} = I_{\text{Solitude},t} - I_{\text{Solitude},t-1} \quad \forall t = 1, \dots, 24$$

$$y_{\text{Brighton},t} + z_{\text{Brighton},t} \leq 1 \quad \forall t = 1, \dots, 24$$

$$y_{\text{Sundance},t} + z_{\text{Sundance},t} \leq 1 \quad \forall t = 1, \dots, 24$$

$$y_{\text{Alta},t} + z_{\text{Alta},t} \leq 1 \quad \forall t = 1, \dots, 24$$

$$y_{\text{Park City},t} + z_{\text{Park City},t} \leq 1 \quad \forall t = 1, \dots, 24$$

$$y_{\text{Solitude},t} + z_{\text{Solitude},t} \leq 1 \quad \forall t = 1, \dots, 24$$

where, the GGDF matrix was introduced in Section 2.2.

Figure 2 is a graphic representation of the optimal solution. Unit commitment without transmission network constraints is solved, and the optimal cost is $229,700. This schedule makes extensive use of the Units 1, 2, and 5 (U1, U2, and U5), but it does not use Unit 4 (U4) at all. U1 reaches its upper limit of generation during all its online periods because it is the cheapest unit; i.e., it is dispatched to generate as much power as often as possible.

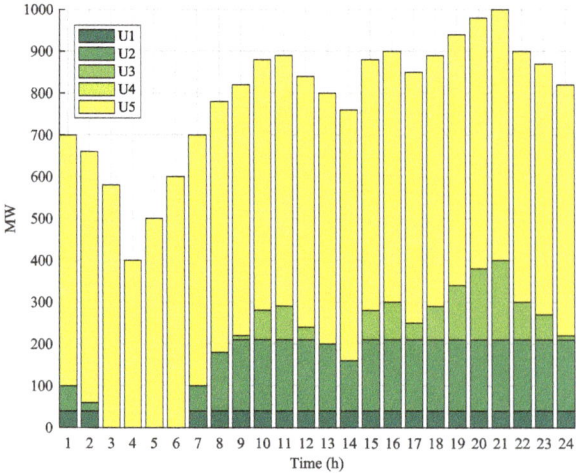

Figure 2. Units' output power without transmission congestion.

Figure 3 displays the UC scheduling and hourly output power generation. For this optimization problem, there is congestion in transmission line between bus D and E during almost all of the 24-hour period, except from hour 2 to hour 6. These power flows reach their limits since U5 produces as much power as possible, while it is constrained by transmission limits.

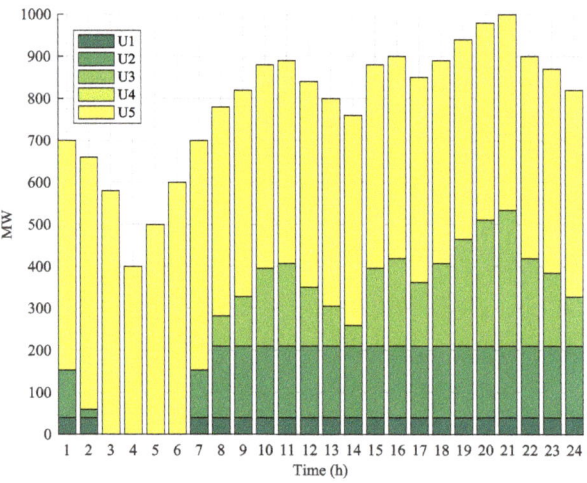

Figure 3. Units' output power with transmission congestion.

Compared the total cost considering both cases, the increased cost due to the transmission network ($267,904.7 − $229,700.0 = $38,204.7) is about 17% of the total cost.

Table 2 shows the computational aspects for the three UC models: (1) DC-, (2) PTDF-, and (3) GGDF-based formulations. The DC-based classical formulation requires a larger number of decision variables. With respect to equality constraints, the PTDF- and GGDF-based formulations require (144) constraints, which is much lower than the classical formulation (240). The inequality constraints are the same for all formulations (2112). Comparing the optimal solution with the classical DC-

and PTDF-based formulations, it is shown that all solutions are the same, which corroborates the optimality and equivalence of the GGDF-based formulation. It also includes the maximum, minimum, and average simulation time considering 100 trials.

Table 2. Computational aspects of the PJM 5-bus system.

Formulation	DC	PTDF	GGDF
Equality constraints	240	144	144
Inequality constraints	2112	2112	2112
Total constraints	2352	2256	2256
Binary variables	360	360	365
Continuous variables	1296	1200	1200
Total variables	1656	1560	1560
Maximum CPU time (seg)	0.218145	0.218145	0.213139
Minimum CPU time (seg)	0.158108	0.147099	0.145098
Average CPU time (seg)	0.167328	0.156782	0.155799

3.2. IEEE 24-Bus Reliability Test System

The IEEE Reliability Test System (RTS) has 33 generating units, 38 transmission lines, and 17 load centers. Bus 1 is selected as the slack bus. The units' minimum on/off times constraints are taken from [31].

The optimal cost without transmission network constraints is $849,359 for the 24-hour period. The operational constraints such as generation limits, minimum up/down time, and initial status of units are verified. Additionally, there is no unserved energy.

For the next simulation, transmission line 14–16 and transmission line 16–17 are limited to 440 MW. The optimal operational cost is $849,365. For this case, the transmission line 14–16 reaches its limit at only one hour ($t = 24$), and there is no loss-of-load for any time period.

To compare the performance of the proposed formulation with both the classical DC- and PTDF-based formulations, the line flow limits on lines 14–16 and 16–17 are set to several maximum values. The second and third columns of Table 3 reports how the operational cost increases caused by the re-dispatch of online units and the commitment of more units in the TCUC problem.

Table 3. Impact of congestion on the operational costs.

P_{mq}^{Max} (MW)	Operating Costs ($)	Cost Increment ($)
-	849,359	-
440	849,365	6
420	849,613	254
400	851,880	2521
380	860,470	11,111
360	884,916	35,557
340	919,852	34,936

It should be mentioned that the three mathematical formulations obtained the same optimal solution. Therefore, the accuracy and optimality of the GGDF-based formulation is also confirmed. It is worth emphasizing that for all simulations there are not unserved energy and all constraints are verified.

When the transmission limits on lines 14–16 and 16–17 are set to 440 MW, only line 14–16 reaches its limit. However, when the limit is set to 420 MW, line 16–17 also reaches its limit. Line 14–16 operates at its maximum limit for 5 h, and line 16–17 operates at its maximum limit from hour 2 to hour 6. As the transmission limit is reduced, lines 14–16 and 16–17 reach their transmission limits for most of the time. When the limit is set to 360 MW, line 14–16 reaches its limit for the 24 h and line 16–17 from hour 1 to hour 8 and at hour 24.

Figure 4 shows that all formulations require more simulation time to determine the optimization problem as the limit of the transmission network is constrained. In other words, the simulation time increases more than five times for the classical DC-based formulation compared with the PTDF- and GGDF-based formulations, when, for example, the transmission power flow constraint is set to 360 MW.

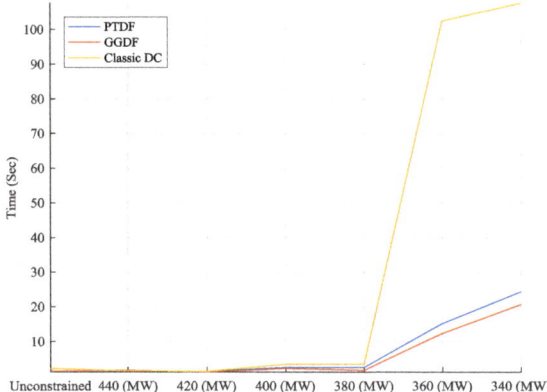

Figure 4. Average run times of TCUC with different transmission network limits.

3.3. IEEE 118-Bus System

The IEEE 118-bus test system has 91 buses with loads, 186 existing branches, and 54 generators. Bus 69 is selected as the slack bus. The system and production cost data are taken from [31].

Table 4 shows that the number of inequality constraints in the classical DC formulation is 4128, whereas the number of equality constraints is 1320 for the PTDF- and GGDF-based formulations. The number of decision variables for the classical formulation is 19,656, whereas the number of decision variables is 16,848 for both (PTDF and GGDF) formulations.

Table 4. Computational aspects for the IEEE 118-bus system.

Formulation	DC	PTDF	GGDF
Equality constraints	4128	1320	1320
Inequality constraints	28,392	28,392	28,392
Total constraints	32,520	29,712	29,712
Binary variables	3888	3888	3888
Continuous variables	15,768	12,960	12,960
Total variables	19,656	16,848	16,848

Considering the lower simulation time obtained by the GGDF-based formulation, we will use this methodology to evaluate the transmission-constrained UC problem using different transmission network conditions.

The optimal cost without the transmission network constraints is $2,246,477 for the 24-hour period. Table 5 reports the impact of congestion in the operational costs for different maximum power flow limits assuming that all transmission lines are limited to the set value. As expected, operational costs increase for each reduction in the maximum power flow limit.

Table 5. Impact of congestion on operating costs.

P_{mq}^{Max} (MW)	Operational Costs ($)	Cost Increment ($)
-	2,246,477	-
420	2,246,510	33
380	2,246,675	165
340	2,247,515	840
300	2,249,704	2189
260	2,253,259	3555
240	2,259,500	11,985
180	2,273,226	13,726
140	2,300,779	27,553

To investigate the GGDF's computational performance, we increase the time period horizon from one day to one week using the same daily system load for every day. In this case, the optimal cost is $15,725,340, and the CPU execution time for solving the proposed UC formulation is 9.6250 sec.

4. Conclusions

This paper presented an alternative formulation to take into account the transmission network constraints in the UC problem. The GGDF-based formulation, similar to the PTDF-based formulation, required fewer variables because it excluded the voltage phase angles as decision variables. The accuracy and performance in the transmission-constrained UC problem using the GGDF-based formulation were compared with the classical DC- and PTDF-based formulations using several test electrical power systems (PJM 5-bus system, IEEE 24-bus RTS system, and IEEE 118-bus system). Simulations accomplished in this study using a commercial solver support the accuracy and superior computational performance of the proposed formulation, with results showing that it can lead to a more suitable methodology applied especially to medium and large-scale electrical power systems without sacrificing optimality.

Future work will incorporate uncertainty in the proposed transmission-constrained UC problem described in this paper. Modeling renewable uncertainty and decomposition techniques (Benders) will also be studied soon.

Author Contributions: In this study, all the authors were involved in the mathematical formulation, simulation, results analysis and conclusions as well as manuscript preparation. All authors have approved the submitted manuscript.

Acknowledgments: This work was supported by the Chilean National Commission for Scientific and Technological Research (CONICYT) under grant Basal FB0008, and by UTFSM, Chile, under project USM PI_M_18_14. Additionally, the authors would like to thank the associate editor and the anonymous reviewers for their valuable comments.

Conflicts of Interest: The authors declare no conflicts of interest.

Appendix A

GGDF is defined as:

$$GGDF = PTDF + GGDF_{m,slack} \tag{A1}$$

In the technical literature, the GGDF of the slack bus can be expressed using the following equation:

$$GGDF_{m,slack} = \frac{P_m - PTDF_{l,g}p_i}{\sum\limits_{i=1}^{Ng} p_i} \quad \forall g \neq slack \tag{A2}$$

Substituting Equation (8) in Equation (A2) yields:

$$GGDF_{m,slack} = \frac{PTDF_l(p_i - Pd_i) - PTDF_{l,g}p_i}{\sum\limits_{k=1}^{N} Pd_k} \tag{A3}$$

Simplifying Equation (A3) yields:

$$GGDF_{m,slack} = -\frac{PTDF_l(Pd_i)}{\sum\limits_{k=1}^{N} Pd_k} \tag{A4}$$

Rearranging Equation (A4), the following equation is accomplished:

$$GGDF = PTDF - \frac{PTDF_l(Pd_i)}{\sum\limits_{k=1}^{N} Pd_k} = PTDF\left(I - \frac{Pd_i}{D}\right) \tag{A5}$$

where D is the total load of the power system, and I is the identity matrix whose dimension depends on the number of buses.

It is very important to highlight that the same GGDF matrix is obtained when any slack bus is selected to compute the PTDF.

References

1. O'Neill, R.P.; Helman, U.; Sotkiewicz, P.M.; Rothkopf, M.H.; Stewart, W.R. Regulatory Evolution, Market Design and Unit Commitment. In *The Next Generation of Electric Power Unit Commitment Models*; Hobbs, B.F., Rothkopf, M.H., O'Neill, R.P., Chao, H., Eds.; Springer: Boston, MA, USA; Volume 36, pp. 15–37.
2. Inostroza, J.; Hinojosa, V.H. Short-term scheduling solved with a particle swarm optimizer. *IET Gener. Transm. Distrib.* **2011**, *5*, 1091–1104. [CrossRef]
3. Fu, Y.; Li, Z.; Wu, L. Modeling and solution of the large-scale security-constrained unit commitment. *IEEE Trans. Power Syst.* **2013**, *28*, 3524–3533. [CrossRef]
4. Fu, Y.; Shahidehpour, M. Fast SCUC for large-scale power systems. *IEEE Trans. Power Syst.* **2007**, *22*, 2144–2151. [CrossRef]
5. Khodaei, A.; Shahidehpour, M. Transmission switching in security-constrained unit commitment. *IEEE Trans. Power Syst.* **2010**, *25*, 1937–1945. [CrossRef]
6. Ostrowski, J.; Wang, J. Network reduction in the transmission-constrained unit commitment problem. *Comput. Ind. Eng.* **2012**, *63*, 702–707. [CrossRef]
7. Pandzic, H.; Qiu, T.; Kirschen, D.S. Comparison of State-of-the-Art Transmission Constrained Unit Commitment Formulations. In Proceedings of the 2013 IEEE Power and Energy Society General Meeting, Vancouver, BC, Canada, 21–25 July 2013.
8. Sahebi, M.M.R.; Hosseini, S.H. Stochastic security constrained unit commitment incorporating demand side reserve. *Int. J. Electr. Power Energy Syst.* **2014**, *56*, 175–184. [CrossRef]
9. Wang, F.; Hedman, K.W. Dynamic reserve zones for day-ahead unit commitment with renewable resources. *IEEE Trans. Power Syst.* **2015**, *30*, 612–620. [CrossRef]
10. Nasrolahpour, E.; Ghasemi, H. A stochastic security constrained unit commitment model for reconfigurable networks with high wind power penetration. *Electric Power Syst. Res.* **2015**, *121*, 341–350. [CrossRef]
11. Li, Z.; Wu, W.; Wang, J.; Zhang, B.; Zheng, T. Transmission-Constrained unit commitment considering combined electricity and district heating networks. *IEEE Trans. Sustain. Energy* **2016**, *7*, 480–492. [CrossRef]
12. Hinojosa-Mateus, V.; Gutiérrez-Alcaraz, G. A computational comparison of 2 mathematical formulations to handle transmission network constraints in the unit commitment problem. *Int. Trans. Electr. Energy Syst.* **2017**, *27*, 1–15.

13. Alvarez, G.E.; Marcovecchio, M.G.; Aguirre, P.A. Security-Constrained unit commitment problem including thermal and pumped storage units: An MILP formulation by the application of linear approximations techniques. *Electr. Power Syst. Res.* **2018**, *154*, 67–74. [CrossRef]

14. Shaw, J.J. A direct method for security-constrained unit commitment. *IEEE Trans. on Power Syst.* **1995**, *10*, 1329–1342. [CrossRef]

15. Ma, H.; Shahidehpour, S.M. Transmission-constrained unit commitment based on Benders decomposition. *Electr. Power Energy Syst.* **1998**, *20*, 287–294. [CrossRef]

16. Tseng, C.-L.; Oren, S.S.; Cheng, C.S.; Li, C.; Svoboda, A.J.; Johnson, R.B. A transmission-constrained unit commitment method in power system scheduling. *Decis. Support Syst.* **1999**, *24*, 297–310. [CrossRef]

17. Kalantari, A.; Restrepo, J.F.; Galiana, F.D. Security-constrained unit commitment with uncertain wind generation: The loadability set approach. *IEEE Trans. on Power Syst.* **2013**, *28*, 1787–1796. [CrossRef]

18. Helseth, A.; Warland, G.; Mo, B. A hydrothermal market model for simulation of area prices including detailed network analyses. *Int. Trans. Electr. Energ. Syst.* **2013**, *23*, 1396–1408. [CrossRef]

19. Hu, B.; Wu, L.; Marwali, M. On the robust solution to SCUC with load and wind uncertainty correlations. *IEEE Trans. Power Syst.* **2014**, *29*, 2952–2964. [CrossRef]

20. Morales-España, G.; Baldick, R.; García-González, J.; Ramos, A. Power-capacity and ramp-capability reserves for wind integration in power-based UC. *IEEE Trans. Sustain. Energy* **2016**, *7*, 614–624.

21. Tejada-Arango, D.A.; Sánchez-Martın, P.; Ramos, A. Security constrained unit commitment using line outage distribution factors. *IEEE Trans. Power Syst.* **2018**, *33*, 329–337. [CrossRef]

22. Ng, W.Y. Generalized generation distribution factors for power system security evaluations. *IEEE Trans. Power Appar. Syst.* **1981**, *3*, 1001–1005. [CrossRef]

23. Zhang, H.; Vittal, V.; Heydt, G.T.; Quintero, J. Mixed-Integer Linear Programming approach for multi-stage security-constrained transmission expansion planning. *IEEE Trans. Power Appar. Syst.* **2012**, *27*, 1125–1133. [CrossRef]

24. Hinojosa, V.; Ticuna, O.; Gutierrez, G. Improving the mathematical formulation of the unit commitment with transmission system constraints. *IEEE Latin Am. Trans.* **2016**, *14*, 773–781. [CrossRef]

25. Hinojosa, V.H.; Gonzales-Longatt, F. Preventive security-constrained DC-OPF formulation using power transmission distribution factors and line outage distribution factors. *Energies* **2018**, *11*, 1497. [CrossRef]

26. Hinojosa, V.H. A generalized stochastic N-m security-constrained generation expansion planning methodology using partial transmission distribution factors. In Proceedings of the IEEE Power and Energy Society General Meeting Conference 2017, Chicago, IL, USA, 16–20 July 2017.

27. Matpower. Available online: http://www.pserc.cornell.edu/matpower (accessed on 18 August 2018).

28. Arroyo, J.M.; Conejo, A.J. Optimal response of a thermal unit to an electricity spot market. *IEEE Trans. Power Syst.* **2000**, *15*, 1098–1104. [CrossRef]

29. Li, F.; Bo, R. Small test systems for power system economic studies. In Proceedings of the IEEE Power and Energy Society General Meeting, Providence, RI, USA, 25–29 July 2010; pp. 1–4.

30. Grigg, C.; Wong, P.; Albrecht, P.; Allan, R.; Bhavaraju, M.; Billinton, R.; Chen, Q.; Fong, C.; Haddad, S.; Kuruganty, S.; et al. The IEEE reliability test system—1996. A report prepared by the reliability test system task force of the application of probability methods subcommittee. *IEEE Trans. power syst.* **1999**, *14*, 1010–1020. [CrossRef]

31. The REAL Lab Library. University of Washington, Department of Electrical Engineering. Available online: http://www.ee.washington.edu/research/real/library.html (accessed on 18 August 2018).

Article

Adaptive Consensus Algorithm for Distributed Heat-Electricity Energy Management of an Islanded Microgrid

Xiaofeng Dong [1,3], Xiaoshun Zhang [2,*] and Tong Jiang [1]

[1] Department of Electrical and Electronic Engineering, North China Electric Power University, Beijing 100000, China; xiaofengdong1984@gmail.com (X.D.); jiangtong@ncepu.edu.cn (T.J.)
[2] College of Engineering, Shantou University, Shantou 515000, China
[3] State Grid Suzhou Power Supply Company, Suzhou 215004, China
* Correspondence: xszhang1990@gmail.com; Tel.: +86-150-1752-7246

Received: 10 July 2018; Accepted: 19 August 2018; Published: 26 August 2018

Abstract: This paper proposes a novel adaptive consensus algorithm (ACA) for distributed heat-electricity energy management (HEEM) of an islanded microgrid. In order to simultaneously satisfy the heat-electricity energy balance constraints, ACA is implemented with a switch between unified consensus and independent consensus according to the dynamic energy mismatches. The feasible operation region of a combined heat and power (CHP) unit is decomposed into eight searching sub-regions, thus its electricity and heat energy outputs can simultaneously match the incremental cost consensus requirement and the heat-electricity energy balance constraints. Case studies are thoroughly carried out to verify the performance of ACA for distributed HEEM of an islanded microgrid.

Keywords: adaptive consensus algorithm; distributed heat-electricity energy management; eight searching sub-regions; islanded microgrid

---

1. Introduction

Over the past decades, microgrids have attracted extensive attention and study as they provide an efficient and flexible way to integrate various distributed energy resources (DERs), local loads, and energy storage devices [1]. In general, a microgrid is a local energy grid which can be operated in either grid-connected or islanded modes [2]. When a microgrid is islanded, it needs to achieve an energy balance between the energy supply and the demand without the adequate power supply from the main grid [3].

In order to handle this issue, the economic dispatch (ED) is usually employed to minimize the total operating cost while satisfying various operating constraints (e.g., energy balance constraints) [4]. So far, ED of an islanded microgrid can be implemented with two frameworks, including the centralized and distributed frameworks. Under the first framework, the energy management system (EMS) needs to collect the operating parameters of all the energy suppliers and consumers [5], then an optimal dispatch scheme can be determined by a centralized optimization method. As a result, it will inevitably result in three critical problems:

- Communication bottleneck [6] due to the great increasing amount of data from the large integration of DERs;
- Expensive computation [7] for the growing controllable variables and operating constraints from the large integration of DERs;
- Low individual privacy and security [8].

Compared with the centralized ED, a distributed ED can automatically address all of the above problems [6]. Owing to this advantage, many distributed optimization techniques have been proposed for a distributed ED in a microgrid, such as distributed λ-iteration [9], population game method [10], and dual decomposition based optimization (DDO) [11]. Among these approaches, consensus-based algorithms were widely studied due to their remarkable self-organizing ability, significant robustness, and easy scalability [12–16]. In [12], a simple consensus-based optimization was designed for optimal resource management in an islanded microgrid. By considering the ramp rate limitations, a novel consensus and innovations [13] were presented for a multistep ED in a microgrid. Moreover, a novel consensus algorithm based ED was proposed by taking the impacts of communication time delays into account [15]. However, all of these consensus algorithms did not address two important issues:

- Multi-energy dispatch: The above ED only considers the electricity energy dispatch, and did not consider the optimal dispatch of other energies, e.g., the heat energy dispatch;
- Tight coupling features among various energies: As the participation of a combined heat and power (CHP) unit, the electricity and heat energy outputs are tightly coupled because of the feasible operation region constraint, which needs to be carefully designed in the distributed ED.

Therefore, this paper proposes a novel adaptive consensus algorithm (ACA) for distributed heat-electricity energy management (HEEM) of an islanded microgrid, which can not only realize the optimal multi-energy dispatch but also consider the tight coupling features between heat and electricity energies.

The remainder of this paper is organized as follows: Section 2 introduces the mathematical model of distributed HEEM, including the objective function, the operation constraints, and a detailed feature analysis of the incremental cost. Section 3 presents the optimization principle of ACA for distributed HEEM, while the detailed solving process is provided. Case studies on a microgrid with ten energy suppliers and seven energy consumers are given in Section 4, in which four optimization methods are introduced for performance comparison with ACA. Finally, Section 5 concludes the paper.

2. Mathematical Model of Distributed HEEM

In this study, the distributed HEEM aims to minimize the total operating cost of the entire islanded microgrid while satisfying the heat and electricity energy balance constraints and other operating constraints, as illustrated in Figure 1. Note that each controllable unit only communicates with the adjacent units during the computation of distributed HEEM, which is the main difference compared with the centralized ED [17].

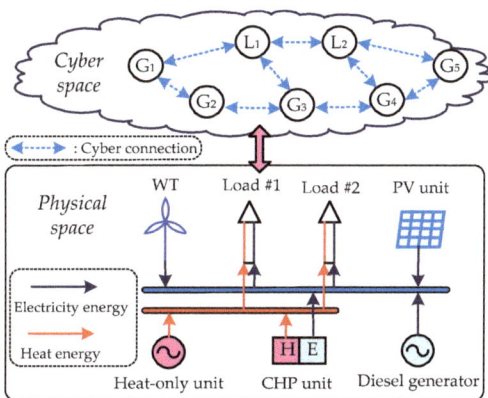

Figure 1. Framework of distributed heat-electricity energy management (HEEM) in an islanded microgrid.

2.1. Objective Function

The total operating cost f_{total} is equal to the sum of all the energy suppliers and consumers, which can be written as:

$$f_{total} = \sum_{i \in \Omega_G} f_i(P_{Gi}, H_{Gi}) + \sum_{i \in \Omega_D} f_i(\Delta P_{Di}), \tag{1}$$

where P_{Gi} and H_{Gi} are the electricity and heat energy outputs of the ith energy supplier, respectively; ΔP_{Di} is electricity energy curtailment of the ith energy consumer which participates in demand response (DR); Ω_G and Ω_D are the sets of the energy suppliers and consumers, respectively; and f_i denotes the operating cost of the ith energy supplier or consumer, which can be calculated as follows [18]:

$$f_i(P_{Gi}, H_{Gi}) = \begin{cases} 0, & \text{for WT or PV unit} \\ \alpha_i + \beta_i P_{Gi} + \gamma_i P_{Gi}^2, & \text{for diesel generator} \\ \alpha_i + \beta_i H_{Gi} + \gamma_i H_{Gi}^2, & \text{for heat} - \text{only unit} , \\ \alpha_i + \beta_i P_{Gi} + \gamma_i P_{Gi}^2 + & \\ \delta_i H_{Gi} + \theta_i H_{Gi}^2 + \xi_i H_{Gi} P_{Gi}, & \text{for CHP unit} \end{cases} \tag{2}$$

$$f_i(\Delta P_{Di}) = \frac{-1}{b_i} \Delta P_{Di}^2 + \frac{P_{Di}^0 - a_i}{b_i} \Delta P_{Di}, i \in \Omega_D, \tag{3}$$

where α_i, β_i, γ_i, δ_i, θ_i, and ξ_i are the operating cost coefficients of the ith energy supplier; a_i and b_i are the operating cost coefficients of the ith energy consumer; WT and PV represent the the wind turbine and photovoltaic unit, respectively; and $P_{Di}{}^0$ is the current initial electricity energy demand of the ith energy consumer.

2.2. Constraints

2.2.1. Energy Balance Constraints

The total energy outputs of all the energy supplier needs to match the total energy demands of all the energy consumers, is as follows:

$$\Delta E = \sum_{i \in \Omega_G} P_{Gi} - \sum_{i \in \Omega_D} \left(P_{Di}^0 - \Delta P_{Di} \right) = 0, \tag{4}$$

$$\Delta H = \sum_{i \in \Omega_G} H_{Gi} - \sum_{i \in \Omega_D} H_{Di} = 0, \tag{5}$$

where H_{Di} is the heat energy demand of the ith energy consumer; ΔE and ΔH are the electricity energy mismatch and heat energy mismatch, respectively, which will be combined into ACA in the latter section.

2.2.2. Lower and Upper Capability Limits

The energy outputs of each energy supplier, and the electricity energy curtailment of each energy consumer should be limited within their lower and upper bounds, as [18,19]:

$$\begin{cases} P_{Gi}^{min} \leq P_{Gi} \leq P_{Gi}^{max}, & \text{for diesel generator} \\ H_{Gi}^{min} \leq H_{Gi} \leq H_{Gi}^{max}, & \text{for heat} - \text{only unit} \\ P_{Gi}^{min}(H_{Gi}) \leq P_{Gi} \leq P_{Gi}^{max}(H_{Gi}), & \text{for CHP unit} \\ H_{Gi}^{min}(P_{Gi}) \leq H_{Gi} \leq H_{Gi}^{max}(P_{Gi}), & \text{for CHP unit} \end{cases} , \tag{6}$$

$$0 \leq \Delta P_{Di} \leq \eta_i P_{Di}^0, i \in \Omega_D, \tag{7}$$

where $P_{Di}{}^{min}$ and $P_{Di}{}^{max}$ are the minimum and maximum electricity energy outputs of the ith energy supplier, respectively; $H_{Di}{}^{min}$ and $H_{Di}{}^{max}$ are the minimum and maximum heat energy outputs of the *i*th energy supplier, respectively; and η_i is the maximum allowable electricity energy curtailment factor of the *i*th energy consumer.

Note that both the WT and PV units are operated at their maximum power points under the current weather conditions [18], thus they do not require a consensus interaction with other controllable devices. Besides, it can be found from Equation (6) that both the lower and upper limits of the electrical energy output of CHP units are determined by different heat energy outputs and vice versa, which indicates that the energy outputs of CHP units should be enclosed by the boundary curve ABCD (i.e., the feasible operating region) [19], as shown in Figure 2.

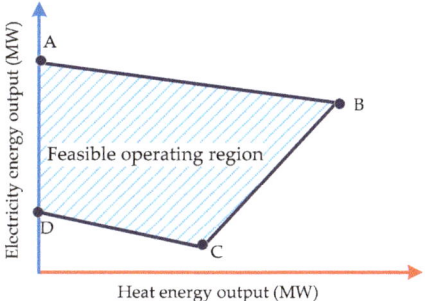

Figure 2. Feasible operating region of a combined heat and power (CHP) unit.

2.3. Feature Analysis

Since only the strictly convex feasible operating region is considered for each CHP unit, the proposed distributed HEEM is a strictly convex optimization with a unique optimum according to the quadratic objective functions Equations (1)–(3) and the linear constraints Equations (4)–(7). Hence, a feasible solution which simultaneously satisfies all the constraints can be regarded as the global optimum of distributed HEEM if all the energy suppliers and consumers can reach a consensus on the incremental cost, as [20]:

$$\frac{\partial f_1(P_{G1})}{\partial P_{G1}} = \cdots = \frac{\partial f_i(P_{Gi})}{\partial P_{Gi}} = \frac{\partial f_i(H_{Gi})}{\partial H_{Gi}} = \cdots = \frac{\partial f_n(\Delta P_{Dn})}{\partial \Delta P_{Dn}} = \lambda, \tag{8}$$

where *n* is the number of controllable devices; and λ is the incremental cost.

Note that such a consensus condition Equation (8) will not hold for a constrained optimization problem, as well as for distributed HEEM. In order to approximate the global optimum, all the constraints Equations (4)–(7) must be satisfied while the consensus condition should be satisfied as much as possible [20].

According to Equation (8), the incremental cost of each agent can be calculated as follows:

$$\begin{cases} \lambda_i^E = 2\gamma_i P_{Gi} + \beta_i, & \text{for diesel generator} \\ \lambda_i^H = 2\gamma_i H_{Gi} + \beta_i, & \text{for heat} - \text{only unit} \end{cases} \tag{9}$$

$$\begin{cases} \lambda_i^E = 2\gamma_i P_{Gi} + \xi_i H_{Gi} + \beta_i, \\ \lambda_i^H = 2\theta_i H_{Gi} + \xi_i P_{Gi} + \delta_i, \end{cases} \text{for CHP unit,} \tag{10}$$

$$\lambda_i^E = \frac{-2}{b_i}\Delta P_{Di} + \frac{P_{Di}^0 - a_i}{b_i}, i \in \Omega_D, \tag{11}$$

where λ_i^E and λ_i^H denote the electricity and heat incremental costs of the *i*th energy supplier or consumer, respectively.

Since all the operating cost coefficients except b_i are the positive constants during each dispatch time interval, the incremental costs of the diesel generator, heat-only unit, and energy consumer, increase linearly with the electricity energy output, heat energy output, and electricity energy curtailment, respectively, as shown in Figure 3. In contrast, the electricity and heat incremental costs of the CHP unit are determined by both the electrical and heat energy outputs, as illustrated in Figure 4. For example, both the electrical and heat incremental costs will increase when the CHP unit is operated from the current operating point to the green region.

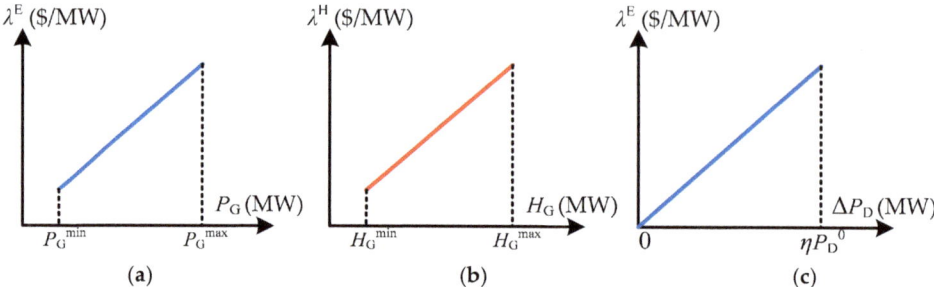

Figure 3. Incremental cost of various units. (**a**) Electrical incremental cost of diesel generator; (**b**) heat incremental cost of heat-only unit; (**c**) electrical incremental cost of energy consumer.

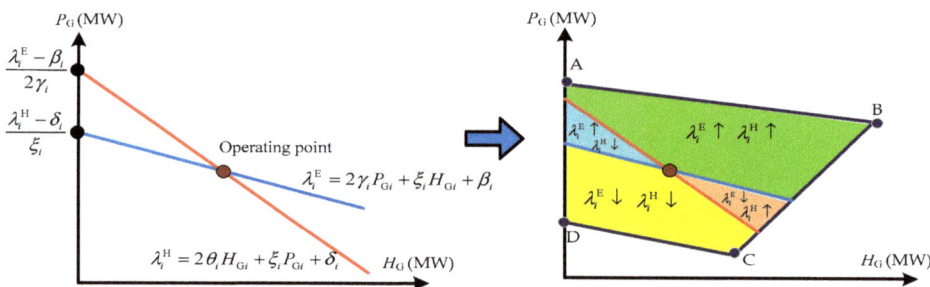

Figure 4. Illustration of incremental costs of a CHP.

3. Design of ACA for Distributed HEEM

3.1. Graph Theory of Interaction Network

The interaction network among different agents can be typically built with a directed graph $G = (V, E, A)$, where $V = \{v_1, v_2, \ldots, v_n\}$ is the set of nodes (agents); $E \subseteq V \times V$ denotes the edges (interactions); and $A = [a_{ij}] \in R^{n \times n}$ is a weighted adjacency matrix [21]. Based on these basic elements, the Laplacian matrix $L = [l_{ij}] \in R^{n \times n}$ and row stochastic matrix $D = [d_{ij}] \in R^{n \times n}$ of the graph G can be determined as follows:

$$\begin{cases} l_{ij} = -a_{ij}, \ \forall i \neq j \\ l_{ii} = \sum_{i=1, i \neq j}^{n} a_{ij} \end{cases}, \tag{12}$$

$$d_{ij}[k] = |l_{ij}| / \sum_{j=1}^{n} |l_{ij}|, \ i = 1, 2, \ldots, n \tag{13}$$

where k is the discrete time index.

In this paper, the weighted adjacency matrix is set to be a simple (0, 1)-matrix, thereby $a_{ij} = 1$ if the ith agent and the jth agent communicate with each other, otherwise $a_{ij} = 0$.

3.2. Adaptive Consensus Algorithm

The basic principle of ACA is that each agent aims to reach a consensus on a specific state with the adjacent agents by regulating its own state based on the current states from the adjacent agents. This process can be described by the first-order consensus, as [20]:

$$x_i[k+1] = \sum_{j=1}^{n} d_{ij}[k]x_j[k], \qquad (14)$$

where x_i is the state of the ith agent, which refers to the incremental cost of each agent for distributed HEEM on the basis of Equation (8).

In this study, each agent will transmit its own energy output or demand to the microgrid EMS at each iteration, then EMS will update ΔE and ΔH, and send them to each agent. In order to satisfy the energy balance constraints Equations (4) and (5), these two mismatches need to be fully considered in the consensus interaction among the agents, which can be achieved as follows:

- Unified consensus: If the signs of ΔE and ΔH are consistent, i.e., $\Delta E \Delta H \geq 0$, then all the agents can update their incremental cost state in a unified interaction network, as

$$\lambda_i[k+1] = \begin{cases} \sum_{j=1}^{n} d_{ij}[k]\lambda_j[k] - \mu\Delta E, & i \in \Omega_E \\ \sum_{j=1}^{n} d_{ij}[k]\lambda_j[k] - \mu\Delta H, & i \in \Omega_H \end{cases}, \qquad (15)$$

- Independent consensus: If the signs of ΔE and ΔH are inconsistent, i.e., $\Delta E \cdot \Delta H < 0$, then the electricity agents and heat agents need to be separated to update their incremental cost state in two independent interaction networks, as:

$$\begin{cases} \lambda_i^E[k+1] = \sum_{j \in \Omega_E} d_{ij}^E[k]\lambda_j^E[k] - \mu\Delta E, & i \in \Omega_E \\ \lambda_i^H[k+1] = \sum_{j \in \Omega_H} d_{ij}^H[k]\lambda_j^H[k] - \mu\Delta H, & i \in \Omega_H \end{cases}, \qquad (16)$$

where Ω_E and Ω_H represent the sets of electricity agents and heat agents, respectively; $d_{ij}{}^E$ is the (i, j) entry of the row stochastic matrix of the interaction network among the electricity agents; $d_{ij}{}^H$ is the (i, j) entry of the row stochastic matrix of the interaction network among the heat agents; and μ denotes the adjustment factor of energy mismatch, $\mu > 0$.

Therefore, each agent will regulate its incremental cost between these two consensus modes according to the sign of $(\Delta E \cdot \Delta H)$, as illustrated in Figure 5. After a series of consensus interactions by Equations (15) and (16), the energy balance constraints Equations (4) and (5) can be satisfied since both the electricity energy mismatch ΔE and heat energy mismatch ΔH will be sufficiently small. It is important that each interaction network should be strongly connected, i.e., any vertex can be realized from any other vertex by a directed path, thereby the consensus convergence can be guaranteed.

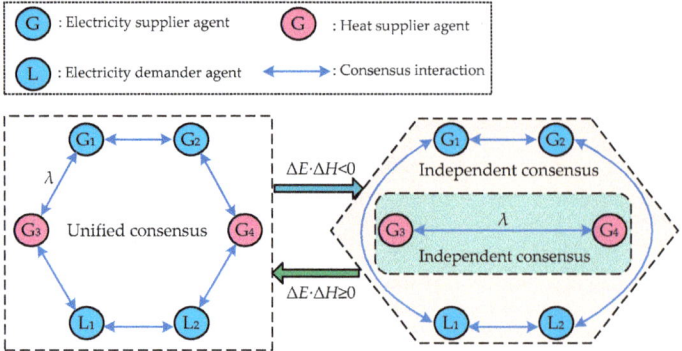

Figure 5. Principle of adaptive consensus algorithm.

3.3. Constraints Handling

Owing to the lower and upper capability limits Equations (6) and (7), all the agents may not reach a consensus on the incremental cost. Hence, a virtual incremental cost [17] is designed in ACA, which corresponds to the actual incremental cost of each agent. Note that each agent is responsible for computing its own incremental cost. More specifically, each agent can update its virtual incremental cost via a consensus interaction with adjacent agents by Equations (15) and (16), which is not limited by the constraints Equations (6) and (7). After updating the virtual incremental cost at each iteration, each agent can calculate its controllable variable by fully considering the constraints, while the actual incremental cost can be determined by Equations (9)–(11). Hence, all the constraints of distributed HEEM can be satisfied, while all the agents can reach a consensus on the incremental cost as much as possible.

1. *Diesel* generator: The electrical energy output can be modified as follows:

$$P_{Gi}^c = \left(\lambda_i^E - \beta_i\right)/2\gamma_i, \tag{17}$$

$$P_{Gi} = \begin{cases} P_{Gi}^{min}, & \text{if } P_{Gi}^c < P_{Gi}^{min} \\ P_{Gi}^c, & \text{if } P_{Gi}^{min} \leq P_{Gi}^c \leq P_{Gi}^{max} \\ P_{Gi}^{max}, & \text{if } P_{Gi}^c > P_{Gi}^{max} \end{cases}, \tag{18}$$

where $P_{Gi}{}^c$ is the consensus value of the electrical energy output of the *i*th energy supplier.

2. *Heat-only unit*: The heat energy output can be modified as follows:

$$H_{Gi}^c = \left(\lambda_i^H - \beta_i\right)/2\gamma_i \tag{19}$$

$$H_{Gi} = \begin{cases} H_{Gi}^{min}, & \text{if } H_{Gi}^c < H_{Gi}^{min} \\ H_{Gi}^c, & \text{if } H_{Gi}^{min} \leq H_{Gi}^c \leq H_{Gi}^{max} \\ H_{Gi}^{max}, & \text{if } H_{Gi}^c > H_{Gi}^{max} \end{cases}, \tag{20}$$

where $H_{Gi}{}^c$ is the consensus value of the heat energy output of the *i*th energy supplier.

3. *Energy consumer*: The electricity energy curtailment can be modified as follows:

$$\Delta P_{Di}^c = \left(P_{Di}^0 - a_i - b_i\lambda_i^E\right)/2, \tag{21}$$

$$\Delta P_{Di} = \begin{cases} 0, & \text{if } \Delta P_{Di}^c < 0 \\ \Delta P_{Di}^c, & \text{if } 0 \leq \Delta P_{Di}^c \leq \eta_i P_{Di}^0 \\ \eta_i P_{Di}^0, & \text{if } \Delta P_{Di}^c > \eta_i P_{Di}^0 \end{cases}, \tag{22}$$

where $\Delta P_{Di}{}^c$ is the consensus value of the electrical energy curtailment of the ith energy consumer.

4. *CHP unit*: Since the electrical and heat energy outputs are highly coupled, the incremental cost should be controlled to meet the energy balance constraints and the feasible operating region constraint. Hence, the feasible operating region is decomposed into eight searching sub-regions, See Figure 6, allowing the CHP unit to adjust its energy outputs based on the current energy mismatches and the consensus value of incremental costs, as given in Table 1.

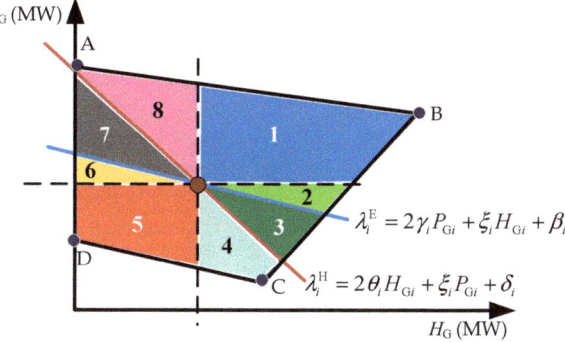

Figure 6. Eight searching sub-regions of the CHP unit.

Table 1. Adjusting rules of energy outputs of CHP unit.

$\Delta E > 0$	$\Delta H > 0$	$\lambda_i^E[k] > \lambda_i^{AE}[k-1]$	$\lambda_i^H[k] > \lambda_i^{AH}[k-1]$	(P_{Gi}, H_{Gi})
True	True	True	True	No adjustment
True	True	True	False	No adjustment
True	True	False	True	No adjustment
True	True	False	False	Sub-region #5
True	False	True	True	Sub-region #2
True	False	True	False	No adjustment
True	False	False	True	Sub-region #3
True	False	False	False	Sub-region #4
False	True	True	True	Sub-region #8
False	True	True	False	Sub-region #7
False	True	False	True	No adjustment
False	True	False	False	Sub-region #6
False	False	True	True	Sub-region #1
False	False	True	False	No adjustment
False	False	False	True	No adjustment
False	False	False	False	No adjustment

Note that the CHP unit does not need to adjust its electrical and heat energy outputs if the consensus requirement and energy balance constraints cannot be satisfied simultaneously. For instance, when both the current virtual incremental costs ($\lambda_i^E[k]$, $\lambda_i^H[k]$) are larger than the last actual incremental costs ($\lambda_i^{AE}[k-1]$, $\lambda_i^{AH}[k-1]$), while both the energy mismatches are positive ($\Delta E > 0$, $\Delta H > 0$), then the CHP unit will readjust the energy balances by increasing its incremental costs, thus its electrical and heat energy outputs will remain unchanged. In addition, when the CHP unit needs to adjust its energy outputs, the electrical and heat energy outputs can be updated according to the energy mismatches, as follows:

$$\begin{cases} P_{Gi}[k+1] = P_{Gi}[k] - \mu_E \Delta E \\ H_{Gi}[k+1] = H_{Gi}[k] - \mu_H \Delta H \end{cases} \qquad (23)$$

where μ_E and μ_H denote the adjustment factors of electrical and heat energy outputs, respectively.

If the operating point of the CHP unit is beyond the corresponding sub-region, then the electrical and heat energy outputs should be modified by the closest point (the point with the shortest Euclidean distance to the updated operating point) within the sub-region.

3.4. Execution Procedure

To sum up, the detailed execution procedure of ACA for distributed HEEM of an islanded microgrid is given in Algorithm 1, where τ is the energy mismatch tolerance, which is set to be 0.001 in this paper.

Algorithm 1. ACA for distributed HEEM.

1: Initial the algorithm parameters;

2: Design the interaction network among different agents;

3: Input the operating data of the current optimization task;

4: Calculate the electricity and heat energy mismatches by Equations (4) and (5);

5: **While** $|\Delta E| > \tau$ or $|\Delta H| > \tau$

6:      **If** $\Delta E \cdot \Delta H \geq 0$ then

7:          Update the virtual incremental cost of each agent by unified consensus Equation (15);

8:      **Else**

9:          Update the virtual incremental cost of each agent by independent consensus Equation (16);

10:      **End If**

11: Calculate the electricity energy output of each diesel generator by Equations (17) and (18);

12: Calculate the heat energy output of each heat-only unit by Equations (19) and (20);

13: Calculate the electricity energy curtailment of each energy consumer by Equations (21) and (22);

14: Modify the energy outputs of each CHP unit based on the adjusting rule in Table 1 and the eight searching sub-regions in Figure 6;

15: Calculate the electricity and heat energy mismatches by Equations (4) and (5);

16: Set $k: = k + 1$;

17: **End While**

Output the optimal energy dispatch strategy of each agent.

---

4. Case Studies

4.1. Simulation Model

In order to test the multi-energy dispatch, the islanded microgrid [18] with three PV units, two WTs, two diesel generators, one heat-only unit, two CHP units, and seven controllable energy consumers, is used for the simulation. Hence, both the electrical and heat parts are simultaneously considered in simulation, where the detailed mathematical model of distributed HEEM can be constructed by acquiring the operating constraints and operating cost function of energy for each supplier or consumer. Furthermore, the operating cost coefficients are given in Table 2; the physical topology is provided in Figure 7, and the interaction network among them is illustrated in Figure 8. In addition, three operating scenarios (i.e., scenarios #1 to #3) with different energy outputs of renewables (i.e., 0.8, 0.6, and 1 MW), instead of a single operating scenario, are designed for evaluating the optimization performance of different algorithms more scientifically. The adjustment factors μ, μ_E, and μ_H are set to be 10, 0.1, and 0.1, respectively. The following simulations will be carried out in

Matlab R2016a by a personal computer with Intel(R) Xeon (R) E5-2670 v3 CPU at 2.3 GHz with 64 GB of RAM.

Table 2. Operating cost coefficients of controllable units.

Type	No.	α_i	β_i	γ_i	δ_i	θ_i	ζ_i
Diesel generator	G_1	10.193	210.36	250.2	-	-	-
	G_2	2.305	301.4	1100	-	-	-
Heat-only unit	G_3	33	12.3	6.9	-	-	-
CHP unit	G_4	339.5	185.7	44.2	53.8	38.4	40
	G_5	100	288	34.5	21.6	21.6	8.8
Energy consumer	L_1	1	−0.002	-	-	-	-
	L_2	1	−0.002	-	-	-	-
	L_3	1	−0.001	-	-	-	-
	L_4	1	−0.001	-	-	-	-
	L_5	1	−0.001	-	-	-	-
	L_6	1	−0.0035	-	-	-	-
	L_7	1	−0.0035	-	-	-	-

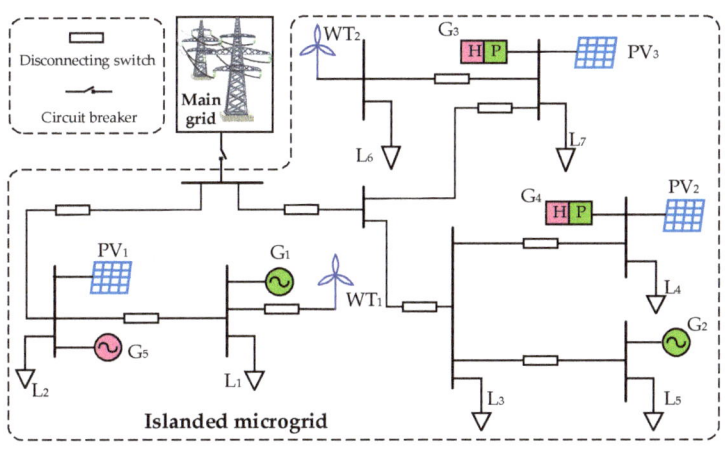

Figure 7. Physical topology of the testing islanded microgrid.

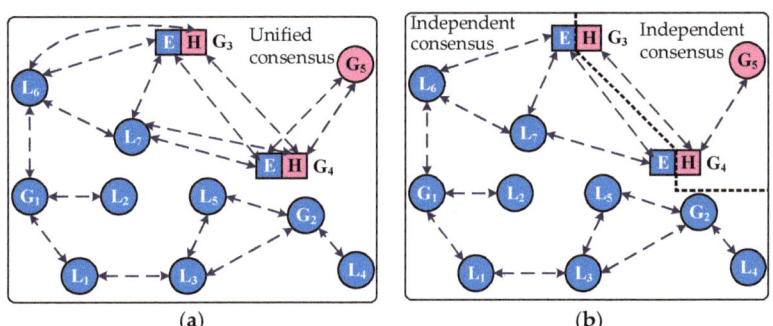

Figure 8. Interaction network of different agents. (**a**) Unified consensus; (**b**) independent consensus.

4.2. Study of Convergence

This case study is executed to reveal the convergence of ACA. Figure 9 shows the convergence process of ACA for distributed HEEM under scenario #1. It can be found from Figure 9a that the virtual incremental cost of each agent will update between unified consensus mode and independent consensus mode according to the dynamic energy mismatches, in which the incremental heat costs cannot reach a consensus with other incremental electrical costs due to the energy balance constraints Equations (4) and (5). Besides, some energy agents have reached their energy capability limits after a few interactions, as shown in Figure 9b. Moreover, two CHP units can adaptively adjust their energy outputs based on the adjusting rule in Table 1, see Figure 9c, where the zero searching sub-region indicates that the energy outputs of the CHP unit remain unchanged. Finally, both the electrical and heat energy mismatches (ΔE and ΔH) can simultaneously satisfy the energy mismatch tolerance after approximately 150 iterations, see Figure 9d, i.e., $|\Delta E| < \tau$ and $|\Delta H| < \tau$. All of this proves that the convergence of ACA can be effectively guaranteed, while the consensus requirement and all the constraints Equations (4)–(7) can be fully satisfied.

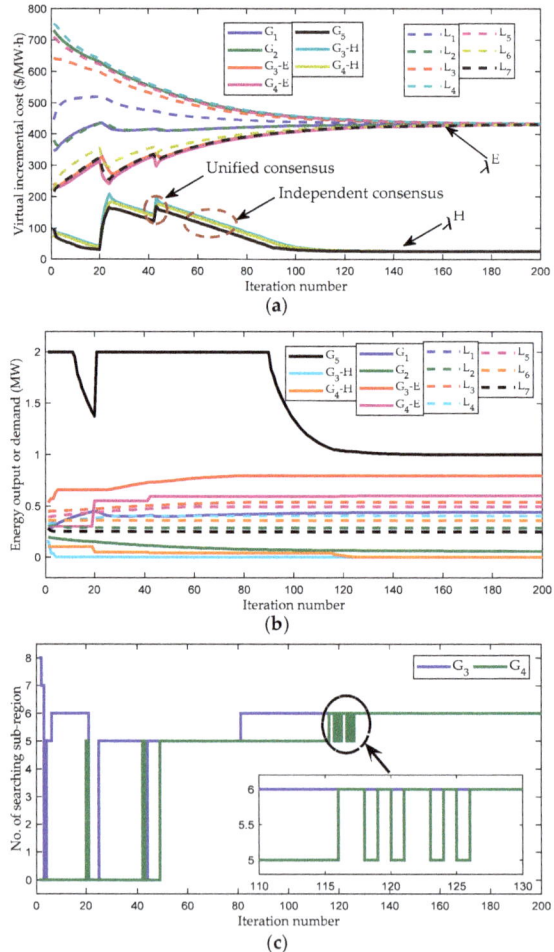

Figure 9. *Cont.*

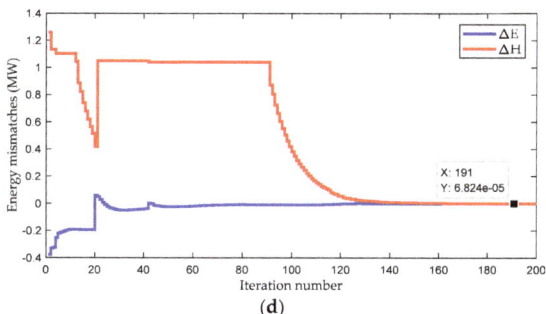

(**d**)

Figure 9. Convergence process of ACA for distributed HEEM. (**a**) Virtual incremental cost; (**b**) energy output or demand; (**c**) no. of searching sub-region of CHP units; (**d**) energy mismatches.

Here, the implementation period of distributed HEEM is set as 2 s for testing the real-time optimization performance of ACA. Note that the total time of each iteration includes the calculation time and information transmission time, which can be set as 1 ms with a conservative estimation. Figure 10 gives the real-time optimization of distributed HEEM obtained by ACA under three different scenarios as the total energy output of PV and WT units varies. It also verifies that ACA can converge to an optimal solution of distributed HEEM, while it can fully satisfy the real-time optimization of distributed HEEM because its convergence time is much shorter than the implementation period. Furthermore, it is clear that the incremental electrical cost decreases with the increasing electrical energy outputs of renewables.

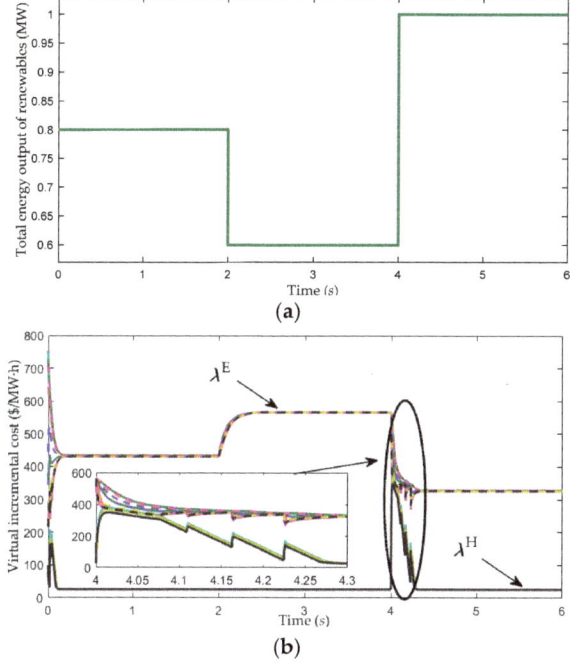

Figure 10. Real-time optimization of distributed HEEM by ACA. (**a**) Electrical energy fluctuation of renewables; (**b**) virtual incremental cost.

4.3. Comparative Results and Discussions

In order to further test the performance of ACA, four optimization algorithms, including genetic algorithm (GA) [22], interior point method (IPM) [23], distributed particle swarm optimization (DPSO) [24], and DDO [11], are introduced for comparisons, where the first two methods are centralized; and the latter two methods are distributed.

Table 3 provides the dispatch strategies obtained by different algorithms under scenario #1. It illustrates that ACA can converge to a high-quality optimum of distributed HEEM, which is very similar to the global optimum obtained by centralized IPM. Furthermore, the quality of the obtained optimum of GA is the lowest due to its premature convergence. This also demonstrates the effectiveness of ACA for distributed HEEM.

Table 3. Obtained dispatch strategies under scenario #1.

No.	Energy Type	Dispatch Strategy (MW)				
		GA	IPM	DPSO	DDO	ACA
G_1	Electrical	0.4013	0.3134	0.4150	0.3178	0.4427
G_2	Electrical	0.1617	0.0492	0.2000	0.0500	0.0602
G_3	Electrical	0.8448	0.9963	1.0000	0.9923	0.7962
	Heat	0.3182	0.0056	0.0000	0.0069	0.0000
G_4	Electrical	0.4588	0.5946	0.6000	0.5941	0.5999
	Heat	0.2952	0.0326	0.0000	0.0354	0.0000
G_5	Heat	0.3858	0.9619	1.0000	0.9577	1.0000
L_1	Electrical	0.3938	0.3771	0.4500	0.3744	0.3600
L_2	Electrical	0.3220	0.3310	0.3600	0.3353	0.2880
L_3	Electrical	0.4736	0.5359	0.5400	0.5360	0.5400
L_4	Electrical	0.3624	0.4030	0.4050	0.4034	0.4050
L_5	Electrical	0.4395	0.4919	0.4950	0.4921	0.4950
L_6	Electrical	0.3922	0.3614	0.4500	0.3605	0.3600
L_7	Electrical	0.2840	0.2531	0.3150	0.2524	0.2520
H_D	Heat	1	1	1	1	1
PV	Electrical	0.3	0.3	0.3	0.3	0.3
WT	Electrical	0.5	0.5	0.5	0.5	0.5
Total operating cost ($/h)		1201.48	1091.38	1153.99	1091.57	1113.91

Table 4 gives the comparison results obtained by three algorithms under different scenarios in 100 runs. It shows that IPM, DDO, and ACA always converge to the same optimum with a given initial solution and parameters as they are essentially the deterministic optimization algorithms. In contrast, both GA and DPSO often search different optimums in different runs due to their random heuristic operators. Furthermore, these two heuristic optimization algorithms also result in a much longer execution time than that of the other three methods, where the execution time of DPSO is shorter than that of GA due to its higher computation efficiency and distributed feature. Besides, the quality of optimum obtained by DDO is only lower than that of IPM, but its execution time is about four times that of ACA. Similarly, the optimum obtained by ACA is similar to the global optimum obtained by IPM, while the execution time is nearly the same. Hence, ACA is very suitable to yield the distributed HEEM because of its excellent performance regarding optimum quality and execution time.

Table 4. Comparison results obtained by five algorithms under different scenarios in 100 runs.

Scenario No.	Algorithm	Type	Execution Time (s)	Total Operating Cost ($/h)		
				Max	Avg	Min
#1	GA	Centr.	9.45	1250.29	1207.94	1148.17
	IPM	Centr.	0.32	1091.38	1091.38	1091.38
	DPSO	Distr.	4.17	1153.99	1151.90	1124.50
	DDO	Distr.	1.27	1091.57	1091.57	1091.57
	ACA	Distr.	0.20	1113.91	1113.91	1113.91
#2	GA	Centr.	9.16	1334.22	1299.12	1251.35
	IPM	Centr.	0.56	1168.46	1168.46	1168.46
	DPSO	Distr.	4.14	1228.44	1228.15	1226.12
	DDO	Distr.	1.49	1198.32	1198.32	1198.32
	ACA	Distr.	0.51	1212.54	1212.54	1212.54
#3	GA	Centr.	9.36	1178.54	1135.87	1083.35
	IPM	Centr.	0.26	1020.67	1020.67	1020.67
	DPSO	Distr.	4.42	1080.39	1078.45	1041.42
	DDO	Distr.	1.24	1020.77	1020.77	1020.77
	ACA	Distr.	0.34	1024.73	1024.73	1024.73

4.4. Scalability Test of ACA

This case study is used for testing the scalability of ACA for a larger scale system. In general, ACA will lead to a slower convergence rate for a larger scale microgrid with more agents. For testing the scalability of ACA, different scales of microgrid are designed based on the presented microgrid with 12 agents, in which the scales are 5, 10, 50, 100, and 500 times of the presented microgrid, respectively. Figure 11 shows the convergence process of ACA for two scales of microgrids under scenarios #3. It can be found that ACA can also converge to the optimal virtual incremental costs when the number of agents increases from 12 to 600. Although the number of agents increases by fifty-fold, the iteration number of convergence only increases from 340 to 488. In addition, Figure 12 provides statistical results of iteration number of convergence under different numbers of agents by ACA. Similarly, it shows that the iteration number of convergence increases marginally as the number of agents increases from 12 to 6000 under different scenarios. More specifically, the iteration number of convergence with 6000 agents is only 2.8 times of that with 12 agents under scenario #1. This reveals that ACA is suitable for real-world application with a high number of agents due to its superior scalability.

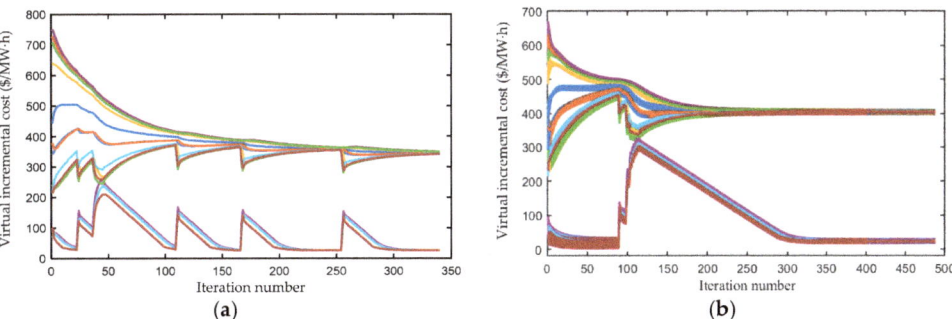

Figure 11. Convergence process of ACA under scenario #3. (**a**) 12 agents; (**b**) 600 agents.

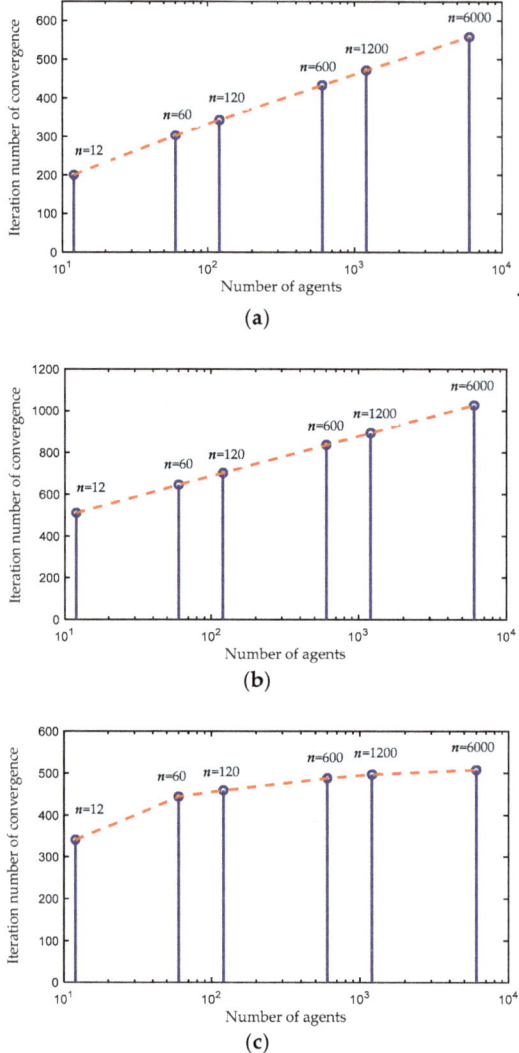

Figure 12. Iteration number of convergence under different numbers of agents by ACA. (**a**) Scenario #1; (**b**) Scenario #2; (**c**) Scenario #3.

5. Conclusions

In summary, this paper presents a novel ACA for distributed HEEM of an islanded microgrid, which has the following contributions:

1. The ACA based distributed HEEM can effectively address the multi-energy dispatch of an islanded microgrid in a simple distributed manner, while various constraints (e.g., the tight coupling features among various energies) can be completely satisfied.
2. The proposed eight searching sub-regions effectively make the CHP unit adaptively adjust its energy outputs to simultaneously meet the consensus requirement and the heat-electricity energy balance constraints.

3. Through the switch between unified consensus and independent consensus, ACA gradually converges to the optimal solution of the whole system according to the dynamic energy mismatches.

4. ACA can not only obtain a high-quality optimum of distributed HEEM, but also guarantee a short execution time. Hence, it can be generalized to be applied to other real-time distributed optimization issues of integrated energy systems.

Our future work will focus on improving the flexibility and generality of ACA by combining the model-free heuristic search or machine-learning mechanisms. Hence, it can handle more complex distributed optimization with high nonlinearity and nonconvexity, discontinuous and nondifferentiable objective functions.

Author Contributions: Data curation, X.D.; Investigation, X.Z.; Methodology, X.Z.; Project administration, X.D. and T.J.; Supervision, T.J.; Writing-original draft, X.Z.

Funding: This research was funded by [State Grid Corporation of China Science and Technology Project] grant number [2017-142].

Acknowledgments: The authors would like to gratefully acknowledge the State Grid Corporation of China.

Conflicts of Interest: The authors declare no potential conflict of interest.

References

1. Alsaidan, I.; Alannazi, A.; Gao, W.; Wu, H.; Khodaei, A. State-of-the-art in microgrid-integrated distributed energy storage sizing. *Energies* **2017**, *10*, 1421. [CrossRef]
2. Kim, H.-M.; Lim, Y.; Kinoshita, T. An intelligent multiagent system for autonomous microgrid operation. *Energies* **2012**, *5*, 3347–3362. [CrossRef]
3. Gao, H.; Chen, Y.; Xu, Y.; Liu, C.-C. Dynamic load shedding for islanded microgrid with limited generation resources. *IET Renew. Power Gener.* **2016**, *10*, 2953–2961. [CrossRef]
4. Lin, W.-M.; Tu, C.-S.; Tsai, M.-T. Energy management strategy for microgrid by using enhanced bee colony optimization. *Energies* **2016**, *9*, 5. [CrossRef]
5. Maghsoodlou, F.; Masiello, R.; Ray, T. Energy management systems. *IEEE Power Energy Mag.* **2004**, *2*, 49–57. [CrossRef]
6. Loia, V.; Vaccaro, A. Decentralized economic dispatch in smart grids by self-organizing dynamic agents. *IEEE Trans. Syst. Man Cybern. Syst.* **2014**, *44*, 397–408. [CrossRef]
7. Zheng, W.; Wu, W.; Zhang, B.; Sun, H.; Liu, Y. A fully distributed reactive power optimization and control method for active distribution networks. *IEEE Trans. Smart Grid* **2016**, *7*, 1021–1033. [CrossRef]
8. Amin, S.M. Smart grid security, privacy, and resilient architectures: Opportunities and challenges. In Proceedings of the 2012 IEEE Power and Energy Society General Meeting, San Diego, CA, USA, 22–26 July 2012; pp. 1–2.
9. Hu, J.; Chen, M.Z.Q.; Cao, J.; Guerrero, J.M. Coordinated active power dispatch for a microgrid via distributed lambda iteration. *IEEE Trans. Emerg. Sel. Top. Circuits Syst.* **2017**, *7*, 250–261. [CrossRef]
10. Mojica-Nava, E.; Barreto, C.; Quijano, N. Population games methods for distributed control of microgrids. *IEEE Trans. Smart Grid* **2015**, *6*, 2586–2595. [CrossRef]
11. Yang, Z.; Wu, R.; Yang, J.; Long, K.; You, P. Economical operation of microgrid with various devices via distributed optimization. *IEEE Trans. Smart Grid* **2016**, *7*, 857–867. [CrossRef]
12. Xu, Y.; Li, Z. Distributed optimal resource management based on the consensus algorithm in a microgrid. *IEEE Trans. Ind. Electron.* **2015**, *62*, 2584–2592. [CrossRef]
13. Hug, G.; Kar, S.; Wu, C. Consensus + innovations approach for distributed multiagent coordination in a microgrid. *IEEE Trans. Smart Grid* **2015**, *6*, 1893–1903. [CrossRef]
14. Han, R.; Meng, L.; Ferrari-Trecate, G.; Coelho, E.A.A.; Vasquez, J.C.; Guerrero, J.M. Containment and consensus-based distributed coordination control to achieve bounded voltage and precise reactive power sharing in islanded AC microgrids. *IEEE Trans. Ind. Appl.* **2017**, *53*, 5187–5199. [CrossRef]
15. Chen, G.; Zhao, Z. Delay effect on consensus–based distributed economic dispatch algorithm in microgrid. *IEEE Trans. Power Syst.* **2018**, *33*, 602–612. [CrossRef]

16. Zhao, T.; Ding, Z. Distributed agent consensus-based optimal resource management for microgrid. *IEEE Trans. Sustain. Energy* **2018**, *9*, 443–452. [CrossRef]

17. Zhang, X.; Xu, H.; Yu, T.; Yang, B.; Xu, M. Robust collaborative consensus algorithm for decentralized economic dispatch with a practical communication network. *Electr. Power Syst. Res.* **2016**, *140*, 597–610. [CrossRef]

18. Liu, N.; Wang, J.; Wang, L. Distributed energy management for interconnected operation of combined heat and power-based microgrids with demand response. *J. Mod. Power Syst. Clean Energy* **2017**, *5*, 478–488. [CrossRef]

19. Zhang, G.; Cao, Y.; Cao, Y.; Li, D.; Wang, L. Optimal energy management for microgrids with combined heat and power (CHP) generation, energy storage, and renewable energy sources. *Energies* **2017**, *10*, 1288. [CrossRef]

20. Zhang, Z.; Chow, M.-Y. Convergence analysis of the incremental cost consensus algorithm under different communication network topologies in a smart grid. *IEEE Trans. Power Syst.* **2012**, *27*, 1761–1768. [CrossRef]

21. Gibbons, A. *Algorithmic Graph Theor*; Cambridge University Press: Cambridge, UK, 1985.

22. Walters, D.C.; Sheble, G.B. Genetic algorithm solution of economic dispatch with value point loading. *IEEE Trans. Power Syst.* **1993**, *8*, 1325–1332. [CrossRef]

23. Duvvuru, N.; Swarup, K.S. A hybrid interior point assisted differential evolution algorithm for economic dispatch. *IEEE Trans. Power Syst.* **2011**, *26*, 541–549. [CrossRef]

24. Erdeljan, A.; Capko, D.; Vukmirovic, S.; Bojanic, D.; Congradac, V. Distributed PSO algorithm for data model partitioning in power distribution systems. *J. Appl. Res. Technol.* **2014**, *12*, 947–957. [CrossRef]

Article

A Hybrid DA-PSO Optimization Algorithm for Multiobjective Optimal Power Flow Problems

Sirote Khunkitti [1], Apirat Siritaratiwat [1], Suttichai Premrudeepreechacharn [2], Rongrit Chatthaworn [1,*] and Neville R. Watson [3]

[1]   Department of Electrical Engineering, Faculty of Engineering, Khon Kaen University,
      Khon Kaen 40002, Thailand; sirote_khunkitti@kkumail.com (S.K.); apirat.siritaratiwat@gmail.com (A.S.)
[2]   Department of Electrical Engineering, Faculty of Engineering, Chiang Mai University,
      Chiang Mai 50200, Thailand; suttic@eng.cmu.ac.th
[3]   Department of Electrical and Computer Engineering, University of Canterbury,
      Christchurch 8140, New Zealand; neville.watson@canterbury.ac.nz
*   Correspondence: rongch@kku.ac.th; Tel.: +66-84-685-2286

Received: 3 August 2018; Accepted: 27 August 2018; Published: 29 August 2018

Abstract: In this paper, a hybrid optimization algorithm is proposed to solve multiobjective optimal power flow problems (MO-OPF) in a power system. The hybrid algorithm, named DA-PSO, combines the frameworks of the dragonfly algorithm (DA) and particle swarm optimization (PSO) to find the optimized solutions for the power system. The hybrid algorithm adopts the exploration and exploitation phases of the DA and PSO algorithms, respectively, and was implemented to solve the MO-OPF problem. The objective functions of the OPF were minimization of fuel cost, emissions, and transmission losses. The standard IEEE 30-bus and 57-bus systems were employed to investigate the performance of the proposed algorithm. The simulation results were compared with those in the literature to show the superiority of the proposed algorithm over several other algorithms; however, the time computation of DA-PSO is slower than DA and PSO due to the sequential computation of DA and PSO.

Keywords: dragonfly algorithm; metaheuristic; optimal power flow; particle swarm optimization

---

1. Introduction

For the past few decades, the optimal power flow (OPF) problem has played an essential role in studying the economy terms of power systems [1,2]. The OPF problem is a nonlinear, nonconvex, large-scale, and static programming problem [3] that optimizes selected objective functions while satisfying a set of equality and inequality constraints. The power balance equations are the equality constraints, and the limits of state and control variables are the inequality constraints of the OPF problem. The state variables consist of slack bus active power generation, load bus voltages, reactive power generation, and apparent power flow. The control variables involve active power generation except at slack bus, generator bus voltages, tap ratios of transformers, and reactive powers of shunt compensation capacitors. In recent years, because of the rise in fuel cost, which increases generation cost, fuel cost has become the objective function to be optimized in the OPF problem. Moreover, due to the release of emissions from thermal power plants into the atmosphere, emissions are yet another concern for power system operation and planning [4]. At the same time, because the demand for electricity has outpaced the expansion of transmission capacity, the inadequate reactive power sources of power systems have increased losses in transmission lines. Thus, emissions and transmission losses must also be considered as part of the objective functions of the OPF problem.

To solve the OPF problem, several traditional optimization techniques, such as nonlinear programming [5], quadratic programming [6], and the interior point method [7], have been successfully applied. However, these algorithms' nonlinear characteristics make them impractical to use in practical systems. The nonlinear characteristics may cause the obtained solutions to be trapped in local optima, and these algorithms require an enormous amount of computational effort and time. Therefore, many optimization methods need to be improved to overcome these shortcomings [8,9]. Recently, several population-based optimization algorithms, including the OPF problem, have been employed to solve a complex constrained optimization problem in the field of power systems. Some of the other proposed techniques include the genetic algorithm (GA) [10], tabu search (TS) [11], differential evolution (DE) [12], evolutionary programming (EP) [13], probabilistic optimal power-flow (P-OPF) [14], preventive security-constrained power flow optimization [15], ant colony optimization (ACO) [16], grey wolf optimizer (GWO) [17], artificial bee colony (ABC) [18], particle swarm optimization (PSO) [19], and the dragonfly algorithm (DA) [20]. Even with the successful optimization of single-objective population-based optimization techniques, minimizing only one objective function is not sufficient in the power system because there are many problems, such as fuel cost, emissions, and transmission losses, which also need to be minimized. Consequently, many objective functions should be considered because this is a multi objective optimization problem. Since there are three independent objective functions in this study (i.e., fuel cost, emissions, and transmission losses), the number of incompatible optimal solutions between the objective functions is infinite, and these optimal solutions are called Pareto optimal solutions [21].

Several optimization algorithms have been proposed and applied to solve the multiobjective OPF (MO-OPF) problem by many researchers. One of these methods was carried out by converting the multiobjective problem into a single-objective problem and then solving the problem by using a single-objective optimizer. However, this method has some drawbacks, such as the limitation of the available choices, the need for weights for each objective, and the requirement of multiple optimizer runs. To overcome these weaknesses, many researchers have proposed multiobjective evolutionary algorithms, such as the improved strength Pareto evolutionary algorithm (ISPEA2) [22], hybrid modified particle swarm optimization-shuffle frog leaping algorithms (HMPSO-SFLA) [23], modified teaching–learning-based optimization (MTLBO) [24], GWO [17], DE [17], multiobjective modified imperialist competitive algorithm (MOMICA) [25], differential search algorithm (DSA) [26], modified shuffle frog leaping algorithm (MSFLA) [27], modified Gaussian bare-bones multiobjective imperialist competitive algorithm (MGBICA) [28], multiobjective harmony search (MOHS) [29], adaptive real coded biogeography-based optimization (ARCBBO) [30], multiobjective differential evolution algorithm (MO-DEA) [31], hybrid modified imperialist competitive algorithm and teaching–learning algorithm (MICA-TLA) [32], etc., to successfully solve the OPF problem. In the past few decades, various well-proposed multiobjective evolutionary algorithms have been successfully applied and improved in many applications; however, most of them have not been extensively investigated in the OPF problem. Moreover, improving the search performance of the multiobjective evolutionary algorithm for solving the OPF problem is also important. In this paper, a hybrid DA-PSO algorithm is proposed to deal with the MO-OPF problem. The concept of the hybrid algorithm is the combination of the exploration and exploitation phases of the DA and PSO algorithms, respectively. The performance of the proposed algorithm was evaluated on the standard IEEE 30-bus and IEEE 57-bus power systems. Three different objective functions—fuel cost, emissions, and transmission losses—were individually and simultaneously considered as parts of the objective function in the OPF problem. The obtained results were compared with other evolutionary algorithms and the traditional DA and PSO.

The rest of the article is classified into five sections as follows. Section 2 introduces the formulation and constraints of the multiobjective optimization. In Section 3, the traditional DA and PSO are explained, and Section 4 depicts the concept of the proposed algorithm. Section 5 presents the optimization results and the comparisons between the solutions from the proposed algorithm and the solution from other algorithms based on IEEE 30-bus and IEEE 57-bus systems. Finally, in Section 6, the conclusions of the simulation results of the proposed algorithm are described.

2. Problem Formulation and Constraints for Multi Objective Optimization for OPF

Multi-objective optimization is a model that optimizes more than one objective function to find optimal control variables while simultaneously satisfying equality and inequality constraints. The compromised solutions, nondominated solutions, which have more than one optimal solution between each objective, are the optimal solutions referred to as the Pareto front. The multiobjective problem is mathematically formulated as follows:

$$\min f = \left\{ f_1(x, u), f_2(x, u), \dots, f_{N_{obj}}(x, u) \right\} \tag{1}$$

subject to

$$g(x, u) = 0 \tag{2}$$

$$h(x, u) \leq 0 \tag{3}$$

where f is a vector of objective functions to be optimized, N_{obj} is the number of objective functions, $g(x,u)$ are the equality constraints, and $h(x,u)$ are the inequality constraints.

x is a vector of state variables including slack bus active power, load bus voltages, generator reactive powers, and apparent power flows, expressed as follows:

$$x = [P_{gslack}, V_{L1}, \dots, V_{LN_L}, Q_{g1} \dots Q_{gN_{gen}}, S_{l1} \dots S_{lN_l}] \tag{4}$$

where P_{gslack} is the active power generation at slack bus, V_{Li} is the load voltage at bus i, N_L is number of load buses, Q_{gi} is the reactive power generation at bus i, N_{gen} is the number of total generators, S_{li} is the apparent power flow at branch i, and N_l is the number of transmission lines.

u is a vector of control variables consisting of active power generations except at slack bus, generator bus voltages, transformer tap ratios, and reactive powers of shunt compensation capacitors, expressed as:

$$u = [P_{gi;i \in PV_{bus}} \dots P_{gN_{gen}}, V_{g1}, \dots, V_{gN_{gen}}, T_1 \dots T_{N_{tran}}, Q_{c1} \dots Q_{cN_{cap}}] \tag{5}$$

where P_{gi} is the active power generation at bus i, PV_{bus} is the set of generator buses except at slack bus, V_{gi} is the generator bus voltage at bus i, T_i is the transformer tap ratio at bus i, N_{tran} is the number of transformer taps, Q_{ci} is the shunt compensation capacitor at bus i, and N_{cap} is the number of compensation capacitors.

2.1. Objective Functions

In this study, the objective functions of the OPF, consisting of fuel cost, emissions, and transmission line losses, are considered as shown below.

2.1.1. Fuel Cost

The total fuel cost of the generators is considered to be minimized and is given as follows:

$$f_C(x, u) = \sum_{i=1}^{N_{gen}} (a_i P_{gi}^2 + b_i P_{gi} + c_i) \tag{6}$$

where f_C is the total fuel cost of generators function (\$/h), and a_i, b_i and c_i are the fuel cost coefficients of the ith generator units.

2.1.2. Emissions

The emissions function can be represented as the sum of all considered emission types, such as sulphur oxides (SO_x), nitrogen oxides (NO_x), thermal emission, etc. However, in the present study, two important emission types, NO_x and SO_x, are taken into account, as expressed below:

$$f_E(\boldsymbol{x}, \boldsymbol{u}) = \sum_{i=1}^{N_{gen}} \left(\gamma_i P_{gi}^2 + \beta_i P_{gi} + \alpha_i + \xi_i \exp(\lambda_i P_{gi}) \right) \tag{7}$$

where f_E is the total emission generations function (ton/h), and γ_i, β_i, α_i, ξ_i and λ_i are emission coefficients of the ith generator units.

2.1.3. Transmission Line Losses

The system active power loss in the transmission line is formulated as follows:

$$f_L(\boldsymbol{x}, \boldsymbol{u}) = \sum_{k=1}^{N_l} g_k (V_i^2 + V_j^2 - 2V_i V_j \cos(\theta_{ij})) \tag{8}$$

where f_L is the total transmission loss function (MW), g_k is the conductance of the kth line, V_i is the voltage at bus i, V_j is the voltage at bus j, and θ_{ij} is the voltage phase angle difference between buses i and j.

2.2. Constraints

2.2.1. Equality Constraints

The OPF equality constraints are the active and reactive power balance constraints, as follows:

$$P_{gi} - P_{di} = \sum_{j=1}^{N_{bus}} V_i V_j (G_{ij} \cos(\theta_{ij}) + B_{ij} \sin(\theta_{ij})) \tag{9}$$

$$Q_{gi} - Q_{di} = \sum_{j=1}^{N_{bus}} V_i V_j (G_{ij} \sin(\theta_{ij}) - B_{ij} \cos(\theta_{ij})) \tag{10}$$

where P_{di} is the active power demand at bus i, N_{bus} is the number of buses, G_{ij} is the transfer conductance between buses i and j, B_{ij} is the transfer susceptance between buses i and j, and Q_{di} is the reactive power demand at bus i.

2.2.2. Inequality Constraints

$$P_{gimin} \leq P_{gi} \leq P_{gimax} \quad i = 1, 2, \ldots, N_{gen} \tag{11}$$

$$Q_{gimin} \leq Q_{gi} \leq Q_{gimax} \quad i = 1, 2, \ldots, N_{gen} \tag{12}$$

$$V_{gimin} \leq V_{gi} \leq V_{gimax} \quad i = 1, 2, \ldots, N_{gen} \tag{13}$$

$$|S_{li}| \leq S_{limax} \tag{14}$$

$$V_{Limin} \leq V_{Li} \leq V_{Limax} \quad i = 1, 2, \ldots, N_L \tag{15}$$

$$Q_{cimin} \leq Q_{ci} \leq Q_{cimax} \quad i = 1, 2, \ldots, N_{cap} \tag{16}$$

$$T_{imin} \leq T_i \leq T_{imax} \quad i = 1, 2, \ldots, N_{tran} \tag{17}$$

where P_{gimin} and P_{gimax} are the minimum and maximum active power generations at bus i, respectively, Q_{gimin} and Q_{gimax} are the minimum and maximum reactive power generations at bus i, respectively,

V_{gimin}, V_{gimax} are the minimum and maximum generator voltage at bus i, respectively, S_{limax} is the maximum apparent power flow at branch i, V_{Limin}, V_{Limax} are the minimum and maximum load voltage at bus i, respectively, Q_{cimin} and Q_{cimax} are the minimum and maximum shunt compensation capacitor at bus i, respectively, T_{imin}, T_{imax} are the minimum and maximum transformer tap-ratio at bus i, respectively.

2.2.3. Constraints Handling

The inequality of dependent variables, including slack bus active power generation, load bus voltage magnitudes, reactive power generations, and apparent power flows, are integrated into the penalized objective function to maintain these variables within their limits and to refuse infeasible solutions. The penalty function can be expressed as follows [27]:

$$J(x, u) = f(x, u) + K_P (P_{gslack} - P_{gslack}^{lim})^2 + K_V \sum_{i=1}^{N_{load}} (V_{Li} - V_{Li}^{lim})^2$$
$$+ K_Q \sum_{i=1}^{N_{line}} (Q_{gi} - Q_{gi}^{lim})^2 + K_S \sum_{i=1}^{N_{line}} (S_{Li} - S_{Li}^{max})^2 \tag{18}$$

where $J(x,u)$ is the penalized objective function, K_p, K_Q, K_V and K_s are the penalty factors, and x^{lim} is the limit value of the dependent variables, determined as follows:

$$x^{lim} = \begin{cases} x^{max} & if & x > x^{max} \\ x & if & x^{min} < x < x^{max} \\ x^{min} & if & x < x^{min} \end{cases} \tag{19}$$

3. Related Optimization Techniques

3.1. DA

DA is a metaheuristic algorithm which was inspired by the static and dynamic swarming behaviors of dragonflies in nature [33]. Dragonflies swarm for two goals: Hunting (static swarm) and migration (dynamic swarm). In the dynamic swarm, many dragonflies swarm when roaming over long distances and different areas, which is the purpose of the exploration phase. In the static swarm, dragonflies move in larger swarms and along one direction with local movements and sudden changes in the flying path, which is suitable in the exploitation phase.

The behavior of dragonflies can be represented through five principles, which are separation, alignment, cohesion, attraction to a food source, and distraction of an enemy. These five behaviors are described and calculated as follows:

Separation, which is the avoidance of the static crashing of individuals into other individuals in the neighborhood, is calculated by Equation (20).

$$S_i = -\sum_{j=1}^{N} X - X_j \tag{20}$$

where S_i is the separation of the ith individual, N is the number of neighboring individuals, X is the position of the current individual, and X_j is the position of jth neighboring individual.

Alignment, which refers to the velocity matching of individuals to the velocity of others in the neighborhood, is computed by Equation (21).

$$A_i = \frac{\sum_{j=1}^{N} V_j}{N} \tag{21}$$

where A_i is the alignment of the ith individual, and V_j is the velocity of jth neighboring individual.

Cohesion, which is the propensity of individuals towards the center of mass of the neighborhood, is formulated by Equation (22).

$$C_i = \frac{\sum_{j=1}^{N} X_j}{N} - X \tag{22}$$

where C_i is the cohesion of the ith individual

Attraction towards a food source computed by Equation (23), should be the main objective of any swarm to survive.

$$F_i = X^+ - X \tag{23}$$

where F_i is the food source of the ith individual, and X^+ is the position of the food source.

Distraction of an enemy, which is computed by Equation (24), is another survival objective of the swarm.

$$E_i = X^- + X \tag{24}$$

where E_i is the position of enemy of the ith individual, and X^- is the position of the enemy source.

To simulate the movement of artificial dragonflies and update their positions, step vector (ΔX) and position vector (X) are considered. The step vector represents the direction of the movement of the artificial dragonflies and is formulated as follows:

$$\Delta X^{t+1} = (sS_i + aA_i + cC_i + fF_i + eE_i) + \omega^t \Delta X^t \tag{25}$$

where ΔX^{t+1} is the step vector at iteration $t + 1$, ΔX^t is the step vector at iteration t, s, a, c, f and e are the separation weight, alignment weight, cohesion weight, food factor and enemy factor, respectively, and ω^t is the inertia weight factor at iteration t and is calculated by Equation (26).

$$\omega^t = \omega_{max} - \frac{\omega_{max} - \omega_{min}}{Iter_{max}} \times Iter \tag{26}$$

where ω_{max} and ω_{min} are set to 0.9 and 0.4, respectively, $Iter$ is the iteration, and $Iter_{max}$ is the maximum iteration.

The position of the artificial dragonflies can be updated by the following equation:

$$X^{t+1} = X^t + \Delta X^{t+1} \tag{27}$$

where X^{t+1} is the position at iteration $t + 1$, and X^t is the position at iteration t.

When the search space does not have a neighboring solution, the artificial dragonflies need to move around the search space by applying random walk (Levy flight) to improve their stochastic behavior. So, in this case, the position of the dragonflies can be calculated by Equation (28).

$$X^{t+1} = X^t + Levy(d) \times X^t \tag{28}$$

where d is the dimension of the position vectors, and the *Levy* is the Levy flight which is computed by Equation (29).

$$Levy(d) = 0.01 \times \frac{r_1 \times \sigma}{|r_2|^{\frac{1}{\beta}}} \tag{29}$$

where r_1 and r_2 are two uniform random values in a range of [0, 1], and σ is calculated by Equation (30).

$$\sigma = \left(\frac{\Gamma(1+\beta) \times \sin\left(\frac{\pi\beta}{2}\right)}{\Gamma\left(\frac{1+\beta}{2}\right) \times \beta \times 2^{\left(\frac{\beta-1}{2}\right)}} \right)^{1/\beta} \tag{30}$$

where β is the constant (which is equal to 1.5 in this work), and $\Gamma(x) = (x - 1)!$.

3.2. PSO

PSO is a population-based stochastic global optimization technique which was first introduced by Eberhart and Kennedy [34]. The idea of PSO came from the flocking behavior of birds or the schooling of fishes in their food hunting. In the PSO system, the population moves around a multidimensional search space where each particle represents a possible solution. Each particle contains the information of control variables and is associated with a fitness value that indicates its performance in the fitness space. Each particle i consists of its position $X_i = (x_{i,1}, x_{i,2}, \ldots, x_{i,Nvar})$, where $Nvar$ represents the number of control variables, velocity $V_i = (v_{i,1}, v_{i,2}, \ldots, v_{i,Nvar})$ and personal best experience $X_{pbesti} = (x_{pbesti,1}, x_{pbesti,2}, \ldots, x_{pbesti,Nvar})$, and a swarm has a global best experience $X_{gbest} = (x_{gbest1}, x_{gbest2}, \ldots, x_{gbestNvar})$. During each iteration, each particle moves in the direction of its own personal best position provided so far as well as in the direction of the global best position obtained so far by particles in the swarm. The particles are operated according to the equations expressed as follows:

$$V_i^{t+1} = \omega^t \times V_i^t + C_1 \times rand_1 \times (X_{pbest_i}^t - X_i^t) + C_2 \times rand_2 \times (X_{gbest}^t - X_i^t) \tag{31}$$

$$X_i^{t+1} = X_i^t + V_i^{t+1} \tag{32}$$

where V_i^{t+1} is the velocity of particle i at iteration $t + 1$, V_i^t is the velocity of particle i at iteration t, C_1 and C_2 are two positive acceleration constants, $rand_1$ and $rand_2$ are two uniform random values in a range of [0, 1], X_{pbesti}^t is the personal best position of particle i at iteration t, X_i^t is the position of particle i at iteration t, X_{gbest}^t is the global best position among all particles at iteration t, and X_i^{t+1} is the position of particle i at iteration $t + 1$.

4. Proposed Hybrid DA-PSO Optimization Algorithm for MO-OPF Problem

Many optimization algorithms have been proposed to overcome the optimization problem of being trapped in the local optima while the algorithms try to find the best solution. PSO has been proven in several works from the literature to find the optimal solution in various problems [35–38]. Because of its equations in finding the optimal solution by using the best experience of the particles, PSO could quickly converge on the optimal solution, i.e., it is good at exploitation. However, PSO is sometimes still trapped in the local optima because it converges on the optimal solution too quickly. In other words, PSO is poor at exploration, which is an important task of the optimization process. In DA, it applies the Levy flight to improve the randomness and stochastic behavior when there is no neighboring dragonfly. This could significantly improve the exploration process of the algorithm. However, the best experience, which is the personal best, of dragonflies is not applied during the operation. This causes the DA to converge on the optimal solution very slowly and can sometimes cause it to be trapped in the local optima. To overcome these problems, a new algorithm is proposed which combines the prominent points of the DA and PSO algorithms, which are the exploration of DA and the exploitation of PSO. At first, the dragonflies in DA are initialized to explore the search space to find the area of the global solution. Then, the best position of DA is obtained. The obtained best position from DA is then substituted as the global best position in the PSO equation (Equation (31)). After that, the PSO algorithm, which is the exploitation phase, operates by using the global best position from DA, allowing it to provide the expected optimal solution. The velocity and position equations of PSO can be modified as follows:

$$V_i^{t+1} = \omega^t \times V_i^t + C_1 \times rand_1 \times (X_{pbest_i}^t - X_i^t) + C_2 \times rand_2 \times (X_{DA}^{t+1} - X_i^t) \tag{33}$$

$$X_i^{t+1} = X_i^t + V_i^{t+1} \tag{34}$$

where X_{DA}^{t+1} is the best position obtained from DA at iteration $t + 1$.

The application of the proposed DA-PSO algorithm for solving the MO-OPF problem can be described as follows:

Step 1. Clarify the system data comprising the fuel cost coefficients of the generators, emission coefficients of the generators, initial values of generator active powers, initial values of generator bus voltages, initial values of transformer tap ratios, initial values of shunt compensation capacitors, upper limit of S_{li}, lower and upper limits of P_{gi}, Q_{gi}, V_{gi}, V_{Li}, Q_{ci}, and T_i, the parameters of DA and PSO, the number of dragonflies and particles, the number of iterations, and the archive size.

Step 2. Generate the initial population of dragonflies and particles.

Step 3. Convert the constrained multi objective problem to an unconstrained one by using Equation (18).

Step 4. Perform the power flow and calculate the objective functions for the initial population of dragonflies.

Step 5. Find the nondominated solutions and save them to the initial archive.

Step 6. Set the fitness value of the initial population as the food source.

Step 7. Calculate the parameters of DA (s, a, c, f, and e).

Step 8. Update the food source and enemy of DA.

Step 9. Calculate the S, A, C, F, and E by Equations (20)–(24).

Step 10. Check if a dragonfly has at least one neighboring dragonfly, then update step vector (ΔX) and the position of dragonfly (X_{DA}) by Equations (25) and (27), respectively, and if each dragonfly has no neighboring dragonfly, then update X_{DA} by Equation (28) and set ΔX to be zero.

Step 11. If any component of each population breaks its limit, then ΔX or X_{DA} of that population is moved into its minimum/maximum limit.

Step 12. Set the best position obtained from DA as the global best of PSO (X^{gbest}).

Step 13. Update the velocity of the particle (V) and the position of the particle (X_{PSO}) by Equations (33) and (34), respectively.

Step 14. If any component of each population breaks its limit, then V or X_{PSO} of that population is moved into its minimum/maximum limit.

Step 15. Calculate the objective functions of the new produced population.

Step 16. Employ the Pareto front method to save the nondominated solutions to the archive and update the archive.

Step 17. If the maximum number of iterations is reached, the algorithm is stopped; otherwise, go to step 7.

The flowchart of the DA-PSO algorithm for the MO-OPF problem is shown in Figure 1.

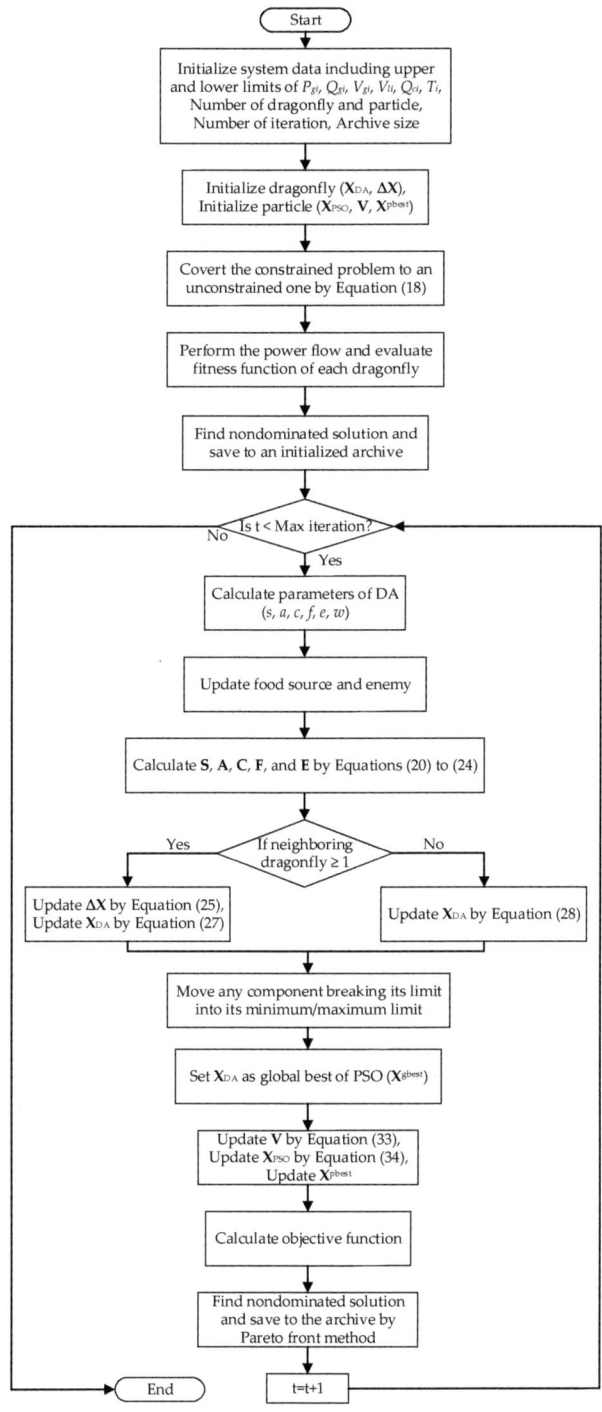

Figure 1. Flowchart of the dragonfly algorithm and particle swarm optimization (DA-PSO) algorithm for solving the multiobjective optimal power flow (MO-OPF) problem.

Energies **2018**, *11*, 2270

5. Simulation Results

To investigate the performance of the proposed algorithm, the IEEE 30-bus and IEEE 57-bus test systems were employed. The proposed algorithm operated for 30 independent runs for each test system. To validate the superiority of the proposed algorithm for solving the economic dispatch optimization problem, the results provided by the proposed algorithm were compared with those of other metaheuristic algorithms from the literature. In order to investigate both the single-objective optimization and multiobjective optimization, the simulation was divided into two cases. Single-objective optimization was evaluated in the first case. In the second case, multiobjective optimization to solve the MO-OPF by using the proposed DA-PSO algorithm was evaluated.

5.1. IEEE 30-Bus Test System

The proposed DA-PSO algorithm was applied to the IEEE 30-bus system to evaluate its performance. The IEEE 30-bus test system was composed of 6 generators at buses 1, 2, 5, 8, 11, and 13, 4 transformers between buses 6 and 9, buses 6 and 10, buses 4 and 12, and buses 27 and 28, and 41 transmission lines. The total system demand was 283.4 MW and 126.6 MVAR. The bus and branch data is given in [39]. The population number and the size of the Pareto archive were set to be 100 and 100, respectively.

5.1.1. Single-Objective OPF

To evaluate the performance of the proposed algorithm for solving the single-objective optimization, three different objective functions consisting of fuel cost, emissions, and transmission loss minimizations were individually considered as part of the objective function. The obtained results by the traditional DA, PSO, and the proposed DA-PSO algorithm for three individual objective functions are shown in Table 1. In Table 2, the best fuel cost of generators provided by the DA-PSO algorithm are compared with other algorithms in the literature, including TS [11], EP [13], ACO [16], SFLA [27], MSFLA [27], improved evolutionary programming (IEP) [40], modified differential evolution optimal power flow (MDE-OPF) [41], stochastic genetic algorithm (SGA) [42], evolutionary-programming-based optimal power flow (EP-OPF) [43], honey bee mating optimization (HBMO) [44], PSO, and DA. The comparison of the best emission values of the DA-PSO algorithm with various algorithms in the literature, including ACO [16], HMPSO-SFLA [23], TLBO [24], MTLBO [24], DSA [26], MSFLA [27], SFLA [27], GA [27], GBICA [28], improved particle swarm optimization (IPSO) [45], PSO, and DA, is shown in Table 3. In Table 4, the best transmission losses provided by DA-PSO algorithm are compared with other algorithms in the literature, including GWO [17], DE [17], MOHS [29], enhanced genetic algorithm with decoupled quadratic load flow (EGA-DQLF) [46], efficient evolutionary algorithm (EEA) [47], enhanced genetic algorithm (EGA) [47], PSO, and DA. It can be seen that the proposed DA-PSO provided better results compared with those of other algorithms for all three objective functions, which can be confirmed by the results in Tables 1–4. However, the computation time of the proposed DA-PSO is much slower than other algorithms in the literature because the proposed algorithm consumed the sequential computation time of DA and PSO as presented in Tables 2–4.

Table 1. Comparison of the simulation results from particle swarm optimization (PSO), dragonfly algorithm (DA), and DA-PSO for IEEE 30-bus system.

Variables	Best Fuel Cost			Best Emission			Best P_{Loss}		
	PSO	DA	DA-PSO	PSO	DA	DA-PSO	PSO	DA	DA-PSO
P_{g1} (MW)	176.2376	176.5128	176.1861	64.1678	64.3407	64.0997	51.6974	51.5987	51.5893
P_{g2} (MW)	48.8432	48.6955	48.8318	67.6692	67.5383	67.6295	80.0000	80.0000	80.0000
P_{g3} (MW)	21.5184	21.4431	21.5119	50.0000	50.0000	50.0000	50.0000	50.0000	50.0000
P_{g4} (MW)	22.1257	22.0995	22.0737	35.0000	35.0000	35.0000	35.0000	35.0000	35.0000
P_{g5} (MW)	12.2000	12.0673	12.2005	30.0000	30.0000	30.0000	30.0000	40.0000	30.0000
P_{g6} (MW)	12.0000	12.0091	12.0000	40.0000	40.0000	40.0000	40.0000	40.0000	40.0000
V_{g1} (p.u.)	1.0500	1.0500	1.0500	1.0500	1.0500	1.0500	1.0500	1.0500	1.0500

Table 1. *Cont.*

Variables	Best Fuel Cost			Best Emission			Best P_Loss		
	PSO	DA	DA-PSO	PSO	DA	DA-PSO	PSO	DA	DA-PSO
V_{g2} (p.u.)	1.0381	1.0379	1.0379	1.0459	1.0472	1.0459	1.0477	1.0476	1.0476
V_{g3} (p.u.)	1.0110	1.0117	1.0109	1.0274	1.0309	1.0277	1.0292	1.0283	1.0292
V_{g4} (p.u.)	1.0194	1.0197	1.0187	1.0353	1.0377	1.0350	1.0366	1.0342	1.0363
V_{g5} (p.u.)	1.1000	1.1000	1.1000	1.1000	1.1000	1.1000	1.1000	1.0387	1.1000
V_{g6} (p.u.)	1.0999	1.0842	1.0828	1.0852	1.0140	1.0713	1.0850	1.0606	1.0712
T_{6-9} (p.u.)	0.9973	1.0318	1.0166	1.0136	1.0017	1.0490	1.0153	1.1000	1.0482
T_{6-10} (p.u.)	0.9000	0.9004	0.9210	0.9000	1.1000	0.9000	0.9000	0.9000	0.9000
T_{4-12} (p.u.)	1.0157	0.9995	0.9980	1.0097	1.0003	0.9954	1.0105	0.9740	0.9962
T_{27-28} (p.u.)	0.9403	0.9501	0.9478	0.9518	1.0136	0.9609	0.9529	0.9647	0.9618
Qc_{10} (MVar)	28.6430	7.0219	10.0521	0.0000	0.0030	5.7174	7.0753	30.0000	5.2416
Qc_{24} (MVar)	0.0000	11.0974	10.6433	30.0000	16.6605	10.6333	17.0085	9.9534	10.6499
Fuel Cost ($/h)	802.5449	802.1299	**802.1241**	945.0484	944.9387	944.7159	968.1335	967.8979	967.8756
Emission (ton/h)	0.363619	0.364411	0.363494	**0.204886**	**0.204861**	**0.204853**	0.207294	0.207280	0.207279
P_Loss (MW)	9.5249	9.4272	9.4041	3.4370	3.4790	3.3292	**3.2974**	**3.1987**	**3.1893**

Table 2. Comparison of the results from DA-PSO and other algorithms when considering only fuel cost as part of the objective function for IEEE 30-bus system.

Algorithms	P_{g1} (MW)	P_{g2} (MW)	P_{g3} (MW)	P_{g4} (MW)	P_{g5} (MW)	P_{g6} (MW)	Emission (ton/h)	Loss (MW)	Cost ($/h)	Time (s)
TS [11]	176.0400	48.7600	21.5600	22.0500	12.4400	12.0000	0.363004	9.4500	802.2900	-
EP [13]	173.8480	49.9980	21.3860	22.6300	12.9280	12.0000	0.357217	9.3900	802.6200	51.40
ACO [16]	181.9450	47.0010	21.4596	21.4460	13.2070	12.0134	0.382000	9.8520	802.5780	-
SFLA [27]	179.0337	49.2580	20.3183	21.3269	11.5420	11.6655	0.372000	9.7444	802.5092	-
MSFLA [27]	179.1929	48.9804	20.4517	20.9264	11.5897	11.9579	0.372300	9.6991	802.2870	-
IEP [40]	176.2358	49.0093	21.5023	21.8115	12.3387	12.0129	0.363610	10.8700	802.4650	99.01
MDE-OPF [41]	175.9740	48.8840	21.5100	22.2400	12.2510	12.0000	0.362900	9.4590	802.3760	23.25
SGA [42]	179.3670	44.2400	24.6100	19.9000	10.7100	14.0900	0.371129	9.5177	803.6990	-
EP-OPF [43]	175.0297	48.9522	21.4200	22.7020	12.9040	12.1035	0.360125	9.7114	803.5710	-
HBMO [44]	178.4646	46.2740	21.4596	21.4460	13.2070	12.0134	0.369212	9.4662	802.2110	28.56
PSO	176.2376	48.8432	21.5184	22.1257	12.2000	12.0000	0.363619	9.5249	802.5449	92.18
DA	176.5128	48.6955	21.4431	22.0995	12.0673	12.0091	0.364411	9.4272	802.1299	103.06
DA-PSO	176.1861	48.8318	21.5119	22.0737	12.2005	12.0000	0.363494	9.4041	**802.1241**	287.13

Table 3. Comparison of the results from DA-PSO and other algorithms when considering only emissions as part of the objective function for IEEE 30-bus system.

Algorithms	P_{g1} (MW)	P_{g2} (MW)	P_{g3} (MW)	P_{g4} (MW)	P_{g5} (MW)	P_{g6} (MW)	Cost ($/h)	Loss (MW)	Emission (ton/h)	Time (s)
ACO [16]	64.3720	72.1604	49.5438	32.9099	28.6113	39.7390	945.5870	3.9368	0.221000	-
HMPSO-SFLA [23]	64.8148	68.0692	50.0000	34.9999	30.0000	40.0000	948.3052	4.4839	0.205200	-
TLBO [24]	63.5221	68.7345	49.9931	34.9894	29.9824	39.9801	947.4392	3.8016	0.205030	-
MTLBO [24]	64.2924	67.6250	50.0000	35.0000	30.0000	40.0000	945.1965	3.5174	0.204930	-
DSA [26]	64.0725	67.5711	50.0000	35.0000	30.0000	40.0000	944.4086	3.2437	0.205826	-
MSFLA [27]	65.7798	68.2688	50.0000	34.9999	29.9982	39.9970	951.5106	5.6437	0.205600	-
SFLA [27]	64.4840	71.3807	49.8573	35.0000	30.0000	39.9729	960.1911	7.2949	0.206300	-
GA [27]	78.2885	68.1602	46.7848	33.4909	30.0000	36.3713	936.6152	9.6957	0.211700	-
GBICA [28]	64.3125	67.4938	50.0000	35.0000	29.9924	40.0000	944.6516	3.3987	0.204900	-
IPSO [45]	67.0400	68.1400	50.0000	35.0000	30.0000	40.0000	954.2480	5.3620	0.205800	-
PSO	64.1678	67.6692	50.0000	35.0000	30.0000	40.0000	945.0484	3.4370	0.204886	91.84
DA	64.0667	67.6897	50.0000	35.0000	30.0000	40.0000	944.8819	3.3564	0.204861	103.20
DA-PSO	64.0997	67.6295	50.0000	35.0000	30.0000	40.0000	944.7159	3.3292	**0.204853**	290.01

Table 4. Comparison of the results from DA-PSO and other algorithms when considering only losses as part of the objective function for IEEE 30-bus system.

Algorithms	P_{g1} (MW)	P_{g2} (MW)	P_{g3} (MW)	P_{g4} (MW)	P_{g5} (MW)	P_{g6} (MW)	Cost ($/h)	Emission (ton/h)	Loss (MW)	Time (s)
GWO [17]	51.8100	80.0000	50.0000	35.0000	30.0000	40.0000	968.3800	0.207310	3.4100	15.90
DE [17]	51.8200	79.9900	49.9900	35.0000	29.9800	40.0000	968.2300	0.207311	3.3800	16.50
MOHS [29]	66.2759	79.6413	46.8835	34.8880	29.1213	30.0558	928.5099	0.212890	3.5165	-
EGA-DQLF [46]	51.6008	80.0000	50.0000	35.0000	30.0000	40.0000	967.8600	0.207281	3.2008	-
EEA [47]	59.3216	74.8132	49.8547	34.9084	28.1099	39.7538	952.3785	0.206735	3.2823	5.72
EGA [47]	51.6740	79.9700	50.0000	35.0000	30.0000	40.0000	967.9300	0.207275	3.2440	29.71
PSO	51.6974	80.0000	50.0000	35.0000	30.0000	40.0000	968.1335	0.207294	3.2974	93.36
DA	51.5941	80.0000	50.0000	35.0000	40.0000	40.0000	967.8869	0.207280	3.1941	102.81
DA-PSO	51.5893	80.0000	50.0000	35.0000	30.0000	40.0000	967.8756	0.207279	**3.1893**	292.33

5.1.2. MO-OPF

In this subsection, the proposed algorithm is investigated as a multiobjective optimization problem, while every two and three objective functions are optimized simultaneously. The best two-dimensional Pareto fronts obtained from the DA, PSO, and DA-PSO algorithms for the IEEE 30-bus system are shown in Figures 2–4. However, DA could not provide the convergent Pareto front when simultaneously considering the emissions and losses as parts of the objective function. This shows that DA is suitable for some objective functions, but that it is not suitable for every objective function for finding optimal solutions. In Figure 5, the Pareto front provided by the DA-PSO algorithm for the three-dimensional Pareto front is shown. For all figures in this system, most of the nondominated solutions obtained by the DA-PSO algorithm are better than those from the DA and PSO algorithms. For instance, at the same level of the fuel cost, the emissions provided by DA-PSO are less than those of DA and PSO. This shows that the new proposed hybrid DA-PSO algorithm, which adopts the exploration phase of the DA and the exploitation phase of the PSO, could improve the performance of the original DA and PSO algorithms.

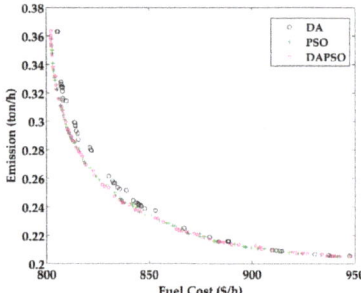

Figure 2. Two-dimensional Pareto fronts when considering fuel cost and emissions as part of the objective function for the IEEE 30-bus system.

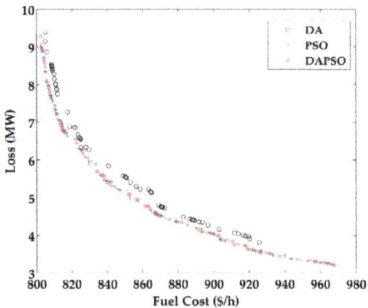

Figure 3. Two-dimensional Pareto fronts when considering fuel cost and transmission losses as part of the objective function for the IEEE 30-bus system.

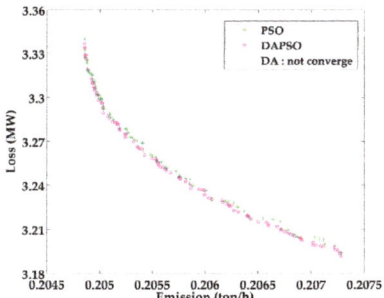

Figure 4. Two-dimensional Pareto fronts when considering emissions and transmission losses as part of the objective function for the IEEE 30-bus system.

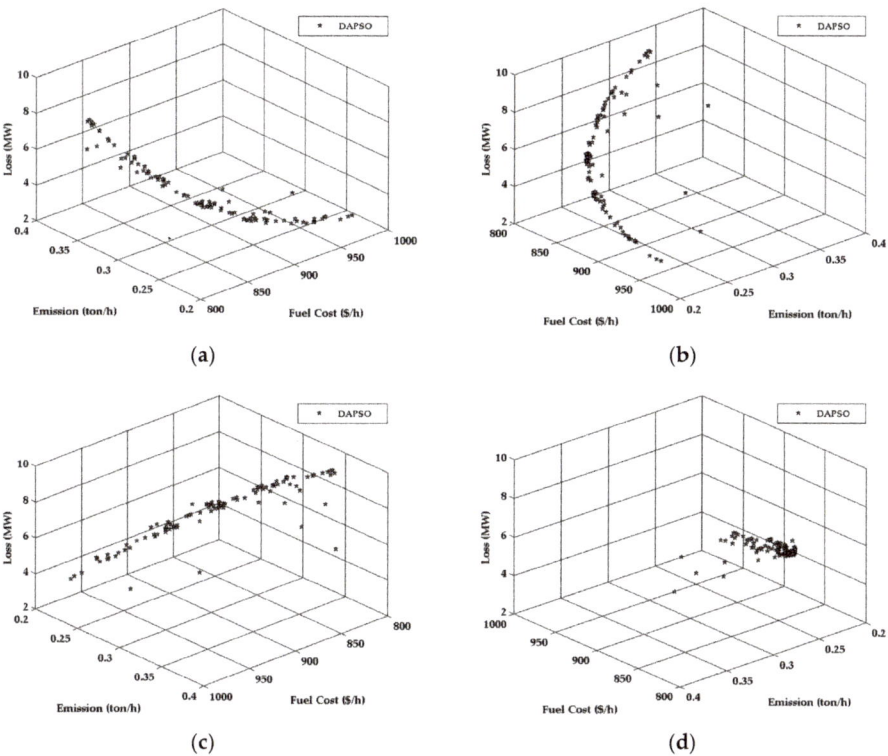

Figure 5. Three-dimensional Pareto fronts when considering fuel cost, emissions, and transmission losses as part of the objective function for the IEEE 30-bus system shown in the different views: (**a**) front view; (**b**) side view; (**c**) back view; (**d**) side view.

5.2. IEEE 57-Bus Test System

The proposed hybrid DA-PSO was also tested on the IEEE 57-bus system to investigate its performance. The system active and reactive power demands were 1250.8 MW and 336.4 MVAR, respectively. It consisted of 7 generators located at buses 1, 2, 3, 6, 8, 9, and 12, 15 transformers, and

80 transmission lines. The detail data were taken from [48]. The population number was 100 and the size of the Pareto archive was 100.

5.2.1. Single-Objective OPF

To verify its performance for solving the single-objective OPF in a larger system, the proposed algorithm was also applied to the IEEE 57-bus test system. Three different objective functions, i.e., fuel cost, emissions, and transmission losses, were individually considered as part of the objective function. The results provided by DA, PSO, and the proposed DA-PSO algorithm for the three individual objectives are shown in Table 5. The best results from DA-PSO are compared with those of: MTLBO [23], DSA [25], GBICA [27], MGBICA [27], ARCBBO [29], MO-DEA [30], MICA-TLA [31], TLBO [48], Levy mutation teaching–learning-based optimization (LTLBO) [49], new particle swarm optimization (NPSO) [50], fuzzy genetic algorithm (Fuzzy-GA) [51], differential evolution pattern search (DE-PS) [52], ABC [53], particle swarm optimization algorithm with linearly decreasing inertia weight (LDI-PSO) [53], evolving ant direction differential evolution (EADDE) [54], gravitational search algorithm (GSA) [55], adaptive particle swarm optimization strategy (APSO) [56], PSO, and DA for the fuel cost objective function; GBICA [27], MGBICA [27], PSO, and DA for the emission objective function; and PSO and DA for the transmission loss objective function—all of which is summarized in Tables 5–8. From these tables, it is obvious that the proposed algorithm could provide more optimized results than the compared algorithms for all three objective functions.

Table 5. Comparison of the simulation results from PSO, DA, and DA-PSO for IEEE 57-bus system.

Variables	Best Fuel Cost			Best Emission			Best P_{Loss}		
	PSO	DA	DA-PSO	PSO	DA	DA-PSO	PSO	DA	DA-PSO
P_{g1} (MW)	142.7472	154.8513	141.4617	236.4846	246.6610	236.4531	193.1342	269.9574	202.6688
P_{g2} (MW)	88.8427	76.6227	87.7806	100.0000	44.6053	100.0000	8.8581	0.2047	0.0000
P_{g3} (MW)	44.9025	49.4440	44.6638	139.9999	140.0000	140.0000	139.9731	60.2481	140.0000
P_{g4} (MW)	70.8490	100.0000	73.6254	100.0000	78.0610	100.0000	100.0000	55.3529	100.0000
P_{g5} (MW)	458.6003	438.7375	458.9904	292.5686	329.7090	292.1457	309.5411	377.9311	308.2507
P_{g6} (MW)	100.0000	100.0000	97.4933	100.0000	82.5653	100.0000	100.0000	90.2309	100.0000
P_{g7} (MW)	360.3487	347.6947	361.7228	298.6306	344.7194	298.4568	410.0000	410.0000	410.0000
V_{g1} (p.u.)	1.0500	1.0500	1.0500	1.0500	1.0500	1.0500	1.0500	1.0500	1.0500
V_{g2} (p.u.)	1.0494	1.0453	1.0488	1.0513	1.0486	1.0506	1.0458	1.0351	1.0450
V_{g3} (p.u.)	1.0479	1.0475	1.0455	1.0532	1.0550	1.0505	1.0528	1.0342	1.0520
V_{g4} (p.u.)	1.0628	1.0755	1.0581	1.0518	1.0303	1.0493	1.0537	1.0454	1.0525
V_{g5} (p.u.)	1.0792	1.0802	1.0745	1.0551	1.0142	1.0506	1.0603	1.0563	1.0566
V_{g6} (p.u.)	1.0455	1.0568	1.0442	1.0266	1.0133	1.0267	1.0349	1.0240	1.0348
V_{g7} (p.u.)	1.0410	1.0629	1.0394	1.0251	1.0556	1.0262	1.0392	1.0119	1.0384
T_{4-8} (p.u.)	0.9429	1.1000	1.0221	0.9629	1.1000	0.9691	0.9625	0.9763	0.9730
T_{4-18} (p.u.)	0.9916	1.0140	0.9953	0.9764	1.0682	0.9870	0.9865	1.0277	1.0275
T_{21-20} (p.u.)	1.0151	1.1000	1.0196	1.0233	1.0113	1.0228	1.0226	1.0286	1.0442
T_{24-25} (p.u.)	0.9000	0.9857	1.0212	0.9082	1.1000	1.1000	0.9111	1.0430	1.0140
T_{24-25} (p.u.)	0.9378	0.9812	0.9896	0.9000	0.9645	0.9609	0.9118	0.9948	1.0145
T_{24-26} (p.u.)	1.0219	1.1000	1.0175	1.0115	0.9981	1.0086	1.0130	1.0312	1.0111
T_{7-29} (p.u.)	0.9901	1.0253	0.9983	0.9791	0.9658	0.9904	0.9826	0.9816	0.9945
T_{34-32} (p.u.)	0.9277	1.0731	0.9582	0.9285	1.0069	0.9682	0.9217	0.9815	0.9577
T_{11-41} (p.u.)	0.9000	0.9879	0.9063	0.9000	0.9062	0.9000	0.9000	1.1000	0.9036
T_{15-45} (p.u.)	0.9667	0.9770	0.9714	0.9750	1.0178	0.9786	0.9779	0.9761	0.9807
T_{14-46} (p.u.)	0.9578	0.9807	0.9616	0.9636	0.9743	0.9569	0.9594	0.9373	0.9616
T_{10-51} (p.u.)	0.9748	0.9899	0.9766	0.9642	0.9824	0.9674	0.9710	0.9395	0.9707
T_{13-49} (p.u.)	0.9300	1.0153	0.9301	0.9257	0.9854	0.9274	0.9292	0.9850	0.9375
T_{11-43} (p.u.)	0.9785	0.9738	0.9756	0.9612	0.9707	0.9647	0.9704	0.9373	0.9767
T_{40-56} (p.u.)	0.9962	1.1000	1.0105	0.9715	0.9460	0.9710	0.9969	1.0457	0.9972
T_{39-57} (p.u.)	0.9692	1.1000	0.9621	0.9728	1.0799	0.9742	0.9629	0.9373	0.9645
T_{9-55} (p.u.)	0.9856	1.0732	0.9988	0.9676	0.9937	0.9810	0.9756	0.9747	0.9842
Q_{c18} (MVar)	18.7450	6.6751	13.2804	0.0000	15.6977	2.4493	10.9247	12.9734	4.5146
Q_{c25} (MVar)	13.8614	8.7220	12.6307	7.3042	22.1164	16.8169	28.7430	12.9949	14.5906
Q_{c53} (MVar)	12.0686	22.2015	13.9725	0.0249	9.3410	12.5551	19.7432	10.7143	12.9250
Fuel Cost ($/h)	**41,698.37**	**41,828.45**	**41,674.62**	45,671.22	45,449.13	45,648.67	44,951.80	43,464.17	45,039.05
Emission (ton/h)	1.9027	1.6883	1.9087	**1.0814**	**1.3097**	**1.0799**	1.3821	1.7562	1.4014
P_{Loss} (MW)	15.4903	16.5502	14.9380	16.8837	15.5210	16.2556	**10.7076**	**13.6430**	**10.1212**

Table 6. Comparison of the results from DA-PSO and other algorithms when considering only fuel cost as part of the objective function for the IEEE 57-bus system.

Algorithms	Cost ($/h)
MTLBO [23]	41,638.3822
DSA [25]	41,686.8200
GBICA [27]	41,740.2884
MGBICA [27]	41,715.7101
ARCBBO [29]	41,686.0000
MO-DEA [30]	41,683.0000
MICA-TLA [31]	41,675.0545
TLBO [48]	41,695.6629
LTLBO [49]	41,679.5451
NPSO [50]	41,699.5163
Fuzzy-GA [51]	41,716.2808
DE-PS [52]	41,685.2950
ABC [53]	41,693.9589
LDI-PSO [53]	41,815.5035
EADDE [54]	41,713.6200
GSA [55]	41,695.8717
APSO [56]	41,713.8868
PSO	41,698.3672
DA	41,828.4473
DA-PSO	**41,674.6209**

Table 7. Comparison of the results from DA-PSO and other algorithms when considering the only emissions as part of the objective function for the IEEE 57-bus system.

Algorithms	Emission (ton/h)
GBICA [27]	1.1881
MGBICA [27]	1.1724
PSO	1.0814
DA	1.3097
DA-PSO	**1.0799**

Table 8. Comparison of the results from DA-PSO and its traditional algorithms when considering only transmission losses as part of the objective function for the IEEE 57-bus system.

Algorithms	Loss (MW)
PSO	10.7076
DA	13.6430
DA-PSO	**10.1212**

5.2.2. MO-OPF

This case proposes a multiobjective optimization problem by using the proposed DA-PSO algorithm to evaluate its performance for the IEEE 57-bus test system. The two-dimensional Pareto fronts provided by the PSO and DA-PSO algorithms for this system are shown in Figures 6–8, while DA could not provide the convergent Pareto fronts for any multiobjective functions in this system. In Figure 9, the three-dimensional Pareto front obtained from DA-PSO is shown. From all figures for this system, the fronts obtained from DA-PSO algorithm are superior to those from PSO, while the fronts obtained from DA could not converge because the best experience of dragonflies in DA is not applied during the operation and the obtained solutions are trapped in the local optima. From the results, it can be seen that the proposed hybrid DA-PSO performs better than the original DA and PSO algorithms once again.

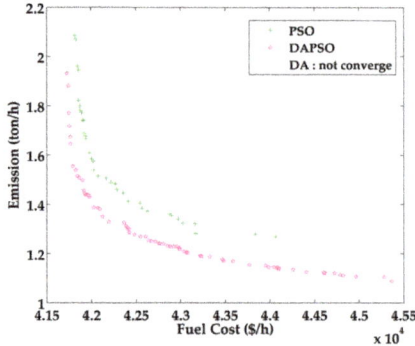

Figure 6. Two-dimensional Pareto fronts when considering fuel cost and emissions as part of the objective function for the IEEE 57-bus system.

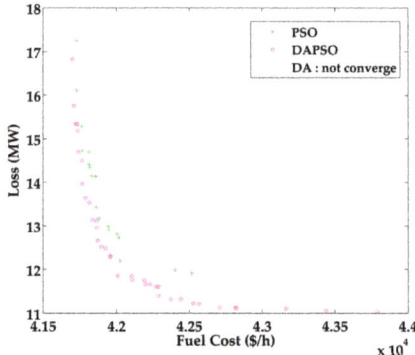

Figure 7. Two-dimensional Pareto fronts when considering fuel cost and transmission losses as part of the objective function for the IEEE 57-bus system.

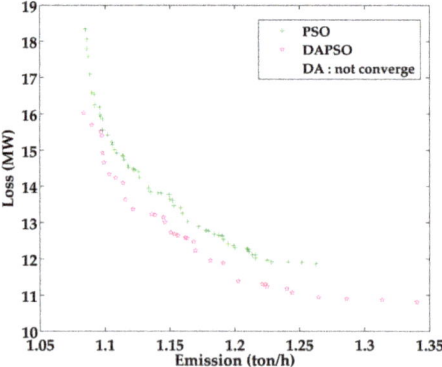

Figure 8. Two-dimensional Pareto fronts when considering emissions and transmission losses as part of the objective function for the IEEE 57-bus system.

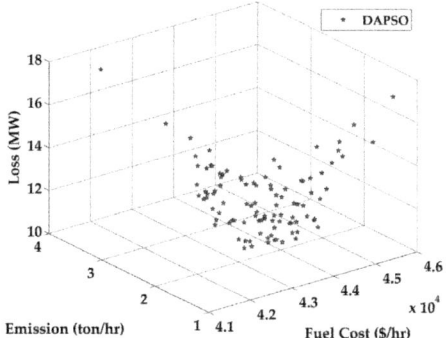

Figure 9. Three-dimensional Pareto fronts when considering fuel cost, emissions, and transmission losses as part of the objective function for the IEEE 57-bus system.

6. Conclusions

In this paper, a hybrid DA-PSO algorithm is proposed to solve the MO-OPF problem in a power system. As the DA is an algorithm that applies Levy flight to improve its randomness and stochastic behavior, this could significantly develop the exploration phase of the algorithm in an optimization. The PSO could quickly converge on the optimal solution because of its equations for finding optimal solutions by using the best experience of the particles. This makes PSO perform well at the exploitation phase in an optimization. The new hybrid DA-PSO algorithm combines the prominent points of these two algorithms, which are the exploration phase of DA and the exploitation phase of the PSO, to improve its performance for finding the optimal solution of the OPF problem. The proposed algorithm was used to minimize fuel cost, emissions, and transmission losses, which are considered to be parts of the objective function. The standard IEEE 30-bus and 57-bus systems were employed to investigate the performance of the proposed algorithm to find the optimal settings of the control variables. In order to investigate the single-objective and multiobjective optimizations, the simulation was divided into two cases. First, the proposed algorithm was used to solve a single-objective function. The results from the proposed algorithm show its superiority over other optimization algorithms in the literature. For the other case, the DA-PSO was successfully employed to solve the MO-OPF problem because the Pareto fronts generated by DA-PSO are better than those obtained by the original DA and PSO algorithms. All simulation results support the applicability, potential, and effectiveness of the proposed algorithm. However, the computation time of the DA-PSO is much slower than other algorithms in the literature because of the sequential computation of DA and PSO.

Author Contributions: Conceptualization, S.K. and R.C.; Methodology, S.K.; Software, S.K.; Validation, S.K., A.S., R.C., and N.R.W.; Formal Analysis, S.K.; Investigation, S.K.; Resources, A.S.; Data Curation, S.K.; Writing-Original Draft Preparation, S.K.; Writing-Review & Editing, S.K., R.C., N.R.W. and S.P.; Visualization, S.K.; Supervision, A.S.; Project Administration, R.C. and A.S.; Funding Acquisition, A.S.

Funding: This research was funded by the Thailand Research Fund through the Royal Golden Jubilee Ph.D. Program (Grant no. PHD/0192/2557) to Sirote Khunkitti and Apirat Siritaratiwat.

Conflicts of Interest: The authors declare no conflict of interest.

Abbreviations

ABC	artificial bee colony
ACO	ant colony optimization
APSO	adaptive particle swarm optimization strategy
ARCBBO	adaptive real coded biogeography-based optimization
DA	dragonfly algorithm
DE	differential evolution
DE-PS	differential evolution pattern search
DSA	differential search algorithm
EADDE	evolving ant direction differential evolution
EEA	efficient evolutionary algorithm
EGA	enhanced genetic algorithm
EGA-DQLF	enhanced genetic algorithm with decoupled quadratic load flow
EP	evolutionary programming
EP-OPF	evolutionary-programming-based optimal power flow
Fuzzy-GA	fuzzy genetic algorithm
GA	genetic algorithm
GSA	gravitational search algorithm
GWO	grey wolf optimizer
HBMO	honey bee mating optimization
HMPSO-SFLA	hybrid modified particle swarm optimization-shuffle frog leaping algorithms
IEP	improved evolutionary programming
IPSO	improved particle swarm optimization
ISPEA2	improved strength Pareto evolutionary algorithm
LDI-PSO	particle swarm optimization algorithm with linearly decreasing inertia weight
LTLBO	Levy mutation teaching–learning-based optimization
MDE-OPF	modified differential evolution optimal power flow
MGBICA	modified Gaussian bare-bones multiobjective imperialist competitive algorithm
MICA-TLA	hybrid modified imperialist competitive algorithm and teaching–learning algorithm
MO-DEA	multiobjective differential evolution algorithm
MOHS	multiobjective harmony search
MOMICA	multiobjective modified imperialist competitive algorithm
MO-OPF	multiobjective optimal power flow
MSFLA	modified shuffle frog leaping algorithm
MTLBO	modified teaching–learning-based optimization
NPSO	new particle swarm optimization
OPF	optimal power flow
P-OPF	probabilistic optimal power flow
PSO	particle swarm optimization
SGA	stochastic genetic algorithm
TS	tabu search

References

1. Sanseverino, E.R.; Buono, L.; Di Silvestre, M.L.; Zizzo, G.; Ippolito, M.G.; Favuzza, S.; Quynh, T.T.T.; Ninh, N.Q. A distributed minimum losses optimal power flow for islanded microgrids. *Electr. Power Syst. Res.* **2017**, *152*, 271–283. [CrossRef]
2. Christakou, K.; Tomozei, D.C.; Le Boudec, J.Y.; Paolone, M. AC OPF in radial distribution networks—Part I: On the limits of the branch flow convexification and the alternating direction method of multipliers. *Electr. Power Syst. Res.* **2017**, *143*, 438–450. [CrossRef]
3. Roy, P.K.; Ghoshal, S.P.; Thakur, S.S. Biogeography based optimization for multi-constraint optimal power flow with emission and non-smooth cost function. *Expert Syst. Appl.* **2010**, *37*, 8221–8228. [CrossRef]

4. Ma, H.; Hart, J.L. Economic Dispatch in View of the Clean Air Act of 1990. *IEEE Trans. Power Syst.* **2000**, *9*, 972–978.

5. Momoh, J.A.; El-Hawary, M.E.; Adapa, R. A review of selected optimal power flow literature to 1993 part I: Nonlinear and quadratic Programming Approaches. *IEEE Trans. Power Syst.* **1999**, *14*, 96–103. [CrossRef]

6. Burchett, R.C.; Happ, H.H.; Vierath, D.R. Quadratically Convergent Optimal Power Flow. *IEEE Trans. Power Appar. Syst.* **1984**, *PAS-103*, 3267–3275. [CrossRef]

7. Yan, X.; Quintana, V.H. Improving an interior-point-based off by dynamic adjustments of step sizes and tolerances. *IEEE Trans. Power Syst.* **1999**, *14*, 709–716. [CrossRef]

8. Ghasemi, M.; Ghavidel, S.; Ghanbarian, M.M.; Habibi, A. A new hybrid algorithm for optimal reactive power dispatch problem with discrete and continuous control variables. *Appl. Soft Comput.* **2014**, *22*, 126–140. [CrossRef]

9. Ghasemi, M.; Ghanbarian, M.M.; Ghavidel, S.; Rahmani, S.; Mahboubi Moghaddam, E. Modified teaching learning algorithm and double differential evolution algorithm for optimal reactive power dispatch problem: A comparative study. *Inf. Sci. (N. Y.)* **2014**, *278*, 231–249. [CrossRef]

10. Lai, L.L.; Ma, J.T.; Yokoyama, R.; Zhao, M. Improved genetic algorithms for optimal power flow under both normal and contingent operation states. *Int. J. Electr. Power Energy Syst.* **1997**, *19*, 287–292. [CrossRef]

11. Abido, M.A. Optimal Power Flow Using Tabu Search Algorithm. *Electr. Power Compon. Syst.* **2002**, *30*, 469–483. [CrossRef]

12. El Ela, A.A.A.; Abido, M.A.; Spea, S.R. Optimal power flow using differential evolution algorithm. *Electr. Power Syst. Res.* **2010**, *80*, 878–885. [CrossRef]

13. Yuryevich, J.; Wong, K.P. Evolutionary Programming Based Optimal Power Flow Algorithm. *IEEE Trans. Power Syst.* **1999**, *14*, 1245–1250. [CrossRef]

14. Deng, X.; He, J.; Zhang, P. A novel probabilistic optimal power flow method to handle large fluctuations of stochastic variables. *Energies* **2017**, *10*, 1623. [CrossRef]

15. Wu, X.; Zhou, Z.; Liu, G.; Qi, W.; Xie, Z. Preventive security-constrained optimal power flow considering UPFC control modes. *Energies* **2017**, *10*, 1199. [CrossRef]

16. Sliman, L.; Bouktir, T. Economic Power Dispatch of Power System with Pollution Control Using Multiobjective Ant Colony Optimization. *Int. J. Comput. Intell. Res.* **2007**, *3*, 145–153. [CrossRef]

17. El-Fergany, A.A.; Hasanien, H.M. Single and Multi-objective Optimal Power Flow Using Grey Wolf Optimizer and Differential Evolution Algorithms. *Electr. Power Compon. Syst.* **2015**, *43*, 1548–1559. [CrossRef]

18. Bai, W.; Lee, D.; Lee, K. Stochastic Dynamic AC Optimal Power Flow Based on a Multivariate Short-Term Wind Power Scenario Forecasting Model. *Energies* **2017**, *10*, 2138. [CrossRef]

19. Abido, M.A. Optimal power flow using particle swarm optimization. *Int. J. Electr. Power Energy Syst.* **2002**, *24*, 563–571. [CrossRef]

20. Bashishtha, T.K. Nature Inspired Meta-heuristic dragonfly Algorithms for Solving Optimal Power Flow Problem. *Int. J. Electron. Electr. Comput. Syst.* **2016**, *5*, 111–120.

21. Shang, R.; Wang, Y.; Wang, J.; Jiao, L.; Wang, S.; Qi, L. A multi-population cooperative coevolutionary algorithm for multi objective capacitated arc routing problem. *Inf. Sci. (N. Y.)* **2014**, *277*, 609–642. [CrossRef]

22. Yuan, X.; Zhang, B.; Wang, P.; Liang, J.; Yuan, Y.; Huang, Y.; Lei, X. Multi-objective optimal power flow based on improved strength Pareto evolutionary algorithm. *Energy* **2017**, *122*, 70–82. [CrossRef]

23. Narimani, M.R.; Azizipanah-Abarghooee, R.; Zoghdar-Moghadam-Shahrekohne, B.; Gholami, K. A novel approach to multi objective optimal power flow by a new hybrid optimization algorithm considering generator constraints and multi-fuel type. *Energy* **2013**, *49*, 119–136. [CrossRef]

24. Shabanpour-Haghighi, A.; Seifi, A.; Niknam, T. A modified teaching-learning based optimization for multi objective optimal power flow problem. *Energy Convers. Manag.* **2014**, *77*, 597–607. [CrossRef]

25. Ghasemi, M.; Ghavidel, S.; Ghanbarian, M.M.; Gharibzadeh, M.; Azizi Vahed, A. Multi-objective optimal power flow considering the cost, emission, voltage deviation and power losses using multi objective modified imperialist competitive algorithm. *Energy* **2014**, *78*, 276–289. [CrossRef]

26. Abaci, K.; Yamacli, V. Differential search algorithm for solving multi objective optimal power flow problem. *Int. J. Electr. Power Energy Syst.* **2016**, *79*, 1–10. [CrossRef]

27. Niknam, T.; rasoul Narimani, M.; Jabbari, M.; Malekpour, A.R. A modified shuffle frog leaping algorithm for multi objective optimal power flow. *Energy* **2011**, *36*, 6420–6432. [CrossRef]

28. Ghasemi, M.; Ghavidel, S.; Ghanbarian, M.M.; Gitizadeh, M. Multi-objective optimal electric power planning in the power system using Gaussian bare-bones imperialist competitive algorithm. *Inf. Sci. (N. Y.)* **2015**, *294*, 286–304. [CrossRef]

29. Sivasubramani, S.; Swarup, K.S. Multi-objective harmony search algorithm for optimal power flow problem. *Int. J. Electr. Power Energy Syst.* **2011**, *33*, 745–752. [CrossRef]

30. Ramesh Kumar, A.; Premalatha, L. Optimal power flow for a deregulated power system using adaptive real coded biogeography-based optimization. *Int. J. Electr. Power Energy Syst.* **2015**, *73*, 393–399. [CrossRef]

31. El-Sehiemy, R.A.; Shaheen, A.M.; Farrag, S.M. Solving multi objective optimal power flow problem via forced initialised differential evolution algorithm. *IET Gener. Transm. Distrib.* **2016**, *10*, 1634–1647. [CrossRef]

32. Ghasemi, M.; Ghavidel, S.; Rahmani, S.; Roosta, A.; Falah, H. A novel hybrid algorithm of imperialist competitive algorithm and teaching learning algorithm for optimal power flow problem with non-smooth cost functions. *Eng. Appl. Artif. Intell.* **2014**, *29*, 54–69. [CrossRef]

33. Mirjalili, S. Dragonfly algorithm: A new meta-heuristic optimization technique for solving single-objective, discrete, and multi objective problems. *Neural Comput. Appl.* **2016**, *27*, 1053–1073. [CrossRef]

34. Eberhart, R.; Kennedy, J. A New Optimizer Using Particle Swarm Theory. In Proceedings of the Sixth International Symposium on Micro Machine and Human Science (MHS'95), Nagoya, Japan, 4–6 October 1995; pp. 39–43.

35. Naderi, E.; Narimani, H.; Fathi, M.; Narimani, M.R. A novel fuzzy adaptive configuration of particle swarm optimization to solve large-scale optimal reactive power dispatch. *Appl. Soft Comput.* **2017**, *53*, 441–456. [CrossRef]

36. Zhou, Y.; Wu, J.; Ji, L.; Yu, Z.; Lin, K.; Hao, L. Transient stability preventive control of power systems using chaotic particle swarm optimization combined with two-stage support vector machine. *Electr. Power Syst. Res.* **2018**, *155*, 111–120. [CrossRef]

37. Mehdinejad, M.; Mohammadi-Ivatloo, B.; Dadashzadeh-Bonab, R.; Zare, K. Solution of optimal reactive power dispatch of power systems using hybrid particle swarm optimization and imperialist competitive algorithms. *Int. J. Electr. Power Energy Syst.* **2016**, *83*, 104–116. [CrossRef]

38. Shihabudheen, K.V.; Mahesh, M.; Pillai, G.N. Particle swarm optimization based extreme learning neuro-fuzzy system for regression and classification. *Expert Syst. Appl.* **2018**, *92*, 474–484. [CrossRef]

39. The University of Washington Electrical Engineering. Power System Test Case Archive, the IEEE 30-Bus Test System Data. Available online: https://www2.ee.washington.edu/research/pstca/pf30/pg_tca30bus.htm (accessed on 9 November 2017).

40. Ongsakul, W.; Tantimaporn, T. Optimal Power Flow by Improved Evolutionary Programming. *Electr. Power Compon. Syst.* **2006**, *34*, 79–95. [CrossRef]

41. Sayah, S.; Zehar, K. Modified differential evolution algorithm for optimal power flow with non-smooth cost functions. *Energy Convers. Manag.* **2008**, *49*, 3036–3042. [CrossRef]

42. Bouktir, T.; Slimani, L.; Mahdad, B. Optimal Power Dispatch for Large Scale Power System Using Stochastic Search Algorithms. *Int. J. Power Energy Syst.* **2008**, *28*. [CrossRef]

43. Sood, Y. Evolutionary programming based optimal power flow and its validation for deregulated power system analysis. *Int. J. Electr. Power Energy Syst.* **2007**, *29*, 65–75. [CrossRef]

44. Niknam, T.; Narimani, M.R.; Aghaei, J.; Tabatabaei, S.; Nayeripour, M. Modified Honey Bee Mating Optimisation to solve dynamic optimal power flow considering generator constraints. *IET Gener. Transm. Distrib.* **2011**, *5*, 989. [CrossRef]

45. Niknam, T.; Narimani, M.R.; Aghaei, J.; Azizipanah-Abarghooee, R. Improved particle swarm optimisation for multi objective optimal power flow considering the cost, loss, emission and voltage stability index. *IET Gener. Transm. Distrib.* **2012**, *6*, 515–527. [CrossRef]

46. Kumari, M.S.; Maheswarapu, S. Enhanced Genetic Algorithm based computation technique for multi objective Optimal Power Flow solution. *Int. J. Electr. Power Energy Syst.* **2010**, *32*, 736–742. [CrossRef]

47. Surender Reddy, S.; Bijwe, P.R.; Abhyankar, A.R. Faster evolutionary algorithm based optimal power flow using incremental variables. *Int. J. Electr. Power Energy Syst.* **2014**, *54*, 198–210. [CrossRef]

48. The University of Washington Electrical Engineering. Power System Test Case Archive, the IEEE 57-Bus Test System Data. Available online: https://www2.ee.washington.edu/research/pstca/pf57/pg_tca57bus.htm (accessed on 9 November 2017).

49. Ghasemi, M.; Ghavidel, S.; Gitizadeh, M.; Akbari, E. An improved teaching-learning-based optimization algorithm using Lévy mutation strategy for non-smooth optimal power flow. *Int. J. Electr. Power Energy Syst.* **2015**, *65*, 375–384. [CrossRef]

50. Niknam, T. A new particle swarm optimization for non-convex economic dispatch. *Trans. Electr.* **2011**, *21*, 656–679. [CrossRef]

51. Hsiao, Y.T.; Chen, C.H.; Chien, C.C. Optimal capacitor placement in distribution systems using a combination fuzzy-GA method. *Int. J. Electr. Power Energy Syst.* **2004**, *26*, 501–508. [CrossRef]

52. Gitizadeh, M.; Ghavidel, S.; Aghaei, J. Using SVC to Economically Improve Transient Stability in Long Transmission Lines. *IETE J. Res.* **2014**, *60*, 319–327. [CrossRef]

53. Rezaei Adaryani, M.; Karami, A. Artificial bee colony algorithm for solving multi objective optimal power flow problem. *Int. J. Electr. Power Energy Syst.* **2013**, *53*, 219–230. [CrossRef]

54. Vaisakh, K.; Srinivas, L.R. Evolving ant direction differential evolution for OPF with non-smooth cost functions. *Eng. Appl. Artif. Intell.* **2011**, *24*, 426–436. [CrossRef]

55. Duman, S.; Güvenç, U.; Sönmez, Y.; Yörükeren, N. Optimal power flow using gravitational search algorithm. *Energy Convers. Manag.* **2012**, *59*, 86–95. [CrossRef]

56. Mahdad, B.; Srairi, K. Hierarchical adaptive PSO for multi objective OPF considering emissions based shunt FACTS. In Proceedings of the 38th Annual Conference on IEEE Industrial Electronics Society IECON 2012), Montreal, QC, Canada, 25–28 October 2012; pp. 1337–1343. [CrossRef]

Article

Dynamic Optimization of Combined Cooling, Heating, and Power Systems with Energy Storage Units

Jiyuan Kuang, Chenghui Zhang, Fan Li and Bo Sun *

School of Control Science and Engineering, Shandong University, Jinan 250061, China;
sdukuangjiyuan@163.com (J.K.); zchui@sdu.edu.cn (C.Z); lifan_12_12@163.com (F.L.)
* Correspondence: sunbo@sdu.edu.cn

Received: 4 August 2018; Accepted: 29 August 2018; Published: 30 August 2018

Abstract: In this paper, a combined cooling, heating, and power (CCHP) system with thermal storage tanks is introduced. Considering the plants' off-design performance, an efficient methodology is introduced to determine the most economical operation schedule. The complex CCHP system's state transition equation is extracted by selecting the stored cooling and heating energy as the discretized state variables. Referring to the concept of variable cost and constant cost, repeated computations are saved in phase operating cost calculations. Therefore, the most economical operation schedule is obtained by employing a dynamic solving framework in an extremely short time. The simulation results indicated that the optimized operating cost is reduced by 40.8% compared to the traditional energy supply system.

Keywords: CCHP system; energy storage; off-design performance; dynamic solving framework

1. Introduction

Combined cooling, heating, and power (CCHP) systems follow the principle of cascade utilization of energy with high energy efficiency and have become a major research focus [1–6]. It is verified that operation optimization can improve their performance to some extent [7–10]. However, fluctuating energy demands might not always fall within the high efficiency region of CCHP systems [11,12]. Satisfactory operation cannot be achieved easily without energy storage units, which can facilitate high-efficiency CCHP system operation and increase the energy conservation rate by approximately 21% [13]. Meanwhile, the introduction of energy storage units makes CCHP system optimization very difficult [14,15].

The most common operating strategy is based on following the electric loads or following the thermal loads [16,17]. Current studies solve the optimal operating strategy of CCHP systems with storage units in the following way: the outputs of different pieces of equipment in each stage are taken as equivalent optimization variables, which are limited by the plant capacity and energy balance. After setting an objective function, various kinds of algorithms are applied with the objective of seeking the optimal operating schedule. The current studies can be separated into the following two general categories based on their algorithms.

Nearly half of the published research papers employ intelligent optimization algorithms, which are mainly genetic algorithms (GAs) and particle swarm optimization (PSO) algorithms, to solve the CCHP system operation optimization problem. Wang et al. employed GA to optimize an electric load-following operating strategy of a CCHP system [18]. Zeng et al. employed GA to determine the optimal operating solution of a CCHP system combined with ground source heat pumps [19]. Wang et al. built a two-time scale optimized model of a CCHP system, and an improved PSO algorithm is proposed [20]. Considering the co-optimization issue of CCHP system with ice-storage air-conditioners, Bao et al. introduced the Improved PSO algorithm to the solution of the day-ahead operating schedule [21].

Numerous examples of linear programming (LP) applications to CCHP system operation optimization can also be found. Shaneb et al. purposed an optimal online operation of residential CCHP systems using LP [22]. Bischia et al. built a detailed nonlinear CCHP system model, which was piecewise approximated as several linear models, and introduced mixed-integer linear programming (MILP) to optimize the operating schedule [23]. Gu et al. built a prediction control model of a CCHP system; its prediction errors and system deviations were corrected online by rolling optimization, and the dispatch schedule in each step of the rolling optimization was determined by using MILP [24]. Luo et al. proposed two-stage optimization and control structure of the CCHP system, and employed MILP to search the operating schedule [25].

GA, PSO, and MILP can easily optimize the CCHP system operation as long as storage units are not introduced. However, the operation optimization of CCHP systems with storage units is more complex than that of systems without storage units [26], and the methods mentioned above cannot handle the optimization of such systems adequately. Difficulties arise not only from the numerous optimization variables corresponding to each stage, but also because of the correlation between adjacent stages due to the existence of storage units [27]. To be more specific, the energy storage state of each stage depends on the energy supply of the previous stage, whereas the energy supply of each stage is influenced by its current energy storage. To describe the correlation between the adjacent stages, complex constraints must be applied.

Hence, it is not certain that GA or PSO can provide optimal solutions. This conclusion is derived from the fact that different results are obtained for the same problem when they are applied repeatedly. MILP is improved to be efficient when the optimization model is considered to be linear. However, to the best of our knowledge, there is no linear CCHP system that has already been developed, so the piecewise linearity model is constantly used when considering off-design performance. As a result, the computation load is large.

Very few studies have employed dynamic programming in CCHP system operating optimization. Facci et al. applied dynamic programming to a no-storage CCHP system. Considering that generator restart would require extra cost, the generator status in terms of starting and stopping was set as a 0–1 state variable and dynamic programming was employed [28]. Based on previous work, Facci et al. built a CCHP system with storage units. Considering the off-design performance, a dynamic model was established. To reduce the difficulties of the non-linear optimizing problem, dynamic programming combined with meta-heuristic optimization is applied [29].

Their study represented a rare example of the application of dynamic programming to CCHP system operation optimization. However, existing studies maintain a relatively simple system structure. The computation will increase significantly as more plants are introduced, particularly storage units. Further research on dynamic programming applications should be conducted for CCHP systems with complex structure.

In summary, the operation optimization of CCHP systems with storage units should be solved dynamically. Traditional methods such as PSO, GA, and MILP cannot be utilized to tackle it successfully. By resolving the dynamic problem in stages, a dynamic solving framework is created. The computation reduction in complex systems needs significant research, though the prospect of dynamic programming has been confirmed preliminarily.

In this paper, a common CCHP system is proposed. The electric demand is supplied by a power generation unit (PGU) and the power grid. The excess electricity can be sold back to grid. The recovered thermal energy is used to satisfy heating and cooling demands. In addition, two separate heat pumps can also be used to satisfy the thermal demands. The difference between thermal the energy demand and supply can be offset using the thermal storage tanks. The state transition equation is extracted according to the dynamic relationship of the energy storage. A dynamic optimization is proposed to determine the most economical operating schedule. The CCHP system operating optimization is divided into small static problems based on the framework of dynamic programming. The economical concept of variable cost and constant cost are introduced to solve static problems, which can be expressed by the same mathematical model and then solved by the same method with very few computations. As the day-ahead optimization simulation shows, significant improvements over the traditional energy system have been achieved.

2. CCHP System Modeling

The structure and energy flux of the CCHP system are depicted in Figure 1. The power generation unit (PGU), which is connected to the grid, consumes natural gas to generate electricity and thermal energy simultaneously. The exhaust heat exchanger transfers heat from the exhaust gas to jacket water. The absorption chiller recovers energy from the jacket water to produce cooling water. Similarly, the domestic hot water heat exchanger recovers energy from the jacket water to produce domestic hot water. The chiller and exchanger are assisted by separate heat pumps. The thermal storage tanks store extra energy and supply it when necessary. In winter, the cooling demand changes to a space heating demand and the original cold storage tank is employed to store heating water. Meanwhile, the absorption chiller functions as a normal heat exchanger to satisfy the heating demand associated with the corresponding heat pump. It must be noted that each of the operations of the equipment obeys the solution for operating optimization.

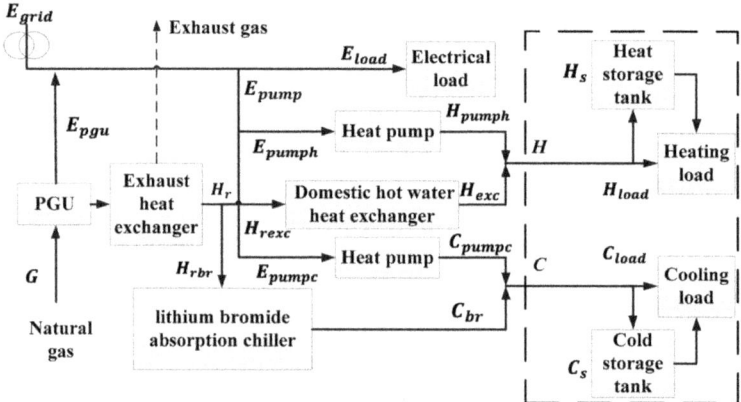

Figure 1. CCHP system.

2.1. State Transition Equation of CCHP System

The components enclosed within a rectangle with dashed borders in Figure 1 constitute the critical section of this system. The state transition equation of the dynamic relationships of the production, load, and stored energy between the kth hour and $(k+1)$th hour can be expressed as follows:

$$
\begin{pmatrix} H_s(k+1) \\ C_s(k+1) \end{pmatrix} = \begin{pmatrix} \eta_h & 0 \\ 0 & \eta_c \end{pmatrix} \cdot \begin{pmatrix} H_s(k) \\ C_s(k) \end{pmatrix} + \begin{pmatrix} 1 & 1 & 0 & 0 \\ 0 & 0 & 1 & 1 \end{pmatrix} \cdot \begin{pmatrix} H_{exc}(k) \\ H_{pumph}(k) \\ C_{br}(k) \\ C_{pumpc}(k) \end{pmatrix} - \begin{pmatrix} 1 & 0 \\ 0 & 1 \end{pmatrix} \cdot \begin{pmatrix} H_{load}(k) \\ C_{load}(k) \end{pmatrix} \quad (1)
$$

$$
f(k+1) = f(k) + v(k) \quad (2)
$$

where η_h is the heat storage efficiency, which represents the proportion of thermal energy remaining after one dissipation stage, and η_c has a similar physical significance; H_s and C_s represent the quantities of stored heating and cooling energy, respectively. The heating and cooling contribution of heat pump are signified as H_{pumph} and C_{pumpc}, respectively. C_{br} and H_{exc} are the chiller and exchanger outputs, respectively. H_{load} and C_{load} are heating and cooling energy demands, respectively. f is the total operating cost and v is the phase cost.

Equation (1) is the core of this paper, based on which the dynamic solving framework is established. Therefore, the huge dynamic problem of CCHP system operating optimization is dynamically broken up into smaller static problems. The operating cost function v is the key of static problem, which will be discussed in chapter three.

2.2. Plant Modeling

The PGU is a gas-fired small internal combustion generating set, whose data is listed in Table 1.

Table 1. Performance of a small naturally aspirated internal combustion engine generator [9].

PLR	η_i	η_g	p_j	p_e	p_l
0.000	0.0000	0.0000	0.5628	0.2764	0.1608
0.100	0.1020	0.7700	0.5227	0.2955	0.1818
0.200	0.1809	0.7800	0.5031	0.3006	0.1963
0.300	0.2250	0.8200	0.4903	0.3097	0.2000
0.400	0.2637	0.8400	0.4865	0.3108	0.2027
0.500	0.2871	0.8600	0.4861	0.3125	0.2014
0.600	0.3085	0.8750	0.4892	0.3237	0.1870
0.700	0.3184	0.8850	0.4818	0.3285	0.1898
0.800	0.3184	0.9000	0.4745	0.3285	0.1971
0.900	0.3039	0.9100	0.4507	0.3169	0.2324
1.000	0.2886	0.9200	0.4336	0.3147	0.2517

Note: *PLR* is the load rate, and η_g and η_i are the efficiencies of the generator and internal combustion engine, respectively. p_j and p_e are the energy ratios corresponding to the jacket water and exhaust, respectively. p_l represents the heat loss rate.

Taking η_{re} and l_{rj} as the exhaust heat exchanger efficiency and the dissipated thermal energy ratio of jacket water in heat exchanging process, respectively, the waste heat recovery ratio η_{rw} is given by:

$$\eta_{rw} = (1 - \eta_{pgu}) \cdot \left[(1 - l_{rj}) \cdot p_j + \eta_{re} \cdot p_e \right] \tag{3}$$

The recovered waste heat H_r is used to drive the absorption chiller and domestic hot water heat exchanger, whose efficiency are η_{br} and η_{exc}, respectively. The contribution of the absorption chiller and heat exchanger are C_{br} and H_{exc}, respectively. The chilling and heating coefficient of performance (COP) of the electric heat pump are COP_c and COP_h, respectively. The heating and cooling contribution of heat pump are H_{pumph} and C_{pumpc}, respectively. The consumed electricity of heat pump is E_{pump}.

3. Methodology

3.1. Optimal Operation Model

The optimization of a CCHP system with storage units is dynamic in nature. Thus, the solution framework is based on dynamic programing. The state variable selection is the most important step in dynamic programing. Energy storage should be chosen because it serves as a link between adjacent stages (see Equations (1)). Thereupon, the optimization model can be established. The state variable discretization is as follows.

s_k (H_s, C_s) denotes the heating and cooling energy storage of stage k, where $0 \leq H_s \leq N_h$ and $0 \leq C_s \leq N_c$. N_h and N_c are the capacities of the heating and cooling storage tanks, respectively. After setting m and n, s_k can be discretized into $(m + 1) \cdot (n + 1)$ state points. The state point $s_k \left(p \frac{N_h}{m}, q \frac{N_c}{n} \right)$ can be expressed simply as $s_k^{p,q}$, where $0 \leq p \leq m$, and $0 \leq q \leq n$. The set of $s_k^{p,q}$ is expressed as s_k. Larger m and n lead to more accurate discretization and more state points.

According to the discretization described above, s_k is arrayed as depicted in Figure 2.

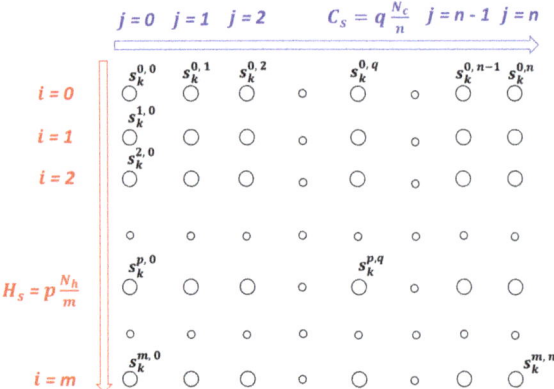

Figure 2. Arrangement of s_k.

$u_k\left(s_k^{p,q}, s_{k+1}^{i,j}\right)$ represents the optimal operation solution of the CCHP system for transferring $s_k^{p,q}$ to $s_{k+1}^{i,j}$. The corresponding cost is expressed as $v_k\left(s_k^{p,q}, s_{k+1}^{i,j}\right)$. The method to solve u_k and v_k is proposed in Section 3.3.

The shortest path model of the CCHP system operation optimization problem is as shown in Figure 3. The energy storage in each stage corresponds to a point set s_{k+1}. Based on the state point $s_k^{p,q}$ selected in the previous stage, the path from $s_k^{p,q}$ to $s_{k+1}^{i,j}$ has a unique length expressed as $v_k\left(s_k^{p,q}, s_{k+1}^{i,j}\right)$. The minimum cost of the CCHP system operating schedule is represented by the length of the shortest path from s_1 to s_{N+1}.

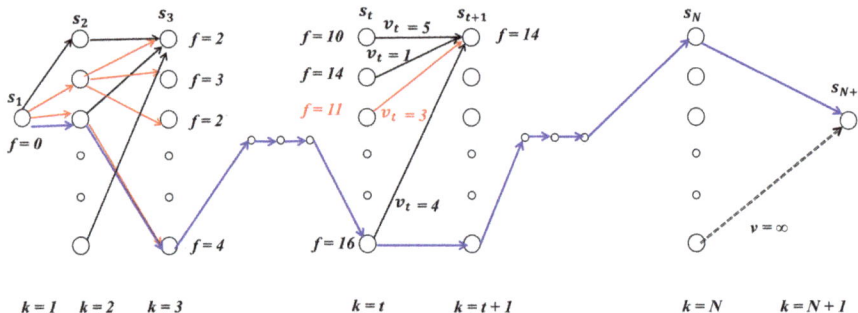

Figure 3. Shortest path model.

The shortest path problem of CCHP system operation can be described as follows. The oriented graph in Figure 3 is represented as $D = (S, A)$, $s_k^{p,q}$ and $s_{k+1}^{i,j}$ (the state points of adjacent stages) are joined by an oriented arc $a\left(s_k^{p,q}, s_{k+1}^{i,j}\right)$, and the weight of the arc is represented as $v(a)$, where $v(a) = v_k\left(s_k^{p,q}, s_{k+1}^{i,j}\right)$. If there is no arc joining $s_k^{p,q}$ and $s_{k+1}^{i,j}$, then $v_k\left(s_k^{p,q}, s_{k+1}^{i,j}\right)$ is set to $+\infty$. Suppose P is a path of D from the initial point s_1 to the end point s_{N+1}, and define the weight of P as the sum of each arc in P, represented as $v(P)$. The objective of this shortest path problem is to find the minimum-weight path P_0 among all of the paths P from s_1 to s_{N+1}, where:

$$v(P_0) = \min_P v(P) \tag{4}$$

P_0 is the shortest path from s_1 to s_{N+1}. The weight of P_0 is the distance from s_1 to s_{N+1}, represented as $f(s_1, s_{N+1})$. For CCHP system operation, $f(s_1, s_{N+1})$ is the minimum operating cost. Thus, the optimization problem can be solved by finding P_0.

3.2. Shortest Path Determination Based on Dynamic Programming

The shortest path search is a multi-stage decision problem. The optimality principle was developed particularly to solve this kind of issue. Moreover, dynamic programming is proposed by transforming the multi-stage process into single stages. The result obtained by dynamic programming is certain to be optimal due to optimality principle. The best methods recognized for solving the shortest path problem involve dynamic programming without exception. The diagram of the dynamic programming flow used in this paper is provided in Figure 4.

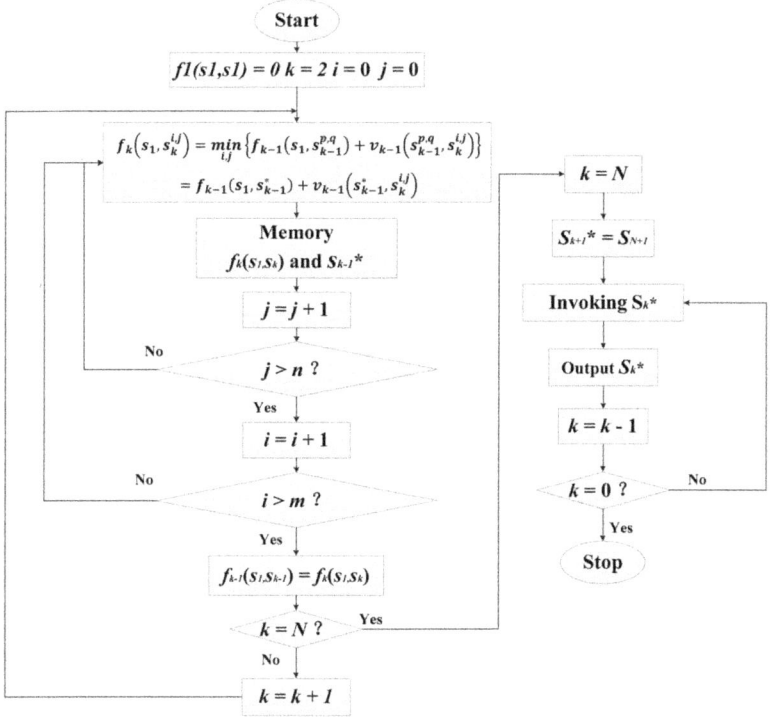

Figure 4. Dynamic programming flow diagram.

The shortest path P_0 from s_1 to s_{N+1} always starts from s_1, passing through one state point $s_N^{i,j}$, and finally arriving at s_{N+1}. According to the optimality principle, the path from s_1 to $s_N^{i,j}$ is the shortest. Hence, the dynamic programming equation of this model is obtained as:

$$f_{N+1}(P_0) = f_{N+1}(s_1, s_{N+1}) = \min_{i,j}\left\{ f_N\left(s_1, s_N^{i,j}\right) + v_N\left(s_N^{i,j}, s_{N+1}\right) \right\} \tag{5}$$

Using s_{k-1}^* to signify the optimal state point selected from s_{k-1}, a more general expressions can be derived as:

$$f_k\left(s_1, s_k^{i,j}\right) = \min_{i,j}\left\{f_{k-1}\left(s_1, s_{k-1}^{p,q}\right) + v_{k-1}\left(s_{k-1}^{p,q}, s_k^{i,j}\right)\right\} = f_{k-1}\left(s_1, s_{k-1}^*\right) + v_{k-1}\left(s_{k-1}^*, s_k^{i,j}\right) \quad (6)$$

and:

$$f_1(s_1, s_1) = 0 \quad (7)$$

As shown in Equations (6) and (7), forward dynamic programming is applied. This problem is solved step by step. Meanwhile, the shortest distance and path selection are recorded. The optimization problem is solved when $f_{N+1}(s_1, s_{N+1})$ is obtained.

In addition, as can be seen in Figure 2, larger m and n lead to larger s_k. To reduce the amount of unnecessary calculations, the discretization is separated into two steps. Firstly, the energy storage is discretized with rough accuracy and dynamic programming is applied to search the shortest path. Secondly, the energy storage is discretized with precise accuracy near the path obtained in the first optimization. The second optimizing result is precise to 1 kW·h.

3.3. Static Problem: Analysis of Stage Cost

The static problem is searching for the minimum cost resulting from the state transition. In other words, its objective is to determine $v_k\left(s_k^i, s_{k+1}^j\right)$ according to state points s_k (H_s, C_s) and s_{k+1} (H_s, C_s).

According to Equation (1), the heating and cooling production of stage k can be represented by the energy storage of stages k and $k + 1$. Based on the required energy production, the most economical dispatch strategy and its corresponding cost can be determined by referring to the operation optimization of a no-storage CCHP system, which is a static problem. Although LP, GA, and PSO can be employed, it is time consuming to calculate the static problems repeatedly in dynamic programming. In this section, the operating cost is solved without any optimizations by introducing the concept of variable cost.

The operating cost of a CCHP system consists of electricity and gas costs. The stage cost v can be calculated as follows:

$$v = E_{price} \cdot E_{grid} + G_{price} \cdot G \quad (8)$$

where G is the consumed natural gas and G_{price} is the gas price. The amount of electricity received from the power grid is given by:

$$E_{grid} = E_{load} - E_{pgu} + \frac{H_{pumph}}{COP_h} + \frac{C_{pumpc}}{COP_c} \quad (9)$$

For the given state points s_k (H_s, C_s) and s_{k+1} (H_s, C_s), the total heating and cooling demand, H and C, respectively, are fixed.

Based on the modeling of the PGU and exhaust heat exchanger given in Section 2.2, E_{pgu} and G can be fitted as polynomial functions of H_r. The required data are listed in Tables 1 and 2. Because H_r is the function of C_{br} and H_{exc}, the conclusion can be drawn that both E_{pgu} and G are functions of C_{br} and H_{exc}. Hence, v can be represented as a function of C_{br} and H_{exc}. Referring to the economics, the operating cost consists of constant cost v' and variable cost Δv:

$$v = v' + \Delta v \quad (10)$$

Assume that all of the heating and cooling energy is provided by the heat pump and that the electrical load is supplied by the power grid. The constant cost v' is determined by E_{price}, E_{load}, H and C. In other words, v' cannot be optimized:

$$v\prime = E_{price} \cdot \left(E_{load} + \frac{H}{COP_h} + \frac{C}{COP_c} \right) \tag{11}$$

Starting the generator results in an extra gas cost, while the produced power offsets the power bought from the grid. The variable cost Δv represents the change in cost resulting from generator operation at different power levels:

$$\Delta v = G_{price} \cdot G(C_{br}, H_{exc}) - E_{price} \cdot \left(E_{pgu}(C_{br}, H_{exc}) + \frac{H_{exc}}{COP_h} + \frac{C_{br}}{COP_c} \right) \tag{12}$$

The domain of this function is:
$$\begin{aligned} 0 \le H_{exc} \le H \\ 0 \le C_{br} \le C. \end{aligned} \tag{13}$$

$v\prime$ has no relationship with C_{br} and H_{exc}. To determine the minimum stage operating cost v, attention should be paid to Δv, which is a function of C_{br} and H_{exc}. Hence, the essence of static problems is searching for the minimum value of Δv. According to the expression for Δv, E_{price} is the most influential parameter. Its influence is shown in Figure 5. For clarity, C_{br} and H_{exc} are combined into H_r.

Table 2. CCHP system plants parameters [30,31].

Parameters	Values
Generator capacity N_{pgu}	90 kW
Efficiency of domestic hot water heat exchanger η_{exc}	0.95
Rated efficiency of absorption chiller η_{br}	0.8
Efficiency of exhaust heat exchanger η_{re}	0.8
Energy loss ratio of jacket water in exhaust heat exchanger l_{rj}	0.2
Heating value of natural gas Q	35,500 kJ/m^3
Price of natural gas G_{price}	2 ¥/m^3

According to the expression for Δv, E_{price} is the most influential parameter. Its influence is shown in Figure 5. For clarity, C_{br} and H_{exc} are combined into H_r.

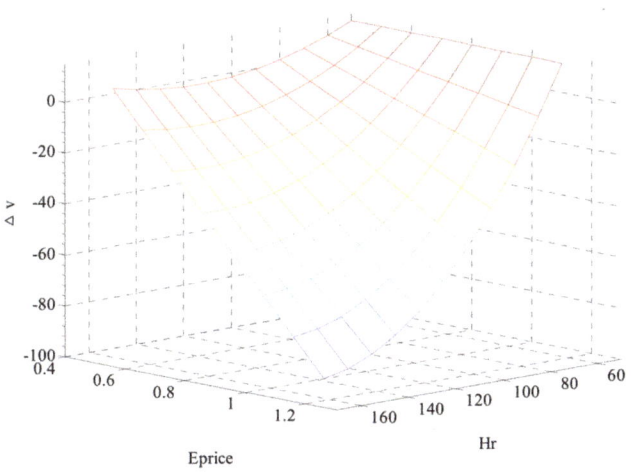

Figure 5. Relationship between H_{re} and Δv for different E_{price}.

When E_{price} is 1.1 ¥/(kW·h), all of the heating and cooling is supplied by the absorption chiller and domestic hot water exchanger. When E_{price} is 0.7 ¥/(kW·h), the heat recovery of 144.7 kW·h corresponds the most economic operating strategy. When E_{price} is 0.4 ¥/(kW·h), the heat recovery of 123 kW·h corresponds to the peak efficiency of the generator. The generator should work to ensure that the Δv as small as possible, so long as Δv is negative. Otherwise, it is more economical to stop the generator when the efficiency is low.

E_{pgu}, H_{pumph}, and C_{pumpc} can be determined based on H_{exc} and C_{br}. Hence, the optimal dispatch strategy can be described as u (C_{br}, H_{exc}, E_{pgu}, H_{pumph}, C_{pumpc}). The operating cost v is also obtained. By referring to the concept of constant cost and variable cost, thousands of repeated computations can be eliminated.

4. Case Study

4.1. Load Description and Basic Data

The energy demands can be obtained from [29,32]. The building was simulated using EnergyPlus (5.0.0.035). The description of the simulated building is given in Table 3. For our test case, we selected two typical days in summer and winter, as reported in Figure 6.

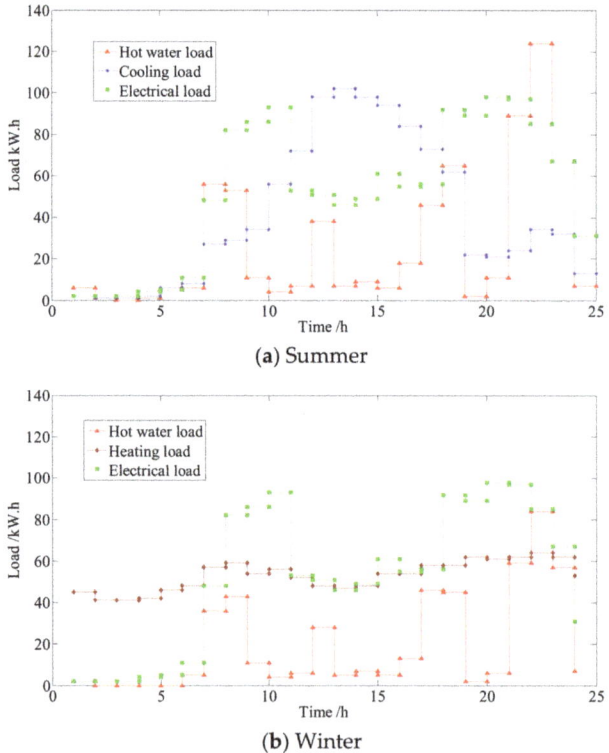

Figure 6. Energy demand time traces.

Table 3. Description of the simulated building [32].

Parameter	Data
Location	Baltimore
Area	4014 m^2
Volume	11,622 m^3
Gross wall area	1695 m^2
Window glass area	184 m^2
Lights (on average)	16.10 W/m^2
Elec plug and process (on average)	12.16 W/m^2
People	254 people

In addition, there is no cooling load in winter. Instead, extra hot water is required by the central air-conditioning system to keep the dormitory warm. This part of the hot water is separated from that consumed by bathing and so on. The cooling storage tank is employed to store this part of the hot water.

The electricity price (in Yuan ¥) per hour refers to [33]:

$$E_{price}\ (k) = \begin{cases} 0.4(k = 1, 2, 3, 4, 5, 6, 24) \\ 0.7(k = 7, 11, 12, 13, 14, 15, 16, 17, 23) \\ 1.1(k = 18, 19, 20, 21, 22) \end{cases} . \tag{14}$$

The CCHP system parameters are listed in Table 4.

Table 4. Constant parameters of the CCHP system [30,31].

Parameters	Values
Rated COP for electrically driven heat pump COP_h, COP_c	3
Cold storage coefficient C_d	0.97
Heat storage coefficient H_d	0.95
Capacity of heat storage unit N_h	150 kW·h
Capacity of cold storage unit N_c	120 kW·h
Rated power of generator P_{rated}	90 kW
Capacity of absorption chiller N_{br}	100 kW

The parameters of a traditional energy system are listed in Table 5. The heating load is supplied by a gas boiler, the electrical load is supplied by the power grid, and the cooling load is supplied by an electrically driven air conditioning system.

Table 5. Constant parameters of a traditional energy system [9].

Parameters	Values
Efficiency of Power Plant	0.35
Efficiency of power transmission	0.92
COP of electrically driven air conditioning system	3
Efficiency of gas boiler	0.88

The fuel parameters employed for the traditional energy system and CCHP systems to calculate the operation targets are listed in Table 6.

Table 6. Parameters of natural gas and coal [9].

Type of Fuel	Heating Value	CO$_2$ Emission When Thoroughly Burned
Coal	29,300 kJ/kg	2.69 kg/kg
Natural gas	35,500 kJ/m^3	1.96 kg/m^3

4.2. Results and Analysis

The state variable discretization process was divided into two steps with accuracy at 10 kW·h and 1 kW·h. According to the discretization method described in Section 3.1, the amount of computations required was reduced by 98%.

The optimal results obtained using the loads in summer are presented in Figures 7 and 8. The negative power grid output values indicate generator feedback power to the grid.

As shown in Figure 7, the energy storage units store energy when the demand is low and then supply a substantial portion of the energy demand during the peak power consumption periods. As depicted in Figure 8, the generator operates at the load rate of about 80%, although the electrical load fluctuates sharply. The storage units serve to reduce the peaks and fill the valleys, which dramatically improves the energy utilization. Nevertheless, the energy demand tendencies remain observable in the generator operation tracking results. Moreover, the power track of the generator follows the power price of the grid. The generator operates at a high load rate when electrical power is expensive. An appropriate load rate is applied when power is modestly priced. The generator would stop at a low power price.

From 8:00 to 10:00, the generator operated at nearly full capacity, and some extra power was sold to the grid. It can be seen from Table 1 that the operating efficiency at full capacity is lower than the maximum efficiency. However, electrical power is so expensive that it is profitable to sacrifice some efficiency. Moreover, the storage units store considerable energy to prepare for the upcoming phase of peak energy consumption.

From 19:00 to 20:00, the thermal energy demand is low. Due to the high electricity price and large amount of electricity demand, the generator operated at nearly full capacity. Meanwhile, large quantity of thermal energy is stored to handle the next peak of thermal energy consumption.

The generator stopped at 23:00. The subsequent thermal demand can be supplied by the energy stored beforehand. If the generator continues operating, the stored energy would remain unutilized. The generator should stop operating although there was little power demand at 23:00.

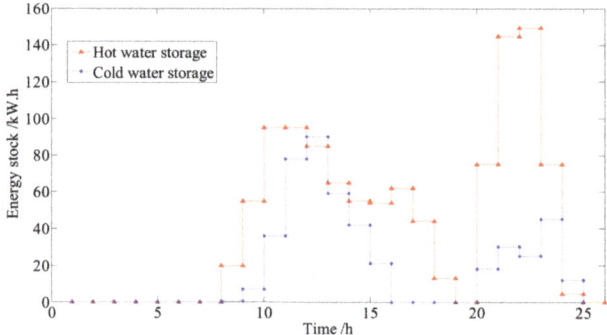

Figure 7. Changes in energy storage under summer conditions.

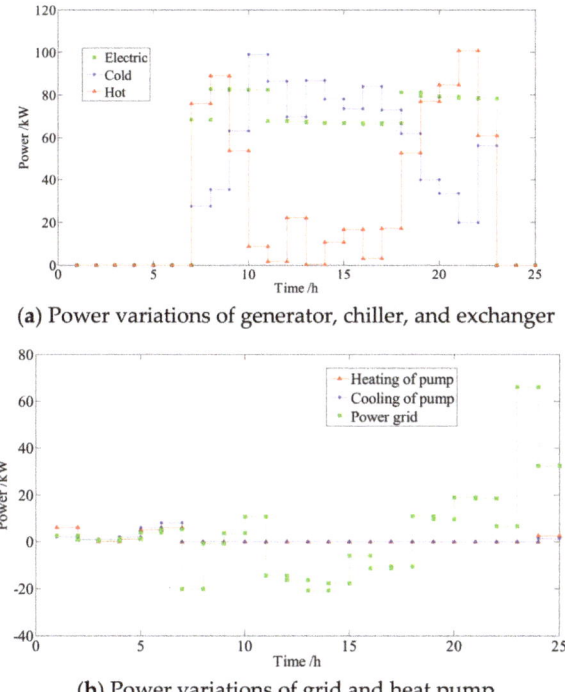

(a) Power variations of generator, chiller, and exchanger

(b) Power variations of grid and heat pump

Figure 8. Power variations of system components under summer conditions.

In summary, the operation optimization is influenced by three main factors. The most important factor is the power demand, which determines the general trend of the optimization results. The next factor is the price of electricity, which strongly affects the operating state of the generator. The last factor is the dissipation of stored energy, which restricts the energy storage time. These three factors jointly determine the optimization results.

Under the energy demands of a typical day in summer, the operating targets of the CCHP system obtained using dynamic programming and the traditional energy system targets are provided in Table 7. The operating cost is converted into dollars.

Table 7. Targets comparison of a whole day of operation in summer.

Operating Target	Operating Cost ($)	CO_2 Emission (kg)	Fuel Consumption (MJ)
Traditional energy system	254.1	2038.7 kg	20,464.92
CCHP system	150.4	924.2 kg	15,953.76
Variation	103.7	1114.5 kg	4511.16
Rate of change	40.8%	54.7%	22.0%

The energy efficiency is the proportion of energy consumed by users and the fossil energy consumed by the power station, gas boiler, and CCHP system. The operating cost of the CCHP system is reduced by 40.8% compared to that in the traditional energy system. Furthermore, the fuel energy saving ratio is 22.0% and the carbon emission is decreased by 54.7% in the CCHP system.

The optimal results obtained considering the loads in winter are presented in Figures 9 and 10.

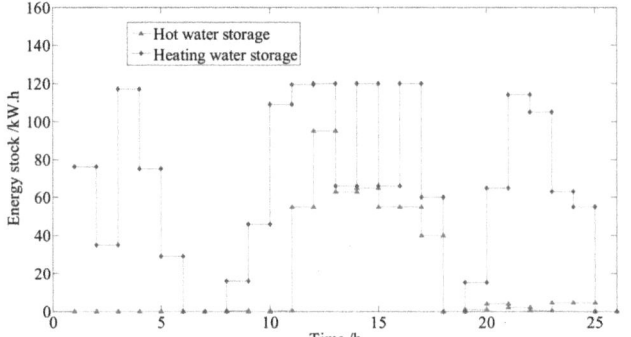

Figure 9. Changes in energy storage under winter conditions.

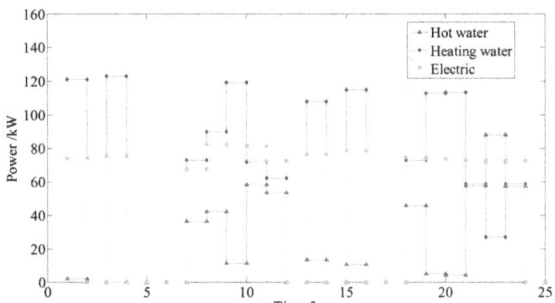

(**a**) Power variations of generator, chiller, and exchanger

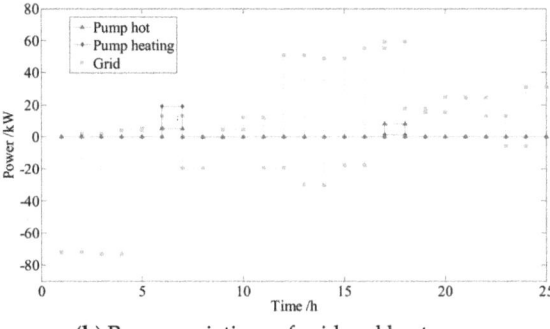

(**b**) Power variations of grid and heat pump

Figure 10. Power variations of system components under winter conditions.

As mentioned previously, the hot water required by the central air conditioning system to keep the dormitory warm was separated from that consumed by bathing and so on, and the cool storage tank was employed to store this part of the hot water.

Generally, the optimization result under winter conditions is influenced by the three factors discussed for summer conditions. However, a significant characteristic occurs at late night. Unlike in the results obtained for summer, the generator starts at night because the heating load is heavy in winter. Because the electricity is cheap late at night, the generator has to operate at the highest efficiency. Otherwise, it has to stop. Hot water is stored to supply heating. Under the energy demands of a typical day in winter. The operating targets are compared in Table 8.

Table 8. Targets comparison of a whole day of operation in winter.

Operating Target	Operating Cost ($)	CO_2 Emission (kg)	Fuel Consumption (MJ)
Traditional energy system	247.4	1895.8	21,289.68
CCHP system	158.9	1147.1	17,753.76
Variation	88.5	748.7	3535.92
Rate of change	35.8%	39.5%	16.7%

When compared with the traditional energy system, the operating cost is reduced by 35.8%, the fuel energy saving ratio is 16.7%, and the carbon emission is decreased by 39.5%.

The conclusion can be drawn that this optimization method not only ensures that the optimal operating cost is achieved, but also obviously improves the fuel energy saving and environment protection. Moreover, all the optimizing results were obtained in less than three seconds.

5. Conclusions

In this paper, a CCHP system with storage units was designed. Due to its complex structure and internal coupling relation, especially considering that its operation progress is essentially dynamic, traditional optimizing algorithms have some insufficiencies in optimizing its operating schedule. Recent research has improved the advantages of dynamic programming applied to CCHP system optimization. However, its application to a CCHP system with complex structure needs efficient planning to reduce computation.

In the proposed method, the optimization problem was split into a dynamic problem and an embedded static problem. The dynamic problem reflects the essence of the optimization problem, while the static problem provides the basis of the dynamic problem. Thousands of repeated computations were eliminated in economical optimization by introducing the concept of constant cost and variable cost. Compared to a traditional energy system, the operating cost was reduced by 40.8%, the fuel energy saving ratio was 22.0%, and the carbon emission was decreased by 54.7%. Moreover, the optimization of the whole day of a CCHP system requires about 3 s on an average desktop computer. This is a very short optimization time for a CCHP system with energy storage units. Thus, dynamic programming can be successfully employed to solve the optimization of CCHP system with complex structure.

In addition, the optimizing methodology applied in this paper implies a stochastic dynamic solving framework, which will probably contribute to CCHP system optimization. We have achieved some breakthrough and are trying to employ it in stochastic optimization of CCHP systems considering off-design performance.

Author Contributions: Conceptualization, J.K.; Methodology, J.K.; Software, J.K., F.L.; Resources, C.Z., B.S.; Writing-Review & Editing, B.S.

Funding: This work was supported by the National Natural Science Foundation of China (grant numbers 61733010, 61320106011, 61573224, 61573223) and the Young Scholars Program of Shandong University (grant number 2016WLJH29).

Conflicts of Interest: The authors declare no conflict of interest.

References

1. Bilgen, S.; Keleş, S.; Kaygusuz, A.; Sarı, A.; Kaygusuz, A. Global warming and renewable energy sources for sustainable development: A case study in Turkey. *Renew. Sust. Energy Rev.* **2008**, *12*, 372–396. [CrossRef]

2. Wu, D.; Wang, R. Combined cooling, heating and power: A review. *Prog. Energy Combust. Sci.* **2006**, *32*, 459–495. [CrossRef]

3. Guo, L.; Liu, W.; Cai, J.; Hong, B.; Wang, C. A two-stage optimal planning and design method for combined cooling, heat and power microgrid system. *Energy Convers. Manag.* **2013**, *74*, 433–445. [CrossRef]

4.	Jradi, M.; Riffat, S. Tri-generation systems: Energy policies, prime movers, cooling technologies, configurations and operation strategies. *Renew. Sustain. Energy Rev.* **2014**, *32*, 396–415. [CrossRef]

5.	Liu, M.; Shi, Y.; Fang, F. Combined cooling, heating and power systems: A survey. *Renew. Sustain. Energy Rev.* **2014**, *35*, 1–22. [CrossRef]

6.	Bellos, E.; Tzivanidis, C. Optimization of a solar-driven trigeneration system with nanofluid-based parabolic trough collectors. *Energies* **2017**, *10*, 848. [CrossRef]

7.	Ju, L.; Tan, Z.; Li, H.; Tan, Q.; Yu, X.; Song, X. Multi-objective operation optimization and evaluation model for CCHP and renewable energy based hybrid energy system driven by distributed energy resources in China. *Energy* **2016**, *111*, 322–340. [CrossRef]

8.	Palomba, V.; Ferraro, M.; Frazzica, A.; Vasta, S.; Sergi, F.; Antonucci, V. Experimental and numerical analysis of a SOFC-CHP system with adsorption and hybrid chillers for telecommunication applications. *Appl. Energy* **2018**, *216*, 620–633. [CrossRef]

9.	Wei, D.; Chen, A.; Sun, B.; Zhang, C. Multi-objective optimal operation and energy coupling analysis of combined cooling and heating system. *Energy* **2016**, *98*, 296–307. [CrossRef]

10.	Zeng, R.; Li, H.; Liu, L.; Zhang, X.; Zhang, G. A novel method based on multi-population genetic algorithm for CCHP–GSHP coupling system optimization. *Energy Convers. Manag.* **2015**, *105*, 1138–1148. [CrossRef]

11.	Song, X.; Liu, L.; Zhu, T.; Zhang, T.; Wu, Z. Comparative analysis on operation strategies of CCHP system with cool thermal storage for a data center. *Appl. Therm. Eng.* **2016**, *108*, 680–688. [CrossRef]

12.	Gopisetty, S.; Treffinger, P. Generic combined heat and power (CHP) model for the concept phase of energy planning process. *Energies* **2016**, *10*, 11. [CrossRef]

13.	Lozano, M.; Ramos, J.; Carvalho, M.; Serra, L. Structure optimization of energy supply systems in tertiary sector buildings. *Energy Build.* **2009**, *41*, 1063–1075. [CrossRef]

14.	Moussawi, H.; Fardoun, F. Louahlia-Gualous, H. Review of tri-generation technologies: Design evaluation, optimization, decision-making, and selection approach. *Energy Convers. Manag.* **2016**, *120*, 157–196. [CrossRef]

15.	Han, B.; Cheng, W.; Li, Y.; Nian, Y. Thermodynamic analysis of heat driven Combined Cooling Heating and Power system (CCHP) with energy storage for long distance transmission. *Energy Convers. Manag.* **2017**, *154*, 102–117. [CrossRef]

16.	Li, G.Z.; Wang, R.; Zhang, T.; Ming, M.J. Multi-objective optimal design of renewable energy integrated CCHP system using PICEA-g. *Energies* **2018**, *11*, 743. [CrossRef]

17.	Sigarchian, S.; Malmquist, A.; Martin, V. Design optimization of a small-scale polygeneration energy system in different climate zones in Iran. *Energies* **2018**, *11*, 1115. [CrossRef]

18.	Wang, J.; Sui, J.; Jin, H. An improved operation strategy of combined cooling heating and power system following electrical load. *Energy* **2015**, *85*, 654–666. [CrossRef]

19.	Zeng, R.; Li, H.; Jiang, R.; Liu, L.; Zhang, G. A novel multi-objective optimization method for CCHP–GSHP coupling systems. *Energy. Build.* **2016**, *112*, 140–158. [CrossRef]

20.	Wang, F.; Zhou, L.; Ren, H.; Liu, X. Search improvement process-chaotic optimization-particle swarm optimization-elite retention strategy and improved combined cooling-heating-power strategy based two-time scale multi-objective optimization model for stand-alone microgrid operation. *Energies* **2017**, *10*, 1936. [CrossRef]

21.	Bao, Z.; Zhou, Q.; Yang, Z.; Yang, Q.; Xu, L.; Wu, T. A multi time-scale and multi energy-type coordinated microgrid scheduling solution. *IEEE Trans. Power Syst.* **2015**, *30*, 2257–2276. [CrossRef]

22.	Shaneb, O.; Taylor, P.; Coates, G. Optimal online operation of residential μCHP systems using linear programming. *Energy Build.* **2012**, *44*, 17–25. [CrossRef]

23.	Bischi, A.; Taccari, L.; Martelli, E.; Amaldi, E.; Manzolini, G.; Silva, P.; Campanari, S.; Macchi, E. A detailed MILP optimization model for combined cooling, heat and power system operation planning. *Energy* **2014**, *74*, 12–26. [CrossRef]

24.	Gu, W.; Wang, Z.; Wu, Z.; Luo, Z.; Tang, Y.; Wang, J. An online optimal dispatch schedule for CCHP microgrids based on model predictive control. *IEEE Trans. Smart Grid* **2017**, *8*, 2332–2342. [CrossRef]

25.	Luo, Z.; Wu, Z.; Li, Z.; Cai, H.; Li, B.; Gu, W. A two-stage optimization and control for CCHP microgrid energy management. *Appl. Therm. Eng.* **2017**, *125*, 513–522. [CrossRef]

26.	Smith, A.D.; Mago, P.J.; Fumo, N. Benefits of thermal energy storage option combined with CHP system for different commercial building types. *Sustain. Energy Technol. Assess.* **2013**, *1*, 3–12. [CrossRef]

Energies **2018**, *11*, 2288

27. Cho, H.; Smith, A.; Mago, P. Combined cooling, heating and power: A review of performance improvement and optimization. *Appl. Energy* **2014**, *136*, 168–185. [CrossRef]

28. Facci, A.; Andreassi, L.; Ubertini, S. Optimization of CHCP (combined heat power and cooling) systems operation strategy using dynamic programming. *Energy* **2014**, *66*, 387–400. [CrossRef]

29. Facci, A.; Ubertini, S. Meta-heuristic optimization for a high-detail smart management of complex energy systems. *Energy Convers. Manag.* **2018**, *160*, 353. [CrossRef]

30. Mago, P.; Fumo, N.; Chamra, L. Performance analysis of CCHP and CHP systems operating following the thermal and electric load. *Int. J. Energy Res.* **2009**, *33*, 852–864. [CrossRef]

31. Mancarella, P.; Chicco, G. Assessment of the greenhouse gas emissions from cogeneration and trigeneration systems. Part II: Analysis techniques and application cases. *Energy* **2008**, *33*, 418–430. [CrossRef]

32. Office of Energy Efficiency and Renewable Energy. Commercial reference buildings. Available online: http://energy.gov/eere/buildings/commercial-reference-buildings (accessed on 13 November 2012).

33. Li, C.; Shi, Y.; Huang, X. Sensitivity analysis of energy demands on performance of CCHP system. *Energy Convers. Manag.* **2008**, *49*, 3491–3497. [CrossRef]

Article

Using Piecewise Linearization Method to PCS Input/Output-Efficiency Curve for a Stand-Alone Microgrid Unit Commitment

Ha-Lim Lee [1],* and Yeong-Han Chun [2]

[1] Smart Distribution Research Center, Advanced Power Grid Research Division, Korea Electrotechnology Research Institute, bulmosan-ro 10beon-gil, Seongsan-gu, Changwonsi, Gyeongsangnam-do 51543, Korea
[2] Power System Lab, Electronic & Electrical Enginerring School, Hong-Ik University, 94, Wauson-ro, Mapo-gu, Seoul 04066, Korea; yhchun@hongik.ac.kr
*   Correspondence: halim@keri.re.kr; Tel.: +82-10-9632-2894

Received: 7 September 2018; Accepted: 13 September 2018; Published: 17 September 2018

Abstract: When operating a stand-alone micro grid, the battery energy storage system (BESS) and a diesel generator are key components needed in order to maintain demand-supply balance. Using Unit Commitment (UC) to calculate the optimal operation schedule of a BESS and diesel generator helps minimize the operation cost of the micro grid. While calculating the optimal operation schedule for the microgrid, it is important that it reflects the actual characteristics of the implanted devices, in order to increase the schedule result accuracy. In this paper, a piecewise linearization, on the actual power conditioning system (PCS) input/output-efficiency characteristic curve, has been considered while calculating the optimal operation schedule using UC. The optimal schedule result calculated by the proposed method has been examined by comparing the schedule calculated by a fixed input/output-efficiency case, which is conventionally used while solving UC for a stand-alone microgrid.

Keywords: battery energy storage system; micro grid; MILP; PCS efficiency; piecewise linear techniques; renewable energy sources; optimal operation; UC

1. Introduction

A microgrid is a power system covering a partial area, which supplies electric power by utilizing a mixture of energy storage systems and distributed generators, such as renewable energy. To increase the usage of renewable energy in a small sized stand-alone microgrid, the capacity of renewable energy should be relatively larger than a normal microgrid. In these power systems, the output fluctuation from the renewable source is large, which causes many electrical problems. To solve and prevent these problems, various issues must be covered [1]. Using a battery energy storage system (BESS) is a possible method to prevent energy waste by charging the over-generated energy produced from the renewable energy sources [2,3]. Therefore, in a stand-alone microgrid, if a BESS is implanted, the availability of renewable sources and the reliability of power system could be increased [4,5]. To maintain this characteristic of the microgrid by using diesel generator and a frequency controlling BESS, an optimal operation method is necessary. In this case, in order to synchronize every device installed in the microgrid, such as diesel generator, wind turbine, photovoltaic device, etc., the BESS must be used as the master source, operating under constant voltage and constant frequency mode [6].

Many studies have been done on the stable operation of stand-alone microgrids, in order to efficiently utilize the implanted BESS. The research conducted in [6,7] have proposed a management system using a decision tree, based on real time measured state of charge (SOC) for a grid-connected microgrid having a BESS. A management system of a microgrid that operates on grid-connected mode at normal occasions, and stand-alone mode when an accident occurs by controlling the BESS, diesel generators, and hydro-generators is proposed in [8]. In the research conducted in [9,10], an unit commitment (UC) is proposed for a grid-connected microgrid having micro turbines and fuel cells. [11] proposed a mixed integer linear program (MILP) based on rolling horizon controllers for a management system of microgrids having a BESS and renewable sources. In [12], a model predictive controller is proposed for a grid-connected microgrid having a BESS as a management system. The research conducted in [13–15] proposed a MILP calculation method considering a modified optimal dispatch strategy as a management system for a micro grid having a BESS. [16] has proposed a MILP approach for residential distributed energy system planning. [17] proposed an energy management strategy for both grid-connected and isolated microgrids, using a simplified MILP formulation as an UC problem. The research conducted in [18] proposed a MILP approach without using complex heuristics or decompositions for the operation scheduling of a microgrid. [19] proposed a linear programming cost minimization model of UC for stand-alone microgrids in the UK. In [20], an improved genetic algorithm-based method is proposed for UC in a stand-alone microgrid. A mathematical model of frequency control in stand-alone microgrids, which is integrated into the UC problem, is proposed in [21]. The research conducted in [22–24] proposed an optimization methodology for a day-ahead UC model in a microgrid.

This paper regards the actual performance characteristics of a power conditional system (PCS) while drawing the microgrid operation schedule, considering the accuracy of modeling and calculations. Therefore, this paper has assumed that drawing the generator and load device's optimal operation schedule is the most economic and safe method for operating the stand-alone microgrid effectively. In particular, using the MILP method to calculate the UC is assumed to be the most accurate after analyzing research on the related issues. The papers mentioned above have assumed that the efficiency of a PCS, which is connected in front of the BESS, is fixed, whether the input/output quantity from the PCS is small or large. However, the actual PCS efficiency is low when the input/output is small, and the efficiency is high when the input/output is large. In other words, when the input/output from the actual PCS is small, the produced electric loss rate is higher during the power conversion process, compared to larger input/output from the PCS. Ignoring such input/output-efficiency characteristics of the PCS, the precise amount of power loss inccurred from the PCS is disregarded during calculation of UC, and also has possibility for operation failure, when the device implanted in the microgrid operate under the calculated schedule. Also, if the accurate amount of power loss caused from the PCS is not reflected, the total operation cost of the diesel generator could increase compared to the schedule, or cause an imbalance between power supply and demand. In order to compensate for such defects, this paper suggests considering PCS efficiency characteristics, by using piecewise linearization on the actual input/output-efficiency curve of PCS, while calculating UC for the optimal operation schedule. This will reduce the calculation difference of the actual power loss occurred from the PCS, which increases the accuracy of the optimal operation schedule calculated by UC. Such concepts have been developed in an effort to increase the accuracy of a calculated UC schedule, by using the actual device performance characteristics.

The remainder the paper is as follows. In Section 2, the objective functions, constraint conditions, and piecewise linearization of PCS input/output-efficiency for stand-alone microgrid UC calculation are explained. Section 3 contains case study results that has been done in order to analyze the effect of piecewise linearization of PCS input/output-efficiency while calculating the microgrid UC, at an equivalent power system of South Korea's first "energy island", Ga-sa Island.

2. Formulation of Microgrid UC for Optimal Operation

In this chapter, the overall outline of this paper will be explained first. The optimization formulation will follow, in order of objective functions and constraints. Applying a piecewise linearization method to the input/output-efficiency curve of PCS is explained in a separate section, as it is the key idea of this paper.

2.1. Overall Outline

The proposed UC assumes that the target power system is a stand-alone microgrid in an island area, composed of a single BESS and diesel generator, several wind turbines, and photovoltaic devices. The BESS runs on constant voltage constant frequency (CVCF) mode, and the diesel generator is a backup source, which runs only when the electric power is insufficient in the grid. In the proposed stand-alone microgrid UC, the optimal operation schedule of the diesel generator and BESS for the next 24 h is calculated based on the forecasting data of renewable energy and load. The purpose is to calculate the UC for every 15 min and renew the 24 h operation schedule. However, in this paper, only a snapshot view is analyzed to concentrate more on the piecewise linearization of the actual PCS efficiency characteristics. The goal of the UC is to optimize the operation schedule of the distributed energy resources to minimize the diesel generator's operation cost using MILP. In the results of the calculation, the on/off status, output power of the diesel generator, and the SOC of the BESS are calculated. Since power loss occurs from the PCS, it has to be considered while calculating UC by also considering the input/output efficiency of the PCS. If the power loss incurred from the PCS is inaccurate, the devices implanted in the microgrid cannot operate according to the calculated schedule. Therefore, to reduce the difference between the calculated operation schedule and actual performance of the PCS and other devices, the input/output-efficiency characteristics must be considered, by applying a piecewise linearization method to the actual PCS efficiency curve. While solving the UC problem, the MILP technique is used, since several binary variables, such as on/off status of the diesel generator, must be optimized. The objective function and constraints of the proposed stand-alone microgrid UC are as follows.

2.2. Objective Function

The objective function of the proposed stand-alone microgrid UC is to minimize the summation of start-up cost and fuel cost of the diesel generator:

$$\min\left\{\sum_{i=1}^{I}\sum_{t=1}^{T}[FLC_{i,t} + STC_{i,t}] - \alpha_i \cdot SOC_{i,T}\right\} \tag{1}$$

Fuel cost function $FLC_{i,t}$ is defined as the following Equation (2).

$$FLC_{i,t} = a_i g_{i,t}^2 + b_i g_{i,t} + c_i \tag{2}$$

The equation above represents the relation between diesel generator's generation amount and fuel cost in a curve format. However, in order to use it for mathematic calculation, the curve must be approximated into a linear format for each section. $g_{i,t}$, the output of diesel generator i, must be divided to match the number of section b as Equation (3).

$$g_{i,t} = g_i^m \cdot u_{i,t} + \sum_{b=1}^{B} g_{i,t,b} \tag{3}$$

The linearized fuel cost for each section is represented as Equation (4).

$$FLC_{i,t} = FLC_i^m \cdot u_{i,t} + \sum_b MC_{i,b} \cdot g_{i,t,b} \tag{4}$$

The start-up cost of the diesel generator is calculated only when the status changes from stop (off) to run (on), according to the following equation.

$$STC_{i,t} = s_{i,t} \times STF_i \tag{5}$$

If the total amount of electric power generated from the renewable sources is greater than the total amount of load for the next 24 h, the result of the MILP optimal calculation gives an infeasible solution. If the final value of the SOC exceeds the maximum BESS capacity, which is usually 100%, the operator should rather reduce the charging power or increase the discharging power of the BESS. However, whereas the SOC remains below the maximum capacity, it is better to charge as much as power in the BESS for the next day's operation schedule, since the possibility of running a diesel generator declines. To ensure such operation, $-\alpha_i \cdot soc_{i,T}$ is added in the objective function. It induces the maximization of the SOC at the final time period by charging the over-generated electric energy generated from the renewable source. α_i, battery i's energy value, which means the unit energy value for the BESS, is calculated by dividing the price of the BESS by the total life cycle. It must be set lower than the fuel cost of the diesel generator, to prevent using the diesel generator to increase the SOC.

2.3. Constraints

The constraints can be classified into diesel generator-related, BESS-related, and the microgrid overall-related.

When the diesel generator is off ($u_{i,t} = 0$), the output must be 0, and when it is on ($u_{i,y} = 1$), the output must be smaller than the maximum output $g_{i,b}^M$.

$$0 \leq g_{i,t,b} \leq g_{i,b}^M \cdot u_{i,t} \tag{6}$$

Since $s_{i,t}$ and $d_{i,t}$, represent whether the diesel generator is running or stopped, they are closely related to $u_{i,t}$, representing whether the diesel generator is on or off. The relation between these variables should be also represented in a constraint format. The diesel generator's on/off status changes only when it starts to run or stop, using the former status of diesel generator at $t-1$.

$$u_{i,t} - u_{i,t-1} = s_{i,t} - d_{i,t} \tag{7}$$

Once the diesel generator starts to run, it has a minimum duration period to stop generating power. To represent such a characteristic, the diesel generator's minimum run time constraint is added as Equation (8). When diesel generator i starts to run ($s_{i,t} = 1$), it should not stop ($d_{i,t} = 1$) before it has been running for the minimum run time (MUT_i).

$$s_{i,t} + d_{i,t} + d_{i,t+1} + \cdots + d_{i,t+MUT_i-1} \leq 1 \tag{8}$$

On the other hand, once the diesel generator has stopped generating power, it could run only after a certain period of rest, represented as minimum stop time (MDT_i) in Equation (9).

$$d_{i,t} + s_{i,t} + s_{i,t+1} + \cdots + s_{i,t+MDT_i-1} \leq 1 \tag{9}$$

The status of the diesel generator is represented as $d_{i,t}$, and $s_{i,t}$. Each becomes '1', when the diesel generator has stopped and is running, respectively.

Must-run or unavailability state is represented as below, by using state variables.

$$u_{i,t} = 0 \ : \text{unavailability} \tag{10}$$

$$u_{i,t} = 1 \ : \text{must} - \text{run} \tag{11}$$

Meanwhile, the pure reserve power ($r_{i,t}$), produced by the diesel generator i, at time t, is only available when it is running. And the amount of reserve power supply from the diesel generator is closely related to the output of the diesel generator. The respective characteristic is represented as below.

$$0 \leq r_{i,t} \leq RC_{i,t} \cdot u_{i,t} \tag{12}$$

$$g_{i,t} + r_{i,t} \leq g_i^M \cdot u_{i,t} \tag{13}$$

$$g_{i,t} - r_{i,t} \geq g_i^m \cdot u_{i,t} \tag{14}$$

The diesel generator output's increase and decrease ratio is represented as below.

$$g_{i,t} - g_{i,t-1} \leq 60 \cdot RUR_i \tag{15}$$

$$g_{i,t-1} - g_{i,t} \leq 60 \cdot RDR_i \tag{16}$$

BESS can both charge and discharge energy, so binary variables are necessary to represent whether it is charging or discharging $(x_{i,t}, y_{i,t})$, and also the output power $(g_{i,t}^x, g_{i,t}^y)$ must be in a constraint format as well. When the BESS is charging, it is handled as a load from the aspect of the power system. Therefore, it must be represented as a negative generation, and the output of the BESS should be shown as below.

$$g_{i,t} = g_{i,t}^x - g_{i,t}^y \tag{17}$$

$$0 \leq g_{i,t}^x \leq g_i^X \cdot x_{i,t} \tag{18}$$

$$0 \leq g_{i,t}^y \leq g_i^Y \cdot y_{i,t} \tag{19}$$

A binary variable representing whether it is charging or discharging is needed as below.

$$x_{i,t} + y_{i,t} \leq 1 \tag{20}$$

The energy capacity of BESS i, must be in the range of minimum capacity and maximum capacity, which is normally 0 (%) and 100 (%).

$$enrg_i^m \leq enrg_{i,t} \leq enrg_i^M \tag{21}$$

In addition, the BESS's initial energy state and final energy demand can be represented as below.

$$enrg_{i,0} = enrg_i^0 \tag{22}$$

$$enrg_{i,T} = enrg_i^T \tag{23}$$

A constraint format representing the BESS's output-reserve power relation is shown in the following equations.

$$0 \leq r_{i,t} \leq RC_{i,t} \tag{24}$$

$$g_{i,t}^x + r_{i,t} \leq g_i^X \tag{25}$$

$$g_{i,t}^y + r_{i,t} \leq g_i^Y \tag{26}$$

The power balance qualification must be equal to the summary of total generation and total load (L_t). In order to satisfy the power balance, using the minimum production of the diesel generator and BESS, the remaining demand (D_t) must be applied, excluding the power generation from the wind turbine and photovoltaic device.

$$\sum_i g_{i,t} = D_T \tag{27}$$

A constraint handling the reserve power can be represented by assuming that a certain amount of reserve power must be procured for every time period.

$$\sum_i r_{i,t} \geq R_t \tag{28}$$

2.4. Considering the Input/Output-Efficiency of PCS Using Piecewise Linearization Method

As mentioned before, solving UC using MILP results in obtaining the optimal operation schedule of the controllable devices planted in the microgrid. In detail, the output of the diesel generator and BESS at every time step is calculated. From the BESS, the electric power is delivered or received through the PCS. Delivering electric power to the grid refers to discharging the BESS, and receiving refers to charging the BESS. The PCS decides the amount of discharge and charge power and also converts DC to AC or vise-versa, during such procedure, power loss occurs [25]. The loss is related to the efficiency of the designated input/output power value. When the input/output power is small, the efficiency is low and the portion of power loss is high. However, if the input/output power is large, the efficiency is high and the portion of the power loss is low. Such characteristics should be considered while solving UC for an accurate calculation result, and moreover, precise control of the planted devices in the micro rid.

If the power loss occurrence from the PCS is neglected or the efficiency is assumed to be fixed regardless of the input/output amount, the drawn operation schedule causes differences between the calculated and actual performance of the BESS. In addition, using such an inaccurate schedule for the BESS operation as a reference, it makes no reason for the microgrid operator to solve UC for operation scheduling, since the actual operation performance differs from the scheduled reference. An imprecise device operation schedule may cause imbalance between power demand and supply, which can also result in blackouts in severe conditions. In a stand-alone microgrid, the biggest issue is to reduce the difference of renewable energy output and load forecast data, because this causes the largest errors. Therefore, much of the research on microgrid operation focuses on prediction models for weather forecasts, of which the main goal is to calculate a precise operation schedule. Considering the input/output-efficiency characteristics of the PCS is another aspect to achieve for the same goal of increasing the accuracy of the operation schedule.

In this paper, in order to improve the operation schedule accuracy, an additional constraint formula has been added in the MILP calculation process when solving the UC problem. A piecewise linear method to the actual input/output-efficiency characteristic curve of the PCS has been proposed. It has been assumed that the actual input/output-efficiency curve of the PCS, shown as the blue curve from Figure 1, has been piecewise linearized as the black dotted line on Figure 1. The input (kW), output (kW), and efficiency (%) for each section is in Table 1. Input (kW) refers to the scheduled injection power heading toward the PCS for both when the BESS is charging or discharging. The amount of the injection power is calculated considering the loss from the PCS. For example, if the BESS is scheduled to discharge 50 kW, the input (kW) must be 55.1 kW, meaning the efficiency is 90.8% for the desired section, causing 5.1 kW of loss.

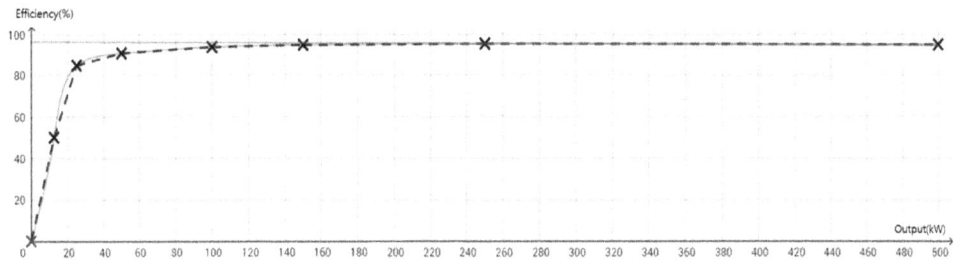

Figure 1. Types of input/output-efficiency characteristic curve of 500 kW power conditioning system (PCS).

Table 1. Input, output, efficiency for each section of a 500 kW power conditioning system (PCS).

Section	Input (kW)	Output (kW)	Eff (%)
1	25.0	12.5	50
2	29.4	25	84.9
3	55.1	50	90.8
4	106.6	100	93.8
5	158.2	150	94.8
6	262.1	250	95.4
7	526.3	500	95

After applying the piecewise linear method on input/output-efficiency curve of the PCS, an additional formulation is necessary to calculate the input (kW) power of the PCS, considering the related section efficiency, in order to calculate BESS *i*'s SOC ($enrg_{i,t}$). However, the efficiency at a certain section is not equal for when the output (kW) is at minimum or maximum value on the related section discretely, instead it is continual. This continuous relation is represented as the section efficiency slope in Equation (29). The variation of input (kW) and output (kW) is used to define the section efficiency slope. As a result, when solving UC for optimal operation scheduling, the SOC of BESS is calculated using Equation (30), which considers the piecewise linearized input/output-efficiency of the PCS.

$$slope_k = \frac{\Delta\text{input(kW)}}{\Delta\text{output(kW)}} \tag{29}$$

$$enrg_{i,t} = enrg_{i,t-1} + \left(g_{i,t}^y - g_{i,t}^x\right) \cdot slope_k \tag{30}$$

3. Case Study

This chapter covers the case studies that have been done. Several assumptions are first explained, and the case study results and analysis follow.

3.1. Assumption

To verify the proposed stand-alone microgrid UC, a case study was done under an equivalent system environment as Ga-sa Island, South Korea, shown in Figure 2. The load of Ga-sa Island is 83~107 kW on average, with a maximum at 167 kW and minimum at 40 kW. There are three 100 kW diesel generators, four 100 kW wind turbines, and a group of photovoltaic devices having total of 320 kW capacity have been installed. A single 3 MWh Li-ion BESS operating in CVCF mode has been installed, connected by three 500 kW PCSs, while PCS #1 is mainly used for normal operation, PCS #2 and #3 are used as back-up under emergency situations. For UC calculation, the initial value of BESS SOC is set at 50%, and the value of SOC at the final time period is also set to be 50% as well. In addition, the input/output-efficiency characteristic curve has been piecewise linearized, equal to the black dotted line from Figure 1 and Table 1.

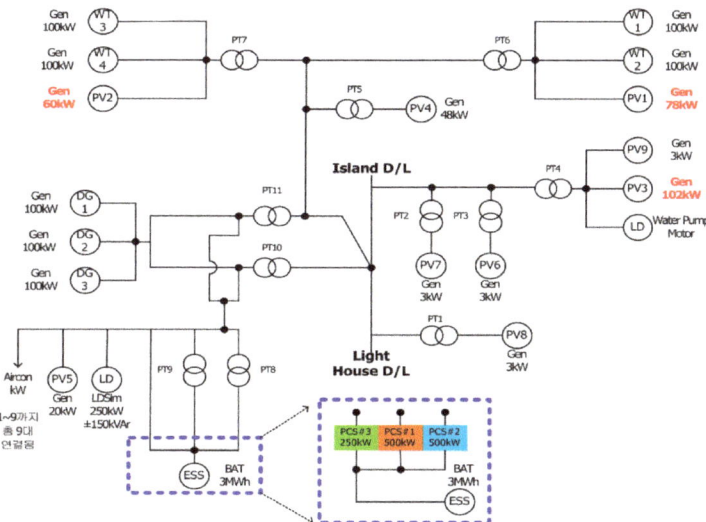

Figure 2. Power System of Ga-sa Island.

The main input data for UC, which is forecast data from the wind turbine and photovoltaic device output and load of Ga-sa Island, is shown in Figure 3. The input forecast data originated from the actual measured data, which was obtained on the 3 April 2015.

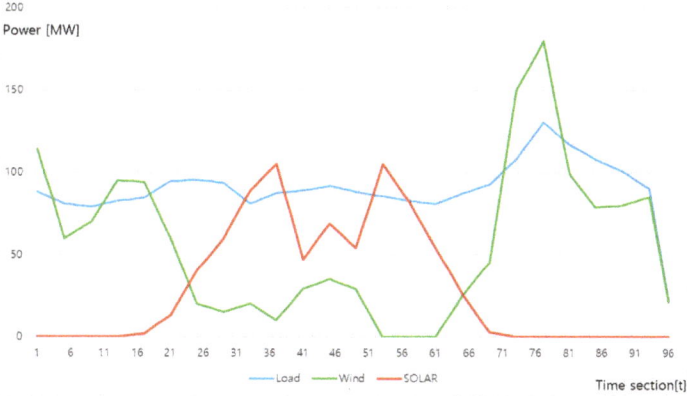

Figure 3. Input forecast data for case study.

To verify the effect of adapting input/output-efficiency of the PCS curve into a linear format by applying a piecewise linear method, a case study was done. Under equal input forecast data and generation device performance, only the efficiency of the PCS is different while solving UC using MILP in the following two cases. CPLEX 2.1 was used as a library function of providing MILP, while the main program was developed using Visual Studio 10.

Case 1: Assuming fixed PCS efficiency (98%) regardless of the input/output value (green line from Figure 1).

Case 2: Assuming the input/output-efficiency of the PCS curve to be piecewise linearized (black dotted line from Figure 1).

257

As a result of calculating UC for both cases, a day-ahead optimal operation schedule for diesel generator output and SOC of the BESS is obtained for a 15 min time period. Since the input forecast data is equal for both scenarios, the result difference is caused by the input/output-efficiency relation of the PCS. In addition, for each case, the actual power loss from the PCS could be calculated by substituting the scheduled PCS input power for every 15 min term into the actual input/output-efficiency curve of PCS, the blue line of Figure 1, or by using the following Equation (31). By this method, the power loss could be measured when assuming the microgrid is running on the calculated schedule. Also the amount of power loss from the PCS could be calculated and compared, which enabled the verification of the effect of piecewise linearization to the input/output-efficiency curve of the PCS. Figure 4 shows the analysis process for the case study.

$$loss_p = \left(g_{i,t}^x, g_{i,t}^y / eff_p \right) \times 100 - g_{i,t}^x, g_{i,t}^y \tag{31}$$

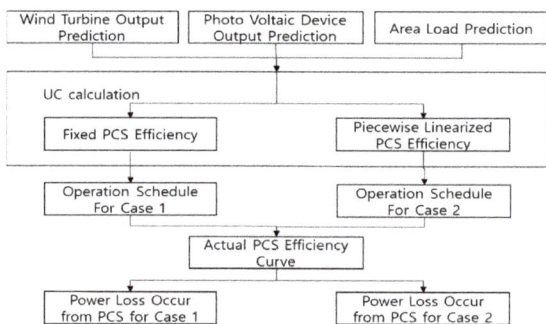

Figure 4. Analysis process for case study.

α_i, battery i's energy value is set to be \$3, after regarding the price of the BESS and its life cycle. Also, the coefficients for the diesel generator, a_i, b_i and c_i, are set to be 0.0001, 0.2, and 10, respectively. The fuel cost for the diesel generator is \$1.5, and the start-up cost is \$4.5, while the ramp rate for both increase and decrease in generation is set as 7.5 MW/min. $soc_{i,T}$, the SOC designed to be set at the final schedule time for the BESS, is set to be larger than 50%. MUT_i and MDT_i, the minimum run time and stop time for the diesel generator, is 0.5 h. The initial energy value of the BESS is assumed to have 50% of the device's capacity.

3.2. Case Study Result

The figures below are the calculation result of the proposed stand-alone microgrid optimal operation schedule by solving UC at the equivalent power system of Ga-sa Island, South Korea.

The red and yellow column from Figures 5 and 6 represent the output from the diesel generator and PCS. When the PCS charges power and stacks SOC for the BESS, it becomes a load for this aspect of the power system. Therefore, the negative expressions in Figures 5 and 6 indicate the charging power from the PCS. The blue line is the SOC. The initial value of SOC is set to be 50%, and the final SOC value is set to be 50% as well.

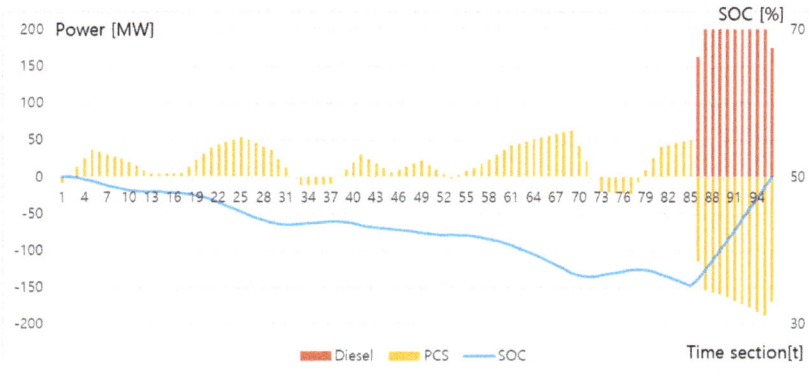

Figure 5. Simulation result from Case 1.

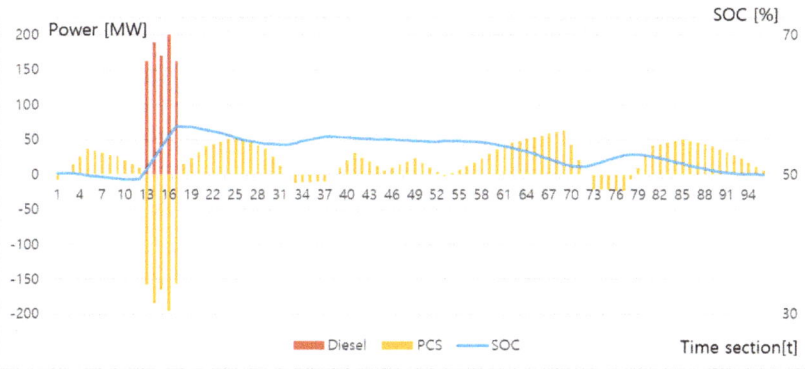

Figure 6. Simulation result from Case 2.

Table 2 is the analysis of case study results. Total generation from diesel and total output from PCS is the summation of each device's calculated result of an optimal operation schedule for 96 time periods. Total loss from PCS is the traced value of loss by substituting each of the calculated optimal operation schedules for the 96 time periods into the actual PCS input/output-efficiency curve.

Table 2. Case Study Analysis.

	Total Generation from Diesel [kWh]	Total Output from PCS [kWh]	Total Loss from PCS [kWh]
Case 1	2137.4	−39.8	1000.42
Case 2	884.1	1213.7	878.06

The result of UC calculation for Case 1 shows that the diesel generator must produce 2137.4 kWh of electric power in order to satisfy the objective function regarding the constraints. For Case 2, the diesel generator must produce 884.1 kWh, which is 1253.3 kWh less than Case 1's result. For the output of the PCS, −39.8 kWh has been supplied to the microgrid, or 39.8 kWh has been absorbed from the microgrid in Case 1, and for Case 2, 1213.7 kWh has been supplied to the microgrid. The loss incurred from the PCS in Case 1 is 1000.42 kWh, and 878.06 kWh for Case 2, which is 122.6 kWh less than Case 1. As expected, one could see that, by calculating the optimal operation schedule while assuming the PCS input/output-efficiency to be fixed, this results in having more loss from the PCS than when the microgrid actually operate as the calculated UC schedule.

If the actual curve of PCS input/output-efficiency is piecewise linearized as in Case 2, the loss from the PCS could be closer to actual performance. Also, the predicted value of the diesel generator will be calculated more precisely, while maintaining the demand-supply balance for the microgrid. In practice, by comparing the results of Case 1 and Case 2, the diesel generator's generation is reduced by 1253.3 kWh, when the piecewise linearized PCS input/output-efficiency is considered, while calculating UC.

4. Conclusions

The main goal of this paper is to increase the results of calculated day-ahead schedules for the implanted devices in a stand-alone microgrid. If a CVCF mode BESS is implemented into a stand-alone microgrid, an energy management system is necessary in order to operate the microgrid under stable and economical conditions. In this paper, a UC was proposed for calculating the implemented diesel generator and BESS's optimal operation schedule, based on the forecast data of renewable source generation output and load. To calculate the operation schedule, the summation of generator's start-up cost and fuel cost must be minimized, and many constraints considering the diesel generator and BESS's characteristics should be considered. Also, the calculation result proved to be closer to the actual performance when the input/output-efficiency of the PCS is piecewise linearized and implemented in the UC constraint. This allows one to predict not only the loss from the PCS, but also allows the diesel generator output to be more accurate, compared to when the PCS efficiency is assumed to be fixed. This leads to an improvement in the efficiency and stability of the microgrid's operation. The proposed stand-alone microgrid UC has been simulated and verified under an equivalent power system on Ga-sa Island, South Korea. Any other microgrid in a similar environment could adapt the main idea of this paper.

The frequency of status alteration, such as changing from charging to discharging or vice versa, at the BESS, is a main concern that limits the lifespan of the device. In future research, a formulation to minimize the status alteration of the BESS will be considered and applied to the proposed idea of this paper. With this, it is expected that the lifespan of one of the most expensive electric device used to construct a microgrid can be extended.

Author Contributions: Software, H.-L.L.; Supervision, Y.-H.C.; Writing—original draft, H.-L.L.

Funding: This work was supported by the Energy & Resource Recycling of the Korea Institute of Energy Technology Evaluation and Planning(KETEP) grant funded by the Korea government Ministry of Knowledge Economy (No. 20171210200830).

Conflicts of Interest: The authors declare no conflict of interest.

Nomenclature

i	subscript representing generator/battery $\in \{1, 2, \cdots, I\}$
t	subscript representing time $\in \{1, 2, \cdots, T\}$
b	subscript representing generator's output section $\in \{1, 2, \cdots, B\}$
a_i, b_i, c_i	coefficient for the diesel generator $i's$ fuel cost function
$u_{i,t}$	diesel generator $i's$ on/off status at time $t \in \{0, 1\}$
$s_{i,t}$	diesel generator $i's$ run status at time $t \in \{0, 1\}$
$d_{i,t}$	diesel generator $i's$ stop status at time $t \in \{0, 1\}$
$x_{i,t}$	battery $i's$ discharge status at time $t \in \{0, 1\}$
$y_{i,t}$	battery $i's$ charge status at time $t \in \{0, 1\}$
$g_{i,t,b}$	diesel generator $i's$ output at output section b at time t [MW]
$g_{i,t}^x$, $g_{i,t}^y$	battery $i's$ discharge/charge output at time t [MW]
$r_{i,t}$	diesel generator/battery $i's$ reserve energy at time t [MW]
$enrg_{i,t}$	battery $i's$ charging energy at time t [MWh]
$g_{i,t}$	diesel generator/battery $i's$ overall output at time t [MW]
$FLC_{i,t}$	diesel generator $i's$ fuel cost function [\$/h]

$FLCg_i^m$	diesel generator $i's$ fuel cos t at minimum output limit [\$/h]
$MC_{i,t}$	diesel generator $i's$ marginal cos t at output sec tion b [\$/MWh]
STC_i	diesel generator $i's$ startup cos t function [\$/h]
STF_i	diesel generator $i's$ startup cos t [\$]
α_i	battery $i's$ energy value
$soc_{i,T}$	battery $i's$ SOC at the final schedule time T [%]
RUR_i	diesel generator $i's$ output increase ratio [MW/min]
RDR_i	diesel generator $i's$ output decrease ratio [MW/min]
MUT_i	diesel generator $i's$ minimum run time [h]
MDT_i	diesel generator $i's$ minimum stop time [h]
g_i^m, g_i^M	diesel generator $i's$ output mimimum/maximum limit [MW]
$g_{i,b}^M$	diesel generator $i's$ maximum output at sec tion b [MW]
g_i^X, g_i^Y	battery $i's$ discharge/charge output maximum limit [MW]
$RC_{i,t}$	diesel generator/battery $i's$ maximum reserve energy at time t [MW]
$enrg_i^m$	battery $i's$ minimum energy level [MWh]
$enrg_i^M$	battery $i's$ maximum energy level [MWh]
$enrg_i^0$	battery $i's$ initial energy level [MWh]
$enrg_i^T$	battery $i's$ final energy level [MWh]
$slope_k$	slope at sec tion k
$loss_p$	electric loss from PCS when output is p [MW]
D_t	remaining/total load at time t
eff_p	actual efficiency when output of BESS is p [MW]

References

1. Li, X.; Hui, D.; Wu, L.; Lai, X. Control strategy of battery state of charge for wind/battery hybrid power system. In Proceedings of the IEEE International Symposium on Industrial Electronics, Bari, Italy, 4–7 July 2010.

2. Le Dinh, K.; Hayashi, Y. Experiment with an OPF controller based on HPSO-TVAC for a PV-supplied microgrid with BESS. In Proceedings of the IEEE PES General Meeting | Conference & Exposition, National Harbor, MD, USA, 27–31 July 2014.

3. Katiraei, F.; Iravani, M.R.; Lehn, P. Microgrid autonomous operation during and subsequent to islanding process. In Proceedings of the IEEE Power Engineering Society General Meeting, Denver, CO, USA, 6–10 June 2004.

4. Vandoorn, T.L.; De Kooning, J.D.M.; Meersman, B.; Guerrero, J.M.; Vandevelde, L. Voltage-Based Control of a Smart Transformer in a Microgrid. *IEEE Trans. Ind. Electron.* **2013**, *60*, 1291–1305. [CrossRef]

5. Schauder, C.; Mehta, H. Vector analysis and control of advanced static VAr compensators. *IEE Proc. C Gener. Transm. Distrib.* **1993**, *140*, 299–306. [CrossRef]

6. Philip, J.; Singh, B.; Mishra, S. Performance evaluation of an isolated system using PMSG based DG set, SPV array and BESS. In Proceedings of the IEEE International Conference on Power Electronics, Drives and Energy Systems (PEDES), Mumbai, India, 16–19 December 2014.

7. Wang, Y.; Tan, K.T.; So, P.L. Coordinated control of battery energy storage system in a microgrid. In Proceedings of the IEEE PES Asia-Pacific Power and Energy Engineering Conference (APPEEC), Kowloon, China, 8–11 December 2013.

8. Piphitpattanaprapt, N.; Bangerdpongchai, D. Optimal dispatch strategy of hybrid power generation with battery energy storage system in islanding mode. In Proceedings of the IEEE Innovative Smart Grid Technologies-Asia (ISGT ASIA), Bangkok, Thailand, 3–6 November 2015.

9. Wu, X.; Wang, X.; Bie, Z. Optimal generation scheduling of a microgrid. In Proceedings of the 3rd IEEE PES Innovative Smart Grid Technologies Europe (ISGT Europe), Berlin, Germany, 14–17 October 2012.

10. Moradi, H.; Esfahanian, M.; Abtahi, A.; Zilouchian, A. Modeling a Hybrid Microgrid Using Probabilistic Reconfiguration under System Uncertainties. *Energies* **2017**, *10*, 1430. [CrossRef]

11. Malysz, P.; Sirouspour, S.; Emadi, A. MILP-based rolling horizon control for microgrids with battery storage. In Proceedings of the IECON 2013—39th Annual Conference of the IEEE Industrial Electronics Society, Vienna, Austria, 10–13 November 2013.

12. Parisio, A.; Rikos, E.; Glielmo, L. A model predictive control approach to microgrid operation optimization. *IEEE Trans. Control Syst. Technol.* **2014**, *22*, 1813–1827. [CrossRef]
13. Moshi, G.G.; Bovo, C.; Berizzi, A. Optimal operational planning for pv-wind-diesel-battery microgrid. In Proceedings of the IEEE Eindhoven PowerTech, Eindhoven, The Netherlands, 29 June–2 July 2015.
14. Hu, C.; Luo, S.; Li, Z.; Wang, X.; Sun, L. Energy coordinative optimization of wind-storage-load microgrids based on short-term prediction. *Energies* **2015**, *8*, 1505–1528. [CrossRef]
15. Sandhya Rani, P.; Sumanth, K. A Model Predictive Control Approach for Efficient Optimization of Microgrid Operations using Mixed Integer Linear Programming. *Int. J. Eng. Res.* **2016**, *5*, 510–524. [CrossRef]
16. Wouters, C.; Fraga, E.S.; James, A.M. An energy integrated, multi-microgrid, milp (mixed-integer linear programming) approach for residential distributed energy system planning-a south Australian case-study. *Energy* **2015**, *85*, 30–44. [CrossRef]
17. Zaree, N.; Vahidinasab, V. An MILP formulation for centralized energy management strategy of microgrids. In Proceedings of the Smart Grids Conference (SGC), Kerman, Iran, 20–21 December 2016.
18. Parisio, A.; Glielmo, L. A mixed integer linear formulation for microgrid economic scheduling. In Proceedings of the IEEE International Conference on Smart Grid Communications (SmartGridComm), Brussels, Belgium, 17–20 October 2011; pp. 505–510. [CrossRef]
19. Hawkes, A.D.; Leach, M.A. Modelling high level system design and unit commitment for a microgrid. *Appl. Energy* **2009**, *86*, 1253–1265. [CrossRef]
20. Liang, H.Z.; Gooi, H.B. Unit commitment in microgrids by improved genetic algorithm. In Proceedings of the IPEC, Singapore, 27–29 October 2010.
21. Farrokhabadi, M.; Cañizares, C.A.; Bhattacharya, K. Unit Commitment for Isolated Microgrids Considering Frequency Control. *IEEE Trans. Smart Grid* **2018**, *9*, 3270–3280. [CrossRef]
22. Deckmyn, C.; Van de Vyver, J.; Vandoorn, T.L; Meersman, B.; Desmet, J.; Vandevelde, L. Day-ahead unit commitment model for microgrids. *IET Gener. Transm. Distrib.* **2017**, *11*, 1–9. [CrossRef]
23. Nemati, M.; Braun, M.; Tenbohlen, S. Optimization of unit commitment and economic dispatch in microgrids based on genetic algorithm and mixed integer linear programming. *Appl. Energy* **2018**, *210*, 944–963. [CrossRef]
24. Jabbari-Sabet, R.; Moghaddas-Tafreshi, S.M.; Mirhoseini, S.S. Microgrid operation and management using probabilistic reconfiguration and unit commitment. *Int. J. Electr. Power Energy Syst.* **2016**, *75*, 328–336. [CrossRef]
25. Kyle, C. *Power Conversion System Architectures for Grid Tied Energy Storage*; National Institude of Standrads and Technology; Technology Administration, U.S. Department of Commerce: Washington, DC, USA, 2012.

Article

Demand Bidding Optimization for an Aggregator with a Genetic Algorithm

Leehter Yao [1,*], Wei Hong Lim [2], Sew Sun Tiang [2], Teng Hwang Tan [2], Chin Hong Wong [3] and Jia Yew Pang [4]

[1] Department of Electrical Engineering, National Taipei University of Technology, Taipei 10608, Taiwan
[2] Faculty of Engineering, Technology and Built Environment, UCSI University,
Kuala Lumpur 56000, Malaysia; limwh@ucsiuniversity.edu.my (W.H.L.);
tiangss@ucsiuniversity.edu.my (S.S.T.); TanTH@ucsiuniversity.edu.my (T.H.T.)
[3] Department of Engineering and Information Technology, UCSI College, Kuala Lumpur 56000, Malaysia;
wongch@ucsicollege.edu.my
[4] School of Engineering and Physical Sciences, Heriot Watt University, Putrajaya 62200, Malaysia;
j.pang@hw.ac.uk
* Correspondence: ltyao@ntut.edu.tw; Tel.: +886-2-771-2171 (ext. 2174)

Received: 8 August 2018; Accepted: 18 September 2018; Published: 20 September 2018

Abstract: Demand response (DR) is an effective solution used to maintain the reliability of power systems. Although numerous demand bidding models were designed to balance the demand and supply of electricity, these works focused on optimizing the DR supply curve of aggregator and the associated clearing prices. Limited researches were done to investigate the interaction between each aggregator and its customers to ensure the delivery of promised load curtailments. In this paper, a closed demand bidding model is envisioned to bridge the aforementioned gap by facilitating the internal DR trading between the aggregator and its large contract customers. The customers can submit their own bid as a pairs of bidding price and quantity of load curtailment in hourly basis when demand bidding is needed. A purchase optimization scheme is then designed to minimize the total bidding purchase cost. Given the presence of various load curtailment constraints, the demand bidding model considered is highly nonlinear. A modified genetic algorithm incorporated with efficient encoding scheme and adaptive bid declination strategy is therefore proposed to solve this problem effectively. Extensive simulation shows that the proposed purchase optimization scheme can minimize the total cost of demand bidding and it is computationally feasible for real applications.

Keywords: demand bidding; demand response; genetic algorithm; load curtailment; optimization

1. Introduction

The electricity market is evolving towards an open-access environment by restructuring the vertically integrated utilities into multiple independent market players including: generator, transmission system operator (TSO), distributor, retailer and aggregator. Under the restructured market, economics and profitability become the main concerns of all market players. Proper implementation of deregulation policy in electricity industry can enhance energy efficiency by incorporating competition between generation and retailing. In addition, the operating costs of electricity customers can be reduced by having more options in purchasing energy [1].

Despite the benefits offered by the deregulated electricity market, some challenges that tend to degrade the reliability of power supply have been encountered by different market players. For instance, network security risk is the main concern for both TSO and distributor because these market players need to prevent the violation of safe operating parameters in the transmission and distribution networks during the contingencies. Retailers and load service entity (LSE) are vulnerable to financial risk because they need to buy electricity from the wholesale market with volatile prices and then resell it to their customers with fixed tariff [2–6]. The sudden increase of peak load or unexpected declination of generation reserve can produce extreme price spikes and incur high energy cost to the retailer and LSE, at some point might lead to the bankruptcy issue [7].

Given the flexibility of customers in consuming electricity, demand response (DR) is recently deployed as a low cost and yet environmentally friendly solution to tackle the abovementioned issues. In general, two types of DR are available in the electricity market. This includes the incentive-based DR and price-based DR that encourage the customers to curtail or delay the usage of electricity by referring to the incentive and market price, respectively [8]. Successful implementation of DR is expected to benefit individual market player and ultimately the entire electricity market [9–12]. For instance, both of TSO and distributor can benefit from DR by using it to relieve the network congestion and enhance the quality of power supply at transmission and distribution levels, respectively. DR can also help the retailer and LSE to cover financial risk caused by spot price volatility by rewarding their customer to curtail energy consumption at the time periods with extreme price spikes. In [13–15], the equivalency between DR at demand side and energy supply at supply side was established and the importance of DR resource to be compensated equally as electricity market price was emphasized.

Considerable efforts were devoted to develop various demand bidding optimization strategies from the perspectives of various deregulated electricity market players by leveraging the flexibility of electricity customers in contributing their DR resources. A distributed DR bidding algorithm was designed in [16] to submit the limit order bids of electrical vehicle (EV) fleet to a day-ahead market in order to minimize the peak demand profiles without affecting the commuting performance of EVs. A multistage stochastic optimization framework was proposed in [17] to help LSE to reduce the energy procurement costs by coordinating the demand bids of flexible loads. A Monte Carlo method was used in [18] to coordinate the coalition of energy consumers that are sensitive to DR programs by optimizing the purchase bidding offers in day-ahead market. The presence of price-sensitive demand was proven to reduce the volatility of electricity prices in peak hours. In [19], an optimal bidding strategy was formulated from the perspective of electricity retailer to minimize the energy procurement costs from different sources by considering the boundary limits of pool prices in electricity market. Procurement strategy obtained using mixed-integer linear programming (MILP) revealed that it is more cost effective for a retailer to procure electricity from the sources with lower price uncertainties. A DR-integrated network-constrained unit commitment model was used in [20] to analyze the impacts of price-based DR on market clearing and location marginal prices (LMP) of power system. It was found that the higher DR participation rate can lead to better payment for DR loads. An optimal bidding strategy of a microgrid in day-ahead market was formulated in [21] using MILP, aiming to minimize its operating cost by considering the uncertainties of renewable sources, storage systems and price-sensitive demands. In [22], both of residential DR and network on-load tap charger were integrated to solve the unbalanced issue in low voltage distribution network with minimum cost without violating the comfort level of users. The over-generation issue of renewable source was mitigated in [23] by designing the reward and penalty functions for DR programs.

While the abovementioned studies modelled DR as a price responsive load, some recent works envisioned DR as a commodity to be traded in energy market. In [24], a market clearing strategy in day-ahead market was solved by MILP to minimize the variability of wind power via the optimal load reduction dispatched by DR resources. A comprehensive framework consisting of four DR bidding strategies was solved in [25] using MILP to decide the optimal DR contribution of aggregator in wholesale market. A demand response exchange (DRX) market that considered DR as a tradable resource between buyers (e.g., TSO, distributor and retailer) and sellers (retailer and aggregator) was proposed in [26]. Under this DRX framework, an aggregator can represent its customers to bargain with the retailers for buying electricity with lower prices. Furthermore, the aggregator can combine the flexible DR loads from its customers so that it can negotiate for more profitable DR selling prices with TSO and distributor. A clearing strategy for pool-based electricity market was developed in [27] using microeconomic theory from the viewpoint of DRX operator, aiming to maximize the overall benefits of DR for all market players. A decentralization concept was adopted by the market clearing strategy proposed in [28]. It was assumed that the market operator can broadcast the price signal to each self-interested market player so that the latter can adjust the DR iteratively through Walrasian auction. A hierarchical market model was proposed in [29] to implement the residential DR mechanism by connecting the upper level of utility operator and lower level of electricity customers through a group of aggregators. All involved entities were considered self-interested because the aggregator aimed for profit maximization while both of utility operator and customers emphasized on cost minimization. Blockchain technology was envisioned in [30] to formulate a peer-to-peer decentralized energy trading mechanism without the presence of any intermediary agent in order to reduce the transaction cost of DR programs. In [31], an optimal purchase strategy was formulated using MILP to facilitate the demand bidding between aggregator and electricity customers, aiming to minimize the total purchase cost required to deliver the promised load curtailment. Extensive efforts were needed in [31] to transform the nonlinear load curtailment constraints into the linear form before the demand bidding optimization problem can be solved using MILP. A decentralized DR framework was developed in [32] from the perspective of an independent system operator (ISO), aiming to minimize the suppliers' generation cost and the customers' discomfort cost simultaneously without violating the confidentiality. The control signals broadcasted by ISO were used to incentivize both supplier and customers to modify their respective generation and demand profiles, hence optimizing their objectives independently.

Given the restriction of minimum power rating imposed in electricity markets [33], small-scale electricity customers are not eligible to directly participate in DR trading. They are unable to produce any noticeable impacts in electricity market and obtain the promising clearing prices. One of the solutions is to allow an aggregator to combine the overall potential of load curtailment offered by all small-scale electricity customers under its service regions and submit an aggregated DR supply curve on behalf of them to market operator. The DR bidding curve submitted by an aggregator for trading in each time slot is represented as a pair of quantity-price bid to indicate the quantity of load curtailment to be offered and the associated selling price. If the DR biddings offered by an aggregator are chosen during the market clearing process, it needs to deliver the promised load curtailments during the demand bidding event which might be scheduled in day, month or year ahead. Although numerous market clearing strategies were proposed in [16–32] to address the demand bidding optimization problems, majority of these studies emphasized at the upper level that involves interaction between the aggregators and different entities of electricity market (e.g., TSO, distributor and retailer) in order to determine the optimal combination of quantity and clearing price of load curtailment. Despite of its importance in practical scenario, limited researches were done at the lower level to investigate the interaction between an aggregator and all customers under its service regions to ensure the delivery of all promised load curtailment bids for market operation.

In this paper, a new insight is proposed to bridge the aforementioned gap by designing a closed bidding process that enables an aggregator to procure load curtailments from all customers under its service regions. Once the market clearing results are announced, the aggregator needs to broadcast a target profile that indicates the total load curtailment to be achieved in each bidding slot and the maximum bidding price per unit of load curtailment to all of its customers. Based on the given information, interested customers can participate in the bidding process by submitting their load curtailment bids and the selling price in each bidding slot to the aggregator. A purchase optimization scheme is then developed from a modified genetic algorithm (MGA) to ensure the mutual interests of both aggregators and its customers are preserved throughout the demand bidding event. From the perspective of an aggregator, the proposed purchase strategy is crucial in searching for the optimal combination of load curtailment bids that can minimize the total bidding cost while satisfying the target load curtailment profiles. For customers, they can perform load curtailment at their preferred time with minimum disruption of comfort levels and be rewarded based on their expected prices. Nevertheless, it is nontrivial to determine the best combination of load curtailment bids given that each customer has a unique load shedding pattern. The inclusion of customers with different load curtailment patterns leads to the formulation of a nonlinear demand bidding optimization problem that requires high computational overhead to solve. A numbers of modifications have been incorporated into MGA to enhance its search efficiency during the demand bidding optimization process.

In general, the technical novelty and contribution of this paper can be presented as follows:

1. A closed demand bidding model is adopted between an aggregator and all customers under its service regions to facilitate the internal trading of load curtailment at lower level so that the aggregator is able to deliver the promised load curtailment for market operation. To the best of authors' knowledge, the proposed work is different with the majority of works in [16–32] because the latter focused on upper level that involves the demand bidding between different entities in electricity market (e.g., TSO, distributor and retailer) with aggregators.

2. A purchase optimization scheme is designed for aggregator to determine the best combination of aggregated load curtailment bids and optimal purchase price, aiming to minimize the total bidding cost while satisfying all load curtailment constraints. Given the unique load shedding patterns of different customers, the demand bidding model formulated is highly nonlinear and difficult to solve. A modified genetic algorithm (MGA) incorporated with a delicate gene encoding scheme and an adaptive search mechanism is designed to solve the nonlinear demand bidding optimization efficiently without having to transform the objective function or constraints into linear forms such as those reported in [19,21,25,31].

3. Simulation results show that the optimal purchase price per unit load curtailment obtained by the proposed method in every bidding slot is different and at least 20% lower than the average bidding price offered by the customers, implying its excellent cost minimization capability. The proposed MGA is also efficient and suitable for practical applications because it incurs low computational times in solving the demand bidding optimization problem.

The remaining parts of this paper are presented in the following sequences. The mathematical formulation of closed demand bidding is explained in Section 2. Section 3 describes the purchase optimization scheme developed using MGA for demand bidding. The detailed descriptions of MGA including its gene encoding scheme and adaptive search mechanism are also provided. Section 4 presents the computer simulation results and the conclusions are drawn in Section 5.

2. Problem Formulation

After the market clearing process has determined the DR capacity to be sold for market operation, a demand trading between customers with large contract capacity and aggregator is conducted in order to fulfill the target load curtailment profile generated. The examples of customers with large contract capacity are the electricity users from commercial and industrial sectors and the load service entities with contract capacity higher than 0.5 MW. All customers that are willing to contribute their DR capacity in the upcoming demand bidding event are encouraged to register with the aggregator so that it can find the best load shedding patterns for customers. Assume that a total of N customers have registered to participate in a demand bidding event and the characteristics of load shedding are unique for all of the N participants. For example, the load of air conditioning systems are the main load for commercial customers, while the main load of industrial customers could be contributed from manufacturing process. Aggregator can suggest the suitable load shedding patterns for every n-th customer after reviewing their track records and load characteristics. The information contained in these recommended load shedding patterns include the most suitable maximum load shedding capacity P_{max}^n and the least load shedding period δ_n, $n = 1,...,N$.

Assume that the proposed demand bidding model focuses on day-ahead operation. An online platform is established to facilitate the broadcasting of upcoming demand bidding event from the aggregator and the submission of bidding from customers through the Internet. The information provided by the aggregator in its bidding request include: (a) the beginning and completion times of the next demand bidding event represented by t^s and t^e, respectively; (b) the maximum purchase price per kWh of load curtailment represented by σ_{max}; and (c) the deadline for customers to respond their participation in next demand bidding event represented by t^d. For the sake of convenience in indexing, the beginning and completion times of demand bidding are designated in hours. Similarly, the time periods of each customer to submit their load curtailment bid are also expressed in hours. Consider M as the number of time slots available for demand bidding, therefore $M = t^e - t^s + 1$.

If the customers do not reply by the deadline t^d, the aggregator assumes that they are not interested to contribute in next demand bidding event. Each of the n-th customer that is interested to contribute for current demand bidding needs to reply to the online platform by submitting the bids containing the information of: (a) the start and end times suitable for load shedding in the upcoming demand bidding represented by $\hat{\tau}_n^s$ and $\hat{\tau}_n^e$, respectively; (b) the quantity of load shedding $\hat{P}_{nm}$ offered at each m-th time slot, $m = 1 \ldots M$; and (c) the selling price $\hat{\sigma}_{nm}$ per kWh of load shedding at each m-th time slot. Notably, the bidding interval submitted by each n-th customer, denoted as $[\hat{\tau}_n^s, \hat{\tau}_n^e]$, is within the interval allowed for demand biding $[t^s, t^e]$ as stipulated by aggregator, i.e.,

$$[\hat{\tau}_n^s, \hat{\tau}_n^e] \subset [t^s, t^e], \; n = 1 \ldots N. \tag{1}$$

Define σ_{max} as the maximum purchase price broadcasted by aggregator to all customers. The purchase price $\hat{\sigma}_{nm}$ bid by every n-th customer in each m-th time slot of demand bidding cannot exceed the threshold σ_{max}, i.e.,

$$\hat{\sigma}_{nm} \leq \sigma_{max}, \; n = 1 \ldots N, \; m = 1 \ldots M. \tag{2}$$

For the convenience of notation, let $\hat{\beta}_n^s$ and $\hat{\beta}_n^e$ be the integer indices transformed from customer's submitted start time $\hat{\tau}_n^s$ and end time $\hat{\tau}_n^e$ allowed for load curtailment, respectively, where $\hat{\beta}_n^s = \hat{\tau}_n^s - t^s + 1$ and $\hat{\beta}_n^e = \hat{\tau}_n^e - t^s + 1$. It is not compulsory for customers to submit their bids in all time slots because $\hat{P}_{nm}$ and $\hat{\sigma}_{nm}$ can be set as null values at the time slots where no demand bids are offered by the customers, i.e.,

$$\hat{P}_{nm} = 0, \; \hat{\sigma}_{nm} = 0, \; \forall m \in \left([1, M] \backslash [\hat{\beta}_n^s, \hat{\beta}_n^e] \right) \tag{3}$$

Furthermore, the amount of load curtailment $\hat{P}_{nm}$ offered by each n-th customer on the online platform at any m-th bidding slot is constrained by the predefined maximum load shedding capacity P_{max}^n, i.e.,

$$\hat{P}_{nm} \leq P_{max}^n, \; n = 1 \ldots N, \; m = 1 \ldots M. \tag{4}$$

After receiving the load demand bids from all interested customers, the aggregator needs to decide an optimal combinations of bids at every m-th time slot in order to satisfy the target load shedding profile Γ_m at the m-th time slot with minimum total purchase cost. Although the proposed purchase optimization strategy is formulated from the viewpoint of an aggregator, the best interest of customers are secured by considering their offers in terms of capacity and selling price per kWh of load curtailment at all bidding slots.

Let $h_{nm} \in \{0, 1\}$ be the purchase status of the bid offered by the n-th customer at the m-th time slot, where $h_{nm} = 1$ shows that the offered bid is purchased by the aggregator, while $h_{nm} = 0$ implies that the bid is rejected. Denote P_{nm} as the load reduction offered by the n-th customer at the m-th time slot, while σ_{nm} as the selling price per kWh of load reduction offered. In this paper, a purchase optimization strategy is to be developed to find the best combination of purchase status $h_{nm}^*, n = 1 \ldots N, m = 1 \ldots M$, so that the total bidding cost can be minimized. Define the overall optimal purchase status as a set $H^* \equiv \{ h_{nm}^* | h_{nm}^* \in \{0, 1\}, n = 1 \ldots N, m = 1 \ldots M \}$, then

$$H^* = \underset{h_{nm}, n=1 \ldots N, m=1 \ldots M}{\text{argmin}} \sum_{n=1}^{N} \sum_{m=1}^{M} h_{nm} \sigma_{nm} P_{nm} \tag{5}$$

$$\text{subject to} \quad \sum_{n=1}^{N} h_{nm} P_{nm} \geq \Gamma_m, \ m = 1 \dots M. \tag{6}$$

Please note that

$$h_{nm} = 0, \forall m \in \left([1, M] \setminus [\hat{\beta}_n^s, \hat{\beta}_n^e]\right) \tag{7}$$

It is noteworthy that the amount of demand purchased from the customer and the price paid to the customer needs to be same as the bid offered by the customer, i.e.,

$$P_{nm} = \hat{P}_{nm}, \ \sigma_{nm} = \hat{\sigma}_{nm}, \ n = 1 \dots N, \ m = 1 \dots M. \tag{8}$$

The number of time slots for the n-th customer to perform load shedding continuously are restricted by the minimum load curtailment periods denoted as δ_n. The value of δ_n is determined by aggregator based on the load characteristic of each n-th customer. Therefore,

$$\sum_{m=1}^{M} h_{nm} \geq \delta_n, \ n = 1 \dots N. \tag{9}$$

Given that the load shedding for the customer with capacity greater than 5000 kW usually requires certain period of process, it takes time and effort to reduce the load to a large enough value. Aggregator usually promises to customer that the load shed request only conducts once in every demand bidding event. Therefore, the optimization in (5) is also constrained by the condition that the duration of load shed cannot be separated. Define $\mathbb{B}(\cdot)$ as an operator that returns true if the operand is 1 and false if the operand is 0, while $\mathbb{I}(\cdot)$ is an operator that returns 1 if the operand is true and 0 if the operand is false. Then, the constraint specifies that load curtailment can only be conducted once for every customer is formulated as shown below:

$$\sum_{m=1}^{M-1} \mathbb{I}\left(\mathbb{B}(h_{nm}) \oplus \mathbb{B}\left(h_{n(m+1)}\right)\right) = 2 \tag{10}$$

where $\oplus$ denotes the Boolean operator exclusive OR.

The optimization of objective function defined in (5) bounded by the constraint functions of (6)–(9) is a non-convex binary optimization problem. A purchase optimization scheme based on modified genetic algorithm will be proposed in the next section to find an optimal bidding purchase solution that can minimize the total biding cost without violating all load curtailment constraints.

3. Proposed Purchase Optimization Scheme

The binary optimization problem to be solved in (5) involves with N customers submitting their bids in M time slots. A total of $2^{N \times M}$ solutions are to be searched if binary search approach is to be used in the optimization. Apparently, binary search is unrealistic because the number of customers N tends to be a large value. Since the demand bidding optimization considered in this paper does not need to be an on-line calculation, a modified genetic algorithm (MGA) is designed as the purchase optimization scheme to solve the demand bidding optimization problem with better efficiency.

3.1. Chromosome Design

The combination of binary purchase status h_{nm} in every m-th time slot for every n-th customer is an intuitive choice of forming a gene. All N genes could be cascaded as a chromosome with $N \times M$ binary values. Nevertheless, this simple and intuitive chromosome design not only suffers from the computational inefficiency due to its excessive length, but it also cannot guarantee to satisfy the constraints in both (9) and (10). For this reason, a unique encoding scheme for chromosome design is proposed in this paper. To include all possible bid purchase status for all participated customers, a table of all possible combinations of $0, 1, 2, \dots, M$ connected time slots are shown in Table 1 where t^s

and t^e are assumed to be 13:00 and 17:00, respectively, for illustration. Denote L as the total number of combinations among M time slots, then

$$L = \frac{M(M+1)}{2} + 1 \tag{11}$$

Every possible combination in Table 1 is represented by an index denoted as $g \in \Omega$ where $\Omega = \{l|l \in I, 0 \leq l \leq (L-1)\}$.

For every pair of submitted start time $\hat{\beta}_n^s$ and end time $\hat{\beta}_n^e$ allowed for load shedding from the n-th customer, there are $\hat{\chi}_n = \hat{\beta}_n^e - \hat{\beta}_n^s + 1$ time slots for potential load curtailment. Among those $\hat{\chi}_n$ time slots, let Φ_n be a set containing the indices of all possible combinations $0, 1, 2, \ldots, \hat{\chi}_n$ connected time slots within the interval between $\hat{\beta}_n^s$ and $\hat{\beta}_n^e$. Then, there are L_n possible combinations, where

$$L_n = \frac{\hat{\chi}_n(\hat{\chi}_n+1)}{2} + 1 \tag{12}$$

Denote $Tra(u)$ as an operator that transforms the interval u into a set containing indices of time slots in the look-up table such as the one in Table 1. Therefore, $Tra([1, M]) = \Omega$ and $Tra([\hat{\beta}_n^s, \hat{\beta}_n^e]) = \Phi_n$, where $\Phi_n \subset \Omega$. For instance, if $\hat{\beta}_n^s$ and $\hat{\beta}_n^e$ are set to be 14:00 and 16:00, respectively, there are in total of 7 possible combinations among L combinations in Table 1 with color blocks located within the interval $[\hat{\beta}_n^s, \hat{\beta}_n^e]$. Table 1 shows that $\Phi_n = \{0, 2, 3, 4, 7, 8, 11\}$. Similarly, if $\hat{\beta}_n^s$ and $\hat{\beta}_n^e$ are set to be 16:00 and 17:00, respectively, there are totally 4 possible combinations among L combinations in Table 1 with color blocks located within the interval $[\hat{\beta}_n^s, \hat{\beta}_n^e]$. Table 1 shows that $\Phi_n = \{0, 4, 5, 9\}$. As soon as every n-th customer makes his or her bid, the set Φ_n is generated and stored in the database in aggregator's main computer. To find the best combination of subintervals within every n-th customer's submitted interval allowed for curtailment, the chromosome is designed to be a string of genes $g_n \in \Phi_n, n = 1 \ldots N$. If the j-th chromosome in the k-th generation is denoted as ζ^{jk}, then

$$\zeta^{jk} = g_1^{jk} \cup g_2^{jk} \ldots \cup g_N^{jk} \tag{13}$$

where $g_n^{jk} \in \Phi_n, n = 1 \ldots N$.

Without loss of generality, the index j numbering the chromosome in a gene pool and the index k numbering the generation are omitted if it is appropriate for the convenience of notations. The purchase status represented by g_n is obtained by referring to the look-up table such as in Table 1. Let $Dec(g_n)$ be an operator to decode the gene g_n into a set $H \equiv \{h_{nm}|h_{nm} \in \{0, 1\}, n = 1 \ldots N, m = 1 \ldots M\}$ containing purchase status of the n-th customer. For instance, if $g_n = 7$ in $\Phi_n = \{0, 2, 3, 4, 7, 8, 11\}$, then $H_n = \{0, 1, 1, 0, 0\}$ whereas if $g_n = 9$ in $\Phi_n = \{0, 4, 5, 9\}$, then $H_n = \{0, 0, 0, 1, 1\}$ based on the look-up table in Table 1.

Table 1. Possible combinations of $0, 1, 2, \ldots, M$ connected time slots for $M = 5$.

Index	Slot 1 (13:00)	Slot 2 (14:00)	Slot 3 (15:00)	Slot 4 (16:00)	Slot 5 (17:00)
0	$h_{i1} = 0$	$h_{i2} = 0$	$h_{i3} = 0$	$h_{i4} = 0$	$h_{i5} = 0$
1	$h_{i1} = 1$	$h_{i2} = 0$	$h_{i3} = 0$	$h_{i4} = 0$	$h_{i5} = 0$
2	$h_{i1} = 0$	$h_{i2} = 1$	$h_{i3} = 0$	$h_{i4} = 0$	$h_{i5} = 0$
3	$h_{i1} = 0$	$h_{i2} = 0$	$h_{i3} = 1$	$h_{i4} = 0$	$h_{i5} = 0$
4	$h_{i1} = 0$	$h_{i2} = 0$	$h_{i3} = 0$	$h_{i4} = 1$	$h_{i5} = 0$
5	$h_{i1} = 0$	$h_{i2} = 0$	$h_{i3} = 0$	$h_{i4} = 0$	$h_{i5} = 1$
6	$h_{i1} = 1$	$h_{i2} = 1$	$h_{i3} = 0$	$h_{i4} = 0$	$h_{i5} = 0$
7	$h_{i1} = 0$	$h_{i2} = 1$	$h_{i3} = 1$	$h_{i4} = 0$	$h_{i5} = 0$
8	$h_{i1} = 0$	$h_{i2} = 0$	$h_{i3} = 1$	$h_{i4} = 1$	$h_{i5} = 0$
9	$h_{i1} = 0$	$h_{i2} = 0$	$h_{i3} = 0$	$h_{i4} = 1$	$h_{i5} = 1$
10	$h_{i1} = 1$	$h_{i2} = 1$	$h_{i3} = 1$	$h_{i4} = 0$	$h_{i5} = 0$
11	$h_{i1} = 0$	$h_{i2} = 1$	$h_{i3} = 1$	$h_{i4} = 1$	$h_{i5} = 0$
12	$h_{i1} = 0$	$h_{i2} = 0$	$h_{i3} = 1$	$h_{i4} = 1$	$h_{i5} = 1$
13	$h_{i1} = 1$	$h_{i2} = 1$	$h_{i3} = 1$	$h_{i4} = 1$	$h_{i5} = 0$
14	$h_{i1} = 0$	$h_{i2} = 1$	$h_{i3} = 1$	$h_{i4} = 1$	$h_{i5} = 1$
15	$h_{i1} = 1$	$h_{i2} = 1$	$h_{i3} = 1$	$h_{i4} = 1$	$h_{i5} = 1$

Coloring in Table 1 denotes the possible continuous time intervals selected for load curtailment.

3.2. Fitness Function

Referring to (5)–(7), the proposed MGA is used to find the optimal bid purchase strategy so that the cost due to bid purchasing can be minimized while fulfilling the target load curtailment committed at every m-th time slot. Let F be the fitness function of MGA and ϑ be a penalty factor, F is then defined as:

$$F = \sum_{n=1}^{N} \sum_{m=1}^{M} h_{nm} \sigma_{nm} P_{nm} + \vartheta \sum_{m=1}^{M} B_m \tag{14}$$

where

$$B_m = \begin{cases} 0, \text{ if } \sum_{n=1}^{N} h_{nm} P_{nm} \geq \Gamma_m; \\ \Gamma_m - \sum_{n=1}^{N} h_{nm} P_{nm}, \text{ otherwise.} \end{cases} \tag{15}$$

The fitness function F is to be minimized by MGA. The penalty factor ϑ is designed to penalize the associated fitness value if the constraint for target load curtailment in every time slot is not fulfilled. However, the first term in (14) is the cost needs to be paid by the aggregator for demand bidding while the second term is the load difference. A guideline for tuning the penalty factor ϑ is thus needed.

Let $\bar{\sigma}$ be the average price offered by all participated customers, i.e.,

$$\bar{\sigma} = \frac{\sum_{n=1}^{N} \sum_{m=1}^{M} \sigma_{nm}}{NM} \tag{16}$$

The first term in (14),

$$\sum_{n=1}^{N} \sum_{m=1}^{M} h_{nm} \sigma_{nm} P_{nm} \cong \bar{\sigma} \sum_{n=1}^{N} \sum_{m=1}^{M} h_{nm} P_{nm} \tag{17}$$

Referring to (15), if $\Gamma_m > \sum_{n=1}^{N} h_{nm} P_{nm}$, assume

$$\Gamma_m = \kappa \sum_{n=1}^{N} h_{nm} P_{nm} \tag{18}$$

where κ is a constant to be adjusted and $\kappa > 1$. The penalty factor ϑ is designed to make both of the first and second terms in (14) roughly have the same weighting despite that both terms have different characteristics. Substituting (15), (17) and (18) into (14), yields

$$\bar{\sigma} \sum_{n=1}^{N} \sum_{m=1}^{M} h_{nm} P_{nm} \cong \vartheta(\kappa - 1) \sum_{n=1}^{N} \sum_{m=1}^{M} h_{nm} P_{nm} \tag{19}$$

Therefore,

$$\vartheta = \frac{\bar{\sigma}}{\kappa - 1} \tag{20}$$

The penalty factor ϑ can thus be adjusted according to (20).

3.3. Crossover, Regeneration and Mutation

From (13), every gene $g_n^{jk} \in \Phi_n$ in the j-th chromosome of the k-th generation represents the submitted interval for curtailment from the n-th customer. It is suitable for applying chromosome crossover operation gene to gene so that every gene has a chance being replaced with a new value as the chromosome evolves. However, the new value obtained by crossover also needs to be an element of Φ_n. All chromosomes are rearranged in the gene pool by sorting corresponding fitness values from small to large. A percentage q of all sorted chromosomes is taken as the parent chromosomes for crossover to generate child chromosomes in the next generation. A pair of chromosomes is randomly selected from the gene pool of parent chromosomes. Assume that both j_1-th and j_2-th chromosomes in the k-th generation are selected as parent chromosomes for crossover, i.e., the crossover is applied to the genes $g_n^{j_1 k}$ and $g_n^{j_2 k}$.

To make sure that the crossover result is still contained in Φ_n, the sequence number of $g_n^{j_1 k}$ and $g_n^{j_2 k}$ in Φ_n are used in crossover. Denote the $\mathbb{S}(g_n)$ as an operator to return the sequence number in Φ_n based on the operand g_n. For instance, if $g_n = 7$ in $\Phi_n = \{0, 2, 3, 4, 7, 8, 11\}$, then $\mathbb{S}(g_n) = 5$. Let $Round(x)$ be an operator rounding the number $x \in R$ and returns the result as an integer, i.e., $Round(x) \in I$. The n-th gene of the child chromosomes in the next $(k + 1)$-th generation are regenerated by applying crossover to the same n-th gene of both parent chromosomes in the current k-th generation as follows.

$$\mathbb{S}\left(g_n^{j(k+1)}\right) = Round\left(b\mathbb{S}\left(g_n^{j_1 k}\right) + (1 - b)\mathbb{S}\left(g_n^{j_2 k}\right)\right) \tag{21}$$

$$\mathbb{S}\left(g_n^{(j+1)(k+1)}\right) = Round\left((1 - b)\mathbb{S}\left(g_n^{j_1 k}\right) + b\mathbb{S}\left(g_n^{j_2 k}\right)\right) \tag{22}$$

where $b \in [0, 1]$ is a random number with uniform distribution and $n = 1 \ldots N$. Although the crossover result is the sequence number $\mathbb{S}(g_n)$, the gene $g_n \in \Phi_n$ associated with this sequence number is actually the one to be achieved.

Referring to Table 1, there is always an element $g_n = 0$ listed in L_n elements of Φ_n corresponding to the situation that this n-th customer is not selected for curtailment. This is especially true when the total amount of demand reduction promised by all customers is much more than the target load curtailment to be achieved. However, it is difficult to have the crossover result of $\mathbb{S}\left(g_n^{j(k+1)}\right) = 1$ that leads to $g_n^{j(k+1)} = 0$ with the crossover operation defined as in (21)–(22). Define the use rate α as the ratio of the target load curtailment through the period of demand bidding event to the total amount demand reduction promised by all participated customers, i.e.,

$$\alpha = \frac{\sum\limits_{m=1}^{M} \Gamma_m}{\sum\limits_{n=1}^{N} \sum\limits_{m=\hat{\beta}_n^s}^{\hat{\beta}_n^e} P_{nm}} \tag{23}$$

Therefore, the probability of directly assigning $\mathbb{S}\left(g_n^{j(k+1)}\right) = 1$ can be designed as proportional to the use rate α. The smaller α is, the more demands are available to choose in order to fulfill the target load curtailment. Let v be the probability assigning $\mathbb{S}\left(g_n^{j(k+1)}\right) = 1$, while v_{max} and v_{min} be the maximum and minimum probability values corresponding to the minimum and maximum use rates of α_{max} and α_{min}, respectively. Then,

$$
v = \begin{cases} v_{max}, & \text{if } \alpha < \alpha_{min} \\ v_{max} - \frac{v_{max} - v_{min}}{\alpha_{max} - \alpha_{min}}(\alpha - \alpha_{min}), & \text{if } \alpha_{min} \leq \alpha < \alpha_{max} \\ v_{min}, & \text{if } \alpha \geq \alpha_{max} \end{cases} \tag{24}
$$

Therefore, the crossover is defined as the operation generating $\mathbb{S}\left(g_n^{j(k+1)}\right) = 1$ with probability v and generating both $\mathbb{S}\left(g_n^{j(k+1)}\right)$ and $\mathbb{S}\left(g_n^{(j+1)(k+1)}\right)$ as defined in (21) and (22), respectively, with probability $(1 - v)$, $n = 1 \dots N$.

Mutation is a scheme randomly perturbing crossover result. It is conducted on the crossover result for every gene with probability v_{mut}. Denote $|\Phi_n|$ as the size of Φ_n. Mutation adds a random number $\mu_n \in [1, |\Phi_n|]$ to the crossover result $\mathbb{S}\left(g_n^{j(k+1)}\right)$. However, it is possible that the addition result is greater than $|\Phi_n|$. A cycling operation is designed as follows to update $\mathbb{S}\left(g_n^{j(k+1)}\right)$ in order to resolve this difficulty.

$$
\mathbb{S}\left(g_n^{j(k+1)}\right) = \begin{cases} \mathbb{S}\left(g_n^{j(k+1)}\right) + \mu_n, & \text{if } \mathbb{S}\left(g_n^{j(k+1)}\right) + \mu_n \leq |\Phi_n|; \\ \mathbb{S}\left(g_n^{j(k+1)}\right) + \mu_n - |\Phi_n|, & \text{otherwise.} \end{cases} \tag{25}
$$

3.4. Extinction and Immigration Mechanisms

One of the common drawbacks frequently encountered by the GA is the loss of population diversity during the search process. It is especially true when the probability of directly assigning $\mathbb{S}\left(g_n^{j(k+1)}\right) = 1$ is high due to low use rate α. As the chromosomes in gene pool tend to be homogeneous, a premature convergence occurs. An extinction and immigration approach is introduced in MGA by reinitializing the gene pool except leaving the best chromosome. The gene pool is then replaced by randomly generated chromosomes.

The extinction and immigration mechanism is applied in MGA when the convergence of fitness value stagnates. Let F_{min}^k and F_{max}^k be the minimum and maximum fitness value in the k-th generation. The extinction and immigration mechanism is applied if $(F_{max}^k - F_{min}^k) \leq \varepsilon_1$ for continuous G_{ext} generations.

3.5. Overall Framework

For the sake of clarity, the flowchart used to describe the mechanism of demand bidding between an aggregator and all customers under its service regions is illustrated in Figure 1a. Meanwhile, the GA-based purchase optimization scheme used to determine the best combination of purchase status for all customers in order to minimize the total bidding cost while fulfilling all load curtailment constraints is summarized using the flowchart presented in Figure 1b. Elitism is used in the proposed MGA. The best chromosome with smallest fitness value is passed to the next generation. If no improvement of the minimum fitness value for certain number of generations, the MGA might have achieved near optimal solution and can be stopped. The stopping criterion for the proposed MGA is set as $\left|F_{min}^k - F_{min}^{k-1}\right| \leq \varepsilon_2$ for continuous G_{stp} generations.

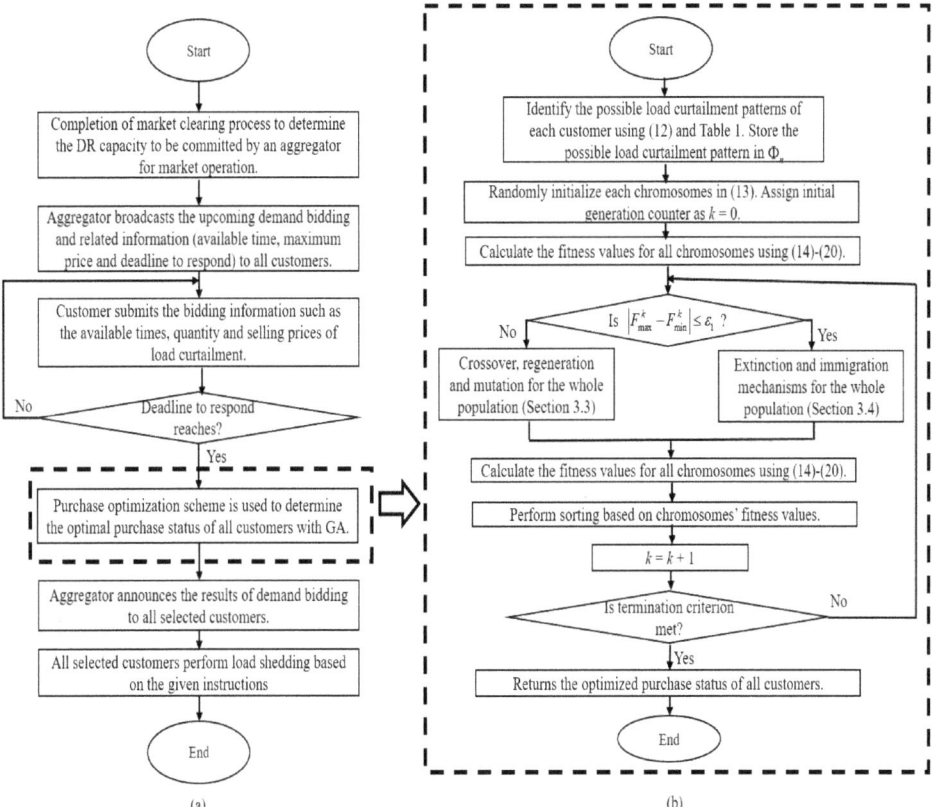

Figure 1. Flowcharts used to describe (**a**) the mechanisms of demand bidding between an aggregator and all customers under its service regions and (**b**) the GA-based purchase optimization scheme used to determine the optimal purchase status of all customers.

4. Simulation Settings and Results

4.1. Simulation Settings

The target load curtailment denoted as Γ_m to be committed at every m-th bidding slots in two case scenarios of demand bidding are summarized in Table 2. Case 1 consists of $M = 5$ bidding slots with $N = 500$ customers are registered as the candidates for demand bidding, while a total of $N = 1000$ customers are considered in the case 2 with $M = 10$ bidding slots. The quantity of load curtailment and bidding price per *kWh* submitted by ten randomly selected customers at each time slot for case 2 are presented in Figures 2 and 3, respectively, to illustrate the characteristic of bidding information. In this paper, it is assumed that the total DR capacity offered by all customers in every bidding slot is much greater than the target load curtailment profile to be achieved. This ensures the aggregator to have greater flexibility in searching for the optimal bidding purchase solution at every bidding slot, hence guaranteeing the feasibility of the demand bidding model. Furthermore, the actual load curtailment delivered by all selected customers during demand bidding cannot be much lesser or much greater than the tolerance ranges of their promised amounts as stipulated by the aggregator in practical scenarios. For the convenience of analysis, the actual amount of load curtailment produced by all selected customers is assumed to be exactly the same as promised.

Table 2. Target load curtailment profiles for two cases of demand bidding.

Bidding Slot	Target Load Curtailment Γ_m (kW)	
	Case 1	Case 2
1	1500	3700
2	2200	4200
3	3500	4600
4	2900	4900
5	2500	5500
6	-	5400
7	-	4700
8	-	4300
9	-	3900
10	-	3200

The proposed purchase optimization scheme designed using MGA is compared with that of developed using regular genetic algorithm (RGA) that did not incorporated with the adaptive bid declination scheme defined in (23)–(24). For the cases 1 and 2 of demand bidding, the population sizes of algorithms are set as 500 and 1000 respectively. The remaining parameters of both MGA and RGA are set to be:

$$\kappa = 1.1, q = 10\%, \alpha_{\max} = 0.4, \alpha_{\min} = 0.05, v_{\max} = 0.4, v_{\min} = 0.01,$$
$$v_{mut} = 0.05, \varepsilon_1 = 2 \times 10^4, G_{ext} = 200, \varepsilon_2 = 1 \times 10^4, G_{stp} = 100$$

In order not to be misled by a single simulation, the simulations results of both compared methods are the average values produced from ten independent simulations. All simulations are run on an Intel ® Core i5-4570 CPU @3.40 GHz, 4.00 GB RAM, and Microsoft Visual Studio 2010.

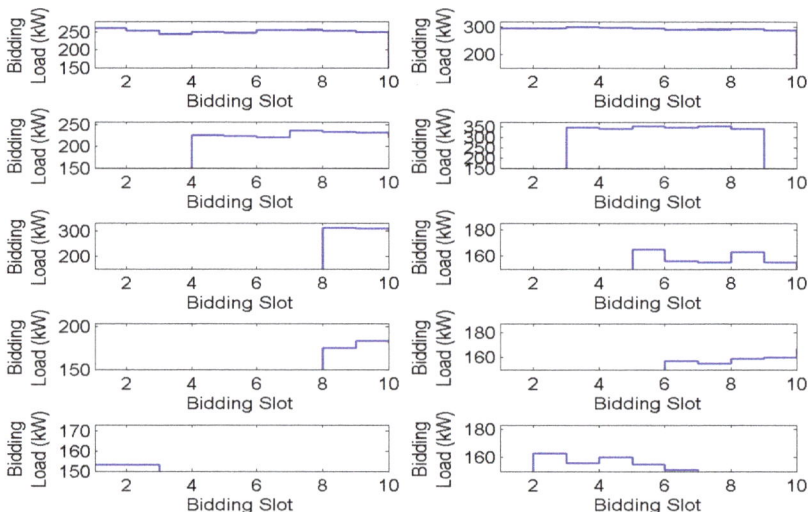

Figure 2. The amount of load curtailment bid by ten randomly selected customers at each bidding slot for case 2 of demand bidding with $M = 10$.

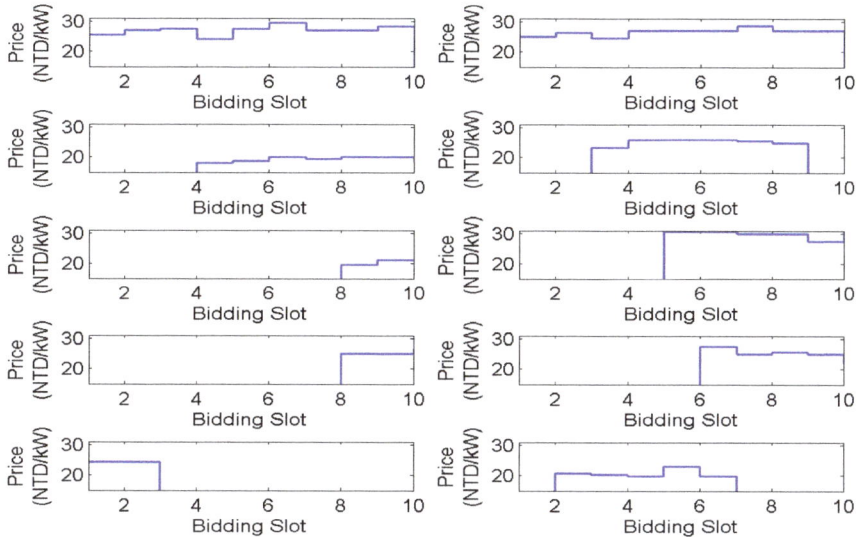

Figure 3. The bidding price per *kWh* offered by ten randomly selected customers at each bidding slot for case 2 of demand bidding with *M* = 10.

4.2. Simulation Results

Denote δ_n as the minimum load curtailment period of each n-th customer, while λ_n represents the actual number of bidding slots purchased by the aggregator from the n-th customer. Let h^*_{nm} be the optimal purchase status of each n-th customer at the m-th bidding slot as determined by the proposed purchase optimization scheme. Then, λ_n is calculated as:

$$\lambda_n = \sum_{m=1}^{M} h^*_{nm}, \; n = 1, \ldots, N. \tag{26}$$

The input load curtailment bids submitted by ten randomly selected customers and the actual demand purchased from these customers determined using the proposed method are shown in Figure 4. The minimum load curtailment period δ_n of these ten customers are set as 1 h. From Figure 4, it is observed that three of these ten randomly selected customers have all of their input load curtailment bids purchased by the aggregator. The demand bids submitted by another six customers are partially selected, while one customer is completely declined for load curtailment. The simulation results in Figure 4 verify that the purchased status of each customer as computed by the proposed work indeed satisfies the constraints (8)–(10) that $\lambda_n \geq \delta_n$ only if $\lambda_n \geq 0$.

Define $\Psi_m = \sum_{n=1}^{N} h^*_{nm} P_m$ as the actual load shedding produced by all selected customers at the m-th bidding slot. The absolute deviation between the actual load shedding Ψ_m and the target load curtailment profile Γ_m at every m-th bidding slot is represented as Δ_m, i.e.,

$$\Delta_m = |\Psi_m - \Gamma_m|, \; m = 1, \ldots, M. \tag{27}$$

Let $\overline{\Delta}$ be a metric used to quantify the average load curtailment difference between the actual quantity of load shedding delivered by the compared methods and the target of load curtailment during demand bidding, that is:

$$\overline{\Delta} = \frac{1}{M} \sum_{m=1}^{M} \Delta_m. \tag{28}$$

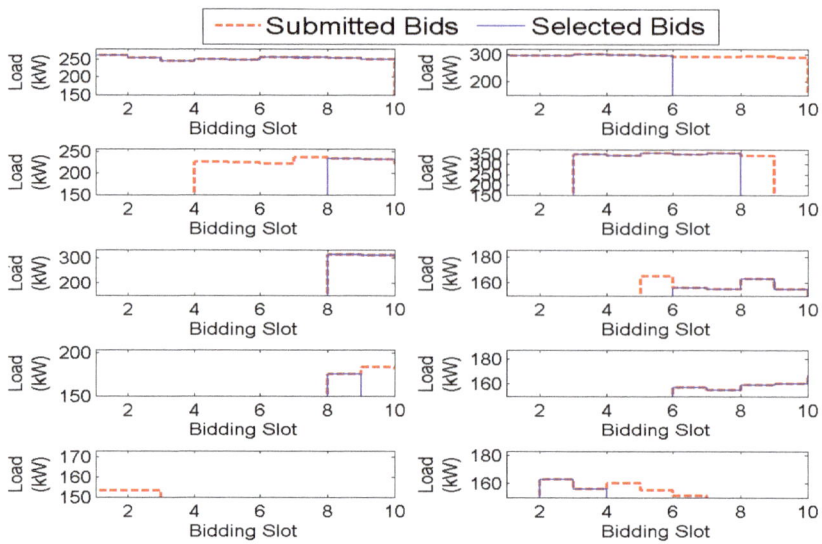

Figure 4. The input load curtailment bids submitted by ten randomly selected customers and the output purchase status computed by the proposed method.

The target load curtailment profile Γ_m a nd the actual load curtailment profile Ψ_m produced by both of the RGA and proposed work for cases 1 and 2 of demand bidding are presented in Figure 5. It is observed that actual load curtailment profiles produced by the proposed method is significantly closer to the target profiles as compared with those of RGA for both cases of demand bidding. The qualitative results shown in Figure 5 are validated by the quantitative results presented in Table 3 because the average load curtailment deviations produced by the proposed work in both cases of demand bidding are at least 10 times lower than those of RGA. The smaller $\overline{\Delta}$ values produced by the proposed work implies its excellent capability in preventing the excessive load curtailment that will lead to the revenue loss of aggregator during demand bidding.

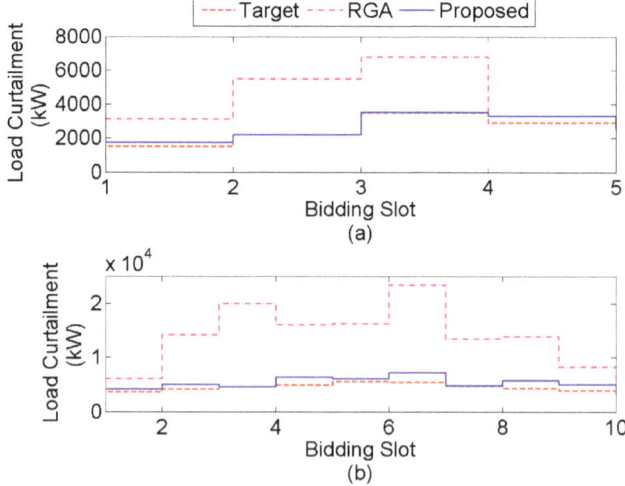

Figure 5. The actual load curtailment profiles produced by both compared methods with respect to the target profile in (**a**) case 1 and (**b**) case 2 of demand bidding.

Table 3. Comparison of the average load difference produced by all compared methods.

	Average Load Curtailment Difference, $\bar{\Delta}$ (kW)	
	RGA	Proposed
Case 1	1814.5	163.0
Case 2	9067.9	810.8

Denote $\bar{\eta}_m$ and $\bar{\eta}_m^*$ as the average bidding price per unit load curtailment submitted by all customers during demand bidding and the optimal purchase price per unit load curtailment as determined by the compared methods, respectively, at the m-th bidding slot. Mathematically,

$$\bar{\eta}_m = \frac{\sum\limits_{n=1}^{N} \sigma_{nm} P_{nm}}{\sum\limits_{n=1}^{N} P_{nm}}, \tag{29}$$

$$\bar{\eta}_m^* = \frac{\sum\limits_{n=1}^{N} h_{nm}^* \sigma_{nm} P_{nm}}{\sum\limits_{n=1}^{N} h_{nm}^* P_{nm}}. \tag{30}$$

Comparisons between $\bar{\eta}_m$ and $\bar{\eta}_m^*$ for cases 1 and 2 of demand bidding are illustrated in Figure 6. It is shown in both Figure 6a,b that the optimal purchase prices $\bar{\eta}_m^*$ produced by both compared methods are different and lower than the average offered price $\bar{\eta}_m$ at every bidding slot for both cases 1 and 2. This observation implies the cost minimization capabilities of the proposed purchase optimization scheme in fulfilling the target load curtailment profiles. Furthermore, the optimal purchase prices $\bar{\eta}_m^*$ produced by the proposed work in all bidding slots are consistently lower than those of RGA because the latter approach tends to purchase more load curtailment than necessary from the customers as shown in Table 3 and Figure 5. This undesirable behavior leads to the increasing of bidding cost of aggregator in purchasing the load curtailment from customers. Let η^* be the total bidding cost incurred by the compared methods in order to satisfy the target load curtailment throughout the demand bidding event, i.e.,

$$\eta^* = \sum_{m=1}^{M} \sum_{n=1}^{N} h_{nm}^* \sigma_{nm} P_{nm} \tag{31}$$

The η^* values produced by RGA and proposed work in both cases of demand bidding are presented in Table 4 to further evaluate the cost minimization capabilities of both compared methods. The proposed work is reported to be more cost effective than RGA in both cases because the optimization results produced by the former method enable the aggregator to deliver the promised load curtailment for market operation with lower financial costs.

Table 4. Comparison of the total bidding cost produced by all compared methods.

	Total Bidding Cost, η^* (NTD'000)	
	RGA	Proposed
Case 1	1253	330
Case 2	3294	520

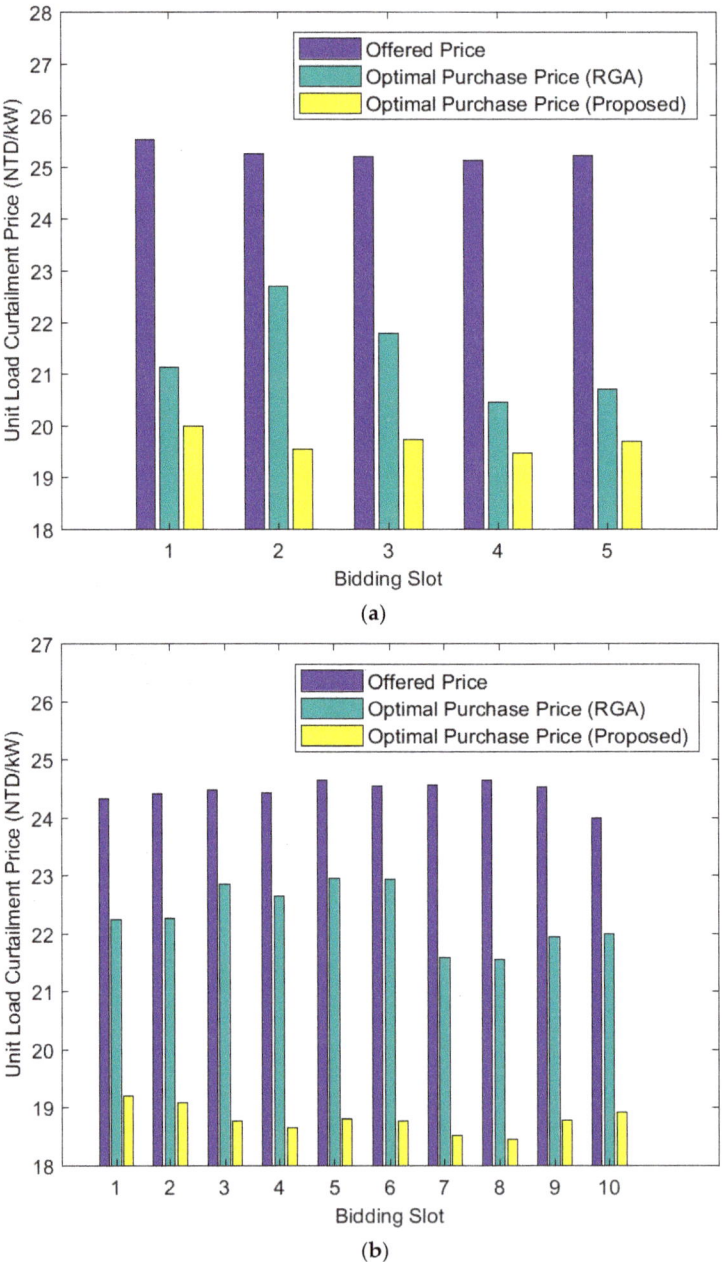

Figure 6. Comparison between the averages offered price per unit load curtailment $\bar{\eta}_m$ and optimal purchase prices per unit load curtailment $\bar{\eta}_m^*$ at every bidding slot in the (**a**) case 1 and (**b**) case 2 of demand bidding.

Table 5 compares the computation time τ^{com} required by both compared methods to solve the demand bidding problems described in cases 1 and 2. The proposed method is reported to incur slightly greater computation times than RGA in both cases. This is because the proposed method is incorporated with an additional adaptive bid declination scheme as described in (23)–(24) to prevent the excessive load curtailment that can lead to the revenue loss of aggregator. It is also observed that case 2 with greater problem complexity has incurred longer computation times for both compared methods. Although the computation times of both compared methods tend to increase with problem complexity, the τ^{com} values obtained for both cases are acceptable because the optimal purchase solutions are computed one day ahead before demand bidding starts. It is also noteworthy that the target load curtailment profiles presented in Table 2 are intentionally made more fluctuating than the practical scenarios with the purpose to evaluate the computational efficiency of the proposed purchase optimization scheme with an additional complexity. The target load curtailment profiles assigned for the practical scenarios are usually constant or less fluctuating than those simulated in this paper. Therefore, the actual computation times incurred can be reasonably shorter that those observed from simulations despite the latter values are short enough for real applications.

Table 5. Comparison of the computational time incurred by all compared methods.

	Computational Time, τ^{com} (s)	
	RGA	Proposed
Case 1	23.44	24.93
Case 2	63.49	68.96

The proposed purchase optimization scheme is evaluated further under different use rates. Let α be the use rate considered when the target load curtailment profile in Table 2 is used for demand bidding. Please note that this use rate can be increased further by multiplying Γ_m in Table 2 with different factors. The simulation results in Table 6 shows that the proposed method outperforms RGA by consistently producing the lower values of $\overline{\Delta}$ and η^* for different use rates. The actual target load curtailment profiles produced by the proposed method in Figure 7 are also closer to the target profiles than those of RGA at majority of the bidding slots, suggesting that the proposed work is less susceptible to the excessive load curtailment issue under different scenarios of use rate.

Table 6. Performance comparisons of all compared methods under different use rates for case 2 of demand bidding.

Use Rate	RGA		Proposed	
	$\overline{\Delta}$ (kW)	η^* (NTD'000)	$\overline{\Delta}$ (kW)	η^* (NTD'000)
1.3α	8128.2	3389	1179.6	1681
1.6α	6530.1	3323	1158.8	2043
2.0α	6275.8	3711	2311.5	2767
2.5α	6023.4	4209	3535.8	3562
5.0α	5772.1	6941	3662.4	6412

Some observations can be highlighted from the simulation results obtained in Table 6 and Figure 7. A significant performance deviation between the proposed method and RGA can be observed when the use rates are relatively low (e.g.,1.3α, 1.6α, 2.0α and 2.5α). The aggregated demand reduction promised by all participated customers in these scenarios is much higher than that of target load curtailment, implying that majority of the submitted input load curtailments bids needs to be declined in order to avoid excessive load curtailment. For RGA that are not equipped with the adaptive bid declination scheme as described in (23)–(24), it is difficult to achieve the crossover results of $\mathbb{S}\left(g_n^{j(k+1)}\right) = 1$ leading to $g_n^{j(k+1)} = 0$ via the crossover operation defined in (21)–(22). On the other hand, the proposed method is able to decline the input bids of the customers with a relatively high probability of v under the low use rate scenarios according to (24). This explains the significant outperformance of the proposed method against the RGA (in terms of $\overline{\Delta}$ and η^*) at the use rates of 1.3α, 1.6α, 2.0α and 2.5α. When the use rate is increased further to 5.0α, smaller performance differences between the proposed method and RGA are observed. In this higher use rate scenario, the probability v assigned to the proposed method becomes much smaller according to (23)–(24). In other words, the adaptive bid declination scheme designed in the proposed method has smaller influence because most of the total demand reduction promised by the customers need to be accepted in order to satisfy the target load curtailment. This justifies the relatively similar performance delivered by both of the proposed method and RGA under the higher use rate scenarios.

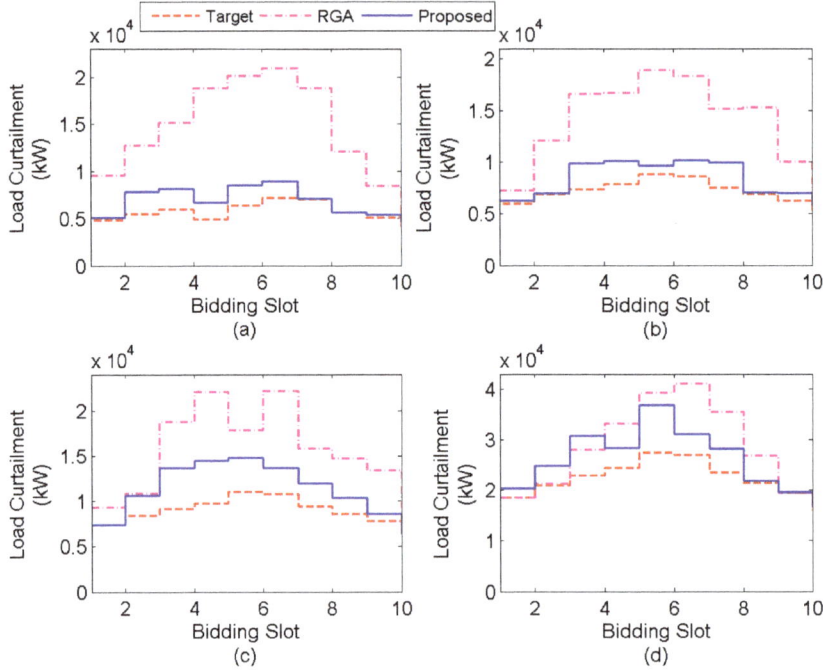

Figure 7. The actual load curtailment profiles produced by both compared methods with respect to the target profile under the use rates of (**a**) 1.3α, (**b**) 1.6α, (**c**) 2.0α, and (**d**) 5.0α.

5. Conclusions

A closed demand bidding model is considered in this paper to facilitate the internal trading of load curtailment between an aggregator and its customers. The proposed work is different with most existing demand bidding models because the latter focused on the interaction between the aggregator and different entities of electricity market (e.g., TSO, distributor and retailer) at upper level to optimize the DR supply curve and market clearing prices. By considering the unique load curtailment patterns of different customers, a purchase optimization scheme is designed to determine the optimal combination of aggregated load curtailment bids and optimal purchase price, aiming to minimize the total bidding cost. A modified genetic algorithm (MGA) incorporated with a delicate gene encoding scheme and an adaptive search mechanism is then proposed to solve the highly nonlinear demand bidding problem efficiently without requiring to linearize the objective function or constraints. Extensive simulations show that the optimal purchase price determined by the proposed purchase optimization scheme via MGA is different in every bidding slot and at least 20% lower than the average bidding prices offered by the customers. This implies the promising performance of proposed method in minimizing the total bidding cost. Furthermore, the computational overhead incurred by the proposed method is low and hence it is suitable for real applications.

Some notable values of the work proposed in this paper are explained as follows. First, it provides an immediate solution for aggregator to deliver its promised load curtailment by establishing a platform to interact with its customers at lower level after the aggregator receives the target load curtailment profiles and clearing prices from the electricity market at upper level. Second, the closed demand bidding model and the information exchange mechanism between seller (i.e., customers) and buyer (i.e., aggregator) are designed to be straightforward and can be easily provided so that the proposed model can be easily deployed for the real application in the field. Third, the proposed purchase optimization scheme is also suitable to be implemented for the utility companies operate in the regulated electricity market in order to maintain the reliability of power system when significant stress conditions are foreseen. For future works, more practical constraints such as regional dispatch constraints can be incorporated between the aggregator and its customers to produce a more realistic demand bidding model that can simulate the practical scenarios better. The penalty mechanisms to be imposed when customers deliver significantly lesser or significantly greater load curtailment than they have promised is another promising direction to extend the current work. Finally, a more comprehensive demand bidding framework can be developed by formulating the purchase optimization scheme as a multi-objective optimization problem. Different objective functions such as comfort level of customers in performing load curtailment can be explored in the multi-objective optimization framework.

Author Contributions: Conceptualization, L.Y.; Methodology, L.Y. and W.H.L.; Software, W.H.L. and S.S.T.; Validation, T.H.T. and C.H.W.; Formal Analysis, T.H.T., C.H.W. and J.W.P.; Writing-Original Draft Preparation, L.Y. and W.H.L.; Writing-Review & Editing, S.S.T., C.H.W. and T.H.T.; Visualization, J.W.P. Supervision, L.Y.

Funding: This work was funded by the Ministry of Science and Technology, Taiwan, R.O.C, under Grant MOST 106-2221-E-027-089.

Conflicts of Interest: The authors declare no conflict of interest.

Nomenclature

Acronyms

DR	Demand response
DRX	Demand response exchange
EV	Electric vehicle
ISO	Independent system operator
LMP	Locational marginal prices
LSE	Load service entity
MGA	Modified genetic algorithm
MILP	Mixed-integer linear programming
RGA	Regular genetic algorithm
TSO	Transmission system operator

Indices and Sets

n	Index of a customer participates in a demand bidding event
l	Index of each possible combination of load shedding pattern
j	Index of chromosome
k	Index of generation
H^*	Set containing the overall optimal purchase status in a demand bidding event
Ω	Set containing the indices of all possible combinations of load shedding patterns in a demand bidding event
Φ_n	Set containing the indices of all possible combinations of load shedding patterns for the n-th customer

Operators

$\mathbb{B}(\cdot)$	An operator that returns true if the operand is 1 and false if the operand is 0
$\mathbb{I}(\cdot)$	An operator that returns 1 if the operand is true and 0 is the operand is false
$\mathbb{S}(\cdot)$	An operator that returns the sequence number of load shedding pattern based on gene value
$Tra(\cdot,\cdot)$	An operator that returns a set containing the possible load shedding patterns of customer based on their available time to perform load shedding
$Dec(\cdot)$	An operator that returns a set containing the purchase status of a customer based on the load shedding pattern encoded in the gene
$Round(\cdot)$	An rounding operator that returns the result an as integer

Parameters and Variables

N	Number of customers participate in a demand bidding event
$P_{\max}^n$	Maximum load shedding capacity of the n-th customer
δ_n	Minimum load shedding period of the n-th customer
t^s	Beginning time of a demand bidding event
t^e	Completion time of a demand bidding event
$\sigma_{\max}$	Maximum purchase price per kWh of load curtailment
t^d	Deadline for customer to respond their participation in a demand bidding event
M	Number of time slots available for demand bidding
$\hat{\tau}_n^s$	Suitable time for the n-th customer to start load shedding
$\hat{\tau}_n^e$	Suitable time for the n-th customer to end load shedding
$\hat{P}_{nm}, P_{nm}$	Quantity of load shedding offered by the n-th customer at each m-th time slot
$\hat{\sigma}_{nm}, \sigma_{nm}$	Selling price per kWh of load shedding offered by the n-th customer at each m-th time slot
$\hat{\beta}_n^s$	Integer index to indicate the start time of n-th customer for load shedding
$\hat{\beta}_n^e$	Integer index to indicate the end time of n-th customer for load shedding
h_{nm}	Purchase status for the bids offered by the n-th customer at the m-th time slot
Γ_m	Target load curtailment profile at the m-th time slot
h_{nm}^*	Optimized purchase status for the bids offered by the n-th customer at the m-th time slot
L	Total number of possible combination of load shedding among M time slots
$\hat{\chi}_n$	Time slots for potential load curtailment of n-th customer
L_n	Total number of possible combination of load shedding for the n-th customer among $\hat{\chi}_n$ time slot
g_n	Gene used to encode the possible load shedding patterns of n-th customer
ξ^{jk}	A chromosome used to encode the possible load shedding patterns of all customer
F	Fitness function value
$F_{\min}^k$	Minimum fitness value at the k-th generation
$F_{\max}^k$	Maximum fitness value at the k-th generation
ϑ	Penalty factor
B_m	Violation of target load curtailment produced at the m-th time slot
$\bar{\sigma}$	Average price offered by all participated customers
κ	A constant with value greater than 1
q	Percentage of fittest chromosomes to be selected as parent chromosomes
α	Use rate of total demand reduction promised by all customers
$\alpha_{\max}$	Maximum use rate
$\alpha_{\min}$	Minimum use rate
b	A random number with uniform distribution between 0 and 1
v	Probability of not selecting customers for load curtailment
$v_{\max}$	Maximum probability of not selecting customer for load curtailment
$v_{\min}$	Minimum probability of not selecting customer for load curtailment
v_{mut}	Mutation probability

μ_n	A random number used to perturb the crossover result of n-th gene
ε_1	Threshold value used to trigger the extinction and immigration mechanism
ε_2	Threshold value used to terminate the search process of genetic algorithm
G_{ext}	Maximum generation number to allow for no improvement in population diversity before triggering the extinction and immigration mechanism
G_{stp}	Maximum generation number to allow for no fitness improvement of best chromosome before terminating the search process of genetic algorithm
λ_n	Actual number of bidding slots purchased by the aggregator from each n-th customer
Ψ_m	Actual load shedding produced by all selected customers at the m-th time slot
Δ_m	Absolute deviation between the target and actual load shedding produced at the m-th time slot
$\overline{\Delta}$	Average load curtailment difference between the actual quantity of load shedding and the target load curtailment during demand bidding
$\overline{\eta}_m$	Average bidding price per unit load curtailment submitted by all customer at the m-th time slot
$\overline{\eta}_m^*$	Optimal purchase price per unit load curtailment determined at the m-th time slot
η^*	Total bidding cost incurred by the aggregator to satisfy the target load curtailment in demand bidding
τ^{com}	Computation time required to solve the demand bidding optimization problem

References

1. Lai, L.L. *Power System Restructuring and Deregulation: Trading, Performance and Information Technology*; Wiley: New York, NY, USA, 2002.
2. Deng, S.; Oren, S.S. Electricity derivatives and risk management. *Energy* **2006**, *31*, 940–953. [CrossRef]
3. Fleten, S.S.; Pettersen, E. Constructing bidding curves for a price-taking retailer in the Norwegian electricity market. *IEEE Trans. Power Syst.* **2005**, *20*, 701–708. [CrossRef]
4. Philpott, A.B.; Pettersen, E. Optimizing demand-side bids in day-ahead electricity markets. *IEEE Trans. Power Syst.* **2006**, *21*, 488–498. [CrossRef]
5. Herranz, R.; Roque, A.M.S.; Campos, F.A. Optimal demand-side bidding strategies in electricity spot markets. *IEEE Trans. Power Syst.* **2012**, *27*, 1204–1213. [CrossRef]
6. Ramachandra, B.; Srivastava, S.K.; Edrington, C.S.; Cartes, D.A. An intelligent auction scheme for smart grid market using a hybrid immune algorithm. *IEEE Trans. Ind. Electron.* **2011**, *58*, 4603–4612. [CrossRef]
7. Weare, C. *The California Electricity Crisis: Causes and Policy Options*; Public Policy Institute of California: San Francisco, CA, USA, 2003.
8. Benefits of Demand Response in Electricity Markets and Recommendations for Achieving Them. Available online: https://emp.lbl.gov/sites/all/files/report-lbnl-1252d.pdf (accessed on 1 June 2018).
9. Sezgen, O.; Goldman, C.A.; Krishnarao, P. Option value of electricity demand response. *Energy* **2007**, *32*, 108–119. [CrossRef]
10. Kirschen, D.S. Demand-side view of electricity markets. *IEEE Trans. Power Syst.* **2003**, *18*, 520–527. [CrossRef]
11. Yao, L.; Yao, L.; Lim, W.H. A soft curtailment of wide-area central air conditioning load. *Energies* **2018**, *11*, 492. [CrossRef]
12. Yao, L.; Chou, Y.C.; Lin, C.C. Scheduling of direct load control using genetic programming. *Int. J. Innov. Comput. I* **2011**, *7*, 2515–2528.
13. Rahimi, F.; Ipakchi, A. Demand response as a market resource under the smart grid paradigm. *IEEE Trans. Smart Grid* **2010**, *1*, 82–88. [CrossRef]
14. Demand Response Compensation in Organized Wholesale Energy Markets. Available online: https://www.ferc.gov/EventCalendar/Files/20110315105757-RM10-17-000.pdf (accessed on 1 June 2018).
15. Wholesale Competition in Regions with Organized Electric Markets. Available online: https://www.ferc.gov/whats-new/comm-meet/2008/101608/E-1.pdf (accessed on 1 June 2018).
16. Vaya, M.G.; Andersson, G. Optimal bidding strategy of a plug-in electric vehicle aggregator in day-ahead electricity markets under uncertainty. *IEEE Trans. Power Syst.* **2015**, *30*, 2375–2385. [CrossRef]
17. Mohsenian-Rad, H. Optimal demand bidding for time-shiftable loads. *IEEE Trans. Power Syst.* **2015**, *30*, 939–951. [CrossRef]

18. Menniti, D.; Costanzo, F.; Scordino, N.; Sorrentino, N. Purchase-bidding strategies of an energy coalition with demand-response capabilities. *IEEE Trans. Power Syst.* **2009**, *24*, 1241–1255. [CrossRef]

19. Nojayan, S.; Mohammadi-Ivatloo, B.; Zare, K. Optimal bidding strategy of electricity retailers using robust optimization approach considering time-of-use rate demand response programs under market price uncertainties. *IET Gener. Transm. Distrib.* **2015**, *9*, 328–338. [CrossRef]

20. Wu, L. Impact of price-based demand response on market clearing and locational marginal prices. *IET Gener. Transm. Distrib.* **2013**, *7*, 1087–1095. [CrossRef]

21. Liu, G.; Xu, Lu.; Tomsovic, K. Bidding strategy for microgrid in day-ahead market based on hybrid stochastic/robust optimization. *IEEE Trans. Smart Grid.* **2016**, *7*, 227–237. [CrossRef]

22. Rahman, M.N.; Arefi, A.; Shafiullah, G.; Hettiwatte, S. A new approach to voltage management in unbalanced low voltage using demand response and OLTC considering consumer preference. *Int. J. Electr. Power* **2018**, *99*, 11–27. [CrossRef]

23. Russell, M.; Sun, Z.; Dagli, C.H. Reward/penalty design in demand response for mitigating overgeneration considering the benefits from both manufacturers and utility company. *Procedia Comput. Sci.* **2017**, *114*, 425–432.

24. Parvania, M.; Fotuhi-Firuzabad, M. Integrating load reduction into wholesale energy market with application to wind power integration. *IEEE Syst. J.* **2012**, *6*, 35–45. [CrossRef]

25. Parvania, M.; Fotuhi-Firuzabad, M.; Shahidehpour, M. ISO's optimal strategies for scheduling the hourly demand response in day-ahead markets. *IEEE Trans. Power Syst.* **2014**, *29*, 2636–2645. [CrossRef]

26. Nguyen, D.T. Demand response for domestic and small business consumers: A new challenge. In Proceedings of the 2010 IEEE PES Transmission and Distribution Conference and Exposition, New Orleans, LA, USA, 19–22 April 2010.

27. Nguyen, D.T.; Negnevitsky, M.; Groot, M.D. Pool-based demand response exchange-concept and modeling. *IEEE Trans. Power Syst.* **2011**, *26*, 1677–1685. [CrossRef]

28. Nguyen, D.T.; Negnevitsky, M.; Groot, M.D. Walrasian market clearing for demand response exchange. *IEEE Trans. Power Syst.* **2012**, *27*, 535–544. [CrossRef]

29. Gkatzikis, L.; Koutsopoulos, I.; Salonidis, T. The role of aggregators in smart grid demand response markets. *IEEE J. Sel. Areas Commun.* **2013**, *31*, 1247–1257. [CrossRef]

30. Pop, C.; Ciora, T.; Antal, M.; Angel, I.; Salomie, I.; Bertoncini, M. Bloackchain based decentralized management of demand response programs in smart energy grids. *Sens.* **2018**, *18*, 162. [CrossRef] [PubMed]

31. Yao, L.; Lim, W.H. Optimal purchase strategy for demand bidding. *IEEE Trans. Power Syst.* **2018**, *33*, 2754–2762. [CrossRef]

32. Bahrami, S.; Hadi Amini, A.; Shafie-khah, M.; Catalão, J.P.S. A decentralized electricity market scheme enabling demand response deployment. *IEEE Trans. Power Syst.* **2018**, *33*, 4218–4227. [CrossRef]

33. Energy & Ancillary Services Market Operations. Available online: http://www.pjm.com/-/media/documents/manuals/archive/m11/m11v87-energy-and-ancillary-services-market-operations-03-23-2017.ashx (accessed on 10 June 2018).

Article

Analyzing of a Photovoltaic/Wind/Biogas/Pumped-Hydro Off-Grid Hybrid System for Rural Electrification in Sub-Saharan Africa—Case Study of Djoundé in Northern Cameroon

Nasser Yimen *, Oumarou Hamandjoda, Lucien Meva'a, Benoit Ndzana and Jean Nganhou

National Advanced School of Engineering, University of Yaoundé I, POB: 8390, Yaounde, Cameroon;
oumahama@yahoo.com (O.H.); jrl67_mevaa@yahoo.com (L.M.); bendzana@yahoo.fr (B.N.);
jean.nganhou@yahoo.com (J.N.)
* Correspondence: nazerois@yahoo.fr; Tel.: +237-670-147-046

Received: 12 September 2018; Accepted: 28 September 2018; Published: 3 October 2018

Abstract: Traditional electrification methods, including grid extension and stand-alone diesel generators, have shown limitations to sustainability in the face of rural electrification challenges in sub-Saharan Africa (SSA), where electrification rates remain the lowest in the world. This study aims at performing a techno-economic analysis and optimization of a pumped-hydro energy storage based 100%-renewable off-grid hybrid energy system for the electrification of Djoundé, which is a small village in northern Cameroon. Hybrid Optimization of Multiple Energy Resources (HOMER) software was used as an analysis tool, and the resulting optimal system architecture included an 81.8 kW PV array and a 15 kW biogas generator, with a cost of energy (COE) and total net present cost (NPC) of €0.256/kWh and €370,426, respectively. The system showed promise given the upcoming decrease in installation cost of photovoltaic systems. It will be viable in parts of SSA region but, significant investment subsidies will be needed elsewhere. The originality of this study can be emphasized in three points: (1) the modelling with the recently introduced pumped-hydro component of HOMER; (2) broadening sensitivity analysis applications to address practical issues related to hybrid renewable energy systems (HRES); and, (3) consideration of the agricultural sector and seasonal variation in the assessment of the electricity demand in an area of SSA.

Keywords: hybrid renewable energy system; pumped-hydro energy storage; off-grid; optimization; HOMER software; rural electrification; sub-Saharan Africa; Cameroon

---

1. Introduction

Energy, especially electricity, is a vital commodity for everyday life in the contemporary world. It is the primary driver for any human, social, or economic development. However, electricity is still a luxury in many places around the world [1]. According to the International Energy Agency (IEA) [2], 1.1 billion people viz. 14% of the world's population did not have access to electricity in 2016. The issue was especially acute in Sub-Saharan Africa (SSA) where 588 million people needed access to electrical energy. The rural electrification rate in the region was only 23%, as compared with 71% in urban areas. This rate was unequally distributed, as illustrated by the electrification rates of selected countries shown in Figure 1. While some countries, such as Ethiopia and Kenya, have experienced rural electrification above 50%, others, such as Chad and Mauritania, have achieved less than 5%. The situation is particularly worrying because over 60% of the regional population lives in rural areas. This is an obstacle to political change, job creation, social welfare, economic growth, the modernization of education, the adoption of modern agricultural technology, and the promotion of gender equality, as demonstrated in [3–6].

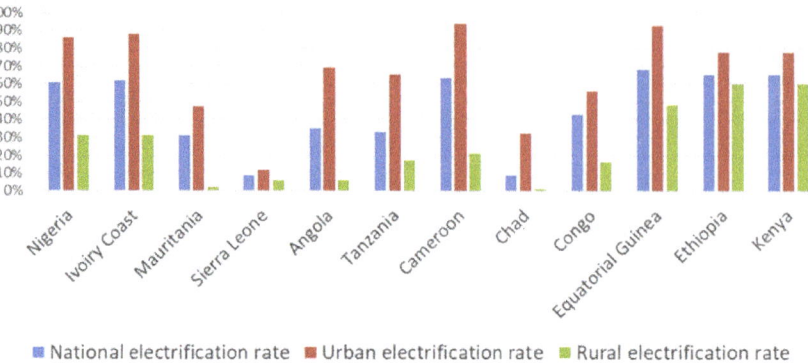

■ National electrification rate   ■ Urban electrification rate   ■ Rural electrification rate

Figure 1. Electrification rates of some Sub-Saharan African (SSA) countries.

The grid extension has long been the primary means of electrification of the region. However, connection to the grid is most of the time practically impossible due to geographical remoteness, thick jungles, rugged terrains, high costs of supply, low household incomes, low consumptions, dispersed settlement of consumers, and inadequate road infrastructures [7]. As a result, decentralized diesel generators are most often used for rural electrification in the region. However, the high costs that are associated with the transportation of fuels and the maintenance of those systems make them unsuitable for rural electrification in developing countries. On a world basis, fossil fuel resources are experiencing a rapid depletion, resulting in an ever-increasing price which tends to make them unaffordable for developing nations. The growing evidence of global warming phenomena due to the release of greenhouse gases when burning those fuels is another critical reason for reducing our dependence on them [8]. Therefore, finding alternative energy sources to meet the growing energy demand while minimizing adverse environmental impacts is becoming an imperative task.

Renewable energy sources, namely solar, wind, biomass, geothermal, and hydro, being inexhaustible, locally available, free, and eco-friendly can constitute potential sources of alternative energy, especially for local power generation in remote rural areas. Increasing interest has been given to their utilization since the oil crises of the 1970s [9].

The main drawbacks associated with the utilization of renewable energy sources are their unreliability and inability to work efficiently due to their intermittent and fluctuating nature, which generally leads to the over-sizing of the system, thereby increasing the investment cost. The hybrid renewable energy systems (HRES) have recently gained popularity as an effective means to deal with the disadvantages that are related to single source based renewable energy systems. A hybrid system is made up of two or more power generation plants fed with appropriate fuels (renewable or fossil fuel) along with energy storage and electronic appliances. The main advantages of hybrid renewable energy systems over single-source systems include [10]:

- higher reliability,
- better efficiency,
- reduced energy storage capacity, and
- lower levelized life-cycle power cost.

For developing countries, the literature on hybrid renewable energy systems is dominated by optimal design studies. The main problem addressed in those studies is to find the appropriate size or number of each component constituting the system, so as to maximize/minimize the objective functions and satisfy all constraints. Many approaches and software tools have been used to handle the issue, as mentioned above.

Reviewing all of the research carried out in this area is beyond the scope of this paper. However, for an indicative purpose, we only mention some, notably the analysis conducted by Nfah and Ngundam [11] on a pico-hydro (PH)/PV hybrid system incorporating a biogas generator and a battery for remote areas of Cameroon. Using HOMER software the authors simulated the system and determined the optimal configurations for localities in the Southern and Northern regions. The cost of energy (COE) and breakeven grid distance were determined at 0.352 €/kWh and 12.9 km and at 0.395 €/kWh and 15.2 km, respectively, for southern and northern locations. Adaramola et al. [12] used HOMER to perform the techno-economic optimization of a solar/wind/Diesel Generator (DG) hybrid energy system in remote areas of Ghana. Considering the levelized cost of electricity (LCOE) and the net present cost (NPC) as the performance criteria, they found out that the optimal system was made up of an 80 kW PV array, a 100 kW wind turbine, and a 600 Surrette 4KS25P, and produced 791.1 MWh of electricity yearly at the cost of $0.281/kWh. Halabi et al. [13] used Homer to model and simulate a PV/diesel/battery HRES to meet domestic electricity needs in Sabah, Malaysia. They found that the optimized system's NPC and COE were $5,571,168 and 0.311 $/kWh, respectively, and its economic and technical performance was better than that of the existing standalone diesel generator and a hypothetical PV/Battery system. Singh et al. [14] used a swarm based artificial bee colony algorithm to find the optimal configuration of a hybrid PV/Wind/Biomass/battery system that is designed to meet the electricity demand of Patiala, an island village of India. They found that the optimal hybrid system was made up of a 250-kW PV array, 18 wind turbines of 1 kW, a biomass generator of 40 kW, and a 1.4-kah battery, and it had an NPC and COE of $7,230,378 and 0.173 $/kWh, respectively. The system configuration analyzed by the previous authors was the focus of the study by Sigarchian et al. [15], who used HOMER software to perform techno-economic feasibility of using a biogas generator fuelled by locally produced biogas as a backup engine in comparison with using a diesel engine for the same purpose. The results showed that the NPC and COE of the optimized hybrid system with a biogas generator were, respectively, 18% and 20% lower than those of the system with a diesel generator. Baghdadi et al. [16] were interested in the design and simulation of a hybrid PV/Wind/Diesel/battery system to satisfy the power requirement of Adrar, a location in southern Algeria. They considered the renewable fraction as the performance criterion to be minimized and used HOMER software to perform the analysis which revealed that the renewable fraction of such a system could reach 70%. Ma et al. [17] investigated on a hybrid PV/Wind system integrated with a pumped hydro storage (PHS) to meet the domestic electricity demand of a hypothetical island in Hong Kong, China. They developed a mathematical model to simulate dozens of cases with different component capacities. They then focused on a technically feasible case made up of 110-kWp PV arrays, two wind turbines of 10.4 kW, and a pumped hydro storage system with a 5106-m^3 upper reservoir. Finally, they concluded that PHS is the best energy storage system for 100% energy autonomy in islanded communities. Kenfack et al. [18] used HOMER software to investigate a hybrid PV/Micro-hydro/Battery system in Batocha, Cameroon. The analysis revealed that the optimal system to meet the electricity demand of the location was made up of a 5-kW PV array, a 2.12-kW Hydro plant, a 1-kW diesel generator and 125 units of a 24-Ah battery. The NPC and COE of the optimized system were estimated at $70,042 and $0.278/kWh, respectively. Singh and Fernandez [19] carried out analyses on a Photovoltaic-Wind-Battery hybrid system in Almora, a remote village of India. They used the MATLAB programming environment to implement Cuckoo Search, a new meta-heuristic algorithm for solving the optimization related problem of the system. They found that the NPC and LCOE of the optimized system were Rs 7.69 lakhs and 18.38 Rs/kWh, respectively. Ahmad et al. [20] performed a techno-economic optimization analysis on a wind-photovoltaic-biomass hybrid renewable energy system for rural electrification of Kallar Kahar, a Pakistani village. The researchers used HOMER for their study and found that the optimal system that was made up of a 15 MW PV array, a 15 MW wind farm, and a 20 MW biomass generator, had an NPC of $180,290,247.40 and an LCOE of 0.05744 $/kWh. Kusakana [21] developed the appropriate mathematical model to simulate and optimize a hydrokinetic/diesel/Pumped-Hydro-Storage hybrid system to meet the energy demand of a hypothetical South African village, while considering daily diesel fuel consumption as the objective function to be minimized. They used MATLAB to implement the model that was developed, and found that such a system could help reduce daily operating costs by 88% when compared to a stand-alone diesel generator. Sawle et al. [22] went further by considering simultaneously technical, economic and social performance criteria through a new multi-objective function in the analysis of a HRES integrating wind, PV, biomass, diesel generator, and battery bank for the electrification of remote Indian areas. The authors used evolutionary optimization techniques and found that the PV/Biomass/Diesel/Battery configuration was the most efficient.

Table 1 summarizes all of the studies mentioned above. This summary highlights the fact that most of the HRES-based studies focused on meeting the demand for domestic and community electricity in rural areas and those that take into account the agricultural sector are rare. To the best of the authors' knowledge, such an application has never been performed in the SSA region, where the agricultural sector accounts for an average of 15% of GDP and employs more than 50% of the labor force, particularly in rural areas [23]. Also, the influence of the changes in weather and seasons on electricity consumption has never been taken into account in an HRES-based study in sub-Saharan Africa. Most SSA countries experience two types of seasons: the rainy season and dry season. Furthermore, it is clear from the literature review that the HOMER software is the most preferred tool for HRES analysis and that battery systems are the most used storage devices. However, battery storage systems have some disadvantages that make them less suitable than pumped hydro storage systems for HRES applications in sub-Saharan Africa. They contain lead and sulphuric acid, which entails risks of explosion and environmental degradation, as well as the need for recycling after use [24]. Besides, a study conducted in [25] shows that PHS systems have a lower lifecycle cost (LCC) than batteries. However practical cases of PHS-based HRESs are extremely rare in the literature. Given the advantages of PHS over batteries mentioned above, PHS-based HRESs may be more technically and economically efficient. Therefore, the modelling and analysing of these systems will help to clarify this hypothesis and promote the use of this energy storage technology in HRES systems. Early versions of HOMER did not include any specific PHS component in the storage component library. Recently, HOMER has introduced a generic 245 kWh PHS component. To the best of our knowledge, up to this point, no study has used this component to model a pumped hydro energy storage system. Besides, most of the previous studies that performed sensitivity analysis did not provide practical interpretations of their results. These interpretations are essential to better understand many aspects of HRESs in order to ensure their sustainable promotion for rural electrification.

The present study intends to fill the previously mentioned literature gaps by using HOMER software to analyze a PV/Wind/Biogas/PHS hybrid renewable energy system to meet domestic, community, commercial, and agricultural electricity needs of Djoundé, a remote location of Northern Cameroon.

The remainder of this paper is structured as follows. The next part, Section 2, introduces the materials and methods adopted to carry out the study. The results obtained are presented in Section 3, followed by Section 4, the discussion part. The paper ends with a conclusion in Section 5.

Table 1. Selected studies on hybrid renewable energy systems (HRES) in developing countries.

S No	Authors/Ref.	Country of Application	Energy Sources	Storage Device	Load Type	Study Objectives	Technique/Software
1	Nfah et al. [11]	Cameroon	MHP-SPV-BGS	Battery	DS	Minimizing NPC	HOMER
2	Adaramola et al. [12]	Ghana	WES-SPV-DG	Battery	DS	Minimizing NPC	HOMER
3	Halabi et al. [13]	Malaysia	SPV-DG	Battery	DS	Minimizing NPC	HOMER
4	Singh et al. [14]	India	SPV-WES-BGS	Battery	DS, C	Minimizing NPC	ABC algorithm
5	Sigarchian et al. [15]	Kenya	SPV-WES-BGS	Battery	DS	Minimizing NPC	HOMER
6	Baghdadi et al. [16]	Algeria	WES-SPV-DG	Battery	DS	Maximise the RF	HOMER
7	Ma et al. [17]	China	WES-SPV	PHS	DS, C	Finding a feasible configuration	Mathematical models
8	Kenfack et al. [18]	Cameroon	SPV-MHS	Battery	DS, C	Minimizing NPC	HOMER
9	Singh and Fernandez [19]	India	WES-PV	Battery	DS, C	Minimizing NPC	MATLAB & Cuckoo Search
10	Ahmad et al. [20]	Pakistan	WES-PV-BMS	No SD	DS, C	Minimizing NPC	HOMER
11	Kusakana [21]	South Africa	HKN-DG	PHS		Maximise the RF	MATLAB
12	Sawle et al. [22]	India	WES-PV-BMS-DG	Battery	DS, C	Multi objective	Genetic algorithm

Note: ABC: Artificial Colony Bee; BGS: Biogas Generating System; BMS: Biomass Generating System; C: Commercial; DG: Diesel Generator; DS: Domestic Sector; MHP: Micro Hydropower; PHS: Pumped Hydro System; RF: Renewable fraction; SD: Storage Device; SPV: Solar Photovoltaic; WES: Wind Energy System.

2. Materials and Methods

2.1. Introduction

Two main approaches for analysing and optimizing HRES have been reported in the literature: optimization techniques and software tools. A thorough review of all these methods is given in [26,27]. Software tools that include, among others, HOMER, HIBRID2, and HOGA, have received increasing attention in the literature given the dramatic improvement in computing power of modern computers. The Hybrid Optimization of Multiple Energy Resources (HOMER), initially developed in 1992 at the National Renewable Energy Laboratory (NREL) in the United States of America (USA), appears to be the most widely used tool in light of the literature review above. It is available in two classes of versions: HOMER Legacy (free) and HOMER Pro (commercial), and it can perform simulations, optimizations, and sensitivity analyses. Besides, its library includes a wide range of technologies and components that make it handy for modelling HRES. These merits have justified the choice of HOMER in its latest version, HOMER Pro Version 3.12.1, as an analysis tool. Further information about its operational mode is provided in [28]. Analyses were performed on Windows 10 Pro 64-bit with 2 GHz Intel Core i7 CPU, 8 GB of RAM, and 3 GB GPU.

The block diagram of the adopted research methodology is illustrated in Figure 2. HOMER software was complemented by a pre-HOMER phase, including a detailed assessment of the village load, available resources, and site layout. During this phase, information collected with through surveys, expert opinions, and literature reviews was analysed to obtain data adapted to HOMER in addition to other technical and economic parameters. Detailed information on the methodology is given in the following subsections.

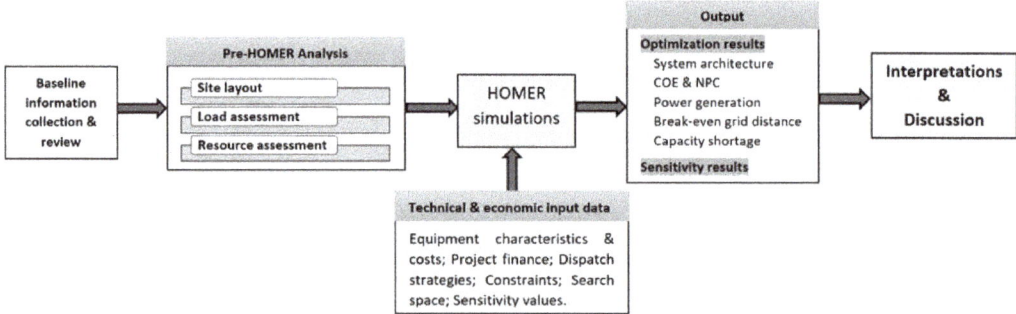

Figure 2. Block diagram of research methodology.

2.2. Study Location

The autonomous hybrid system to be designed was intended to meet the electricity needs of. The village is one of the localities of Cameroon not yet connected to the national electricity grid. Figure 3 shows the geographical situation of the study location, while the related background information is displayed in Table 2.

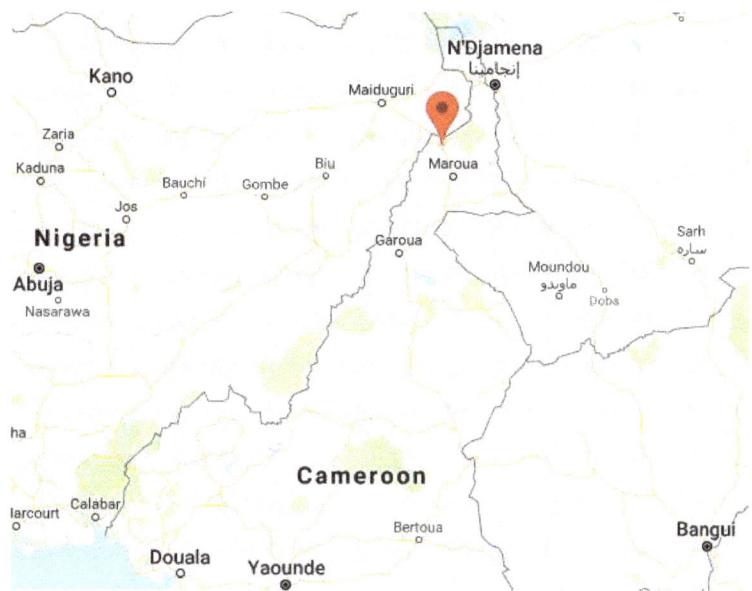

Figure 3. Geographical situation of the study location.

Table 2. General information about the study location.

Particulars	Details
Country	Cameroon
Region	Far North
Division	Mayo-Sava
Name of the municipality	Mora
Latitude	11°03′00″ North
Longitude	14°18′00″ East
Elevation above sea level	100 m
Number of households	180
Nearest power transformer	Mora, 18 km
Main socio-economic activities	Agriculture, small business, and crafts

2.3. Load Assessment

The electricity needs of remote rural areas are generally lower than those of urban areas. In this study, the demand for electrical energy at the study site was assessed on the basis of a survey taking into account the future needs of the village as well as expert opinion and previous cases implemented in Pakistan and India [29,30]. The rating and the number of energy-consuming appliances needed for the 180 households in the village, as well as all other sectors considered are shown in Table 3. The two seasons that prevail in the study areas, namely the rainy season (May to September) and the dry season (October to April), affect the energy consumption of some devices, such as fans and irrigation pumps. In fact, the temperatures in the rainy season are lower than those in the dry season, so that most of the time, the fans are not used during the rainy season. Moreover, thanks to the rain that falls during the rainy season, less water is required for irrigation. Consequently, the hourly power demand of the study location was evaluated separately for the two seasons, as presented in Table 4. The daily electricity demand of the study area during dry and rainy seasons were determined at 381.07 kWh/day and 302.23 kWh/day, respectively, for an annual average of 348.02 kWh/day. The annual electricity demand was evaluated at 127,027 kWh/year. Day-to-day variability of 20% and time-step-to-time-step variability of 15% were considered to make the load profile more realistic. The peak power and load factor were, respectively, 57.88 kW and 0.25. The seasonal load profile is shown in Figure 4.

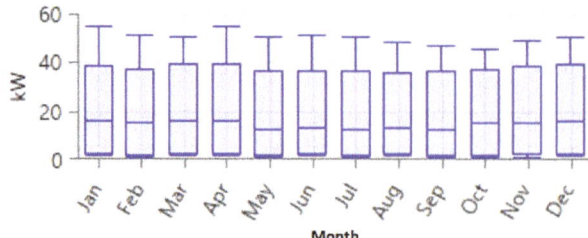

Figure 4. The seasonal load profile of the study location.

Table 3. Appliances' requirement and rating for different sectors of energy consumption.

Load Type	Appliances	Rating (W)	Total Quantity
A-Domestic			
	CFL	15	360
	TV	65	180
	Radio	12	180
	Mobile Charger	12	180
	Fan	40	360
	Water pump	450	9
B-Commercial			
	CFL	15	12
Shops	Fan	40	12
	Refrigerator	500	1
Mini dairy	-	3000	1
Flour mill	-	4800	1
C-Agricultural			
Water irrigation pumps	-	2200	3
Cutting machine	-	1500	2
Threshing machine	-	4000	2
D-Community			
School	CFL	15	20
	Fan	40	2
	CFL	15	5
Health centre	Fan	40	6
	Refrigerator	500	1
Street lights	CFL	100	20

Note: CFL: Compact Fluorescent Lamp.

Table 4. Hourly electricity demand of the study location during dry and rainy seasons.

Hour	Domestic load (kW)				Commercial Load (kW)						Agricultural Load (kW)				School		Community Loads — Health Centre			STL	Electricity Demand (kW)	
	CFL	TV	Radio	Fan DS/RS	WP	MC	CFL	Fan DS/RS	RF	FM	MD	WIP DS/RS	TM	CM	CFL	Fan DS/RS	CFL	Fan DS/RS	RF	CFL	Dry Season	Rainy Season
01:00									0.5								0.075		0.5	2	3.075	3.075
02:00									0.5								0.075		0.5	2	3.075	3.075
03:00									0.5								0.075		0.5	2	3.075	3.075
04:00	5.4		2.2						0.5								0.075		0.5	2	10.675	10.675
05:00	5.4		2.2			2.2			0.5		3						0.075		0.5	2	15.875	15.875
06:00			2.2			2.2			0.5		3			3			0.075		0.5		11.475	11.475
07:00						2.2			0.5		3			3	0.6	0.8/0	0.075		0.5		10.675	9.875
08:00									0.5					3	0.6	0.8/0			0.5		5.4	4.6
09:00									0.5						0.6	0.8/0			0.5		2.4	1.6
10:00									0.5						0.6	0.8/0			0.5		2.4	1.6
11:00				7.2/0					0.5	4.8					0.6	0.8/0		0.24/0	0.5		14.64	6.4
12:00			2.2	7.2/0				0.48/0	0.5	4.8					0.6	0.8/0		0.24/0	0.5		17.32	8.6
13:00			2.2	7.2/0				0.48/0	0.5			6.6/6.6			0.6	0.8/0		0.24/0	0.5		19.12	10.4
14:00		11.7	2.2	7.2/0				0.48/0	0.5			6.6/0			0.6	0.8/0		0.24/0	0.5		30.82	15.5
15:00		11.7		7.2/0				0.48/0	0.5				8		0.6	0.8/0		0.24/0	0.5		30.02	21.3
16:00		11.7		7.2/0				0.48/0	0.5				8					0.24/0	0.5		28.62	20.7
17:00		11.7		7.2/0				0.48/0	0.5									0.24/0	0.5		20.62	12.7
18:00	5.4	11.7		7.2/0			0.18	0.48/0	0.5								0.075		0.5	2	28.035	20.355
19:00	5.4	11.7	2.2		9		0.18	0.48/0	0.5								0.075		0.5	2	32.035	31.555
20:00	5.4	11.7	2.2		9		0.18	0.48/0	0.5								0.075		0.5	2	32.035	31.555
21:00	5.4	11.7	2.2		9		0.18	0.48/0	0.5								0.075		0.5	2	32.035	31.555
22:00	5.4	11.7					0.18	0.48/0	0.5								0.075		0.5	2	20.835	20.355
23:00							0.18	0.48/0	0.5								0.075		0.5	2	3.735	3.255
00:00									0.5								0.075		0.5	2	3.075	3.075

Note: CFL: Compact Fluorescent Lamp; CM: Cutting machine; DS: Dry season; FM: Flour mill; MC: Mobile charger; MD: Mini dairy; RF: Refrigerator; RS: Rainy season; STL: Street lights; TM: Treshing machine; TV: Television; WIP: Water irrigation pump; WP: Water pump; WS: Wet season.

2.4. System Configuration

The schematic diagram of the system involved in this study is presented in Figure 5. The system is equipped with three power generators (PV array, biogas generator, and wind turbine); an energy storage device, the pumped hydro storage (PHS); a converter system; a control station; and, a load.

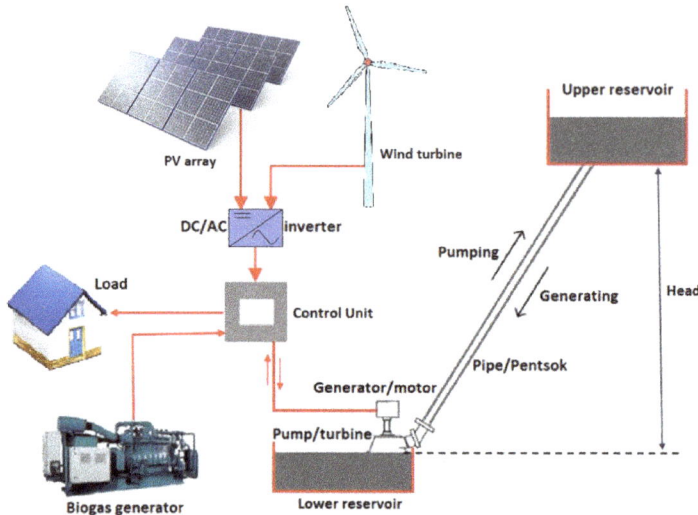

Figure 5. The schematic diagram of the hybrid renewable energy system.

The PHS stores excess electricity from intermittent sources in the system (PV and wind) for use during periods of insufficient generation to meet the demand for electrical load. The operating principle of the pumped hydro system can be explained briefly, as follows. During a period of excess energy supply, the surplus wind or PV power is used to pump and raise water from the lower reservoir to the upper reservoir. Later, when a supply-demand imbalance occurs, the stored water is allowed to return to the lower reservoir, thus enabling electricity production through a turbine/generator unit [31]. The biogas generator is used as a backup power source to be activated in case of insufficient combined production from PV array, wind turbine and PHS. The proposed system as implemented by the HOMER software is illustrated in Figure 6. The pumped hydro storage, which is in fact an AC component, is connected to the DC bus in HOMER. Indeed, HOMER models the PHS component as a special battery with an initial and minimum state of charge of 100% and 0%, respectively. The consequence of this modelling is the need for a rectifier in the HOMER model, although this is not necessary in the actual configuration of the system. The aim of the rectifier in HOMER's model is to convert the surplus AC output from the biogas generator to DC current to be stored by the storage device in case of cycle charging dispatch strategy. The system is an off-grid system, i.e., it is not connected to the grid. The grid component on the HOMER's schematic presentation was introduced for the purpose of comparing the proposed autonomous system with the grid extension.

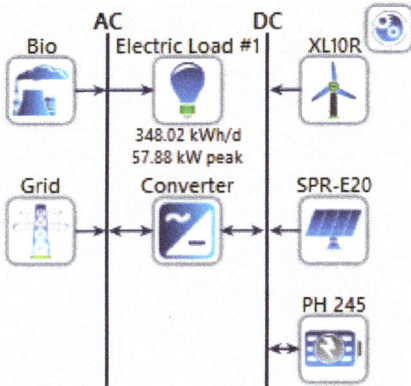

Figure 6. The proposed hybrid system in Hybrid Optimization of Multiple Energy Resources (HOMER).

2.5. Assessment of Available Energy Resources

2.5.1. Available Solar and Wind Resources

Data on solar radiation and wind speed in Djounde, the study location, were collected from the NASA Surface meteorology and Solar Energy (SSE) database [32] at the coordinates of Mora (11°03'00" N, 14°18'00" E), the closest data location. The annual average solar radiation was found to be 5.82 kWh/m^2/day, while the average clearness index was 0.6. The maximum solar irradiation, 6.67 kWh/m^2/day, is that of March whereas the minimum, 4.77 kWh/m^2/day, is that of August. Figure 7 displays the monthly daily average solar radiation and the clearness index of the selected location. It was generated by HOMER after SSE data entry. It highlights the high potential of the area for solar energy that can be used to generate electricity through PV arrays.

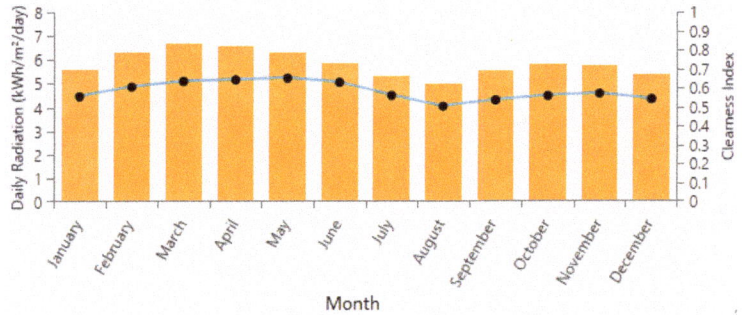

Figure 7. The monthly daily average solar radiation and clearness index for Djoundé.

Monthly data on mean wind speed are shown in Figure 8. The average annual wind speed at the site was 4.95 m/s at the 50 m anemometer. A study by Kidmo et al. [33] showed that the high altitude above the sea level makes the site fruitful for wind power production, despite the relatively low wind speed recorded.

Figure 8. The average monthly wind speed measured at 50 m in Djoundé.

2.5.2. Biomass Resources

Biomass refers to all organic materials that can be converted into energy. It includes both materials of plant origin (agricultural residues, leaves, wood) and those of animal origin (animal and human wastes, living beings of the soil, animal corpses). In this study, livestock manure was considered to be the only biomass resource for power generation. The data filled in HOMER was the average daily manure available for biogas production. Based on the survey conducted in the study area, the total livestock population was 811, consisting of cows (213), horses (12), mules (29), and goats (557). Table 5 provides a detailed assessment of the potential for generating biogas and electricity from on-site livestock manure. The evaluation of the biogas was carried out on the basis of manure yield. For that, we assumed a daily manure production of 10 kg/day for any cow/mule/horse and 1 kg /day for any goat [34]. When considering a recovery factor of 0.7 and a "gas yield per kg of wet manure" of 0.036 m^3/kg [35], the total biogas yield was determined at 78 m^3/day. Finally, given that 0.73 m^3 of biogas is needed to produce 1 kWh of electricity [36], the total potential for producing electricity from the livestock manure produced on site was 107 kWh/day. As inputs for biomass resources, HOMER requires the cost of biomass, gasification ratio (GR), and lower heating value (LHV). The gasification ratio indicates the amount of biogas produced per unit mass of biomass, while the LHV is the amount of energy contained in 1 kg of biogas available to feed the biogas generator. In this study, the GR and LHV were 0.05 kg/kg and 5.5 MJ/kg, respectively [37]. The cost of biomass was set at 0 €/t.

Table 5. Electricity potential from biomass of the study area.

Livestock	Population	Dung Availability (kg/head/day)	Total Dung (kg/day)	Total Dung (Recovery Factor = 0.70)	Total Gas Yield (m^3/day)	Potential Power Yield (kWh/day)
Cows	213	10	2130	1491	53.7	74
Horse	12	10	120	84	3	4
Mule	29	10	290	203	7,3	10
Goat	557	1	557	390	14	19
Total	811	-	3097	2168	78	107

2.6. System Analysis

2.6.1. PV Array

The model of PV solar module adopted in this study was SPR-E20-327, a monocrystalline module manufactured by Sunpower. The module has a rated power of 327 W$_p$ and it can produce a maximum voltage of 600 V DC. Table 6 presents the technical specifications of the selected PV module. The following equation is used by HOMER to calculate the output of a PV array [38]:

$$P_{output} = Y_{PV} f_{PV} \left(\frac{G_T}{G_{T,STC}} \right) [1 + \alpha_P (T_c - T_{c,STC})],$$ (1)

where, f_{PV} is the PV derating factor (%), Y_{PV} the rated capacity of the PV array (kW), G_T the global solar radiation incident on the surface of the PV array (kW/m^2), and $G_{T,STC} = 1$ kW/m^2 is the standard amount of incident radiation at the standard test condition (25 °C), α_p is the temperature coefficient of power (%/°C), T_c is the PV cell temperature, and $T_{c,STC}$ is the PV cell temperature under the standard test condition. HOMER uses the Graham and Hollands algorithm to generate hourly global solar radiation from the monthly average global solar radiation [39]. The temperature coefficient of power for the selected module is −0.38%/°C, as shown in Table 6 of the technical specifications of the said PV module. A derating factor of 0.95 was considered in this study. The lifespan of the PV generator is assumed to be 25 years. The total capital cost, replacement cost, and operation and maintenance cost for the PV installation were estimated at 3000 €/kW, 3000 €/kW, and 10 €/kW/year [40].

Table 6. PV module specifications [41].

Item	Specification
Manufacturer	Sunpower
PV Module type	Mono-si
Module number	SPR-E20-327-C-AC
Module efficiency	20.4%
Power capacity	327 W
Power tolerance	+5/−0%
Rated voltage (Vmpp)	54.7 V
Rated current (Impp)	5.98 A
Open-Circuit Voltage (VoC)	64.9 V
Short-Circuit Current (ISC)	6.46 A
Maximum system voltage	DC 600 V
Power Temp Coef	−0.38%/°C
Volt Tem coef	−175 mV/°C
Current Temp Coef	3.5 mA/°C
Dimensions	46 mm × 1559 mm × 1046 mm
Operating temperature	−40 °C to +85 °C
Area	1.63 m^2
Weight	18.60 kg

2.6.2. Wind Turbine

The wind turbine model that is considered for this study is the bergey excel 10-R model, manufactured by Bergey Windpower. Its nominal power is 10 kW at 12 m/s. The technical specifications of the turbine are presented in Table 7, while its power curve is shown in Figure 9. The power law was used to calculate wind speed at the hub height [12]:

$$U_{hub} = U_{anem} \cdot \left(\frac{Z_{hub}}{Z_{anem}} \right)^{\alpha}, \tag{2}$$

where, U_{hub} and U_{anem} are the wind speeds at the hub and anemometer height (Z_{hub}, and Z_{anem}), and α is the power law exponent whose the typical value for low roughness site is 0.14 [15]. The total initial cost, replacement cost, and operation and maintenance cost were estimated at 50,000 €/unit, 30,000 €/unit, and 200 €/year [40]. The turbine lifespan was assumed to be 20 years.

Table 7. Technical specifications of the selected wind turbine model [42].

Item	Specification
Manufacturer	Bergey WindPower
Model	Bergey excel 10-R
Nominal power	10 kW at 12 m/s
Cut-in Wind Speed	2.5 m/s
Cut-Out Wind Speed	None
Furling Wind Speed	14–20 m/s
Max. Design Wind Speed	60 m/s
Temperature range	−40 to + 60 °C
Hub height	30 m
Type	3 Blade Upwind

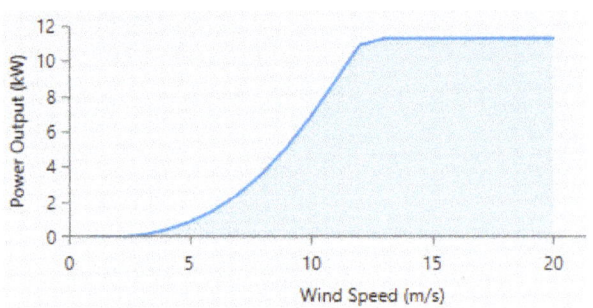

Figure 9. Power curve of the selected wind turbine.

2.6.3. Biogas Generator

A generic biogas generator connected to an AC output is considered for this study. HOMER takes into account the available biogas when sizing the generator. The capital cost, replacement cost, and maintenance costs of a 1-kW biogas generator were set at €1500, €1200, and €0.1/h [15], respectively. The generator lifespan was set at 15,000 h of operation. The minimum load ratio was assumed to be 30% of the capacity. The generator was off for two hours (from 06:00 to 09:00) every weekend for maintenance operations, as shown in Figure 10.

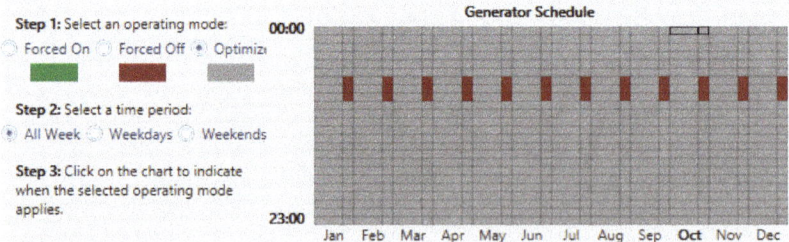

Figure 10. The generator schedule.

2.6.4. Pumped-Hydro Storage

The generic 245 kWh pumped-hydro component recently introduced into the HOMER component library was used to model the PHS station in this study. The system can store up to 1000 m^3 of water dischargeable over a 12-h period [43], resulting in a discharge flow rate (Q) of 0.0231 m^3/s and a capacity power (P_C) of 20.4 kW. Moreover, the head height of the system (H) was determined at 100 m from the following relationship [17]:

$$P_C = \rho g Q H \eta, \tag{3}$$

where: P_C = power capacity (2,0400 W); ρ= mass density of water (1000 kg/m^3); g = acceleration due to gravity (9.8 m/s^2); Q = discharge flow rate (0.0231 m^3/s); and, η = PHS efficiency when discharging (assumed at 90%).

The costs of pumped-hydro energy storage systems are provided by the Electricity Storage Association and range from €440/kW to €1320/kW [44]. Accordingly, the initial cost of the system can range between €9000 and €27,000. For this study, the initial cost, replacement cost, and O&M cost were taken as €15,000, €10,000, and €300/year. The lifespan of the system was assumed to be 25 years.

2.6.5. Converter

A power converter system is necessary to ensure the continuity of energy flow between the DC and AC electrical components of the system. In this study, a generic system converter comprising an inverter and a rectifier to perform a bidirectional AC-DC conversion was considered. The capital cost, replacement and O&M costs for 1 KW were assumed to be €650, €600, and €0, respectively [15]. The inverter efficiency is considered to be 95%, while the rectifier efficiency is set at 100% to take into account the fact that a rectifier, although required in HOMER implementation, is non-existent in the real system. A converter lifespan of 15 years was considered.

2.7. Simulation and Optimization

2.7.1. The Assessment Criteria

The HOMER's assessment criteria considered for optimal system design is the Total Net Cost (NPC). It is the performance criterion by which all feasible system configurations are ranked in the optimization results. The NPC is the sum of all discounted values of costs and revenues related to the system. In the present case, the costs included the upfront costs, replacement costs, operating costs, and fuel cost, while the revenues involved salvage values of the system components. The NPC of an HRES is expressed, as follows [38]:

$$\text{NPC} = \frac{C_{ann,tot}}{CRF(d, N)}, \tag{4}$$

where $C_{ann,tot}$ = total annualized cost of the system, CRF= Capital recovery factor, N = project lifespan, and d = discount rate.

The total annualized cost of the system is given as:

$$C_{ann,tot} = C_{ann,cap} + C_{ann,rep} + C_{ann,O\&M} + C_{ann,fuel} - R_{ann,salv} \tag{5}$$

where $C_{ann,cap}$, $C_{ann,rep}$, and $C_{ann,O\&M}$ are, respectively, the annualized capital, replacement, and maintenance costs of all components of the system; $C_{ann,fuel}$ is the annualized cost of fuels used to feed the power generators, and $R_{ann,salv}$ represents the annualized total savage value of all system components. A Capital Recovery Factor (CRF) converts a present value into a uniform annual cash flow series over the project lifespan (N) at a specified discount rate (d). CRF formula is given as:

$$CRF = \frac{d(1+d)^N}{(1+d)^N - 1}. \tag{6}$$

In this study, the life of the project was set at 20 years. The discount rate is calculated on the basis of the following equation [15]:

$$d = \frac{i - f}{1 + f}, \tag{7}$$

where f is the annual inflation and i the nominal interest rate. The annual inflation rate and nominal interest rate were considered at 3% and 8%, respectively.

Besides the NPC, HOMER also calculates the levelized cost of energy (COE). This is the average cost of production by the system of one kWh of electricity. COE is expressed as the ratio of the total annualized cost of the system to the total annual useful electricity output of the system. The formula of COE is given as [34]:

$$\text{COE} = \frac{C_{ann,tot} - C_{boiler} \cdot H_{thermal}}{E_{served}} \qquad (8)$$

where $C_{ann,tot}$ = the total annualized cost of the system (€/year), C_{boiler} = boiler marginal cost (€/kWh), H_{served} = total thermal load served (kWh/year), and E_{served} = total electrical load served (kWh/year). In this study, thermal load is not served, hence $H_{served} = 0$.

2.7.2. Dispatch Strategy

A dispatch strategy is a set of rules governing the operation of the generator(s) and the storage device(s). HOMER software can model two dispatch strategies: load following (LF) and cycle charging (CC) [38]. Under the load following strategy, whenever a generator is activated, it only produces the power that is needed to meet the demand; while under the cycle charging strategy, each time a generator is turned on, it runs at full capacity, the surplus power being stored by the power storage device. Both LF and CC strategies were considered in this study.

2.7.3. Optimization Variables and Search Space

An optimization variable, also referred to as decision variable, is a variable that can be controlled by the system designer and for which HOMER can take into account several possible values in its optimization process. Table 8 displays the optimization variables involved in this study and related values to each one. They include the size of the PV array (seven values), the number of wind turbines (10 values), the number of pumped hydro storage stations (six values), the size of the biogas generator (eight values), and the size of the bidirectional converter (nine values). The search space is the set of all possible system configurations in which HOMER searches for the optimal solution. For this study, each combination of the five optimizations variables was simulated for each of the two dispatch strategies considered (LF and CC). The number of configurations simulated by the HOMER software was therefore $7 \times 10 \times 6 \times 8 \times 9 \times 2 = 60,480$ configurations.

Table 8. Optimization variables of the study model.

Optimization variable	PV Array Size (kW)	Number of WT	Number of Pumped Hydro Storage (PHS) (number)	Biogas Generator Size (kW)	Converter Capacity (kW)
Maximum	98.1	9	5	17.5	80
Minimum	0	0	0	0	0
Step	16.35	1	1	2.5	10

2.7.4. Constraints

Constraints are conditions that system configurations must satisfy. A configuration that does not meet one of the specified constraints is considered an unfeasible configuration and is not ranked by HOMER after the simulation and optimization process. Two types of constraints are considered in the present case study:

- The constraint that is related to the capacity shortage is defined by the maximum annual capacity shortage, which was set at 5% in this study. This means that HOMER discarded any system that did not meet at least 95% of the annual electrical load plus the operating reserve.
- Constraints related to the operating reserve are those that impose excess operating capacity to ensure the reliability of the system in the event of a sudden increase in load or a reduction in

renewable energy production. HOMER defines the required operating reserve using four inputs, two of which are as a percentage of the variability of the electricity load: load in current time step and annual peak load; and two as a percentage of renewable energy production: solar power output and wind power output. In this case study, the operating reserve percentages that are associated with the load in current time step, annual peak load, solar power output, and wind power wind output were set at 10%, 15%, 20%, and 50%, respectively.

2.8. Sensitivity Analysis

Sensitivity analysis aims at dealing with uncertainty by investigating the effects of changes in specific parameters on the performance of the system. A parameter is any HOMER's numerical input data that is not a decision variable. Parameters that are involved in a sensitivity analysis are sometimes referred to as sensitivity variables. For each sensitivity variable, a range of values, sensitivity values, is entered into HOMER by the designer. The sensitivity variables considered in this study included wind speed, solar radiation, biomass price, biomass availability, and maximum capacity shortage. All of these sensitivity variables and their associated values are listed in Table 9.

Table 9. Sensitivity variables and associated values.

Sensitivity Variable	Values
Wind speed (m/s)	3.5, 4.95, 8
Solar radiation (kWh/m^2/day)	3.8, 5.82, 7, 8
Capital cost multiplier of PV	0.5, 1, 1.5, 2
Capital cost multiplier of PHS	0.5, 1, 1.5, 2
Biomass price (€/t)	0, 0.2, 0.4, 0.6
Biomass availability (t/day)	0.2, 2.2, 4.5, 7, 9.5
Maximum capacity shortage (%)	0, 2.5, 5, 7.5, 10, 12.5

For speeding up purposes, the sensitivity analyzes were performed sequentially, instead of running them all at the same time. Sensitivity analysis of wind speed and solar radiation was first performed to account for the wide variability in the availability of these two resources, and hence better understand the performance of the proposed system across the sub-Saharan region. The sensitivity analysis of the availability and cost of biomass resources was then conducted to determine the viability of their transport or purchase in the event of unavailability or insufficient production at the site. This was followed by the sensibility analysis of the capital costs of PV and PHS, using cost multipliers, with three objectives: (1) to understand how PV-based HRESs are promising in taking up rural electrification challenges in SSA; (2) to assess the impact of the adoption of PV investment subsidies on the viability of HRESs; and, (3) to evaluate the effect of change in pumped hydro investment cost due site morphology. Finally, the sensitivity analysis of the maximum capacity shortage was run to better manage the trade-off between the reliability and cost of the proposed system.

2.9. The grid Extension

As mentioned above, grid extension is one of the most common rural electrification solutions in sub-Saharan Africa. In this study, it was considered to be an alternative to the proposed off-grid hybrid system. HOMER software when compared both methods by calculating the break-even grid extension distance (BGED) which is the distance from the grid to which the NPCs of the grid extension and the optimized off-grid system are equal. Beyond this distance, the off-grid system is preferable, while closer to the grid, the grid extension is the best solution. The required input parameters for performing the BGED calculation in HOMER are the capital cost per km, annual O&M cost per km, and the grid power price.

The capital cost per km and the O&M cost of the grid extension were, respectively, estimated at €10.000/km and €200/year/km in [18,45]. Given inflation, they were taken at €14.000/km and €300/year/km in this analysis. The average price of electricity from the grid in Cameroon is 0.1 €/kWh.

3. Results

The results of the analyses described above are presented in this section. The optimization results are first analyzed, followed by the description of the sensitivity analyses outcomes.

3.1. Optimization Results

The results of HOMER simulation and optimization processes showed that among the 60.480 system configurations of the HOMER search space, only 11,560 were feasible and classified according to the system architecture in five categories, namely: category 1 (PV/Biogas/PHS), category 2 (PV/Wind/Biogas/PHS), category 3 (PV/PHS), category 4 (PV/Wind/PHS), and category 5 (Wind/Biogas/PHS). The details of the components, as well as the technical and economic specifications of the best hybrid system in each category, are presented in Table 10. The best hybrid system in Category 1, which was the overall optimal hybrid system, was made up of an 81.8 kW PV array, a 15 kW biogas generator, two 245 kWh pumped hydro storage stations, and a 40 kW bi-directional converter with a dispatch strategy of load following. No wind turbine is included in that configuration. Its cost of energy (COE) and total net present cost (NPC) were €0.256/kWh and €370 426 respectively.

The breakdown by component and cost type of this NPC, as presented in Figure 11, shows that it was 87% dominated by the total capital cost of the system. The PV array was the most important component in terms of costs and accounted for 76% of the total capital cost and 66% of the NPC of the system. During the life of the project, only the biogas generator and system converter were replaced for a total replacement cost of €28,066. For this configuration, the total annual electricity production was 159,840 kWh/year, 89% dominated by the production of photovoltaic panels.

Figure 12, which displays the monthly distribution of electrical generation shows that the power generation from biogas generator was higher during the dry season (October to April) than the rainy season. The excess energy and the unmet load for that configuration were, respectively, 13,670 kWh/year and 1972 kWh/year, i.e., 8.6% and 1.6% of total production. An excess electricity is an unused and dumped power once the electrical load demand is met and the upper reservoir of the PHS station is full, while an unmet load is a load that cannot be met due to a gap between the total electrical demand and the total electrical generating capacity. The system's capacity shortage was 6071 kWh/year, i.e., 4.8% of the demand load, which is less than the maximum shortage capacity of 5%, specified as a constraint.

The PV array output throughout the year, shown in Figure 13, reveals that the PV power production took place between 06:00 and 18:00 and it was more likely to reach its maximum (75.7 kW) between 10:00 and 14:00. Furthermore, the total annual PV electrical production was 141,046 kWh/year, which corresponded to a capacity factor of the system of 19.7%.

Figure 14 shows the performance of the biogas generator throughout the year. During the rainy season (May to September), it was more likely that the generator was switched on between 18:00 and midnight, while, during the dry season (October to April), it was from 14:00 to midnight. For both seasons, the biogas generator was likely to deliver its maximal electrical output (14.8 kW) between 18:00 and 00:00. The annual power production from biogas generator was 17,794 kWh/year, representing a capacity factor of the system of 13.5%.

Table 10. Optimization results.

Specification Category	Specification	Unit	Best Hybrid System Per Category				
			Category 1	Category 2	Category 3	Category 4	Category 5
System architecture	PV array 5SPR-E20	kW	81.8	65.4	98.1	81.8	0
	Wind turbine (XL10R)	Number	0	1	0	1	8
	Biogas gen.	kW	15	15	0	0	12.5
	Pumped Hydro (PH 245)	Number	2	2	3	3	4
	Converter	kW	40	40	50	50	70
	Dispatch strategy	LF or CC	LF	LF	CC	CC	CC
Cost	LCOE	€/kWh	0.256	0.260	0.261	0.265	0.417
	NPC	€	370,426	375,945	379,257	383,757	598,368
	Total O&M cost	€/year	4031	4425	644	950	6400
	Total capital cost	€	323,750	324,700	371,800	372,750	524,250
Power production	PV array	kWh/year	141,046	112,837	169,255	141,046	152,249
	Wind turbine	kWh/year	0	19,031	0	19,031	17,323
	Biogas Generator	kWh/year	17,794	18,508	0	0	0
	Total electricity production	kWh/year	158,840	150,376	169,255	160,077	169,572
	Primary load consumption	kWh/year	125,056	124,916	125,225	124,888	123,901
	Capacity shortage	kWh/year (%)	6.071 (4.8)	6430 (5)	4756 (3.7)	6269 (4.9)	6377 (5)
	Unmet load	kWh/year (%)	1972 (1.6)	2112 (1.7)	1503 (1.2)	2140 (1.7)	3127 (2.4)
	Excess electricity	kWh/year (%)	13,670 (8.6)	7158 (4.8)	19,249 (11.4)	12,532 (7.8)	27,693 (15.9)
Capacity factor	PV array 5SPR-E20	%	19.7	19.7	19.7	19.7	0
	Wind turbine (XL10R)	%	0	21.7	0	21.7	21.7
	Biogas gen.	%	13.5	14.1	0	0	18.8

The results of our model also show that the PHS total input and output power were respectively 78,501 kWh/year and 64,050 kWh/year, the difference resulting from the system losses (14,968 kWh/year) and the depletion of storage (508 kWh/year). This value of losses correspond to a conversion efficiency of 20%, which is in the 65%–80% range of round-trip energy efficiency of pumped-hydro storage systems [24]. Losses in pumped-hydro storage systems are mainly made up of pipe friction losses and pump/turbine unit losses [46].

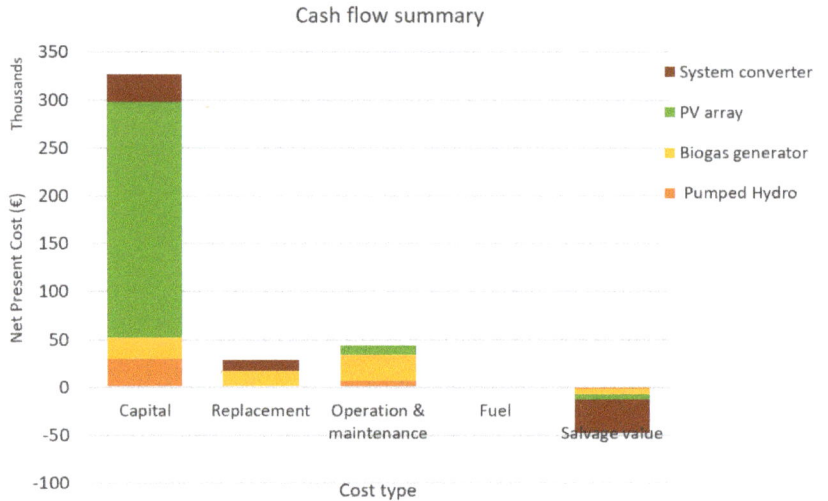

Figure 11. Cash flow summary based on the optimised architecture.

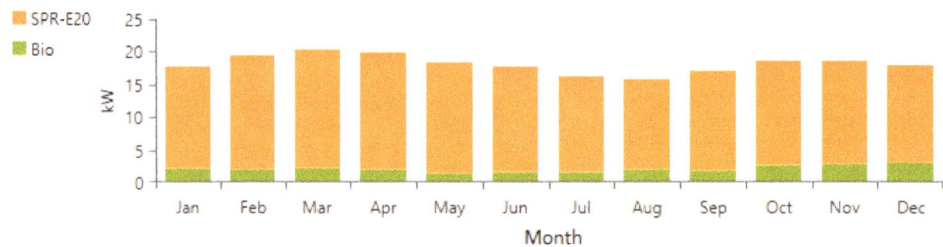

Figure 12. Monthly average electrical output from the optimal configuration system.

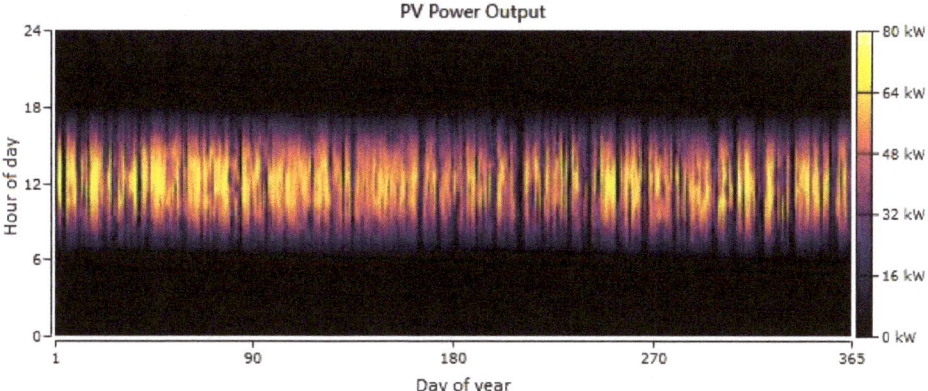

Figure 13. The PV array output.

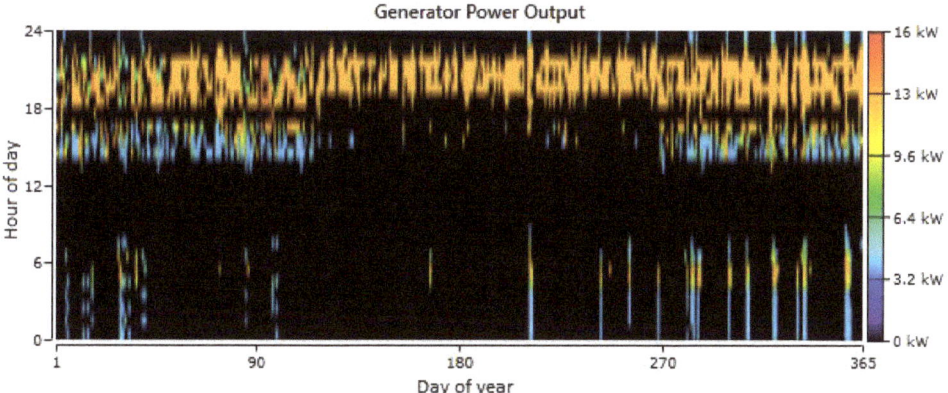

Figure 14. Biogas generator output.

Figure 15, which displays the state of charge of the pumped hydro storage station of the present case, shows that the upper reservoir was relatively more filled during the rainy season than the dry season. This was due to the relatively higher demand for electrical energy during the dry season than during the rainy season, which implied a higher probability of occurrence of unmet load during the dry season.

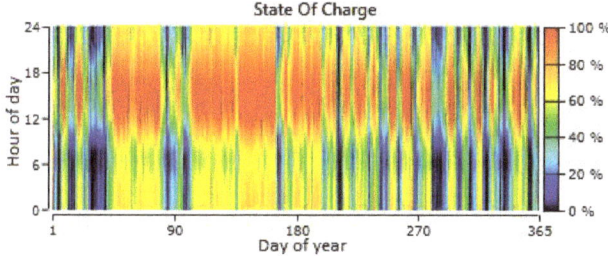

Figure 15. State of charge of the Pumped-hydro storage station.

Figure 16 shows the operating schedules and energy flow of optimized system components over a 72-hour period (29, 30 and 31 December) during the dry season. During the day (8:00–17:00) due to the intense sunlight, the PV output was high and intended to meet the load demand and charge the PHS station. From 8:00 to 12:00, the pumped hydro charge power was likely to peak due to low load demand, and excess power production was likely to occur, as was the case on 29 December. From 12:00 to 18:00, the PV array output might become insufficient to meet the demand load and charge the PHS station, requiring the activation of the biogas generator to fill the gap, as was the case on 29, 30 and 31 December. During the night and early in the morning (17:00–8:00), because of the absence of sunshine, the PV array output was zero so that the demand for electrical energy was mainly satisfied by the power output from the PHS station. However, from 18:00 to 0:00, after peaking, this power became insufficient to cover the load demand, requiring the activation of biogas generator to fill the gap. The latter might in turn also peak without filling the gap for which it had been activated, resulting in an unmet load, as was the case on 30 and 31 December From 0:00 to 8:00, the low load demand was exclusively satisfied by the PHS output.

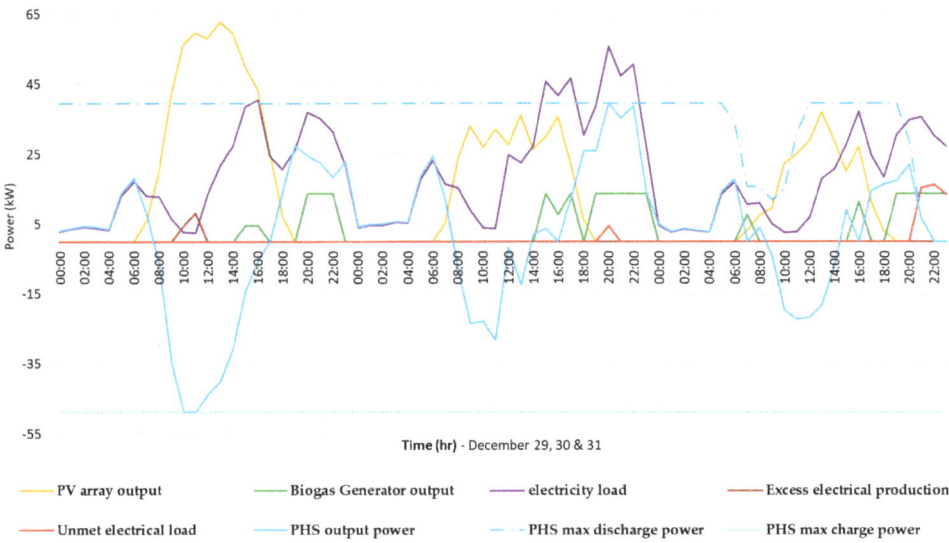

Figure 16. Operations schedules and energy flow of the system components over a 72-hour period.

Figure 17 shows the result of the comparison between the NPCs of the autonomous system designed and the grid extension for the purpose of electrification of Djoundé. This indicates a break-even grid extension distance of 12.78 km, which led to the conclusion that the system designed was the better solution for Djoundé's electrification, with the nearest power transformer being located in Mora, 18 km away.

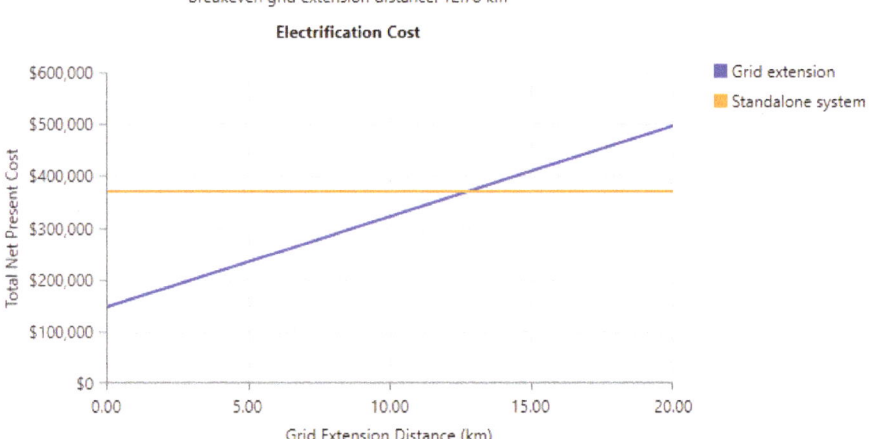

Figure 17. Cost of electrification options of Djoundé.

3.2. Sensitivity Results

The sensitivity result for wind speed and solar radiation is presented in Figure 18. It shows that, for low wind speed values, the optimal system type would be PV/Biogas/PHS; hence, the COE and NPC of the system would not be sensitive to wind speed variation, as no wind turbine would be part of the system. However, for each solar radiation value, increasing, the wind speed would reach a threshold value above which the optimal system type would be PV/Wind/Biogas/PHS. By increasing further above that threshold value, the wind speed would reach another threshold value, above which the optimal system type would be Wind/Biogas/PHS. The higher the solar radiation, the higher the two wind speed threshold values mentioned above. For all wind speed values that are below 6m/s, the PV array would be part of the optimal-system components, and the higher the solar radiation, the lower the NPC and COE. For very high wind speed values, the PV array would not be part of the system; hence the system performances were not sensitive to variations in solar radiation. Therefore, if only the changes in solar radiation and wind speed were taken into account, the optimal system type would be PV/Biogas/PHS and it would have COEs greater than €0.3/kWh in parts of SSA such as Gabon, Equatorial Guinea, Southern Cameroon, Southern Nigeria and Congo, which experience average solar radiations below 5 kWh/day/m^2, and wind speed less than 4.5 m/s [47,48]. The optimal system type would remain PV/Biogas/PHS, but COEs would be reduced to less than 2.5 €/kWh in some such places as Northern Chad and Northern Niger. The wind turbines would be part of the optimal system in the places with higher wind speed, which would help to reduce the COE up to 2 €/kWh in the regions such as the East African and South African coasts that experience wind speed greater than 7 m/s.

Figure 19 presents the result of the sensitivity analysis of the capital costs of PV and PHS systems. The capital cost of PHS showed a less potential impact on the optimal system type and COE. For example, doubling it would not change the optimal system type and it would increase its COE by only 8%. On the other hand, for a 50% reduction in capital cost of PV, the optimal system would change from PV/Biogas/PHS type to PV/PHS type, and its COE would decrease by 37%. For a 50% increase in the PV capital cost, wind turbines would be integrated into the optimal system of which the COE would increase by 29%. Therefore, PV-based HRESs are promising for addressing the challenges of rural electrification in sub-Saharan Africa Given that PV capital cost is expected to decrease by 50% by 2040 [49]. The result of this analysis also highlighted the relevance of government subsidies to the investment costs of photovoltaic technology for PV-HRESs to be viable. On the other hand, the morphology of the site, although having a significant impact on the investment cost of PHS systems, does not affect the viability of the latter as an energy storage device for HRESs in the region.

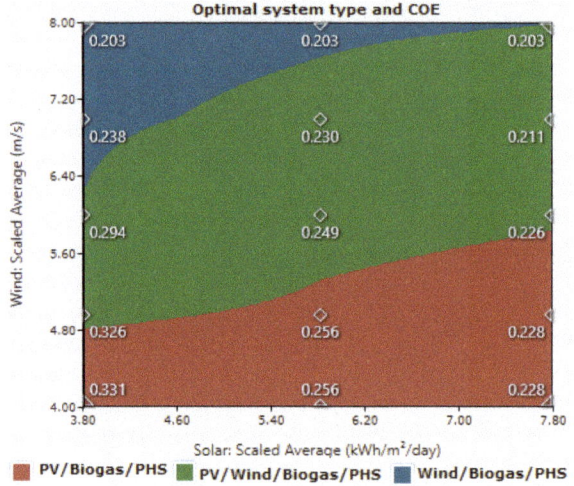

Figure 18. Result of sensitivity analysis of wind speed and solar radiation.

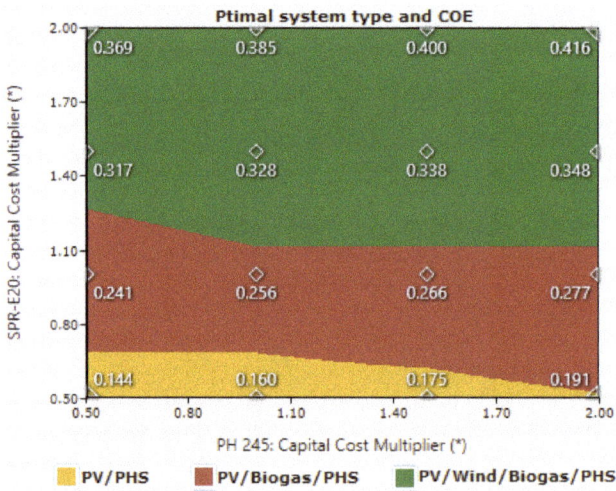

Figure 19. Sensitivity analysis result of PV and PHS capital costs.

Figure 20 presents the result of the sensitivity analysis of the availability and cost of biomass resources. It shows that in the case of deficient biomass resource production, the use of the biogas generator as the backup engine would be infeasible. However, increased availability above the calculated value would not have a significant impact on the system performance. For example, doubling the value of the base scenario would result in only a 0.2% decrease in the system's COE. On the other hand, the result shows that, if the biomass availability value of the base scenario is considered, the biogas generator would no longer be viable if the biomass price was greater than €0.13/t. Thus, in the event of unavailability at the site, the transport or the purchase of biomass resource would not be viable. However, that unavailability would only increase the COE by 2%.

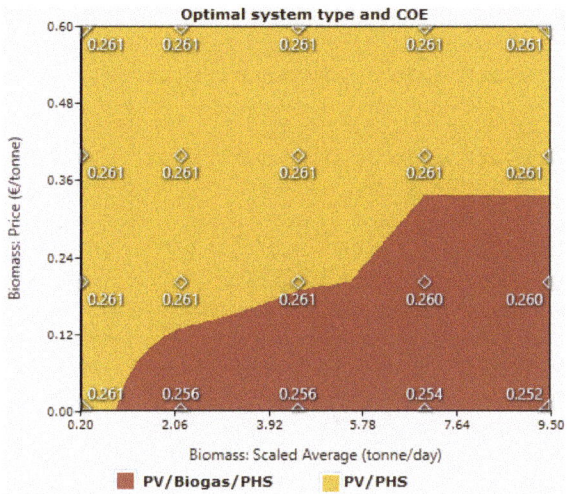

Figure 20. Sensitivity analysis result of the price and availability of biomass.

The result of the sensitivity analysis of maximum capacity shortage is illustrated in Figure 21. It shows that an improvement of the reliability of the system by lowering the maximum annual capacity shortage from 5% to 2.5% would result in NPC and COE increases of only 0.6% and 0.4%, respectively, while an improvement to 0% would result in increases of 9% and 7%, respectively. On the other hand, the degradation of the system reliability by setting its maximum annual capacity shortage at 10% would decrease its NPC and COE by 4% and 3%, respectively. This result reveals that the system could achieve better reliability without substantially increasing the COE.

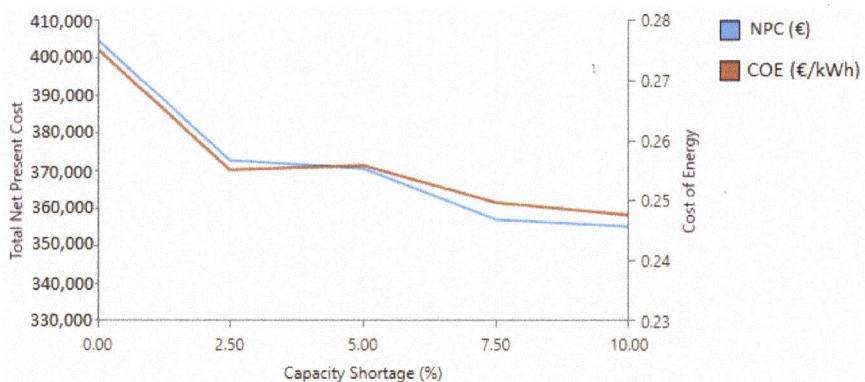

Figure 21. Sensitivity analysis result of the maximum capacity shortage.

4. Discussion

The results of the analyses that are presented above clearly show that the proposed system could help to meet the demand for electricity in a remote village at a lower COE than in any of the previous cases in sub-Saharan Africa cited in the literature review of this article [11,12,15,18]. Each of these cases had a battery system as an energy storage device, which demonstrates to the point that PHS-based hybrid renewable energy systems are technically and economically better than battery-based systems, and confirms the hypothesis defined in the introductory part. Moreover, unlike previous studies, the analysis of the system was preceded by a thorough assessment of electricity demand taking into account all electricity consumption sectors, including agriculture, the primary sector of employment in rural areas of SSA. This implies a more realistic and reliable assessment of load demand, which advances the literature on HRES applications in sub-Saharan Africa.

At the global level, the novelty of this study lies on two points. First, the authors have successfully modelled, simulated, and optimized a PHS based HRES using the 245 kWh PHS component recently introduced by HOMER. The roundtrip efficiency of the modelled PHS system was within the typical value range of real production environments. Other authors have attempted to solve the same problem, namely Ma et al. [17], who developed mathematical models to model and simulate dozen feasible configurations a PHS-based PV/Wind hybrid system to meet the electricity demand of a remote island in Hong Kong. The main limitation of the study by Ma et al. is that the designed models were not able to achieve the optimization of the system. To fill the absence of a PHS component in the HOMER library, Canales and Beluco [50] proposed a method of modelling a pumped-hydro energy storage system with HOMER by making certain adjustments on a battery component for that it represents a PHS system. This approach was then implemented in [51] to model, simulate, and optimize a PHS based energy system to meet the electrical load of a village in South Africa. Although Canales and Beluco's proposed approach allows for optimizing PHS based on HRESs, it has the disadvantage of requiring prior adjustments to specific HOMER components. Also, the introduction of the 245 kWh PHS component in HOMER library has reduced its usefulness. Besides, models developed by Kusakana [21], cited and described in the introductory part of this paper presents less relevance in the context of SSA as cost reduction was not the primary objective of the analysis. When compared to the three preceding approaches, the method devised in this study has the advantage of providing simple modelling, simulation, and cost-based optimization of PHS based HRESs using HOMER software.

The second novelty of this paper lies with providing possible interpretations of sensibility analysis results other than those that highlight the effect of the change in key parameters on the system performance. Among the studies reviewed in this paper, those that performed a sensitivity analysis [13–15,19,20] failed in broadening the interpretation of their analysis results. Unlike previous those studies, this article was able to interpret the result of the sensitivity of wind speed and solar radiation to provide insight into the performance of the proposed system throughout the SSA region. Such information is essential for determining the level of a renewable energy policy to improve the viability of the system in a particular location. Then, the result of the sensitivity analysis of capital costs of PV and PHS systems provides insight into the potential effect of the government's PV capital cost subsidies, as well as the promising prospects of PV based HRESs giving the upcoming decrease in PV capital costs. Such information is essential for the government in designing appropriate policies for microgrid technology. Finally, the sensitivity analysis of the availability and cost of biomass resources provided an overview of the relevance of transportation or purchase of these resources. This information is essential in a resource-limited environment. Indeed, Biomass resources may not be available at the site or may be sought for other uses, such as cooking, soil fertility, or agricultural traction [52]. For illustrative purposes: the biomass available in the baseline scenario could produce 78 m^3 of biogas per day, enough to meet the daily cooking needs of nearly 344 people in rural areas of sub-Saharan Africa, considering that the daily amount of biogas that is required for cooking per person in rural areas is 0.227 m^3/day [53].

Primary beneficiaries of the implementation of this research will be the local populations of Djoundé who will benefit from the project in three points. First, the provision of electricity to the agricultural sector will help solve the crucial problem of poor agricultural performance in the area through the use of electrical machinery in agricultural production, full mechanization, and processing of agricultural products. Indeed, the lack of energy has been reported as the leading cause of the low productivity of the agricultural sector in sub-Saharan Africa, where only 2% of final electricity consumption is devoted to agriculture, as compared to 18% in India [2]. Second, the project is expected to promote the development of small-scale industry and commerce, which will help increase productivity and lead to job creation and poverty reduction [3,5]. Finally, the implementation of the project will contribute to improving the quality of life, health outcomes, gender equality, education, and ending migration and deforestation [4,6].

The barriers to the implementation of the research envisioned in this paper are: high investment costs; lack of a legal, regulatory and institutional framework; lack of funding; and, unrealistic pricing. Designing appropriate renewable energy incentive policies is a critical step in addressing these challenges and then promoting hybrid renewable energy systems for rural electrification in SSA.

The main tool, HOMER Pro, used to perform this research is a Windows application requiring Windows 7, 8, 8.1, or higher.

The originality of this study can be emphasized in three points, namely: (1) the use of HOMER software to model and simulate a pumped-hydro energy storage based HRES; (2) consideration of the agricultural sector and the seasonal variation in the assessment of electricity demand of a rural area of sub-Saharan Africa; and, (3) broadening sensitivity analysis applications to address practical issues that are related to HRESs.

Additional studies are needed to address the limitations of this study, the main ones being: (1) the failure to take into account the increase in energy demand over time due to population growth and technological development; (2) the non-consideration of social and environmental factors in the selection of the best system configuration; and, (3) the failure to account for the losses imposed on the surplus power of the PV array and the wind turbine due to the HOMER modelling of the PHS system as a battery connected to the DC bus.

5. Conclusions

In response to the challenges of rural electrification in sub-Saharan Africa, a 100% renewable hydro-pumped off-grid hybrid energy system, consisting of wind turbines, PV array, and a biogas generator, has been proposed to meet the demand for electricity in Djoundé, a remote village on northern Cameroon. By designing and applying an original approach, we achieved the modeling, simulation, and optimization of the proposed system. The results of the study highlighted the cost-effectiveness and environmental benefits of the proposed system when compared to previous cases in sub-Saharan Africa. Therefore, PHS-based HRESs can be part of solution in achieving "access to affordable, reliable and modern energy for all by 2030", in line with the Millennium Development Goals (MDGs) for energy. Therefore, sub-Saharan African countries are called upon to develop appropriate policies to address hindering factors to the implementation of HRESs.

Author Contributions: Conceptualization, N.Y., O.H. and L.M.; Methodology, N.Y., B.N. and O.H.; Modelling, N.Y.; Validation, O.H., L.M. and B.N.; Formal Analysis, N.Y., O.H. and L.M.; Investigation, L.M. and B.N.; Writing-Original Draft Preparation, N.Y.; Writing-Review & Editing, L.M., O.H., B.N. and J.N.; Supervision, J.N.

Funding: No external funding was provided for this research.

Conflicts of Interest: The authors declare no conflict of interest.

References

1. Riva, F.; Ahlborg, H.; Hartvigsson, E.; Pachauri, S.; Colombo, E. Electricity access and rural development: Review of complex socio-economic dynamics and casual diagrams for more appropriate energy modelling. *Energy Sustain. Dev.* **2018**, *43*, 203–223. [CrossRef]
2. World Energy Outlook Special Report. Energy Access Outlook 2017: From Poverty to Prosperity. Available online: https://www.iea.org/publications/freepublications/publication/WEO2017SpecialReport_EnergyAccessOutlook.pdf (accessed on 10 September 2018).
3. Adams, S.; Klobodu, E.K.M.; Opoku, E.E.O. Energy consumption, political regime and economic growth in sub-Saharan Africa. *Energy Policy* **2016**, *96*, 36–44. [CrossRef]

4. Trotter, P.A. Rural electrification, electrification inequality and democratic institutions in sub-Saharan Africa. *Energy Sustain. Dev.* **2016**, *34*, 111–129. [CrossRef]

5. Feder, G.; Savastano, S. Modern agricultural technology adoption in sub-Saharan Africa: A four-country analysis. In *Agriculture and Rural Development in a Globalizing World*; Routledge: Abingdon-on-Thames, UK, 2017; pp. 11–25.

6. Orlando, M.B.; Janik, V.L.; Vaidya, P.; Angelou, N.; Zumbyte, I.; Adams, N. Getting to Gender Equality in Energy Infrastructure: Lessons from Electricity Generation, Transmission, and Distribution Projects. 2018. Available online: http://documents.worldbank.org/curated/en/930771499888717016/pdf/117350-ESMAP-P147443-PUBLIC-getting-to-gender-equality.pdf (accessed on 12 August 2018).

7. Dagnachew, A.G.; Lucas, P.L.; Hof, A.F.; Gernaat, D.E.; de Boer, H.-S.; van Vuuren, D.P. The role of decentralized systems in providing universal electricity access in Sub-Saharan Africa–A model-based approach. *Energy* **2017**, *139*, 184–195. [CrossRef]

8. McCright, A.M.; Dunlap, R.E. The politicization of climate change and polarization in the American public's views of global warming, 2001–2010. *Sociol. Q.* **2011**, *52*, 155–194. [CrossRef]

9. Lauber, V.; Jacobsson, S. The politics and economics of constructing, contesting and restricting socio-political space for renewables–The German Renewable Energy Act. *Environ. Innov. Soc. Transit.* **2016**, *18*, 147–163. [CrossRef]

10. Sinha, S.; Chandel, S. Review of software tools for hybrid renewable energy systems. *Renew. Sustain. Energy Rev.* **2014**, *32*, 192–205. [CrossRef]

11. Nfah, E.; Ngundam, J. Feasibility of pico-hydro and photovoltaic hybrid power systems for remote villages in Cameroon. *Renew. Energy* **2009**, *34*, 1445–1450. [CrossRef]

12. Adaramola, M.S.; Agelin-Chaab, M.; Paul, S.S. Analysis of hybrid energy systems for application in southern Ghana. *Energy Convers. Manag.* **2014**, *88*, 284–295. [CrossRef]

13. Halabi, L.M.; Mekhilef, S.; Olatomiwa, L.; Hazelton, J. Performance analysis of hybrid PV/diesel/battery system using HOMER: A case study Sabah, Malaysia. *Energy Convers. Manag.* **2017**, *144*, 322–339. [CrossRef]

14. Singh, S.; Singh, M.; Kaushik, S.C. Feasibility study of an islanded microgrid in rural area consisting of PV, wind, biomass and battery energy storage system. *Energy Convers. Manag.* **2016**, *128*, 178–190. [CrossRef]

15. Sigarchian, S.G.; Paleta, R.; Malmquist, A.; Pina, A. Feasibility study of using a biogas engine as backup in a decentralized hybrid (PV/wind/battery) power generation system–Case study Kenya. *Energy* **2015**, *90*, 1830–1841. [CrossRef]

16. Baghdadi, F.; Mohammedi, K.; Diaf, S.; Behar, O. Feasibility study and energy conversion analysis of stand-alone hybrid renewable energy system. *Energy Convers. Manag.* **2015**, *105*, 471–479. [CrossRef]

17. Ma, T.; Yang, H.; Lu, L.; Peng, J. Technical feasibility study on a standalone hybrid solar-wind system with pumped hydro storage for a remote island in Hong Kong. *Renew. Energy* **2014**, *69*, 7–15. [CrossRef]

18. Kenfack, J.; Neirac, F.P.; Tatietse, T.T.; Mayer, D.; Fogue, M.; Lejeune, A. Microhydro-PV-hybrid system: Sizing a small hydro-PV-hybrid system for rural electrification in developing countries. *Renew. Energy* **2009**, *34*, 2259–2263. [CrossRef]

19. Singh, S.S.; Fernandez, E. Modeling, size optimization and sensitivity analysis of a remote hybrid renewable energy system. *Energy* **2018**, *143*, 719–731. [CrossRef]

20. Ahmad, J.; Imran, M.; Khalid, A.; Iqbal, W.; Ashraf, S.R.; Adnan, M.; Ali, S.F.; Khokhar, K.S. Techno economic analysis of a wind-photovoltaic-biomass hybrid renewable energy system for rural electrification: A case study of Kallar Kahar. *Energy* **2018**, *148*, 208–234. [CrossRef]

21. Kusakana, K. Optimization of the daily operation of a hydrokinetic–diesel hybrid system with pumped hydro storage. *Energy Convers. Manag.* **2015**, *106*, 901–910. [CrossRef]

22. Sawle, Y.; Gupta, S.; Bohre, A.K. Socio-techno-economic design of hybrid renewable energy system using optimization techniques. *Renew. Energy* **2018**, *119*, 459–472. [CrossRef]

23. Aoudji, A.K.; Kindozoun, P.; Adegbidi, A.; Ganglo, J.C. Land Access and Household Food Security in Kpomassè District, Southern Benin: A Few Lessons for Smallholder Agriculture Interventions. *Sustain. Agric. Res.* **2017**, *6*, 104. [CrossRef]

24. Luo, X.; Wang, J.; Dooner, M.; Clarke, J. Overview of current development in electrical energy storage technologies and the application potential in power system operation. *Appl. Energy* **2015**, *137*, 511–536. [CrossRef]

25. Ma, T.; Yang, H.; Lu, L. Feasibility study and economic analysis of pumped hydro storage and battery storage for a renewable energy powered island. *Energy Convers. Manag.* **2014**, *79*, 387–397. [CrossRef]

26. Connolly, D.; Lund, H.; Mathiesen, B.V.; Leahy, M. A review of computer tools for analysing the integration of renewable energy into various energy systems. *Appl. Energy* **2010**, *87*, 1059–1082. [CrossRef]

27. Luna-Rubio, R.; Trejo-Perea, M.; Vargas-Vázquez, D.; Ríos-Moreno, G. Optimal sizing of renewable hybrids energy systems: A review of methodologies. *Sol. Energy* **2012**, *86*, 1077–1088. [CrossRef]

28. Bahramara, S.; Moghaddam, M.P.; Haghifam, M. Optimal planning of hybrid renewable energy systems using HOMER: A review. *Renew. Sustain. Energy Rev.* **2016**, *62*, 609–620. [CrossRef]

29. Shahzad, M.K.; Zahid, A.; ur Rashid, T.; Rehan, M.A.; Ali, M.; Ahmad, M. Techno-economic feasibility analysis of a solar-biomass off grid system for the electrification of remote rural areas in Pakistan using HOMER software. *Renew. Energy* **2017**, *106*, 264–273. [CrossRef]

30. Sen, R.; Bhattacharyya, S.C. Off-grid electricity generation with renewable energy technologies in India: An application of HOMER. *Renew. Energy* **2014**, *62*, 388–398. [CrossRef]

31. Mahlia, T.; Saktisahdan, T.; Jannifar, A.; Hasan, M.; Matseelar, H. A review of available methods and development on energy storage; technology update. *Renew. Sustain. Energy Rev.* **2014**, *33*, 532–545. [CrossRef]

32. NASA Surface Meteorology and Solar Energy. Available online: https://power.larc.nasa.gov/ (accessed on 10 September 2018).

33. Kidmo, D.K.; Deli, K.; Raidandi, D.; Yamigno, S.D. Wind energy for electricity generation in the far north region of Cameroon. *Energy Procedia* **2016**, *93*, 66–73. [CrossRef]

34. Bhatt, A.; Sharma, M.; Saini, R. Feasibility and sensitivity analysis of an off-grid micro hydro–photovoltaic–biomass and biogas–diesel–battery hybrid energy system for a remote area in Uttarakhand state, India. *Renew. Sustain. Energy Rev.* **2016**, *61*, 53–69. [CrossRef]

35. Hasan, A.M.; Khan, M.F.; Dey, A.; Yaqub, M.; Al Mamun, M.A. Feasibility study on the available renewable sources in the island of Sandwip, Bangladesh for generation of electricity. In Proceedings of the 2nd International Conference on the Developments in Renewable Energy Technology (ICDRET 2012), Dhaka, Bangladesh, 5–7 January 2012.

36. Mandal, S.; Yasmin, H.; Sarker, M.; Beg, M. Prospect of solar-PV/biogas/diesel generator hybrid energy system of an off-grid area in Bangladesh. *AIP Conf. Proc.* **2017**, *1919*. [CrossRef]

37. Wang, L.; Shahbazi, A.; Hanna, M.A. Characterization of corn stover, distiller grains and cattle manure for thermochemical conversion. *Biomass Bioenergy* **2011**, *35*, 171–178. [CrossRef]

38. Lambert, T.; Gilman, P.; Lilienthal, P. Micropower system modeling with HOMER. *Integr. Altern. Sources Energy* **2005**, 379–418.

39. Graham, V.; Hollands, K. A method to generate synthetic hourly solar radiation globally. *Sol. Energy* **1990**, *44*, 333–341. [CrossRef]

40. Distributed Generation Renewable Energy Estimate of Costs. Available online: Https://www.nrel.gov/analysis/tech-lcoe-re-cost-est.html (accessed on 10 September 2018).

41. SunPower SPR-E20-327. Available online: https://www.energysage.com/panels/SunPower/SPR-E20-327/ (accessed on 10 September 2018).

42. The excel 10 kW Wind Power. Available online: Availableat:bergey.com/products/wind-turbines/10kw-bergey-excel (accessed on 10 September 2018).

43. Pumped Hydro—Homer Energy. Available online: https://www.homerenergy.com/products/pro/docs/3.11/pumped_hydro.html (accessed on 10 September 2018).

44. Deane, J.P.; Gallachóir, B.Ó.; McKeogh, E. Techno-economic review of existing and new pumped hydro energy storage plant. *Renew. Sustain. Energy Rev.* **2010**, *14*, 1293–1302. [CrossRef]

45. Zachary, O.; Authority, R.E. *Rural Electrification Programme in Kenya*; Rural Electrification Authority: Nairobi, Kenya, 2011.

46. Kaldellis, J.; Kapsali, M.; Kavadias, K. Energy balance analysis of wind-based pumped hydro storage systems in remote island electrical networks. *Appl. Energy* **2010**, *87*, 2427–2437. [CrossRef]

47. Fadare, D.; Irimisose, I.; Oni, A.; Falana, A. Modeling of solar energy potential in Africa using an artificial neural network. *Am. J. Sci. Ind. Res.* **2010**, *1*, 144–157. [CrossRef]

48. Mentis, D.; Hermann, S.; Howells, M.; Welsch, M.; Siyal, S.H. Assessing the technical wind energy potential in Africa a GIS-based approach. *Renew. Energy* **2015**, *83*, 110–125. [CrossRef]

49. Energiewende, A.; Mayer, J.N.; Philipps, S.; Saad, N.; Hussein, D.; Schlegl, T.; Senkpiel, C. Current and future cost of photovoltaics. *Berl. Agora Energiewende* **2015**.

50. Canales, F.A.; Beluco, A. Modeling pumped hydro storage with the micropower optimization model (HOMER). *J. Renew. Sustain. Energy* **2014**, *6*, 043131. [CrossRef]

51. Kusakana, K. Feasibility analysis of river off-grid hydrokinetic systems with pumped hydro storage in rural applications. *Energy Convers. Manag.* **2015**, *96*, 352–362. [CrossRef]

52. Bettencourt, E.M.V.; Tilman, M.; Narciso, V.; Carvalho, M.L.S.; Henriques, P.D.S. The livestock roles in the wellbeing of rural communities of Timor-Leste. *Rev. Econ. E Sociol. Rural* **2015**, *53*, 63–80. [CrossRef]

53. Shahzad, K.; Nasir, A.; Shafiq Anwar, M.; Farid, M.U. Bioenergy Prospective and Efficient Utilization Pattern for Rural Energy Supply in District Pakpattan. *J. Agric. Res.* **2017**, *55*, 101–113.

Article

Building Energy Management Strategy Using an HVAC System and Energy Storage System

Nam-Kyu Kim [1], Myung-Hyun Shim [2] and Dongjun Won [2,*]

[1] Hyosung Corporation, 74, Simin-daero, Dongan-gu, Anyang-si, Gyeonggi-do 14080, Korea;
    namkyukim@hyosung.com

[2] Department of Electrical Engineering, Inha University, 100, Inha-ro, Michuhol-gu, Incheon 22212, Korea;
    gildongh66@gmail.com

*   Correspondence: djwon@inha.ac.kr; Tel.: +82-32-860-7404; Fax: +82-32-863-5822

Received: 15 September 2018; Accepted: 4 October 2018; Published: 10 October 2018

Abstract: Recently, a worldwide movement to reduce greenhouse gas emissions has emerged, and includes efforts such as the Paris Agreement in 2015. To reduce greenhouse gas emissions, it is important to reduce unnecessary energy consumption or use environmentally-friendly energy sources and consumer products. Many studies have been performed on building energy management systems and energy storage systems (ESSs), which are aimed at efficient energy management. Herein, a heating, ventilation, and air-conditioning (HVAC) system peak load reduction algorithm and an ESS peak load reduction algorithm are proposed. First, an HVAC system accounts for the largest portion of building energy consumption. An HVAC system operates by considering the time-of-use price. However, because the indoor temperature is constantly changing with time, load shifting can be expected only immediately prior to use. Therefore, the primary objective is to reduce the operating time by changing the indoor temperature constraint at the forecasted peak time. Next, numerous research initiatives on ESSs are ongoing. In this study, we aim to systematically design the peak load reduction algorithm of ESS. The structure is designed such that the algorithm can be applied by distinguishing between the peak and non-peak days. Finally, the optimization scheduling simulation is performed. The result shows that the electricity price is minimized by peak load reduction and electricity usage reduction. The proposed algorithm is verified through MATLAB simulations.

Keywords: building energy management system; HVAC system; energy storage system

1. Introduction

Over the past 40 years, the global consumption of primary energy has increased by approximately 2.4 times. Primary energy consumption has steadily increased, except for certain periods, such as the oil shock and financial crisis. Global energy demand is expected to continue to increase in the upcoming years. By 2040, the world energy demand is expected to increase by 30%. In response to the rapidly changing climate, the 21st Conference of the Parties to the United Nations Framework Convention on Climate Change was held in Paris at the end of 2015, with the aim of reducing greenhouse gas emissions. This move to control greenhouse gas emissions is occurring worldwide to raise the awareness about indiscriminate energy consumption. Hence, renewable energy generation such as solar power generation and wind power generation are spreading worldwide. However, the growth rate of new energy sources is limited and high investment costs are required. Therefore, technologies for efficiently using the existing energy are emerging. According to the data published by the U.S. Energy Information Administration in 2017, the heating, ventilation, and air-conditioning (HVAC) system is a heavy load, which accounts for approximately 25% of the total building load [1]. A typically used HVAC system operates to maintain the indoor temperature constant with the set temperature constant. Because of this feature, the HVAC system produces a constant output even at the peak load time, which is a burden to the user in terms of economy. In addition, on the grid side, HVAC systems can increase the required power plant capacity, generation and operating reserve capacity owing to increased peak loads. However, the HVAC system can control the set temperature; therefore, if it is controlled according to the electricity price and outdoor temperature by time, it has the potential to reduce the power consumption or peak load [2]. Another method to efficiently utilize the energy that is being used is via an energy storage system (ESS). The use of ESSs in buildings, which accounts for a significant portion of power consumption, is becoming a necessity for efficient demand management. Recently, the need for energy management has increased. HVAC system and ESS optimization scheduling has been studied extensively. Most of the studies using the HVAC system thus far have only considered reducing the load by modifying the indoor temperature constraint during the demand response time, and simulated by simplifying the thermal model. Thus, controlling the HVAC system only in the demand response time does not account for the economic benefits obtained from other times [3]. Further, a disadvantage is that the thermal model is simple and not similar to the actual temperature change pattern [4]. Significant work has been performed to manage the peak load using the ESS. Electricity costs can be divided into two categories: demand cost and energy cost. Strategies to minimize both the demand cost and energy cost are different. Previous studies have adopted necessary strategies, but the overall process was difficult to understand [5,6]. In the current work, we study the peak load reduction algorithm with the two systems above. Energy management using the HVAC system was designed considering the thermal model and user convenience, and the primary theme was to perform load shifting simultaneously with peak load reduction. The energy management algorithm using the ESS was used to design the structure, such that the overall process of minimizing the demand cost and energy cost can be organized by a single algorithm.

We herein introduce the uptime optimization scheduling of the HVAC system, and the output power optimization scheduling algorithm of the ESS for an optimal energy management. In the first of the two energy management strategies, the HVAC system schedules an uptime to reduce the day's peak load. This does not simply stop the system at the peak time, but aims to maintain the proper indoor temperature considering user convenience. The change in the indoor temperature was implemented using the thermal model provided by MATLAB (R2017a, The MathWorks Inc., Natick, MA, USA) to demonstrate a similar temperature change pattern. HVAC system optimization scheduling was performed by applying a genetic algorithm (GA). Next, the energy management algorithm using ESS is divided into two stages, and ESS output scheduling is implemented through the appropriate stage depending on the load at that time. ESS power optimization scheduling was performed by applying linear programming. Energy management simulations were performed using MATLAB.

Section 2 presents the algorithm for energy management using the HVAC system. Section 3 presents the algorithm for energy management using the ESS. The purpose of each stage and the formulae used are shown in detail. Section 4 shows the simulation results of the proposed algorithms using various cases. Finally, Section 5 presents the conclusions and the next steps of the presented study.

2. Energy Management Using HVAC System

2.1. Thermal Model and User Convenience in HVAC System

In HVAC systems, the indoor and outdoor temperatures are closely related to the load. The outdoor temperature is estimated at the meteorological office site, and it is important to model the indoor temperature with characteristics similar to those of actual HVAC systems [7]. In particular, the indoor temperature changes must be considered simultaneously with changes in HVAC system operation and outdoor temperature. In addition, the indoor temperature is affected by various factors, such as the structure of the building and the specifications of the HVAC system. To reflect these characteristics, the thermal model provided by MATLAB is applied to the HVAC system energy optimization algorithm.

The thermal model is constructed considering the characteristics of the building and the characteristics of the HVAC system, and various parameters can be changed according to the site. The thermal model provided by MATLAB is shown in Figure 1. The model consists of Simulink, and the initial model is a heating system. In this study, a Simulink model is constructed as a script to implement the HVAC system optimization algorithm, and a heating system model is constructed by partially modifying the cooling system model [8]. The formulae for the thermal model are as follows.

$$\frac{dQ_{cool}(t)}{dt} = (T_{in}(t) - T_{cool}) \times (t_{int} \times 3600) \times M_{dot} \times c \ [J/sec] \tag{1}$$

$$\frac{dQ_{heat}(t)}{dt} = (T_{heat} - T_{in}(t)) \times (t_{int} \times 3600) \times M_{dot} \times c \ [J/sec] \tag{2}$$

$$\frac{dQ_{loss}(t)}{dt} = (T_{out}(t) - T_{in}(t))/R_{eq} \ [J/sec] \tag{3}$$

$$P_{cool}(t) = \frac{dQcool(t)}{dt}/1000 \ [kW] \tag{4}$$

$$P_{heat}(t) = \frac{dQheat(t)}{dt}/1000 \ [kW] \tag{5}$$

$$\frac{dT_{in}(t)}{dt} = \begin{cases} \frac{1}{M_{air} \times c} \times \left(\frac{dQ_{loss}(t)}{dt} - \frac{dQ_{cool}(t)}{dt} \right), \text{ when cooling system} \\ \frac{1}{M_{air} \times c} \times \left(\frac{dQ_{loss}(t)}{dt} + \frac{dQ_{heat}(t)}{dt} \right), \text{ when heating system} \end{cases} \tag{6}$$

$$T_{in}(t+1) = T_{in}(t) + t_{int} \times \frac{dT_{in}(t)}{dt} \tag{7}$$

Thermal Model of a House

Note: Time given in units of hours

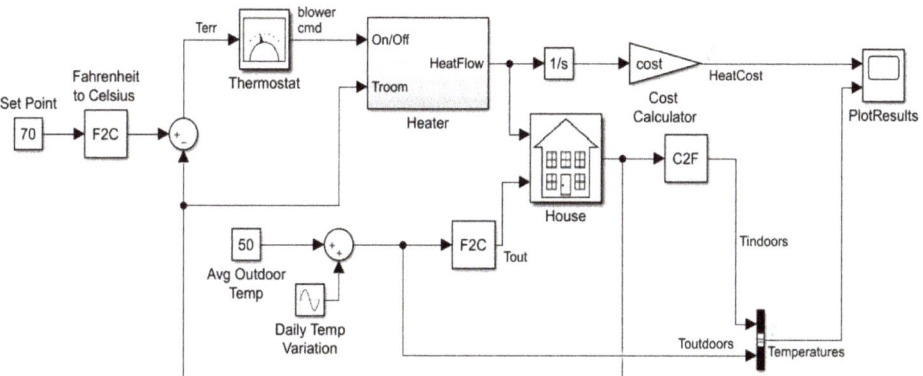

Figure 1. Thermal model of MATLAB simulink.

Equations (1) and (2) represent the HVAC system heat flow in the cooling and heating modes, respectively. Here, t_{int} denotes the algorithm control period, and in the subsequent simulation, the time interval is set to 5 min and its value is 1/12 (h); T_{in} denotes the indoor temperature by the control cycle; T_{cool}, T_{heat} denote the cooling supply temperature, the heating supply temperature; M_{dot}, c denote the HVAC system supply air mass and air heat capacity, respectively. Equation (3) shows the heat loss owing to the outdoor temperature and shows the effect on the indoor temperature change. Here, R_{eq} denotes the building equivalent heat resistance. Equations (4) and (5) show the output in the cooling and heating modes in kW, respectively. Equation (6) is the variation of the indoor temperature considering heat flow by the HVAC system and heat loss by the outdoor temperature. In Equation (7), the indoor temperature is updated according to the value from Equation (6) [9,10].

Once the optimal value is determined through the thermal model, the optimization scheduling of the HVAC system proceeds. The indoor temperature that the HVAC system must control is the most direct parameter that determines the user's comfort. Hence, it is essential to consider user convenience in the energy optimization algorithm using the HVAC system. Figure 2 shows the range in which the user sees comfort in the psychrometric chart. The psychrometric chart shows the relationship among the dry bulb temperature, wet bulb temperature, absolute humidity, relative humidity, water vapor pressure, and enthalpy under atmospheric pressure [11,12]. The two graphs in Figure 2a,b demonstrate the range in which 90% and 80% of the users can feel comfortable, respectively. Because the 90% acceptability level is narrow, it is difficult to expect a load shifting effect to reduce the peak load at a specific time, or to reduce the energy cost through the HVAC operation. Meanwhile, the 80% acceptability level has a relatively wide tolerance range; therefore, the load peak or energy cost reduction can be expected through the proper operation of the HVAC. In the subsequent simulation, the relative humidity was set at 50% and the indoor temperature operating range is at the 80% acceptability level.

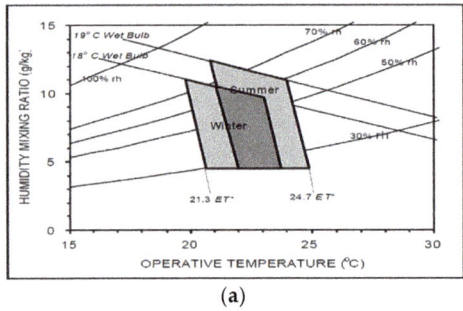 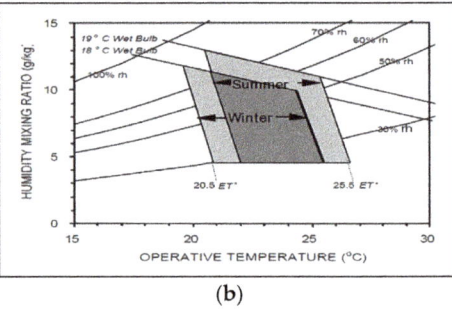

(a)   **(b)**

Figure 2. Seasonal comfort zone in the psychrometric chart. (**a**) 90% acceptability level; (**b**) 80% acceptability level.

2.2. HVAC System Energy Management Optimization Algorithm

The objective function of the algorithm is to minimize the energy cost. The thermal model and user convenience described above are considered. The uptime of the HVAC system is regulated according to the electricity price. The HVAC system is operated at a time when the electricity price is low, thus reducing the operating number of HVAC systems at a relatively high electricity price. However, the pattern of the peak shifting is limited because the indoor temperature is continuously changed by the outdoor temperature. That is, even if the setting temperature is changed by operating at a specific time, it can only affect the next time. The objective function and the constraint conditions of the energy management optimization algorithm using the HVAC system are as follows.

$$\text{Min}\left\{ \sum_{t=1}^{24/t_{int}} (\rho_e(t) \times t_{int} \times u(t) \times P_{HVAC}(t)) \right\} \tag{8}$$

$$P_{HVAC}(t) = u_c(t) \times P_{cool}(t) + u_h(t) \times P_{heat}(t) \tag{9}$$

$$u(t) = u_c(t) + u_h(t) \tag{10}$$

$$u_c(t) + u_h(t) \leq 1 \tag{11}$$

$$T_{in,min}(t) \leq T_{in}(t) \leq T_{in,max}(t) \tag{12}$$

$$T_{work,min}(t) \leq T_{in}(t) \leq T_{work,max}(t) \tag{13}$$

Equation (8) adjusts the operating time such that the energy cost is minimized as an objective function. Here, ρ_e, u, P_{HVAC} denote the electricity price, the HVAC system on/off state, and the HVAC system power, respectively. Equations (9) and (10) represent the values of P_{HVAC} and u of the objective function, respectively; P_{cool} and P_{heat} are the cooling/heating power calculated in the thermal model; u_c and u_h are the binary variables indicating the cooling/heating state of the specific time, respectively. Equation (11) allows only one of the cooling and heating modes to operate at a specific time. Equation (12) represents the indoor temperature range constraint when the time is not the operating time or the peak load time, and this range can be changed according to the condition of the building. Equation (15) shows the indoor temperature range constraint considering user convenience at the operating time. The energy management algorithm of the HVAC system is optimized using the GA [13]. The optimal temperature range is set as a constraint condition considering user convenience, and the change in the indoor temperature is based on a thermal model [14,15]. Basically, it schedules the uptime according to the electricity price. If the peak load time of the day can be determined through the demand prediction, the peak load can be reduced by changing the indoor temperature range limit of the time [16,17].

3. Energy Management Using ESS

ESS can be used to reduce the electricity price and the peak load through charging and discharging. Thus, the cost of electric energy can be reduced. ESS can be more active in responding to peak control than the HVAC system load, and significant savings can be achieved if operated with proper output scheduling. The electricity costs are generally divided into the demand cost and energy cost. In this study, the process of minimizing the power consumption to determine the demand cost, as well as the power consumption to determine the energy cost are both structured systematically. Figure 3 shows the flow chart of the energy management optimization algorithm using the ESS. The algorithm is divided into Stage 1 and Stage 2. Stage 1 performs power scheduling to minimize the peak load. Stage 2 is developed to perform power scheduling to minimize the energy cost while not exceeding the peak load determined at Stage 1.

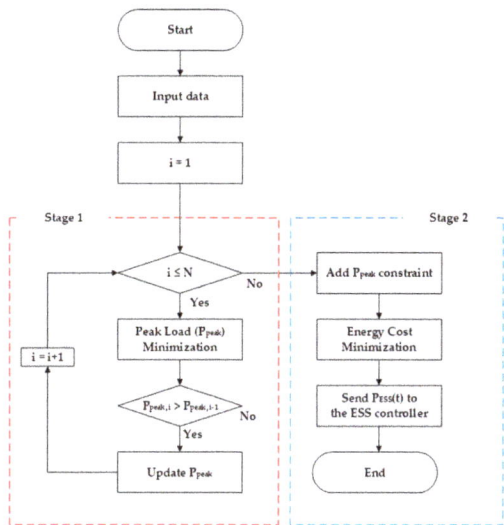

Figure 3. Flowchart of the proposed energy storage system (ESS) optimization algorithm.

In Stage 1, peak load reduction scheduling using the ESS is performed using the predicted monthly load data. Simultaneously, the peak load (with ESS) is repeatedly compared to update the peak load that determines the monthly demand cost. The ESS output is determined according to the peak load reduction scheduling result when the peak load that determines the monthly demand cost is generated in the current month. In Stage 1, the objective function and constraint are as follows.

$$\text{Min}\left\{\rho_s \times P_{peak}\right\} \tag{14}$$

$$-P_{ch,max} \times x(t) \leq P_{ESS(t)} \leq P_{dch,max} \times y(t) \tag{15}$$

$$x(t) + y(t) \leq 1 \tag{16}$$

$$SOC(t) = SOC(t-1) - (y(t) \times P_{ESS}(t) \times \eta_d^{-1} + x(t) \times P_{ESS}(t) \times \eta_c) \tag{17}$$

$$SOC_{lb} \leq SOC \leq SOC_{ub} \tag{18}$$

$$SOC_{final} = SOC_{initial} \tag{19}$$

$$Load(t) - P_{ESS}(t) \leq P_{peak} \tag{20}$$

Equation (14) is an objective function and consists of a base rate. ESS power scheduling is performed to lower the peak load. Here, ρ_s, P_{peak} denote the price per kW of the demand cost, i.e., the peak load. Equation (15) is a charging/discharging power constraint. The charging power has a negative value and the discharging power has a positive value. Here, $P_{ch,max}$, P_{ESS}, $P_{dch,max}$ denote the ESS charging rated power, the ESS power and the ESS discharging rated power, respectively. Equation (16) allows only one of the charging and discharging modes to operate at a specific time. Further, x, y denote the ESS charging and discharging states, respectively. Both values are binary variables. Equation (17) is a formula for calculating the state of charge (SOC) for each control cycle. Here, η_d and η_c denote the ESS charging and discharging efficiencies, respectively. Equation (18) limits the operating range of the SOC of the ESS, and the final SOC is set equal to the initial SOC, as shown in Equation (19). Here, SOC_{lb} and SOC_{ub} denote the SOC lower and upper limits, respectively. Equation (20) shows that P_{peak} has the largest value among differences between the load and the ESS power.

When the peak load that determines the monthly demand cost is determined in Stage 1, Stage 2 adds the constraint using this value. In the objective function, the ESS performs power scheduling considering only the electricity price, excluding the demand cost portion. The objective function and the additional constraint condition in Stage 2 that minimizes the energy cost are as follows.

$$\text{Min}\left\{\sum_{t=1}^{96}\left(\rho_e(t) \times 0.25 \times (Load(t) - P_{ESS}(t))\right)\right\} \tag{21}$$

$$P_{peak} \leq P_{peak,limit} \tag{22}$$

According to Equation (21), the ESS performs optimization scheduling to charge at a low electricity price, and to discharge at a high electricity price considering the electricity price. Here, $\rho_e(t)$ denotes the electricity price. Equation (22) sets the peak load that determines the monthly demand cost as a constraint and limits the occurrence of new peak loads owing to the excessive charging of the ESS according to the electricity price. Here, $P_{peak,limit}$ means a peak load limit. The day when the peak load occurs in a month is preferentially scheduled in Stage 1 by the peak-reduction algorithm, and subsequently rescheduled to Stage 2.

4. Simulation Results

4.1. Simulation of HVAC System Optimization Algorithm

Our simulation is performed by modeling a real building and HVAC system using MATLAB. The building is modeled as a thermal model assuming a building size of 30 m × 10 m. The actual outdoor temperature is based on the actual data from 6 July 2015, in South Korea. The electricity cost is calculated using the actual hourly electricity price in Korea. The hourly electricity price is shown in Figure 4. First, it is assumed that the HVAC system does not operate, and the variation in the indoor temperature is confirmed when the indoor temperature changes only by the external temperature. A comparison of the indoor and outdoor temperatures is shown in Figure 5. The configuration data of the buildings are summarized in Table 1.

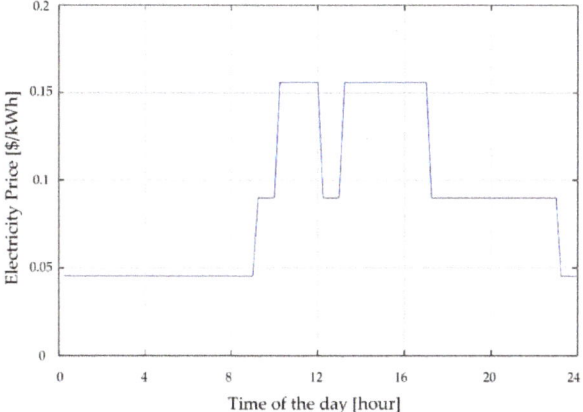

Figure 4. Electricity price applied to the simulation.

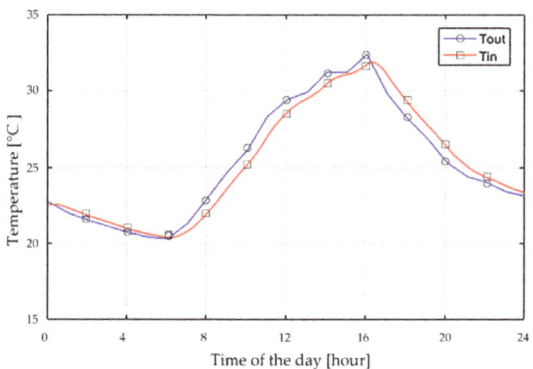

Figure 5. Comparison of indoor and outdoor temperatures (without a heating, ventilation, and air-conditioning (HVAC) system).

Table 1. Configuration data of the building.

Parameter	Definition	Value	Parameter	Definition	Value
lenBuilding	Building Length	30 [m]	widWindows	Window Width	1 [m]
widBuilding	Building Width	10 [m]	htWindows	Window Height	1 [m]
htBuilding	Building Height	4 [m]	Req	Building Equivalent Heat Resistance	1.73×10^{-8} [sec·°C/J]
numWindows	Number of Widows	6 [ea]	M	Indoor Air Mass	1470 [kg]

In this study, the simulation is performed for the summer. One day is composed of non-work time and work time, and the time from 2:00 p.m. to 3:00 p.m. is assumed to be the time when the peak load occurs. Scenarios for the HVAC system simulation are shown in Table 2.

Table 2. Scenarios for the heating, ventilation, and air-conditioning (HVAC) system simulation.

Scenario	Setting Temperature (°C) (Non-Work Time/Work Time)	Indoor Temperature Range (°C) (Non-Work Time/Work Time)	Indoor Temperature Range at Peak Time (°C)
Case 1	27.5/23.5	–	–
Case 2	Scheduling	19.5~27.5/21.5~25.5	–
Case 3			19.5~27.5

4.1.1. Case 1

Case 1 is performed to observe the variation in the indoor temperature when the HVAC system is operated for the summer outdoor temperature change in the building model above. At this time, the HVAC system operates only when the set temperature and indoor temperature differ by 1 °C between 9:00 a.m. and 6:00 p.m., which is the work time. The control period is set to 5 min. The HVAC system data are summarized in Table 3.

Table 3. Configuration data of the HVAC system.

Parameter	Definition	Value	Parameter	Definition	Value
T_{cool}	Cooling supply temperature	14 [°C]	$T_{work,max}$	Work time indoor temperature maximum	25.5 [°C]
M_{dot}	HVAC supply air mass	1 [kg/s]	$T_{in,min}$	Non-work time indoor temperature minimum	19.5 [°C]
$T_{work,min}$	Work time indoor temperature minimum	21.5 [°C]	$T_{in,max}$	Non-work time indoor temperature maximum	27.5 [°C]

We confirmed that the HVAC system is turned on/off as the daytime outdoor temperature increases. Therefore, the indoor temperature is maintained within a certain range. Case 1 is a general HVAC system because it operates at a constant temperature setting. The HVAC system output in Figure 6d is displayed every 5 min. Because the actual output is measured in units of 15 min, the unit load is obtained by averaging over a 15-min period. In the simulation, it is assumed that the peak of the total load occurs between 2:00 p.m. and 3:00 p.m. Therefore, the peak load of the HVAC system is as shown in Table 4, summarizing the results, such as the number of on/off, power consumption, and energy cost.

Table 4. Summary of Case 1 simulation results.

Scenario	Number of On/Off	Peak Load (2:00 p.m.~3:00 p.m.)	Power Consumption	Energy Cost
Case 1	33 [times]	7.36 [kW]	30.01 [kWh]	4.91 [$]

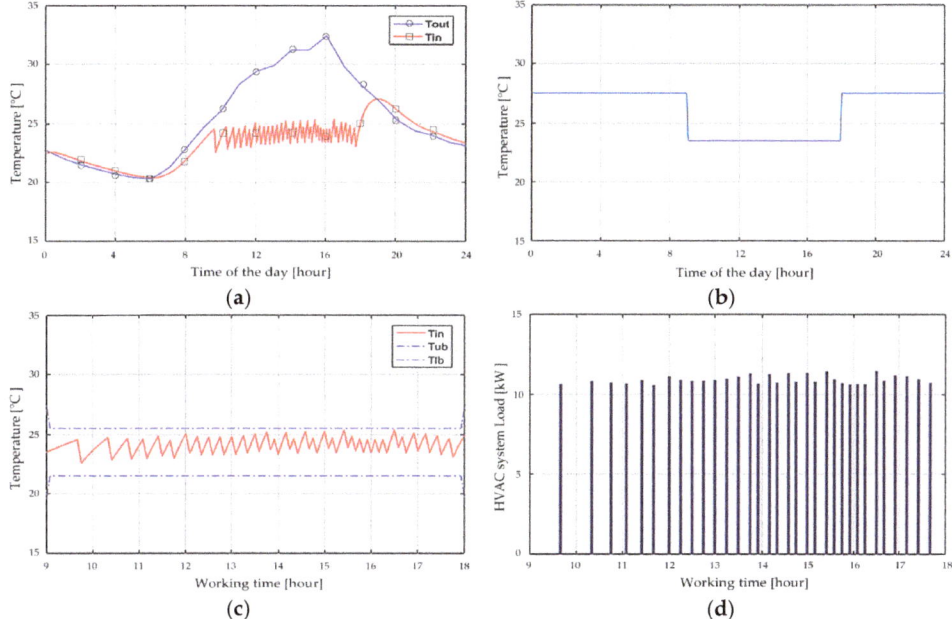

Figure 6. Case 1 simulation results. (**a**) Outdoor temperature and room temperature; (**b**) setting temperature; (**c**) indoor temperature and constraints range (work time); (**d**) HVAC system power (work time).

4.1.2. Case 2

Case 2 uses the GA method to set the set temperature at which the energy cost is minimized. The simulation is performed using the same data as in Case 1 for the building data, HVAC system data, outdoor temperature data, and hourly electricity price data. Figure 7b shows the scheduled setting temperature. The HVAC system is turned on/off by comparing the setting temperature with the indoor temperature. To reduce the energy cost as much as possible, we confirmed that the cost is minimized by lowering the indoor temperature to as low as possible at 12:00 p.m. to 1:00 p.m., and increasing the indoor temperature at an expensive time. In Case 2, we confirmed that power consumption is increased compared to Case 1. This is because the difference between the indoor temperature of the current time and the outdoor temperature of the next time is increased as the indoor temperature is lowered as much as possible at a low electricity price. However, because the indoor temperature is changed according to the electricity price, the total energy cost is reduced by approximately $0.09 compared to Case 1. Table 5 summarizes the results of the number of on/off, peak load, power consumption, and energy cost.

Table 5. Summary of Case 2 simulation results.

Scenario	Number of On/Off	Peak Load (2:00 p.m.~3:00 p.m.)	Power Consumption	Energy Cost
Case 2	35 [times]	6.5 [kW]	30.43 [kWh]	4.82 [$]

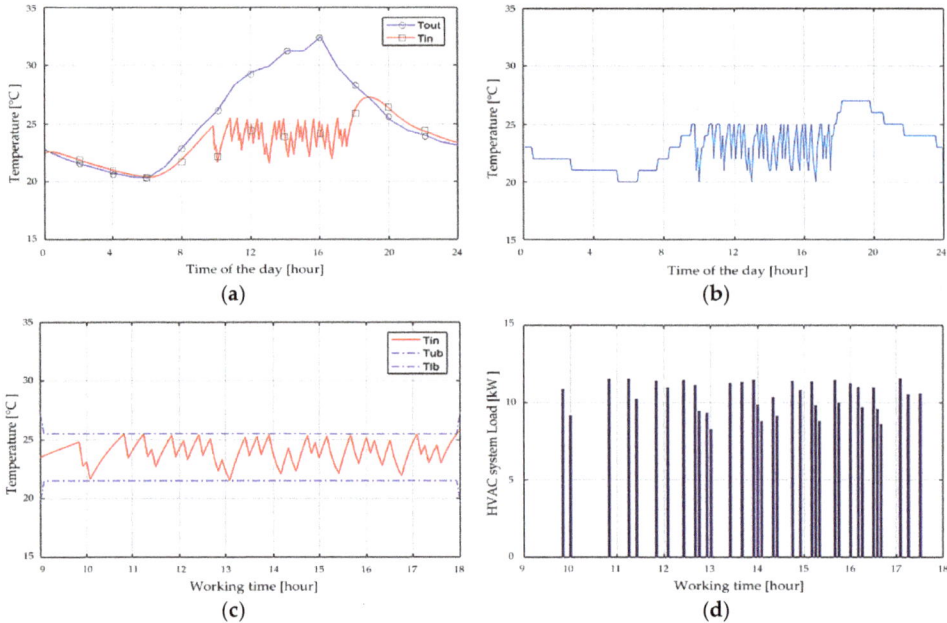

Figure 7. Case 2 simulation results. (**a**) Outdoor temperature and room temperature; (**b**) setting temperature; (**c**) indoor temperature and constraints range (work time); (**d**) HVAC system power (work time).

4.1.3. Case 3

Case 3 controls the HVAC system at the setting temperature at which the energy cost is minimized using the GA method, as in Case 2. However, by maximizing the room temperature range from 2:00 p.m. to 3:00 p.m., which is assumed to be the peak load time, it is a simulation that reduces the peak load and minimizes the energy cost.

The data used is the same as those in Case 2, and the indoor temperature limit at 2:00 p.m. to 3:00 p.m. is changed as shown in Table 6. In Case 3, the HVAC system also operates within a certain temperature range. Figure 8c shows the change in indoor temperature when the indoor temperature range from 2:00 p.m. to 3:00 p.m. is increased. In Case 3, the setting temperature is changed such that the energy cost is the lowest. In Case 3, the indoor temperature is increased to the maximum because the temperature range is increased from 2:00 p.m. to 3:00 p.m. Therefore, when cooling after 3:00 p.m., the HVAC peak load at every 5 min is higher than that of Case 2 to reduce the high temperature. Because the actual load is measured in units of 15 min, we confirmed that the peak load decreases from Case 2. In addition, because the setting temperature is maintained high when the electricity price is expensive, the energy cost is reduced by approximately $0.19 as compared to Case 2. Table 7 summarizes the results of the number of on/off, peak load, power consumption, and energy cost.

Table 6. Indoor temperature constraints between 2:00 p.m. and 3:00 p.m.

Parameter	Definition	Value	Parameter	Definition	Value
$T_{work,min}$	Work time indoor temperature minimum	19.5 [°C]	$T_{work,max}$	Work time indoor temperature maximum	27.5 [°C]

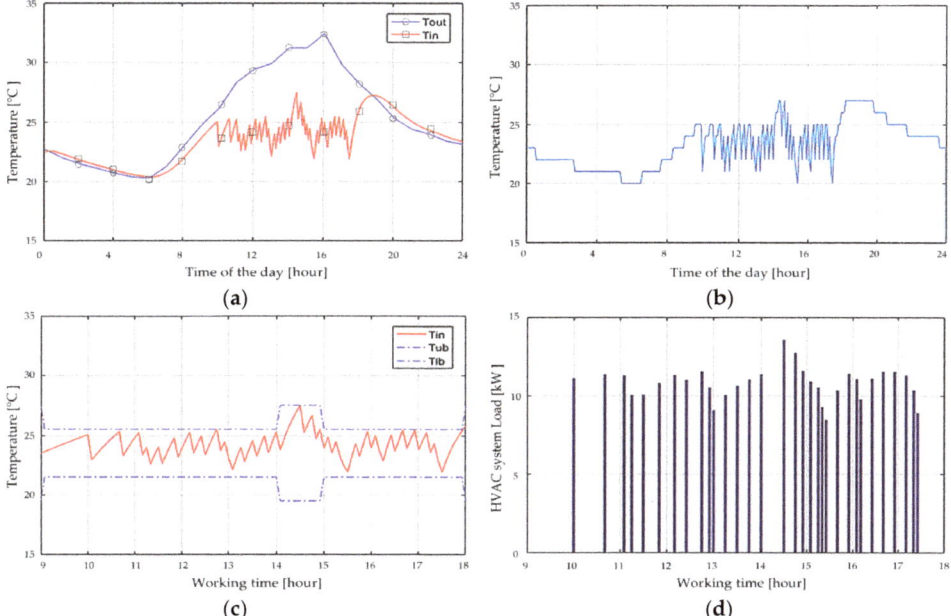

Figure 8. Case 3 simulation results. (**a**) Outdoor temperature and room temperature; (**b**) setting temperature; (**c**) indoor temperature and constraints range (work time); (**d**) HVAC system power (work time).

Table 7. Summary of Case 3 simulation results.

Scenario	Number of On/Off	Peak Load (2:00 p.m.~3:00 p.m.)	Power Consumption	Energy Cost
Case 3	32 [times]	4.52 [kW]	28.83 [kWh]	4.64 [$]

4.2. Simulation of ESS Optimization Algorithm

The ESS optimization algorithm is implemented and simulated using MATLAB. The actual load data from Inha University from February 2017 to August 2017 are used as the simulation load data. As a result of simulating all the dates by applying the algorithm of Stage 1, we confirmed that a peak load occurs on 7 August 2017. Hence, August 2017 is set as the simulation target. The rated power of the ESS is set to 2 MW, and the rated capacity to 2 MWh. The charging and discharging efficiencies are both set to 90%. The initial SOC and final SOC are both set at 50%, and the SOC operating range is set at 10 to 90%. The scenarios for the ESS simulation are shown in Table 8. The reason for configuring the scenario as shown below is that it can demonstrate all the situations that can occur.

Table 8. Scenarios for ESS simulation.

Scenario	ESS Configuration Data	Load Data	Characteristic
Case 1		7 August 2017	Peak load (with ESS) Occurs
Case 2	2 MW/2 MWh Initial SOC: 50% Charging/Discharging efficiency: 90%	9 August 2017	Peak load (without ESS) of Case 2 exceeds peak load (with ESS) of Case 1
Case 3		12 August 2017	Peak load (without ESS) of Case 3 less than peak load (with ESS) of Case 1

4.2.1. Case 1

In Case 1, 7 August 2017 is the day when the peak load (including ESS) occurs. Hence, Stage 1 scheduling is performed to minimize the peak load (with ESS). Because the peak load reduction is the primary goal, the SOC is set to the maximum value before the peak time, and the discharging is continued from 1:00 p.m. to 5:00 p.m. Figure 9a shows that 479.4 kW is reduced at the original peak load using the ESS. Because the pattern of the original load is flat at the peak, the peak load reduction is not large. The results of Case 1 are summarized in Table 9, and the peak load that determines the demand cost in August is set at 5420.8 kW.

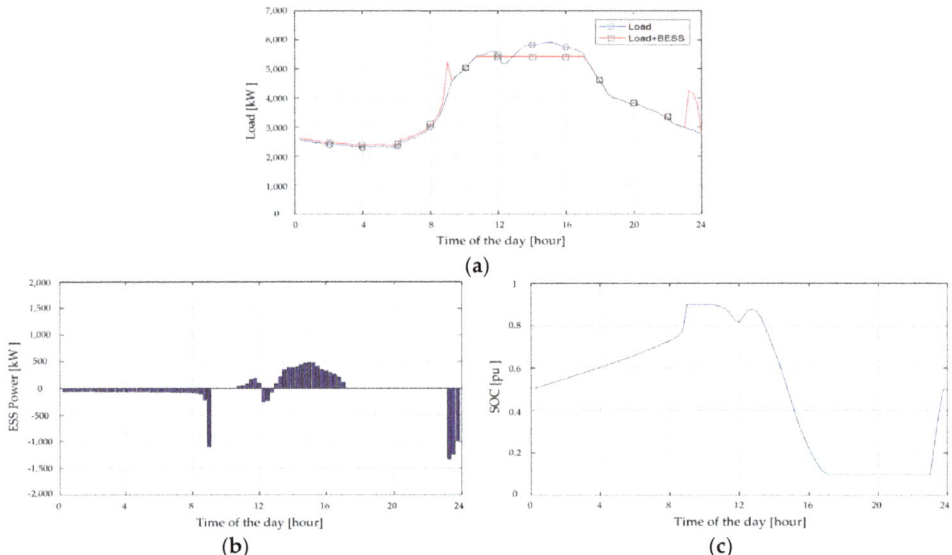

Figure 9. Case 1 simulation results. (**a**) Comparison of the original load and the load after ESS operation; (**b**) ESS power; (**c**) State of charge (SOC).

Table 9. Summary of Case 1 simulation results.

Scenario	Original Peak Load	New Peak Load (with ESS)	Demand Cost Reduction	Daily Energy Cost Reduction
Case 1	5900.2 [kW]	5420.8 [kW]	3097.9 [$]	136.8 [$]

4.2.2. Case 2

Case 2 uses the load data of 9 August 2017. Stage 1 is assessed as not the peak load that determines the demand cost through the scheduling of minimizing the peak load; therefore, it goes to Stage 2 of the flowchart. The purpose of the simulation in Case 2 is to confirm the power scheduling of the ESS when the peak load exceeds the peak load that determines the demand cost. In Figure 10, because the demand cost is not included in the objective function, discharging is performed at a time when the electricity price is high to minimize the energy cost. Therefore, the ESS is charged at 12:00–1:00 p.m., owing to the low electricity price. However, it is not possible to increase the SOC to the maximum owing to the constraint of the peak load that determines the demand cost in Case 1. The simulation results demonstrate that the ESS charges from 11:00 p.m. to 12:00 a.m. to set the final SOC to the same value as the initial SOC. Because the simulation is performed only for one day, the ESS is charged from 11:00 p.m. to 12:00 p.m., and the lowest price is shown in the evening. The results of Case 2 are summarized in Table 10.

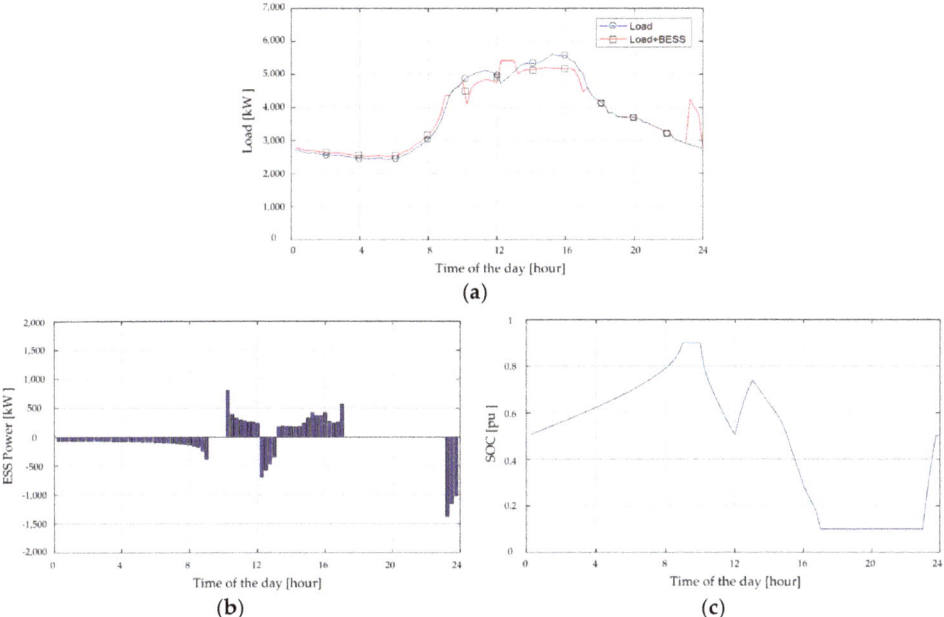

Figure 10. Case 2 simulation results. (**a**) Comparison of the original load and the load after ESS operation; (**b**) ESS power; (**c**) SOC.

Table 10. Summary of Case 2 simulation results.

Scenario	Original Peak Load	New Peak Load (with ESS)	Daily Energy Cost Reduction
Case 2	5614.1 [kW]	5420.8 [kW]	150.9 [$]

4.2.3. Case 3

In Case 3, the load data of 12 August 2017 are used. The peak is not high because Case 3 is a weekend load. The purpose simulating Case 3 is to confirm the power scheduling of the ESS when the peak load does not exceed the peak load that determines the demand cost. In Figure 11b,c, the ESS output power and SOC are similar to the patterns in Case 2. However, because the original peak load is low, the energy cost is minimized without violating the constraints. The peak load is increased by 539.7 kW when operating the ESS compared to the original peak load; however, it does not affect the demand cost. The results of Case 3 are summarized in Table 11.

Table 11. Summary of Case 3 simulation results.

Scenario	Original Peak Load	New Peak Load (with ESS)	Daily Energy Cost Reduction
Case 3	3246.7 [kW]	4929.6 [kW]	193.1 [$]

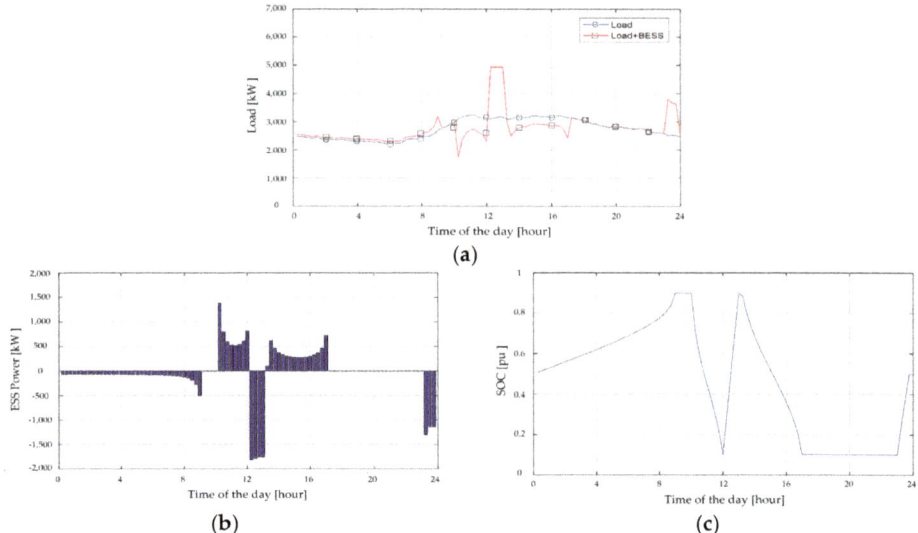

Figure 11. Case 3 simulation results. (**a**) Comparison of the original load and the load after ESS operation; (**b**) ESS power; (**c**) SOC.

5. Conclusions

We herein proposed a method to reduce the peak load by adjusting the operation time of the HVAC system at the peak load time. Algorithms were designed to receive the peak load time period and the electricity price as the input data; thus, the HVAC system performed uptime scheduling to minimize the peak load and energy cost. The thermal model provided by MATLAB was applied, and user convenience was considered. Through simulations, the operation of the algorithm was verified and the simulation results were analyzed. By operating the cooling in advance at a time when the electricity price was low, the indoor temperature was maintained within an appropriate range and the power consumption was reduced. In addition, by changing the room temperature limit during peak hours, we confirmed that the HVAC operation time and the peak of the entire load was reduced. In another energy management strategy, the ESS optimization algorithm was designed to schedule the outputs of the ESS systematically. This algorithm was divided into Stage 1 to reduce the peak load corresponding to the demand cost, and Stage 2 to minimize the overall energy cost. We confirmed that the scheduling was performed by distinguishing the stages through three simulations. The proposed algorithms demonstrated energy management strategies in a single building. The contribution of this paper is that different types of technologies can be controlled for the same purpose, peak load reduction and the energy cost saving. The HVAC system does not simply turn off at peak hours, but it can save the electricity cost while meeting user comfort. The ESS can reduce both the energy cost and the peak load by charging energy at the low price time and discharging at the high price time.

As future work, we will further study the design of an integrated aggregator for the participation of the electricity providers and consumers in many buildings applying the proposed algorithms.

Author Contributions: N.-K.K. carried out the main research tasks and wrote the full manuscript, and M.-H.S. provided technical support to verify the proposed algorithm in simulation software. D.W. validated and double-checked the proposed algorithm, the results, and the whole manuscript.

Funding: This research received no external funding.

Acknowledgments: This work was supported by INHA UNIVERSITY Research Grant.

Conflicts of Interest: The authors declare no conflicts of interest.

Energies **2018**, *11*, 2690

References

1. OECD. *OECD Factbook 2015–2016, Economic, Environmental and Social Statistics*; OECD Publishing: Paris, France, 2016.
2. Erdinc, O.; Tascikaraoglu, A.; Paterakis, N.G. End-User Comfort Oriented Day-Ahead Planning for Responsive Residential HVAC Demand Aggregation Considering Weather Forecasts. *IEEE Trans. Smart Grid* **2017**, *8*, 362–372. [CrossRef]
3. Ali, M.; Safdarian, A.; Lehtonen, M. Demand Response Potential of Residential HVAC loads Considering Users Preferences. In Proceedings of the IEEE PES Innovative Smart Grid Technologies Europe (ISGT Europe), Istanbul, Turkey, 12–15 October 2014.
4. Nguyen, D.T.; Le, L.B. Joint Optimization of Electric Vehicle and Home Energy Scheduling Considering User Comfort Preference. *IEEE Trans. Smart Grid* **2014**, *5*, 188–199. [CrossRef]
5. Mohsenian-Rad, H. Optimal Bidding, Scheduling, and Deployment of Battery Systems in California Day-Ahead Energy Market. *IEEE Trans. Power Syst.* **2016**, *31*, 442–453. [CrossRef]
6. He, G.; Chen, Q.; Kang, C.; Pinson, P.; Xia, Q. Optimal Bidding Strategy of Battery Storage in Power Markets Considering Performance-Based Regulation and Battery Cycle Life. *IEEE Trans. Smart Grid* **2016**, *7*, 2359–2367. [CrossRef]
7. Ji, Y.; Xu, P. A bottom-up and procedural calibration method for building energy simulation models based on hourly electricity submetering data. *Energy* **2015**, *93*, 2337–2350. [CrossRef]
8. Dejvises, J.; Tanthanuch, N. A Simplified Air conditioning Systems Model with Energy Management. *Procedia Comput. Sci.* **2016**, *86*, 361–364. [CrossRef]
9. Lin, Y.; Barooah, P.; Mathieu, J.L. Ancillary Services through Demand Scheduling and Control of Commercial Buildings. *IEEE Trans. Power Syst.* **2017**, *32*, 186–197. [CrossRef]
10. Hao, H.; Sanandaji, B.M.; Poolla, K.; Vincent, T.L. Aggregate Flexibility of Thermostatically Controlled Loads. *IEEE Trans. Power Syst.* **2015**, *30*, 189–198. [CrossRef]
11. Good, N.; Karangelos, E.; Navarro-Espinosa, A.; Mancarella, P. Optimization under Uncertainty of Thermal Storage-Based Flexible Demand Response with Quantification of Residential Users' Discomfort. *IEEE Trans. Smart Grid* **2015**, *6*, 2333–2342. [CrossRef]
12. De Dear, R.; Brager, G.; Cooper, D. *Developing an Adaptive Model of Thermal Comfort and Preference*; RP-884; ASHRAE: Atlanta, GA, USA, 1997.
13. Arabali, A.; Ghofrani, M.; Etezadi-Amoli, M.; Fadali, M.S.; Baghzouz, Y. Genetic-Algorithm-Based Optimization Approach for Energy Management. *IEEE Trans. Power Deliv.* **2013**, *28*, 162–170. [CrossRef]
14. Rahmani-Andebili, M. Scheduling deferrable appliances and energy resources of a smart home applying multi-time scale stochastic model predictive control. *Sustain. Cities Soc.* **2017**, *32*, 338–347. [CrossRef]
15. Rahmani-Andebili, M.; Shen, H. Price-Controlled Energy Management of Smart Homes for Maximizing Profit of a GENCO. *IEEE Trans. Syst. Man Cybern. Syst.* **2017**. [CrossRef]
16. Rahmani-Andebili, M. Cooperative Distributed Energy Scheduling in Microgrids. In *Electric Distribution Network Management and Control*; Springer: Singapore, 2018; pp. 235–254.
17. Rahmani-Andebili, M.; Shen, H. Energy Management of End Users Modeling their Reaction from a GENCO's Point of View. In Proceedings of the International Conference on Computing, Networking and Communications (ICNC), Silicon Valley, CA, USA, 26–29 January 2017.

Article

PLDAD—An Algorihm to Reduce Data Center Energy Consumption

Joao Ferreira [1,*,†,‡], Gustavo Callou [2,‡], Dietmar Tutsch [3,‡] and Paulo Maciel [1,†,‡]

1   Informatics Center, Federal University of Pernambuco, Recife 50740-560, Brazil; prmm@cin.ufpe.br
2   Departament of Computing, Federal Rural University of Pernambuco, Recife 52171-900, Brazil; gustavo.callou@ufrpe.br
3   Automation Technologye, Bergische Universität Wuppertal, D-42119 Wuppertal, Germany; tutsch@uni-wuppertal.de
*   Correspondence: jfsj3@cin.ufpe.br; Tel.: +55-81-2126-8430
†   Federal University of Pernambuco, Informatics Center, Cidade Universitária-50740-560-Recife/PE-Brazil.
‡   These authors contributed equally to this work.

Received: 1 August 2018; Accepted: 17 September 2018; Published: 19 October 2018

Abstract: Due to the demands of new technologies such as social networks, e-commerce and cloud computing, more energy is being consumed in order to store all the produced data. While these new technologies require high levels of availability, a reduction in the cost and environmental impact is also expected. The present paper proposes a power balancing algorithm (power load distribution algorithm-depth (PLDA-D)) to optimize the energy distribution of data center electrical infrastructures. The PLDA-D is based on the Bellman and Ford–Fulkerson flow algorithms that analyze energy-flow models (EFM). EFM computes the power efficiency, sustainability and cost metrics of data center infrastructures. To demonstrate the applicability of the proposed strategy, we present a case study that analyzes four power infrastructures. The results obtained show about a 3.8% reduction in sustainability impact and operational costs.

Keywords: energy flow model; dependability; sustainability; data center; power architectures; optimization

1. Introduction

Social awareness has influenced the way the world works and how people live. Widely available Internet access, the growing mobile market and advances in cloud computing technology are generating a huge amount of data, thus entailing unprecedented demands on energy consumption. The digital universe corresponds to 500 billion gigabytes of data [1], for which only 25% of the world's population is on-line [2].

Data center power consumption has increased significantly over recent years, influenced by the increasing demand for storage capacity and data processing [3–5]. In 2013, data centers in the U.S. consumed 91 billion kilowatt-hours of electricity [6], and this is expected to continue to rise. In addition, critical elements in the performance of daily tasks, such as social networks, e-commerce and data storage, also contribute to the rise in energy consumption across these systems.

Data center infrastructures require electrical components, many of which may directly affect system availability. Fault-tolerant mechanisms are key techniques for handling equipment with limited reliability. The Uptime Institute [7] is an institution that classifies the infrastructure of the data center based on the architectures and characteristics of redundancy and fault tolerance (Tier I, Tier II, Tier III and Tier IV). In this paper, four data centers were analyzed, considering different tiers of architectures for the power subsystem. The power subsystem electrical flow is represented by the energy flow model (EFM) [8].

The algorithm proposed in this paper, named power load distribution algorithm-depth (PLDA-D), improves the results presented in our previous work when considering the operational cost and energy efficiency of data centers [8–10]. Now, we have obtained the shortest path, using Bellman algorithm instructions, considering the energy cost as the main metric and the maximum energy flow (Ford–Fulkerson), considering the energy efficiency of each component of the data center's electrical infrastructure. Thus, we propose this new algorithm that uses two criteria of different classes that complement each other.

Both the proposed algorithm and the EFM model are supported in the Mercury modeling environment (see Section 4). In addition to the EFM model and the proposed algorithm, the Mercury environment also supports reliability block diagrams (RBD) [11], Markov chain [12] and stochastic Petri nets (SPN) [13] modeling, which are an essential part of the analysis. As such, the impact on the power subsystem reliability and availability was included.

The paper is organized as follows. Section 2 presents studies related to this research field. Section 3 introduces the basic concepts of the data center tier classification, sustainability and dependability. Section 4 presents an overview of the Mercury evaluation platform. Section 5 describes the energy flow model (EFM). Section 6 explains the PLDA-D. Section 7 describes the basic models adopted. Section 8 presents a case study, and finally, Section 9 concludes the paper and suggests directions for future work.

2. Related Works

Over the last few years, considerable research has been conducted into energy consumption in data centers. This section presents studies related to this research field.

Al-Fares [14] proposed an engine, called Hedera, for dynamic re-routing of the traffic of networking switch topologies that compose data center infrastructures. The main goal is to optimize the network bandwidth utilization with the proposed scheduler engine that has a minimal overhead on the available flows. Following the proposed approach by the authors, the bandwidth utilization was optimized to over 113% in relation to the static load-balancing strategies.

Dzmitry [15] proposed a methodology, named "Data center energy-efficient network-aware scheduling" (DENS) to manage job performance, energy consumption and data traffic. This proposed strategy is able to dynamically analyze the network feedback and make decisions to improve performance, energy consumption and the traffic. Therefore, the goal is to conduct the balance between those metrics, as well as to minimize the number of computing servers required on the data center that provide support to the services contracted.

These two papers are complementary to ours, as we propose a solution to reduce the energy consumption through the IT equipment of a data center, and those papers reduce the energy consumption by improving the network utilization.

Doria [16] extended the PowerFarm software [17] concept by adding an online monitor of loads and the correspondent power consumption. Additionally, the proposed EnergyFarm tool is able to turn off/on servers as needed according to the demands and respecting logical and physical dependencies. Therefore, the EnergyFarm turns off a set of servers to satisfy the required demand for storage on the data center, which reduces the overall system energy consumption, CO_2 emissions and the respective associated cost.

Heller et. al. [18] proposed an engine, named ElasticTree, for managing the power consumption of computer networks. The ElasticTree is able to dynamically adjust switches in order to couple with the changes of the traffic loads of data centers. The main goal is, besides reducing the energy consumption, to improve performance and the fault tolerance of the system under analysis as well. To accomplish this, methods (e.g., formal models, greedy bin-packer, heuristic and prediction methods) are proposed to decide which links and/or switches must be used.

Neto et. al. [19] proposed an algorithm, named MtLDF, to improve the load balance of fog systems considering performance metrics such as delay and priority. The authors have shown, through applied case studies, that the proposed method is able to reduce the energy consumption by improving the load distribution.

These previous related works are similar to the proposed strategy of our paper. However, none of them propose our own algorithm to minimize the energy consumption of the data center. The PLDA-D proposes the use of a new algorithm, based on the classic algorithms of minimum and maximum flow, making a mixture of both and obtaining a great result, with the search in depth.

3. Basic Concepts

This section discusses the basic concepts needed for a better understanding of the paper and presents an overview of the data center tier classification, followed by concepts regarding sustainability and combinatorial and state-based models. Finally, the concepts of Mercury environment and energy flow model are introduced.

3.1. Tier Classification

A data center infrastructure can be classified based on its redundancy features and fault tolerance ability [7]. This classification provides metrics to data center designers that identify the performance of the electrical infrastructureand strategies adopted. The following lines provide an overview of the four-tier classification.

3.1.1. Data Center Tier I (Basic)

This is a data center that does not offer redundant power and cooling infrastructures. A Tier I data center provides infrastructure to support information technology beyond office hours. Its infrastructure includes a dedicated area for the IT subsystem; a power subsystem with one uninterruptible power supply (UPS) to cope with power spikes and short outages; a dedicated cooling subsystem that does not shut down during office hours; and a generator to protect IT subsystem outages. Figure 1 depicts an example of the power system infrastructure for the Tier I data center.

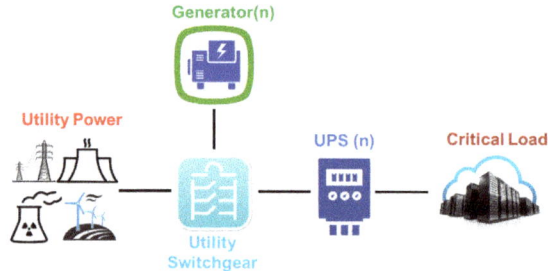

Figure 1. Tier I data center power subsystem.

We discuss how to manage schemas and their evolution for the last two scenarios (static schema management is straightforward and ignored here).

3.1.2. Data Center Tier II (Redundant Components)

A Tier II data center incorporates redundant critical power and cooling components, but with a single power distribution infrastructure. This infrastructure supports planned maintenance activities without interrupting the service, reducing as a result the system downtime. The redundant components include power and cooling equipment, such as UPS, chillers, pumps and engine generators. Figure 2 depicts an example of the power subsystem infrastructure assuming the Tier II classification.

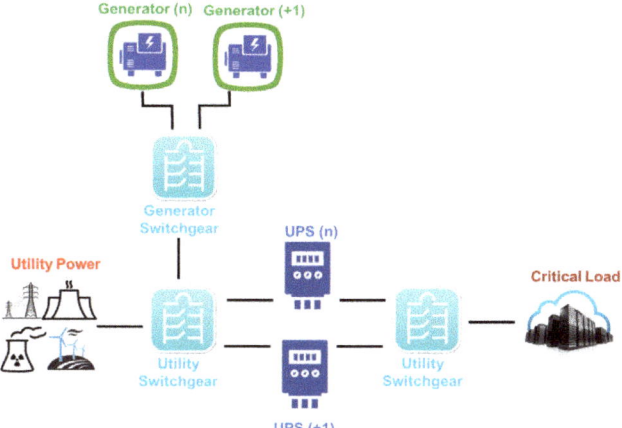

Figure 2. Tier II data center power subsystem.

3.1.3. Data Center Tier III (Simultaneous Maintenance and Operation)

A Tier III data center does not require shutdowns for equipment replacement or maintenance. The Tier III configuration considers the Tier II arrangement including a redundant independent power path (as shown through Figure 3). Therefore, each power component may be shutdown for maintenance without impacting the IT system's operation. Similarly, a redundant cooling subsystem is also provided. These data centers are not susceptible to downtime for planned activities and accidental causes. Planned maintenance activities may be carried out using the redundant components and capabilities of the reference distribution so as to ensure the safe operation of the remaining components.

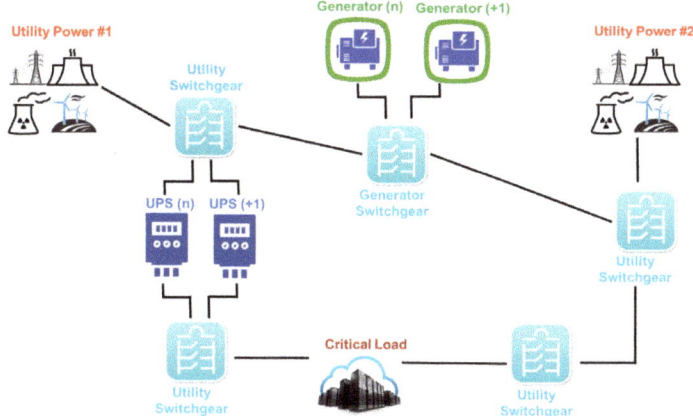

Figure 3. Tier III power system from utility to IT equipment.

3.1.4. Data Center Tier IV (Fault-Tolerant Infrastructure)

A Tier IV adopts the Tier III infrastructure by adding a fault-tolerant mechanism, in which independent systems (electrical and cooling) are present. This tier classification is suitable for international companies that provide 24/7 customer services (as shown through Figure 4).

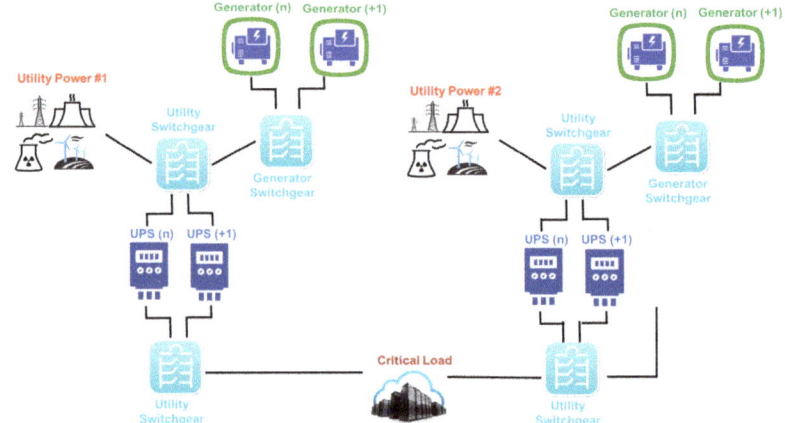

Figure 4. Tier IV power system from utility to IT equipment.

3.2. Sustainability

The concept of the green data center is related to electricity consumption and CO_2 emissions, which depend on the utility power source adopted. For example, in Brazil, 73% of electrical power is derived from clean electricity generation [8], whereas in the USA, 82.1% of generated electricity comes from petroleum, coal or gas [20]. Figure 5 depicts the relationship between the type of material used for power generation in Brazil and the USA.

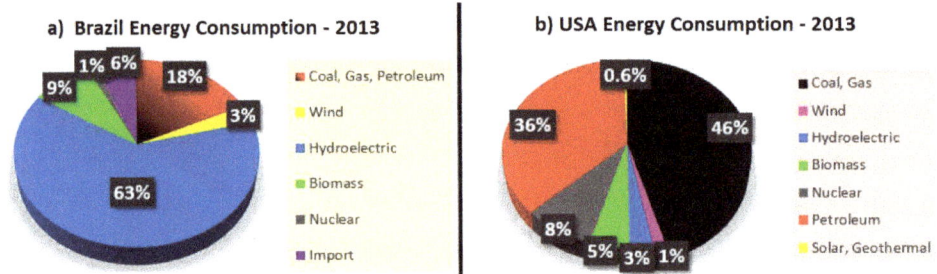

Figure 5. Energy Consumption: Brazil vs. USA.

Several methods and metrics are available for comparing equipment from a sustainability viewpoint.

Exergy is a metric that estimates the energy consumption efficiency of a system. It is defined as the maximal fraction of latent energy that can be theoretically converted into useful work [21].

$$Exergy = Energy \times F \tag{1}$$

where F is a quality factor represented by the ratio of $Exergy/Energy$. For example, F is 0.16 for water at 80 °C, 0.24 for steam at 120 °C and 1.0 for electricity [21].

The PUE (power usage efficiency) is defined as the total load of the data center ($C_{infrastructure}$) divided by the total load of the IT equipment installed (C_{TI}).

$$PUE = \frac{C_{infrastructure}}{C_{TI}} \tag{2}$$

3.3. Combinatorial and State-Based Models

RBD [22], fault trees [11], SPNs [23] and Continuous Time Markov Chains (CTMC) [12] have been used to model fault-tolerant systems and to evaluate some dependability measures. These model types differ in two aspects, i.e., simplicity and respective modeling capability. RBD and fault trees are combinatorial models, so they capture conditions that make a system fail in the structural relationships between the system components. They are more intuitive to use, but do not allow one to express dependencies between system's components. CTMC and SPN models represent the system behavior (failures and repair activities) by its states and event occurrence expressed as labeled state transitions.

These state-based models enable the representation of complex relations, such as active redundancy mechanisms or resource constraints [22,24]. The combination of both types of models is also possible, allowing one to obtain the best of both worlds, via hierarchical modeling. Different model types can be combined with different levels of comprehension, leading to composite hierarchical models. Heterogeneous hierarchical models are being used to deal with the complexity of systems in other domains, such as sensors networks, telecommunication networks and private cloud computing environments.

3.3.1. Reliability Block Diagram

The reliability block diagram (RBD) [25] is a technique for computing the reliability of systems, using intuitive block diagrams. The RBD is able to represent the component's interaction and to verify the relationship over the failed and active status of elements that keeps the system operational.

Figure 6a depicts a series relationship, where the system fails by the failure of a single component. Considering n independent components, the reliability is obtained by Equation (3)

$$P_s = \prod_{i=1}^{m}(P_i) \tag{3}$$

where P_i is the reliability—$R_i(t)$ (instantaneous availability ($A_i(t)$) or steady state availability (A_i))—of block b_i.

Figure 6b shows a parallel arrangement, where the system continues to be operational, even with the failure of a single component. Considering n independent components, the reliability is obtained by Equation (4):

$$P_p = 1 - \prod_{i=1}^{m}(1 - P_i) \tag{4}$$

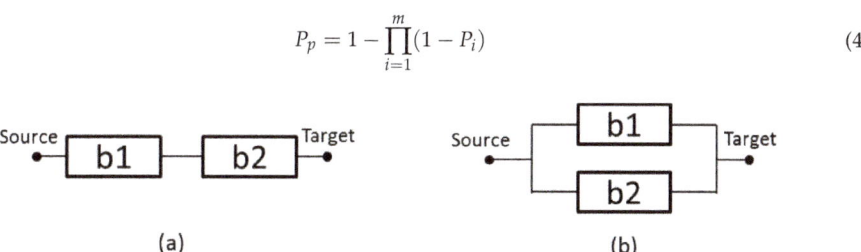

(a)   (b)

Figure 6. (**a**) Serial arrangement; and (**b**) parallel configuration.

For other examples and closed-form equations, the reader should refer to [11].

3.3.2. Stochastic Petri Nets

The Petri net (PN) [26] is able to represent concurrency, communication mechanisms, synchronization and a natural representation of deterministic and probabilistic systems. PN is a graph, in which places are represented by circles and transitions are shown as rectangles. Directed arcs are used to connect places and transitions and vice versa.

This paper considers stochastic Petri nets for conducting dependability analysis of data center power architectures. Figure 7 represents the SPN model of a "simple component", where the places' states are X_ON (activity) and X_OFF (inactivity). When the number of tokens (#) in the place X_ON is greater than zero, this means the component is operational. Otherwise, the component has failed. MTTFand MTTRof the system are used to compute the availability, and these parameters are not shown in the figure, but are associated with the transitions $X\_Failure$ and X_Repair.

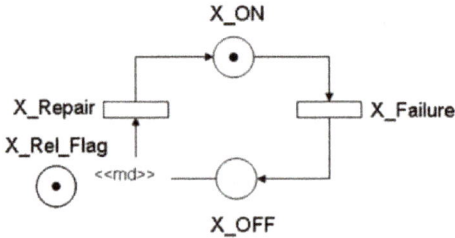

Figure 7. Simple component model.

The expression IF ($\#X\_Rel\_Flag = 1$) : 2 ELSE 1 defines the multiplicity (« md »), represented by the arc from X_OFF to X_Repair. The place X_Rel_Flag is adopted to let one conduct the evaluation of availability or reliability according to the marking of the place p. $\#XRel\_Flag = 1$ means the reliability model is set; otherwise, we have the availability model.

If the number of tokens in the place is zero (X_Rel_Flag and ($\#X\_Rel\_Flag = 0$)), the probability $P\#X\_ON > 0$ computes the component's availability. Otherwise, ($\#X\_Rel\_Flag = 1$); then the probability $P\#X\_ON > 0$ allows one to compute the component's reliability. That enables us to parameterize the model, allowing the system evaluation, considering or not the repair.

3.3.3. Continuous Time Markov Chains

Markov chains can be adopted to analyze various types of systems. A Markov process does not have memory; therefore, it has no influence from the past. The current state is enough to know the future steps. A Markov chain occurs when the process has a discrete state space. These states represent the different conditions that the system may be in. The events are represented by the transitions between the states.

In Figure 8, a new task is represented by the arc with rate λ. The arc with rate μ represents the server. This model depicts a system with two servers that compute received jobs. Considering the number of busy servers as a time function, it is possible to assume the function $X(t)$ or a random variable. The state $X_n(t)$ is named as any modification of X over (t). The state space of the model is the set of all possible states. Therefore, we can compute the transition probabilities from a state to its successor $X_{n+1}(t)$.

In order to accomplish this, it is necessary to define the probability distribution function of $Xn(t)$. Stochastic processes are these random functions of time, where this variable changes its state over time [22].

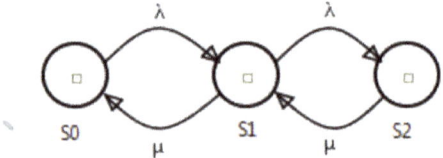

Figure 8. Example of a Continuous Time Markov Chains (CTMC) model.

4. Mercury

The Mercury environment [27,28] was developed by the MoDCS[28] research group for building and evaluating performance and dependability models. The proposed environment can be adopted as a modeling tool for the following formalisms: CTMC [12], RBD [11], EFM [9] and SPN [13,29,30].

Mercury offers useful features that are not easily found in other modeling environments, such as:

- More than 25 probability distributions supported in SPN simulation;
- Sensitivity analyses of CTMC and RBD models;
- Computation of reliability importance indices; and
- Moment matching of empirical data

Figure 9 details the functionalities available in the Mercury environment. The optimization module is able to evaluate the supported models (SPN, RBD, CTMC and EFM) through optimization techniques. In our previous study, we implemented GRASP (Greedy Randomized Adaptive Search Procedure) [31] and PLDA [9]. This paper proposes the PLDA-D as a great improvement over the PLDA. This is because in a single search, the PLDA-D considers two criteria for stopping, i.e., minimum flow for the lowest cost (Bellman) and maximum flow for energy (Ford–Fulkerson), with a scan of the graph in depth for each possible path.

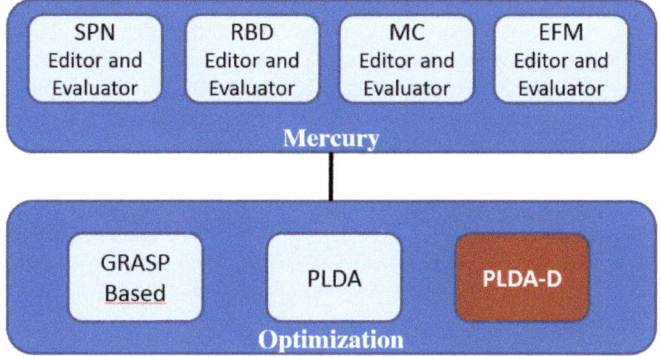

Figure 9. Evaluation environment. SPN, stochastic Petri nets; RBD, reliability block diagrams; EFM, energy-flow models. PLDA-D, power load distribution algorithm-depth.

5. Energy Flow Model

The EFM represents the energy flow between the components of a cooling or power architecture, considering the respective efficiency and energy that each component is able to support (cooling) or provide (power). The EFM is represented by a directed acyclic graph in which components of the architecture are modeled as vertices and the respective connections correspond to edges [8,32]. For more details about the formal definitions of the EFM, the reader is redirected to [32].

An example of EFM is shown in Figure 10. The rounded rectangles equate to the type of equipment, and the labels name each item. The edges have weights that are used to direct the energy that flows through the components. For the sake of simplicity, the graphical representation of EFM hides the default weight of one.

TargetPoint1 and *SourcePoint1* represent the IT power demand and the power supply, respectively. The weights of the edges, i.e., 0.7 and 0.3, are the energy flows via the uninterrupted power supply (UPS) units, UPS1 at 70% and UPS2 at 30%, respectively, for meeting the power demand from the IT system.

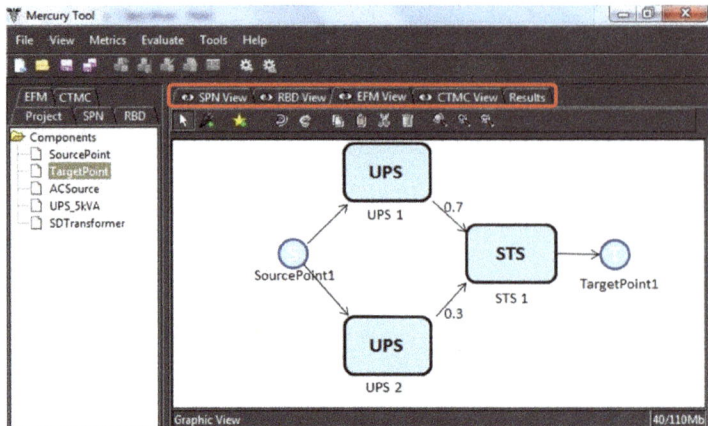

Figure 10. EFM example in the Mercury tool. STS, static transfer switch.

The EFM is employed to compute the overall energy required to provide the necessary energy at the target point. If we consider that the demand from the data center computer room is 100 kW, this value is thus associated with *TargetPoint1*. Assuming that the efficiency of STS1 (static transfer switch) is 95%, the electrical power that the STS component receives is 105.26 kW.

A similar strategy is adopted for components UPS1 and UPS2, however now, dividing the flow according to the associated edge weights, 70% (73.68 kW) for UPS1 and 30% (31.27 kW) for UPS2. Thus, the UPS1 needs 77.55 kW, considering 95% efficiency, and UPS2 needs 34.74, considering 90% electrical efficiency. The *Source Point 1* accumulates the total flow (112.29 kW).

The edge weights are specified by the designer of the model, and there is no guarantee that the best values for the distribution were defined; as a result, higher power consumption may be reached. This work aims at solving such an issue by automatically setting the edge's weight distribution of the EFM model with the PLDA-D algorithm. Therefore, our approach is able to achieve lower power consumption for the system.

Cost

In this paper, the operational cost considers the data center operation period, energy consumed, energy cost and the data center availability. Expression (5) denotes the operational cost.

$$OpCost = P_{Input} \times C_{Energy} \times T \times (A + \alpha(1 - A)) \tag{5}$$

P_{input} is the power supply input; C_{energy} is the energy cost per energy unit; T is the considered time period; A is the system availability; α is the energy percentage that continues to be consumed when the system fails.

6. Power Load Distribution Algorithm in Depth Search

The power load distribution algorithm-depth (PLDA-D) is proposed to minimize the electrical energy consumption of the system represented through EFM models [32]. PLDA-D is a depth search extension of PLDA [9,10]. The Bellman–Ford algorithm [33] is used for searches of the smaller path in a weighted digraph, whose edges have a weight, including a negative one. The Ford–Fulkerson algorithm [34] is used when it is desired to find a maximum flow that makes the best possible use of the available capacities of the network in question. The PLDA-D is a blend of these algorithms since it uses the characteristics of Bellman–Ford to choose the lowest cost, considering the weights of each node and the attributes of Ford–Fulkerson to pass the more significant amount of energy by a specific path, considering the energy efficiency of each piece of equipment.

The time and space analysis of depth-first-search (DFS) differs according to its application area. DFS traverses an entire graph with time Θ ($|V| + |E|$), linear in the graph size. In the worst case, it adopts the space O($|V|$) to store the set of vertices on the actual search path like the stack of vertices visited [35]. Thus, in this setting, the time and space bounds are the same as for breadth-first search, and the choice of which of these two algorithms to use depends less on their complexity and more on the different properties of the vertex orderings the two algorithms produce.

In this case, for the properties of a data center's electrical infrastructure, the depth search implemented in PLDA-D offers an optimal solution, whereas the width search performed in PLDA guarantees only a good solution.

The PLDA-D is divided into three phases: initialization, kernel calculations and search.

6.1. Initialization

This phase initializes variables, calls PLDA-D and computes the input power assigned to the EFM. In Algorithm 1 (initialize PLDA-D), the power infrastructure is represented by graph G (EFM model). Variable R stores a copy of G, so the original EFM is preserved (Line 1). The accumulated cost (*AccumCost*) of the variables, the capacities of the equipment (ccu_v), edge weights ($weigh_e$) and the input power (*inputPower*) are initialized with values of zero (Lines 2–11).

Algorithm 1 Initialization PLDA-D (G).

1: $R = G$;
2: **for** $v \in R$ **do**

3:     $ccu_v = 0$;
4:     $ActualCost_v = \infty$
5:     $AccumCost_v = 0$;
6:     $Child_v = null$
7: **end for**
8: **for** $e \in R$ **do**

9:     $weigh_e = 0$;
10: **end for**
11: $inputPower = 0$;
12: **for** $t \in V_{target}$ **do**

13:     $R = PLDADKernel(R, f_d(t), s)$;
14: **end for**
15: $setUpdateWeight(R)$;
16: **return** R;

$ActualCost_v$ of each node is initialized with a symbol denoting an infinite quantity (Line 4), and the variable *Child* is initialized with a null value. This variable is adopted to create a relationship between nodes (Line 6).

Lines 12–14 call the PLDA-D function for each target node vertex (if there is more than one target in the EFM). The number of calls corresponds to the number of target nodes on G. If there is more than one target node, the energy flow will be distributed considering each.

The EFM edge weights are updated considering the accumulated flow of each component (Line 13).

6.2. Kernel Calculations

The kernel calculations, depicted in Algorithm 2 (Algorithm 2: PLDA-D kernel calculations), execute a loop with two stop criteria. First, it is checked if the demand is higher than zero ($f_d(t) > 0$) and if there is a valid path from $Target(t)$ node to $Source(s)$ node (isPathValid(R,t,s)), where t is target node and s is the source. A valid path is a path from one node to another where the electrical capacity of all components in this path is respected.

Algorithm 2 PLDA-D kernel calculations (R, f_d(t), t, s).

1: **while** $(f_d(t) > 0)$ & (isPathValid(R, t, s)) **do**

2:     $P = getElementsFromBestPath(t)$;

3:     $pf = \infty$;

4:     **for** $i \in P$ **do**

5:         $pf = getMinimumCapacity(pf, fc_{(i)} - ccu_{(i)})$;

6:     **end for**

7:     **for** $i \in P$ **do**

8:         $ccu_i = ccu_i + pf$;

9:         **if** $(fc_{(i)} - ccu_{(i)} = 0)$ **then**

10:             $i.reachedLimit()$

11:         **end if**

12:     **end for**

13:     $f_d(v) = f_d(t)$ - pf;

14: **end while**

15: **return** R

The function $getElementsFromBestPath(t)$ aims at finding the best path from the target to the source node, according to the efficiencies and respecting the capacity of each element. This function is explained in the following section.

The path is stored in list P (Line 2), then the infinity symbol is assigned to the variable pf (possible flow) (Line 3), which stores the possible energy flow in the path. In the first loop (Lines 4–6), pf receives the value returned from getMinimumCapacity() for each path (Line 5), which returns the lower value between the actual possible flow (pf) and the difference between the flow capacity supported by the node $fc_{(i)}$ and the actual flow $ccu_{(i)}$.

The smallest possible value is added to each node of the path. The second loop (Lines 7–12) stores the accumulated flow (ccu_i). The limit of each piece of equipment is respected ($i.reachedLimit()$) (Line 10). In the next valid path query, those that possess a selected node as a limit reached will be disregarded.

The demanded energy of the target node ($f_d(t)$) is updated (Line 13), subtracting the previously transmitted flow from its values. The previous steps are repeated until all valid paths have been analyzed or there is no demand. Finally, residual graph R is returned (Line 15), and only the edge weights are changed from the original graph G.

6.3. Search

We proposed our own version of an algorithm to compute maximum and minimum flows, which was implemented based on the Bellman [33] and Ford and Fulkerson [34] algorithms. More detail is provided in this section.

All paths are traversed from the target to source node. A cost is assigned to each node (component); the lower the value, the better the path. Once the cost associated with each path is calculated, it is possible to direct the flow to the better paths in relation to the electrical energy consumption.

Algorithm 3 (Algorithm 3: best patch choice) shows a function, called "$getElementsFromBestPath$", responsible for identifying the best path through the nodes *Target* to *Source*. In the first execution, the value passed as a parameter (CurrentVertices) is the *Target* node. *ListOfParents* stores the list of nodes with one level of the current node precedence, i.e., the list of parents (Line 1).

Line 2 starts a loop to each node of the list of parents. The first step of the loop chooses one item of equipment from a list of parents to begin the procedure (Line 3). The order of choice does not influence the search. The limit of capacity is verified in Line 4. If the node has reached its capacity limit, the algorithm proceeds looking for other paths available.

Algorithm 3 *Search getElementsFromBestPath(CurrentVertices).*

1: $ListOfParents = getListofParentsFrom(CurrentVertices)$
2: **for** $i \in ListOfParents.size$ **do**

3:    $CurrentParent = ListOfParents[i]$
4:    **if** $CurrentParent.LimitNotReached()$ **then**

5:       $newCost = 1/(CurrentVertices.getAccumCost * (CurrentParent.efficiency/100))$
6:       **if** $CurrentVertices = Target$ **then**

7:          $CurrentParent.setActualCost(newCost)$
8:          $CurrentParent.setChild(CurrentVertices)$
9:          $CurrentParent.setAccumCost($
10:         $CurrentParent.efficiency/100)$
11:       **else**

12:         **if** $CurrentParent.ActualCost > CurrentVertices.getActualCost + newCost$ **then**
13:         $CurrentParent.setActualCost($
14:         $CurrentVertices.getActualCost + newCost)$
15:         $CurrentParent.setChild(CurrentVertices)$
16:         $CurrentParent.setAccumCost($
17:         $CurrentParent.efficiency/100) * CurrentVertices.getAccumCost$
18:       **end if**
19:       $getElementsFromBestPath(CurrentParent)$
20:    **end if**
21: **end for**
22: $ListOfElements.addBestChild(Source)$
23: **return** $(ListOfElements)$

Line 5 computes the cost of each component node, in which the shortest value represents the best path choice. Line 6 verifies if the vertex under analysis (CurrentVertices) is the *Target*. In this case, the *CurrentParent* cost receives *newCost*; the *CurrentVertices* is assigned as the *Child* of *CurrentParent*; and the accumulated cost is computed (Lines 7–9). The accumulated cost represents the cost of the node multiplied by the cost of the path that precedes it. This step draws the best path.

Assuming the *CurrentVertices* are not the *Target*, Line 11 conducts a check that is only satisfied when there is at least one path with a lower cost to be reached.

In this case, the *CurrentParent* cost is updated to the sum of the cost of the *CurrentVertices* plus the *newCost*. The *CurrentVertices* will be the *"best child"* for the *CurrentParent*, and the *CurrentParent* cost is updated considering this new path (Lines 12–14).

In Line 19, a list of elements is filled with the children of the *Source* node, which corresponds to the best path from the *Target* to the *Source* node, according to the expression of Line 5. Finally, a list with the elements of the best path is returned (Line 20).

6.4. PLDA-D Execution

Figure 11 illustrates the step-by-step execution of the PLDA-D. Figure 11a shows an EFM composed of three electrical components *A*, *B*, and *C*, with an efficiency of 80, 90 and 95%, respectively. *S* is the *Source* node, representing an electrical utility, and *T* is the *Target* node, representing a computer room.

The demand (*Dem*) and the efficiency (*Ef*) values are specified by the data center designer. The *Target* node *Acc* value is set to one. The other accumulated costs (*Acc*) are set to zero, and the edge weights are set to the default value, respectively, as depicted in Figure 11a. Phase 1 of the PLDA-D is represented by Figure 11b, where all variables of all vertices are initialized.

Phase 2 starts in Figure 11c following until Figure 11h. The best path is selected according to the efficiency of each component. In Figure 11c, the values of the *ActualCost* and *AccumulatedCost* are computed, and the best child is chosen according to the lowest value of the variable *ActCost*.

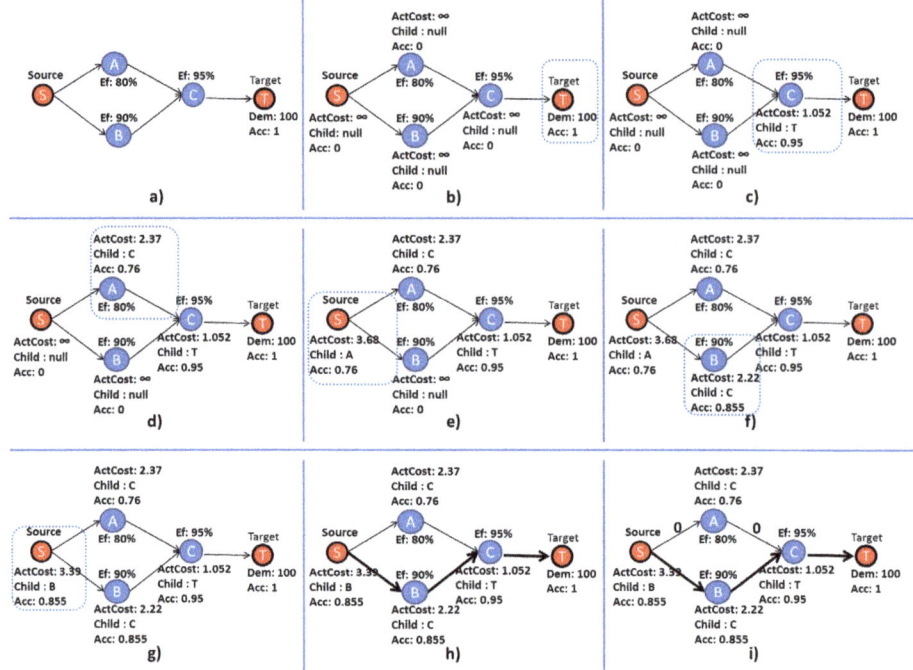

Figure 11. Example PLDA-D execution; blue rectangles highlight the nodes under analysis.

Next, one of the parents of node *C* is chosen; the order of choice does not influence the search. Node *A* was selected, and the values of *Acc* and *ActCost* were computed; the best child is node C; see Figure 11d.

The values of *ActCost* and *Acc* were computed according to Equations (6) and (7), as described in Lines 5 and 14 of Algorithm 3.

$$ActCost = ActCost + \frac{1}{ActCost \times \frac{efficiency}{100}} \tag{6}$$

$$Acc = Acc \times \frac{efficiency}{100} \tag{7}$$

Figure 11e shows the algorithm step in which *Acc* and *ActCost* of all variables were computed and the best child selected. The *Source* is a terminal node and has no parent; thus, the algorithm returns to the node *C* that has two parents. Node *C* has not been thoroughly researched, because there is an unvisited parent node *B*. Figure 11f shows the algorithm step once the variables for node B have been computed.

Figure 11g depicts the step after calculating variables *Acc* and *ActCost* and verifies that the *ActCost* for the current path (3.39) was less than the *ActCost* of the previous path (3.68) for reaching the *Source* node. Thus, the *Source* node has changed the values of its variables, and the best *Child* is now node *B* and no longer node *A*. In other words, the path passing through the node B represents a better choice than passing through the node A.

Figure 11h represents the end of Phase 3, which returns the best path to Phase 2. The best path from the target to source node is: *Target*, *C*, *B*, *Source*. Figure 11i presents the flow distributed by Phase 2. After that, the EFM computes the minimum possible value for the input power.

7. Basic Models

This section presents the analysis of the proposed models for representing the previous four-tier configurations. The baseline architecture is modeled with RBD; however, RBD models cannot completely represent complex systems with dependencies between components.

State-based methods can represent these dependencies, thereby allowing the representation of complex redundant mechanisms. The Achilles heel for state-based methods is the exponential growth of the state space as the problem becomes large, which can either increase the computation time or make the problem mathematically intractable. However, strategies for hierarchical and heterogeneous modeling (based on states and combinatorial models) are essential to represent large systems with complex redundancy mechanisms [22]. MC, SPN, RBD and EFM models were utilized to evaluate the four tiers. The availability was obtained by the RBD, MC or SPN model. The other metrics (cost, PUE, input power) were achieved through the EFM evaluation.

7.1. Tier I Models

Figures 12 and 13 depict the RBD models for power and cooling architectures of Tier I, respectively. The power and cooling architectures were evaluated separately.

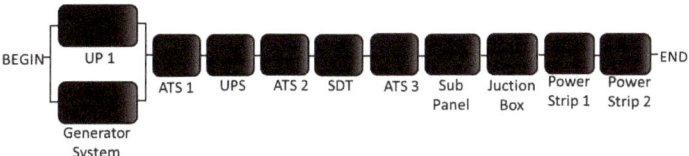

Figure 12. RBD model of the Tier I power infrastructure.

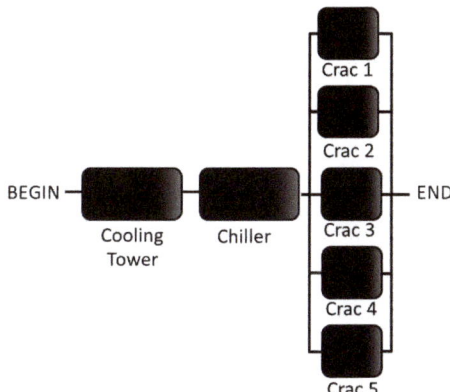

Figure 13. RBD model of the Tier I cooling infrastructure.

After that, we assumed that the system was only operational once both the cooling and power system were working. Therefore, the previous availability results were put together in a serial relationship, meaning that the failure of an electrical device would also affect the cooling equipment. Moreover, the system availability was compared with the Up Time Institute [7], in which there can be no doubt that the results achieved are equivalent.

Once the availability was computed, the EFM shown in Figure 14 was adopted for computing, for instance, cost and operational exergy. Only the electrical infrastructure was consider in the EFM model.

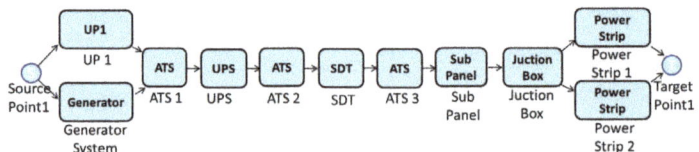

Figure 14. EFM model of Tier I.

7.2. Redundancy N + 1

Redundancy N + 1 is adopted in utility power and generator systems for Tiers II, III and IV. This redundancy is a form of ensuring system availability in the event of component failure. Components (N) have at least one independent backup component (+1). This paper considers redundancy $N + 1$ (generator and UPS), as there is a demand for at least two pieces of equipment. One machine works with a spare backup; thus, N is assumed to be two.

The RBD model is used to obtain the dependability metrics of the electrical infrastructure of data center Tiers II, III and IV. However, due to the system complexity of the redundancy ($N + 1$), the utility power and generator subsystem were modeled in SPN (Figure 15 depicts the corresponding SPN model for that system). This model represents the operational mode of the utility power and generator system, in which the system is operational if the power supply utility (#$C\_UP = 1$) and the two main generators are operating (#$G12\_Up = 2$) or if one main generator and one backup is running, i.e., ((#$G12\_Up = 1$) and (#$Gb\_Up = 1$)).

In this SPN model, the transaction that activate Generators 1 and 2 (G12 Act) is only fired when the power supply utility has failed. Similarly, the transaction *Gb Act* is able to fire once the power supply utility and at least one main generator have failed.

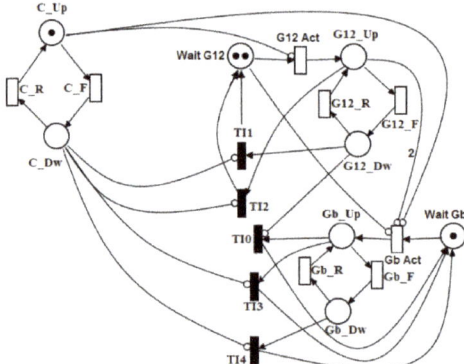

Figure 15. SPN model for the utility power and generator system (UP + GS).

The availability expression obtained by the SPN model is:

$$A = (C_Up = 1)OR(G12_Up = 2)OR((G12_Up = 1)AND(Gb_Up = 1))$$ (8)

The UPS system is modeled with redundancy ($N + 1$), assuming a cold standby strategy. A cold standby redundant system considers a non-active spare component that is only activated when the main active component fails. The components of the UPS system are based on a non-active redundant module that expects to be active when the main module fails. The operational mode of this system considers that at least two UPSs must be active. Figure 16 depicts the Markov chain model adopted to evaluate the availability of the UPS system with redundancy ($N + 1$) in cold standby.

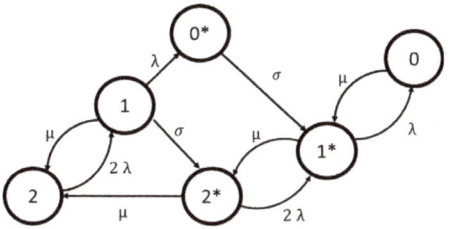

Figure 16. CTMC model for the UPS cold standby system (UPS system).

In Figure 16, State 2 represents the two standard UPSs operating and the backup waiting. State 1 shows the detection of a fault in one UPS. State 2* represents two UPSs operating (one standard UPS and one backup). State 1* represents a fault in the standard or backup UPS. State 0 represents the fault of all the UPS's. State 0* shows the fault of two standard UPSs and the operating of the backup. The failure rate is represented by λ; μ is the repair rate; σ is the mean time to activate the backup UPS.

The availability expression obtained by the CTMC model is A'/A'', where:

$$A' = (\mu\sigma(4\lambda^3 + 4\lambda^2(\mu + \sigma) + \mu^2(\mu + \sigma) + \lambda\mu(3\mu + 2\sigma))) \tag{9}$$

$$A'' = (2\lambda^2\mu^3 + (\lambda + \mu)(2\lambda + \mu)(2\lambda^2 + \mu^2)\sigma + (4\lambda^3 + 4\lambda^2\mu + 2\lambda\mu^2 + \mu^3)\sigma^2) \tag{10}$$

7.3. Tier II Models

Availability results are obtained through the evaluation of these SPN models, as well as the RBD and MC. We use two-level hierarchical models in which RBD is used to represent the overall system on the upper level, and SPN and MC are used to capture the behavior of the subsystem on the lower level, as power and UPS systems. Figure 17 depicts the RBD model adopted to represent the power infrastructure.

The values of the GS + UP1 (Generator_System and UtilityPower1) and the UPS_System used in the RBD models of Tiers II, III and IV are computed through the SPN and MC models in Figures 15 and 16, respectively. The availabilities of the SPN and MC models are computed and inserted into each block of the RBD (e.g., UP1 + GS) models.

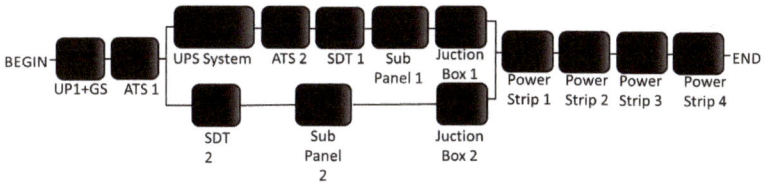

Figure 17. RBD model of Tier II.

Figure 18 depicts the EFM model of the electrical infrastructure of data center Tier II. As the reader may observe, there is a difference between the representation of dependability models and the electrical flow to the power strip component.

At first, representation in series signifies that the failure of one component affects the operation of the data center. In the second, the parallel representation signifies that the electrical flow is distributed by all power strip devices.

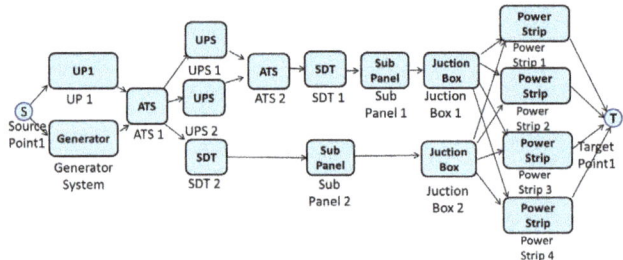

Figure 18. EFM model of Tier II.

7.4. Tier III Models

The data center Tier III model uses hierarchy to represent the UPS system and power generation system. The Tier III model is divided into subsystems; two of them represent the power and UPS systems previously presented (Figures 15 and 16). One path of the electrical flow uses the UPS system with redundant components (Subsystem X), and the other path has no redundant components (Subsystem Y). Both provide possible paths to the set of power strip components (Subsystem P). The availability algebraic expressions of each subsystem is shown in Equations (11)–(13).

$$SubsystemX = ATS1 \times UPSSystem \times ATS2 \times SDT1 \times SubPanel1 \times JuctionBox1 \tag{11}$$

$$SubsystemY = SDT1 \times SubPanel2 \times JuctionBox2 \tag{12}$$

$$SubsystemP = \prod_{i=1}^{n}(PowerStrip_{(n)}) \tag{13}$$

where n is six in this model. Equation (14) shows the algebraic availability expressions of all subsystem (X, Y, P) that compose Tier III.

$$TierIII = (1 - (1 - (UPS\_GS) \times (ATS1 \times UPSSystem \times ATS2 \times SDT1 \times \\ SubPanel1 \times JuctionBox1)) \times (1 - (UP2) \times (SDT1 \times SubPanel2 \times JuctionBox2))) \times \\ (\prod_{i=1}^{n}(PowerStrip_{(n)})) \tag{14}$$

Once availability is computed, the EFM model can be analyzed to provide cost and operational exergy, as well as to ensure that the power restrictions of each device are respected. Figure 19 presents the EFM model adopted for Tier III.

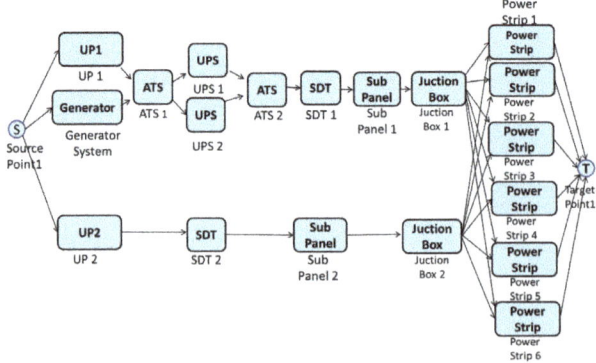

Figure 19. EFM model of Tier III.

7.5. Tier IV Models

Tier IV is the highest level of assurance that a data center can offer. This data center category is fully redundant in terms of electrical circuits (see Figure 4).

The RBD of Tier IV is modeled using a similar approach to Tier III, with hierarchical models. Five subsystems are used, two representing the power and UPS system (Figures 15 and 16). There are two redundant paths of electrical flow, both with redundant UPS systems. One path, named Subsystem Z (see the availability algebraic expression in Equation (15)), is composed of ATS1, UPS System 1, ATS2, SDT1, SubPanel 1and JunctionBox1. The set of power strips for data center Tier IV is present in Equation (16), where m is eight.

$$SubsystemZ = ATS1 \times UPSSystem1 \times ATS2 \times SubPane1 \times JuctionBox1 \tag{15}$$

$$SubsystemPS = \prod_{i=1}^{m}(PowerStrip_{(m)}) \tag{16}$$

There are two utility powers, each with a backup generator system (UtilityPower1 + GeneretorSystem1 and UtilityPower2 + GeneretorSystem2). The availability algebraic expression of Tier IV is presented in Equation (17).

$$TierIV = (1 - (\prod_{i=1}^{n}(1 - (UtilityPower\_GeneretorSys)_{(n)} \times (ATS1 \times UPSSystem1 \newline \times SubPane1 \times ATS2 \times JuctionBox1)_{(n)})) \times (\prod_{i=1}^{m}(PowerStrip_{(m)})) \tag{17}$$

After computing the availability value of Tier IV, the EFM depicted in Figure 20 is adopted.

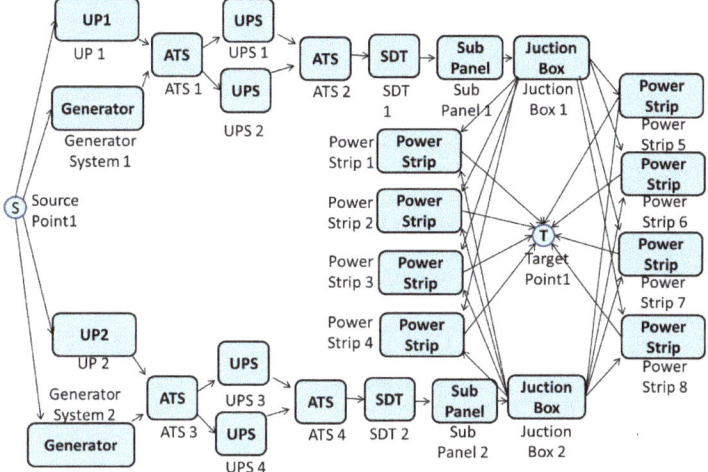

Figure 20. EFM model of Tier IV.

8. Case Study

The main goal of this case study is to validate the proposed models and to show the applicability of the PLDA-D algorithm, considering the data center power infrastructure of Tiers I, II, III and IV. To conduct the evaluation, the environment Mercury was adopted. In addition to computing the dependability metrics, Mercury is adopted for estimating the cost and sustainability impact, as well as to conduct the energy flow evaluation and propose a new one, according to the optimization of the PLDA-D algorithm. Figure 21 depicts the connections between cooling components and electrical infrastructure.

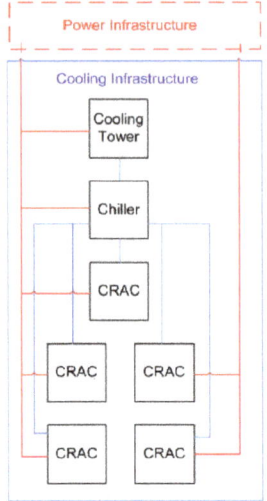

Figure 21. Cooling connections to power the infrastructure.

To validate the Tier I model, the cooling and power infrastructure were evaluated together. The value of the availability proposed for Tier I according to the Up Time Institute is 0.9967. The availability obtained from the RBD model of Figures 12 and 13 was 0.9952. To validate the proposed model, the relative error was used, to compare the difference between the results. Considering the relative error (presented in Equation (18)), the value of 0.0015 was reached.

$$RelativeError = \frac{(Theoretical - Experimental)}{Experimental} \tag{18}$$

A very small value for the relative error was found; therefore, we consider the proposed model to be an accurate representation of the Tier I model. The same strategy was adopted to validate the other proposed models of Tiers II, III and IV.

Table 1 shows the MTTF and MTTR values for each device. These values were obtained from [36].

Table 1. MTTRand MTTRvalues.

Equipment	MTTF (h)	MTTR (h)
AcSource	5380.0	8
Generator	3190.0	8
ATS	282,581	8
UPS	60,000	8
SDT	1,412,908	8
Subpanel	404,000	8
Junction Box	5,224,000	8
Power Strip	215,111	8
Cooling Tower	24,816	48
Chiller	18,000	48
CRAC	37,059	8

To show the applicability of the PLDA-D, four data center power infrastructure tiers were evaluated considering the following metrics: (i) total cost; (ii) operational exergy; (iii) availability; and (iv) PUE (power usage efficiency). These metrics were computed over a period of five years (43,800 h). Each metric was computed before and after the PLDA-D execution.

The electrical flow in a data center starts from a power supply (i.e., utility power), passes through uninterruptible power supply units (UPSs), the step down transformer (SDT), power distribution units (PDUs) (composed of a transformer and an electrical subpanel) and, finally, to the rack. According to the adopted tier configuration, different redundant levels were considered, which impact the metrics computed for this case study. Table 2 presents the electrical efficiency and maximum capacity of each device.

Table 2. Capacity and efficiency. SDT, step down transformer.

Equipment	Efficiency (%)	Capacity (kW)
Utility Power 1	95.3	10,000
Utility Power 2	90.0	10,000
STS 1	99.5	1500
STS 2	98.0	1500
SDT (or transformer) 1	98.5	5000
SDT (or transformer) 2	95.0	5000
Subpanel 1	99.9	1500
Subpanel 2	95.0	1500
UPS 1	95.3	5000
UPS 2	90.0	5000
Junction Box 1	99.9	1500
Junction Box 2	98.0	1500
Power Strip 1	99.5	5000
Power Strip 2	95.1	5000

Table 3 summarizes the results for each power infrastructure of data center Tiers I–IV. Row *Before* presents the results obtained before executing the PLDA-D; row *After* presents the results after PLDA-D execution; *Improvement* (%) is the improvement achieved as a percentage; *Oper. Exergy* is the operational exergy in gigajoules (GJ) (considering five years); *Total Cost* is the sum of the acquisition cost with the operation cost in USD (for five years); *Availability* is the availability level; PUE is the power usage efficiency as a percentage, which corresponds to the total load of the data center divided by the total load of the IT equipment installed.

Table 3. Results of PLDA-D execution with improvement in %. Operational Exergy, Total Cost; Availability and PUE.

Tiers	-	Oper. Exergy	Total Cost	Availability	PUE
	Before	4688	1,173,593	0.999605271	86.84 (%)
Tier I	After	4418	1,165,323	0.999605271	87.50 (%)
	Improvement (%)	6.13	0.71	0	0.77
	Before	10,127	2,289,365	0.9997510814	85.94 (%)
Tier II	After	8837	2,249,961	0.9997510814	87.50 (%)
	Improvement (%)	14.59	1.75	0	1.82
	Before	13,242	3,347,222	0.9999999380	87.52 (%)
Tier III	After	11,077	3,281,087	0.9999999380	89.34 (%)
	Improvement (%)	19.54	2.02	0	2.08
	Before	16,252	4,452,412	0.99999993803	88.39 (%)
Tier IV	After	11,000	4,291,925	0.99999993803	91.84 (%)
	Improvement (%)	47.75	3.74	0	3.9

We apply the PLDA-D algorithm to each EFM architecture, and as a result, the weights presented on each edge of the EFM model are updated, improving the energy flow. The lowest value of the input power is reached, and thus, all metrics related to energy consumption are improved.

From the aforementioned table, the first observation to be noted is the improvement obtained after using the PLDA-D algorithm. The metrics of sustainability, energy consumption and cost are all improved. For instance, even in the data center of Tier I, where no redundant components are considered, improvements were achieved. For instance, the operational exergy was reduced by 6.13% and the total cost by 0.71% (which corresponds to 8720 USD savings), and the PUE metric was also improved by around 0.77%.

Tier II presents an improvement in cost and sustainability metrics. For example, operational exergy was reduced from 10,127 to 8,837 GJ and PUE from 85.94 to 87.50 (%), and the cost improved by 1.75 (%), which would be 39,404 USD. Assuming Tier III, a reduction of almost 20% was observed in operational exergy and 2% in total cost, which in financial resources equates to 66,135 USD. The PUE was improved by 2.08%, reaching 89.34%, a considerable improvement.

The data center classified as Tier IV is the most complete in redundancy and security levels. The values achieved were significant, with a reduction of almost 50% in operating exergy and almost 160,500 USD in five years. The PUE was improved by 3.9%. Figure 22 presents the increase of the total cost and PUE.

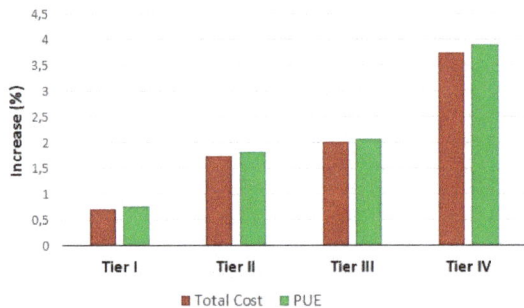

Figure 22. Comparison before and after PLDA-D execution.

Although the improvements to the algorithm seem slight, the long-term values are high. For instance, the total cost of Tier II was 1.75 (%), which means USD 39,403 over five years. Resources from these energy savings could be used for hiring employees, team qualification or acquiring equipment. In order to do this, it is sufficient to adopt a new method for distributing the electrical flow.

Furthermore, the UPS system is responsible for maintaining the IT infrastructure; then, there is a relationship between the tier classification and the capacity of the UPS system. The average power consumption of a computer room according to the tier level is shown in Table 4 [37].

Table 4. Relation between cost/kW before/after PLDA and PLDA-D.

Tier Classification	Cost/kW	After PLDA	After PLDA-D
Tier 1	10,000	9923	9923
Tier 2	11,000	10,870.2	10,870.2
Tier 3	20,000	19,679	19,584
Tier 4	22,000	21,337	21,142

The columns "After PLDA" and "After PLDA-D" represent the results achieved after running the PLDA and PLDA-D algorithms, in which a reduction in comparison with the average power consumption (column "Cost/kW") can be noticed.

To compare the improvement of PLDA-D also with its predecessor, PLDA, we have included the results after the execution of both. For the first two tiers, there was no change in the result, showing that both have good solutions (in this case, optimal); however, as the complexity of the graph increases, PLDA-D continues to offer an optimal result, unlike PLDA, which returns a good solution. For Tier III, the use of PLDA implies a reduction of 1.63%, while with PLDA-D 2.12%. For Tier IV, the improvement is even more significant, since with PLDA 2.21% and PLDA-D, we achieved a 4.05% reduction in the cost/kW.

Therefore, this case study has shown that the proposed approach can be adopted for reducing the cost for a company. In this specific case, we have reduced the cost associated with the electricity consumed through the improvement of the electrical flow inside the data center system infrastructure.

9. Conclusions

The present paper has proposed an algorithm, named the power load distribution algorithm in depth search (PLDA-D), to reduce the electrical energy consumption of data center power infrastructures.

The main goal of the PLDA-D algorithm is to allocate more appropriate values to the edge weights of the EFMs automatically. Such an optimization-based approach was evaluated through a case study, which validated and demonstrated that the results obtained after the execution of the PLDA-D were significantly improved.

For all the architectures of the case study, the results for sustainability impact (operational exergy and PUE) were improved. Power consumption and total cost were also improved. Companies are always looking at reducing costs and their environmental footprint, which has been demonstrated for data centers by optimizing the power load distribution using PLDA-D in the Mercury environment.

For future work, we plan to integrate the PLDA-D with the use of artificial intelligence to predict the energy consumption of data centers, taking into account historical data that date back several years and estimating the environmental impact.

Author Contributions: J.F. conceived of of the presented idea, developed the theory, implemented the algorithms, proposed the formal models and performed the computations. G.C. verified the analytical models and algorithms and revised all the paper. P.M. encouraged J.F. to investigate maximum and minimum flow and to propose a new solution to data centers' electrical power. P.M. and D.T. supervised and revised the findings of this work. All authors discussed the results and contributed to the final manuscript.

Funding: This study was financed in part by Conselho Nacional de Desenvolvimento Científico e Tecnológico (CNPq), Fundação de Amparo a Ciência e Tecnologia de PE (FACEPE) and Bergische Universitat Wuppertal.

Conflicts of Interest: The authors declare no conflict of interest.

References

1. Hallahan, R. Technical information from Cummins Power Generation. In *Data Center Design Decisions and Their Impact on Power System Infrastructure*; Power Generator, 2011. Available online: https://power.cummins.com/system/files/literature/brochures/PT-9020-Data-Ctr-Design-Decisions.pdf (accessed on 15 July 2017)
2. Uhlman, K. Data Center Forum. In *Architecture Solutions for Your Data Center*; Eaton: Natick, MA, USA, 2009.
3. Environmental Protection Agency. Report to Congress on Server and Data Center Energy Efficiency. Available online: http://www.energystar.gov/ia/partners/prod_development/downloads/EPA_Datacenter_Report_Congress_Final1.pdf (accessed on 23 May 2016).
4. Abbasi, Z.; Varsamopoulos, G.; Gupta, S.K. Thermal Aware Server Provisioning and Workload Distribution for Internet Data Centers. In Proceedings of the ACM International Symposium on High Performance Distributed Computing (HPDC10), Chicago, IL, USA, 21–25 June 2010; pp. 130–141.
5. Al-Qawasmeh, A. *Heterogeneous Computing Environment Characterization and Thermal-Aware Scheduling Strategiesto Optimize Data Center Power Consumption*; Colorado State University: Fort Collins, CO, USA, 2012.
6. Delforge, P. America's Data Centers Consuming–and Wasting–Growing Amounts of Energy. Available online: http://switchboard.nrdc.org (accessed on 1 May 2017).
7. The Up Time Institute. Available online: https://uptimeinstitute.com/ (accessed on 9 July 2016).

8. Callou, G.; Maciel, P.; Tutsch, D.; Ferreira, J.; Araújo, J.; Souza, R. *Estimating Sustainability Impact of High Dependable Data Centers: A Comparative Study between Brazilian and US Energy Mixes*; Springer: Vienna, Austria, 2013.
9. Ferreira, J.; Callou, G.; Maciel, P. A power load distribution algorithm to optimize data center electrical flow. *Energies* **2013**, *6*, 3422–3443. [CrossRef]
10. Ferreira, J.; Callou, G.; Dantas, J.; Souza, R.; Maciel, P. An Algorithm to Optimize Electrical Flows. In Proceedings of the 2013 IEEE International Conference on Systems, Man, and Cybernetics, Manchester, UK, 13–16 October 2013; pp. 109–114.
11. Kuo, W.; Zuo, M.J. *Optimal Reliability Modeling–Principles and Applications*; John Wiley and Sons: New York, NY, USA, 2003.
12. Trivedi, K. *Probability and Statistics with Reliability, Queueing, and Computer Science Applications*; Wiley Interscience Publication: New York, NY, USA, 2002.
13. Molloy, M.K. Performance Analysis Using Stochastic Petri Nets. *IEEE Trans. Comput.* **1982**, *9*, 913–917.
14. Al-Fares, M.; Radhakrishnan, S.; Raghavan, B.; Huang, N.; Vahdat, A. Hedera: dynamic flow scheduling for data center networks. In Proceedings of the 7th USENIX Conference on Networked Systems Design and Implementation, San Jose, CA, USA, 28–30 April 2010.
15. Kliazovich, D.; Bouvry, P.; Khan, S.U. *DENS: Data Center Energy-Efficient Network-Aware Scheduling*; Cluster computing; Springer: New York, NY, USA, 2013
16. Ricciardi, S.; Careglio, D.; Sole-Pareta, J.; Fiore, U.; Palmieri, F. Saving energy in data center infrastructures. In Proceedings of the 2011 First International Conference on Data Compression, Communications and Processing (CCP), Palinuro, Italy, 21–24 June 2011.
17. Doria, A.; Carlino, G.; Iengo, S.; Merola, L.; Ricciardi, S.; Staffa, M.C. PowerFarm: A power and emergency management thread-based software tool for the ATLAS Napoli Tier2. In Proceedings of the Computing in High Energy Physics (CHEP), Prague, Czech Republic, 21–27 March 2009.
18. Heller, B.; Seetharaman, S.; Mahadevan, P.; Yiakoumis, Y.; Sharma, P.; Banerjee, S.; McKeown, N. Elastictree: Saving energy in data center networks. In Proceedings of the 7th USENIX Conference on Networked Systems Design and Implementation, San Jose, CA, USA, 28–30 April 2010.
19. Neto, E.C.P.; Callou, G.; Aires, F. An Algorithm to Optimise the Load Distribution of Fog Environments. Available online: https://ieeexplore.ieee.org/abstract/document/8122791/ (accessed on 15 July 2017).
20. Institute for Energy Research. Energy Encyclopedia. Available online: http://instituteforenergyresearch.org/topics/encyclopedia/ (accessed on 29 June 2016).
21. Dincer, I.; Rosen, M.A. *Exergy: Energy, Environment and Sustainable Development*; Elsevier Science: New York, NY, USA, 2007.
22. Maciel, P.; Trivedi, K.S.; Matias, R.; Kim, D.S. Premier Reference Source. In *Performance and Dependability in Service Computing: Concepts, Techniques and Research Directions*; Igi Global: Hershey, PA, USA, 2011.
23. Ajmone Marsan, M.; Conte, G.; Balbo, G. A class of generalized stochastic Petri nets for the performance evaluation of multiprocessor systems. *ACM Trans. Comput. Syst.* **1984**, *2*, 93–122. [CrossRef]
24. Muppala, J.K.; Fricks, R.M.; Trivedi, K.S. Techniques for System Dependability Evaluation. In *Computational Probability*; Kluwer Academic Publishers: Dordrecht, The Netherlands, 2000.
25. Xie, M.; Dai, Y.S.; Poh, K.L. *Computing System Reliability: Models and Analysis*; Springer Science & Business Media: New York, NY, USA, 2004.
26. Murata, T. Petri Nets: Properties, Analysis and Applications. *Proc. IEEE* **1989**, *77*, 541–580. [CrossRef]
27. Silva, B.; Matos, R.; Callou, G.; Figueiredo, J.; Oliveira, D.; Ferreira, J.; Dantas, J.; Junior, A.L.; Alves, V.; Maciel, P. Mercury: An Integrated Environment for Performance and Dependability Evaluation of General Systems. In Proceedings of the IEEE 45th Dependable Systems and Networks Conference (DSN-2015), Rio de Janeiro, Brazil, 22–25 June 2015.
28. Mercury Tool. Available online: http://www.modcs.org/?page_id=1397 (accessed on 13 April 2017).
29. German, R. *Performance Analysis of Communication Systems with Non-Markovian Stochastic Petri Nets*; John Wiley & Sons, Inc.: New York, NY, USA, 2000.
30. Marsan, M.A.; Balbo, G.; Conte, G.; Donatelli, S.; Franceschinis, G. ACM SIGMETRICS Performance Evaluation Review. In *Modelling with Generalized Stochastic Petri Nets*; ACM Press: New York, NY, USA, 1998.
31. Callou, G.; Ferreira, J.; Maciel, P.; Tutsch, D.; Souza, R. An Integrated Modeling Approach to Evaluate and Optimize Data Center Sustainability, Dependability and Cost. *Energies* **2014**, *7*, 238–277. [CrossRef]

32. Callou, G.; Maciel, P.; Tutsch, D.; Araujo, J. Models for Dependability and Sustainability Analysis of Data Center Cooling Architectures. In Proceedings of the 2012 IEEE International Conference on Dependable Systems and Networks (DSN), Boston, MA, USA, 25–28 June 2012.

33. Bellman, R. *On a Routing Problem*; DTIC Document: Bedford, MA, USA, 1956.

34. Ford, L.R., Jr.; Fulkerson, D.R. *Flows in Networks*; Princeton University Press: Princeton, NJ, USA, 1962.

35. Cormen, T.H.; Leiserson, C.E.; Rivest, R.L.; Stein, C. *Introduction to Algorithms*; MIT Press: Cambridge, MA, USA, 2009.

36. IEEE Standards Association. IEEE Gold Book 473. In *Design of Reliable Industrial and Commercial Power Systems*; IEEE Standards Association: Piscataway, NJ, USA, 1998.

37. Turner, I.V.; Pitt, W.; Brill, K.G. *Cost Model: Dollars Per kW Plus Dollars Per Square Foot of Computer Floor*; white paper; Uptime Institute: Seattle, WA, USA, 2008.

Article

Hierarchical Scheduling Scheme for AC/DC Hybrid Active Distribution Network Based on Multi-Stakeholders

Chang Ye, Shihong Miao *, Yaowang Li, Chao Li and Lixing Li

State Key Laboratory of Advanced Electromagnetic Engineering and Technology, Hubei Electric Power Security and High Efficiency Key Laboratory, School of Electrical and Electronic Engineering, Huazhong University of Science and Technology, Wuhan 430074, China; seee_yechang@hust.edu.cn (C.Y.); yw_li@hust.edu.cn (Y.L.); lx_lichao47@hust.edu.cn (C.L.); D201577339@hust.edu.cn (L.L.)
* Correspondence: shmiao@hust.edu.cn; Tel.:+86-13971604685

Received: 4 September 2018; Accepted: 17 October 2018; Published: 19 October 2018

Abstract: This paper presents a hierarchical multi-stage scheduling scheme for the AC/DC hybrid active distribution network (ADN). The load regulation center (LRC) is considered in the developed scheduling strategy, as well as the AC and DC sub-network operators. They are taken to be different stakeholders. To coordinate the interests of all stakeholders, a two-level optimization model is established. The flexible loads are dispatched by LRC in the upper-level optimization model, the objective of which is minimizing the loss of the entire distribution network. The lower-level optimization is divided into two sub-optimal models, and they are carried out to minimize the operating costs of the AC/DC sub-network operators respectively. This two-level model avoids the difficulty of solving multi-objective optimization and can clarify the role of various stakeholders in the system scheduling. To solve the model effectively, a discrete wind-driven optimization (DWDO) algorithm is proposed. Then, considering the combination of the proposed DWDO algorithm and the YALMIP toolbox, a hierarchical optimization algorithm (HOA) is developed. The HOA can obtain the overall optimization result of the system through the iterative optimization of the upper and lower levels. Finally, the simulation results verify the effectiveness of the proposed scheduling scheme.

Keywords: AC/DC hybrid active distribution; hierarchical scheduling; multi-stakeholders; discrete wind driven optimization

1. Introduction

The active distribution network (ADN) is considered an effective way to address the issues caused by the large-scale integration of distributed energy resources (DERs). For this reason, it has aroused great attention in both research and industry fields, and a great number of studies have been done [1–4]. With the large-scale integration of distributed generation (DG), flexible loads, energy storage and new types of power electronic control devices, a large amount of DC equipment has been installed in AC distribution networks. Due to the large number of electrical energy conversion links, using AC distribution networks for power supply only will influence both the cost and the efficiency [5]. The best topology that can accommodate both AC and DC technologies with less need for such conversion is a hybrid one [6]. The idea is to merge the AC and DC distribution networks through a bidirectional converter and establish an AC/DC hybrid ADN in which AC- or DC-type energy sources and loads can be flexibly integrated into the distribution networks such that power can smoothly flow between them [7]. Furthermore, flexible DC technology is applied to construct the AC/DC hybrid ADN, which can achieve energy scheduling over a wide area. The AC/DC hybrid ADN is controllable and allows DC loads and various DERs to plug-and-play easily; it will be an important implementation form of the future distribution network [8].

The characteristics of the AC/DC hybrid ADN provide an efficient means for the integration of DERs with minimum modifications of the current distribution grid [9]. At present, most research on AC/DC hybrid ADNs is focused on the optimal power flow (OPF) problem and coordinated control methods. An efficient OPF algorithm for hybrid AC/DC grids with discrete control devices is presented in [10]. On this basis, the corrective security constraint is taken into account in OPF for a meshed AC/DC power transmission network [11]. Meanwhile, second-order cone programming (SOCP) is applied, and [12] presents a SOCP formulation of the OPF problem for AC/DC systems with voltage source converter (VSC) technology. In terms of coordinated control methods, Ref. [13] presents an overview of various control schemes used for voltage and frequency regulation in standalone and transition mode operation of the hybrid micro-grid. Ref. [14] proposes an instantaneous power-based current control scheme for reactive power compensation in hybrid AC/DC networks, while Ref. [15] presents a novel three-phase reactive power and voltage distributed control method for AC/DC hybrid ADNs based on model predictive control. Furthermore, in order to ensure the safety of the converters and the grid facility, a flexible control strategy is proposed for the AC/DC hybrid grid in [16]. These studies provide strong support for the optimal scheduling and operation of the AC/DC hybrid ADN.

In general, the core technologies for operation of AC/DC hybrid ADN contain flexible DC technology and scheduling technology [17]. Flexible DC technology can provide a better interface for DGs, energy storage, electric vehicles, and other devices [18–20]. Meanwhile, the ADN scheduling technology can flexibly dispatch various types of flexible resources in the distribution network, and ultimately realize the optimal operation of the entire system. Using a hybrid-type distribution system can eliminate several conversion stages and thus improve efficiency and reduce investment costs [21].

Recently, several studies have involved the optimization of the scheduling strategies of the AC/DC hybrid ADN. Ref. [22] builds a system structure for an AC/DC hybrid ADN and then establishes a multi-time scale optimal dispatch model based on a multi-agent system (MAS). The model is designed to reduce the operation cost and energy loss on the premise of absorbing the most DERs. As the AC and DC sub-networks perform power interaction through an interlink convertor (ILC), a hierarchical planning method is introduced into the scheduling process. Ref. [23] focuses on power management in the AC/DC microgrid, and the combination of PV, fuel cells, wind, and battery storage with adjustable parameters is analyzed. An optimization model is investigated to reduce the implementation costs by using a multi-objective particle swarm optimization (MOPSO) algorithm. For an ADN with a single DC section and multiple AC sections, [24] proposes a hierarchical and distributed coordinated multi-source optimal scheduling strategy which divides the dispatch area into a local level and an area level. In [25], a two-stage stochastic centralized dispatch scheme is presented, where the first stage produces day-ahead dispatch decisions for the dispatchable DG units, while the second stage determines appropriate corrective decisions for a set of possible scenarios. On this basis, a robust optimal method is introduced. Reference [26] proposes a bi-level two-stage robust optimal scheduling model for AC/DC hybrid multi-microgrids. In this model, the system is divided into the utility level and the supply level, and two-stage robust optimization is carried out. A multi-interval-uncertainty constrained robust dispatch model is proposed in [27], in order to deal with the uncertainties of renewable energy generation and load power in an AC/DC hybrid microgrid.

The above studies made great efforts towards developing optimal operation models for AC/DC hybrid ADN; however, there are still some persisting problems that need to be addressed. Firstly, as different stakeholders, the AC and DC subgrids should have their own scheduling objectives. Meanwhile, the distribution network operator (DNO) needs to regulate the power grid globally. Few studies consider their scheduling strategy based on multi-stakeholders of an AC/DC hybrid ADN. Secondly, most of the present studies pay little attention to the coordination of various schedulable resources in an AC/DC hybrid ADN. In addition, the interaction between the AC and DC sides still deserves further study. Thus, this paper constructs a coordinated optimization framework for AC/DC hybrid ADN based on multi-stakeholders, and a hierarchical multi-stage scheduling model is established. The upper level is optimized by the load regulation center (LRC), taking the minimum loss of the entire distribution network as the scheduling objective. Meanwhile, the lower level is optimized by the AC and DC sub-networks. Therefore, lower-level optimization is divided into two suboptimal problems, which have objective functions related to their own stakeholders. The advantages of using the above hierarchical optimization scheduling model include: (a) avoiding the difficulty of solving multi-objective optimization; and (b) ensuring that all stakeholders pay attention to their own benefits, and obtain the optimal operating state of the system through mutual cooperation and coordination. Meanwhile, considering the combination of the proposed DWDO algorithm and the YALMIP toolbox, the model is solved by a hierarchical optimization algorithm. The simulation results show that the proposed strategy can minimize the operating costs of AC/DC sub-networks while reducing system losses.

In summary, the contributions of this paper are as follows: (a) a hierarchical multi-stage scheduling model for AC/DC hybrid and is proposed, to avoid the difficulty of solving multi-objective optimization; (b) AC and DC sub-networks are regarded as different stakeholders so that their interaction characteristics are analyzed in a scheduling problem; and (c) a hierarchical optimization algorithm comprising the DWDO algorithm and YALMIP toolbox is developed to solve the hierarchical optimization model.

The rest of this paper is organized into the following sections. Section 2 describes the network structure of the AC/DC hybrid ADN and proposes a hierarchical multi-stage scheduling framework based on multi-stakeholders. Section 3 details the mathematical formulation of the hierarchical optimization scheduling model. In Section 4, a hierarchical optimization algorithm is proposed and researched. In Section 5, case studies and discussions are carried out. Finally, Section 6 put forwards the conclusions that can be drawn.

2. System Framework

As mentioned above, various types of distributed scheduling resources are integrated into the AC/DC hybrid ADN. When the distribution network is in operation, the schedulable resources will interact with each other. Scheduling only a single or partial resource makes it difficult to achieve overall system optimization, and easily leads to redundancy and waste of resources. Thus, global optimization of AC/DC hybrid ADN is necessary. On the other hand, each of the AC and DC areas constitutes a sub-network, which can be scheduled as an independent stakeholder. In view of this, the network structure is shown in Figure 1. The arrows in the figure show the positive power direction.

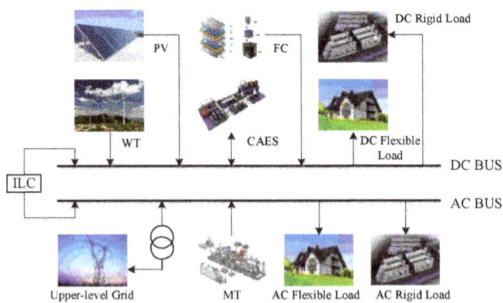

Figure 1. Network structure of the AC/DC hybrid ADN.

In Figure 1, the AC/DC hybrid ADN is divided into an AC section and a DC section. The main components of the DC section include photovoltaics (PVs), wind turbines (WT), fuel cells (FC), compressed air energy storage (CAES), and DC loads, which are connected to the DC bus. The AC section mainly constitutes a micro-gas turbine (MT) and AC loads, with the upper-level power grid (UPG) connected to it. The upper-level power grid supplies electricity based on time-of-use price. Meanwhile, the DC section interacts with the AC section through ILC, which can transmit power in both directions. Additionally, in order to reflect the operating characteristics of the ADN, flexible loads are considered to participate in the scheduling.

In general, the power on both the AC and DC sides influences each other, and it is difficult to directly obtain the optimal operating status of the entire system. Thus, a hierarchical multi-stage scheduling framework based on multi-stakeholders is proposed, as shown in Figure 2. It is necessary to consider both active and reactive power in the optimal power flow of the distribution network [28]. However, in this paper, the main purpose of the lower-level optimization is to determine the active economic dispatch of the power system. Meanwhile, considering the complexity of this model, only active power is analyzed in both the AC and DC sections.

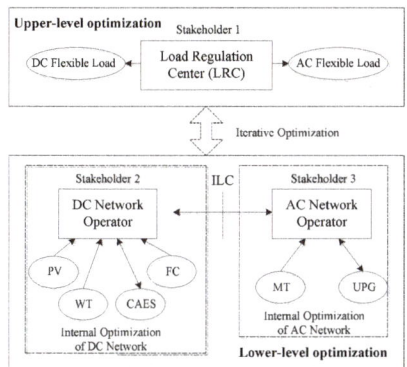

Figure 2. Hierarchical multi-stage scheduling framework based on multi-stakeholders.

In Figure 2, the stakeholder of the upper-level optimization is the load regulation center (LRC). All flexible loads of DC and AC sections are managed by the LRC, in order to reduce the losses of the distribution network. Then, the scheduling results of the flexible loads are transferred to the lower-level optimization, which is comprised of the internal optimization of both the DC network and the AC network. The DGs and CAES are scheduled by the lower-level optimization, as well as the transmission power of the UPG. The results are also transferred to the upper level, and the system performs repeated iterative optimization until it reaches a dynamic balance. Through the hierarchical optimization method, optimal scheduling results for the entire AC/DC hybrid ADN are expected to finally be obtained. The results not only reflect the global optimization of the power grid, but also reflect the local optimization of the AC and DC power grids.

In the next section, we will establish the hierarchical optimization model, and give a detailed solution process for it.

3. Hierarchical Optimization Model

3.1. Optimization Model of the Upper Level

In the upper-level optimization, flexible loads of the entire network are scheduled. The scheduling objective is to reduce system losses, which mainly include line loss and loss of ILC. Therefore, the objective function is expressed in Equation (1).

$$\min F_{\text{ul}} = \sum_{t=1}^{T} (c_{\text{g},t} P_{loss\_1,t} + c_{loss\_ILC} P_{loss\_ILC,t}) \Delta t \tag{1}$$

where T is the scheduling period.

The detailed expression of the above power loss can be expressed as Equations (2) and (3).

$$P_{\text{loss\_1},t} = \sum_{l_1 \in N_{line}^{AC}} P_{loss,l_1,t} + \sum_{l_2 \in N_{line}^{DC}} P_{loss,l_2,t} = \sum_{l_1 \in N_{line}^{AC}} \frac{(P_{l_1,t}^2 + Q_{l_1,t}^2) R_{l_1}}{U_{e\_AC}^2} + \sum_{l_2 \in N_{line}^{DC}} \frac{P_{l_2,t}^2 R_{l_2}}{U_{e\_DC}^2} \tag{2}$$

$$P_{\text{loss\_ILC},t} = (1 - \eta_{\text{ILC}}) \cdot |P_{\text{ILC},t}| \tag{3}$$

where l_1 and l_2 are lines of the AC and DC networks, respectively, and N_{line}^{AC} and N_{line}^{DC} are the total numbers of lines in AC and DC networks, respectively.

The active and reactive power of the network lines are obtained using the existing power flow calculation method. In the optimization process, the node voltage amplitude constraint of the network is considered to be as in Equation (4).

$$V_{s,\min} - \varepsilon_{1,s,t} \leq V_{s,t} \leq V_{s,\max} + \varepsilon_{2,s,t}, s \in N_{\text{node}} \tag{4}$$

where N_{node} is total network node of the system. $\varepsilon_{1,s,t}$ and $\varepsilon_{2,s,t}$ are relaxation variables. In practical calculation, voltage constraints are added as penalty terms to the objective function to avoid the occurrence of voltage over-limit conditions.

In this model, the flexible loads are mainly transferable loads (TLs), which can be expressed as Equation (5).

$$\begin{cases} \sum\limits_{t=1}^{T} \alpha_{j,t} = X_j \\ P_{\text{TL}j,t} = P_{\text{TL},j} \cdot \alpha_{j,t}, j \in N_{\text{TL}} \end{cases} \tag{5}$$

where $\alpha_{j,t} \in \{0,1\}$ are the actual operating state of transferable load j at time t, 1 is operation and 0 is outage. N_{TL} is the total number of transferable loads in the system.

Equation (5) is the constraint condition for the flexible load scheduling. Through the upper-level optimization, the flexible load scheduling result is obtained. Then it is passed to the lower level. Thus, the actual loads of DC and AC sections can be expressed as Equation (6).

$$\begin{cases} P_{\text{L\_DC},t} = P_{\text{CL\_DC},t} + \sum\limits_{j_1 \in N_{\text{TL\_DC}}} P_{TL\_DCj,t} \\ P_{\text{L\_AC},t} = P_{\text{CL\_AC},t} + \sum\limits_{j_2 \in N_{\text{TL\_AC}}} P_{TL\_ACj,t} \end{cases} \tag{6}$$

3.2. Optimization Model of the Lower Level

The lower-level optimization is divided into two parts: the DC section and the AC section, which have their own optimization objectives. As different stakeholders, they are committed to minimizing operating costs in their respective section.

3.2.1. Optimization Model of the DC Section

For the DC section, the scheduling costs mainly include the maintenance cost of generators and energy storage, the fuel cost of FC and CAES, and the electricity transaction cost. The objective function is shown in Equation (7).

$$\min F_{\text{ll\_DC}} = \sum_{t=1}^{T} (C_{PV,t} + C_{WT,t} + C_{CAES,t} + C_{FC,t} + C_{ILC\_DC,t}) \tag{7}$$

The detailed expressions of the costs are expressed in Equation (8).

$$\begin{cases} C_{\text{PV},t} = \sum_{i_{PV} \in N_{PV}} [c_{o,PV} P_{PV,i_{PV},t} + c_{aban,PV} (\overline{P}_{PV,i_{PV},t} - P_{PV,i_{PV},t})] \Delta t \\ C_{\text{WT},t} = \sum_{i_{WT} \in N_{WT}} [c_{o,WT} P_{WT,i_{WT},t} + c_{aban,WT} (\overline{P}_{WT,i_{WT},t} - P_{WT,i_{WT},t})] \Delta t \\ C_{\text{CAES},t} = \sum_{i_{CAES} \in N_{CAES}} \left(c_{CAES} P_{CAES,i_{CAES},t} u_{CAES\_G,t} + c_{o,CAES} P_{CAES,i_{CAES},t} \right) \Delta t \\ C_{\text{FC},t} = \sum_{i_{FC} \in N_{FC}} \left(c_{FC} P_{FC,i_{FC},t} + c_{o,FC} P_{FC,i_{FC},t} \right) \Delta t \\ C_{\text{ILC\_DC},t} = c_{g,t} P_{\text{ILC\_DC},t} \Delta t \end{cases} \tag{8}$$

where N_{PV}, N_{WT}, N_{CAES}, N_{FC} are the total number of PV, WT, CAES and FC in the system. The second items in the right of $C_{\text{PV},t}$ and $C_{\text{WT},t}$ are the penalty cost of PV and wind power curtailment. These can ensure that the power of PV and WT is curtailed as little as possible.

Since CAES that has been applied on a large scale at this stage is mostly supplementarily fired, it needs to consume natural gas to supplement combustion during discharge stage. This cost has been considered in Equation (8). The simplified power model of CAES can be described as Equation (9).

$$\begin{cases} \dot{p}_{\text{CAES\_C},t} = k_{\text{CAES\_C}} P_{\text{CAES\_C},t} \\ \dot{p}_{\text{CAES\_G},t} = k_{\text{CAES\_G}} P_{\text{CAES\_G},t} \end{cases} \tag{9}$$

where $k_{\text{CAES\_C}}$ and $k_{\text{CAES\_G}}$ are the operation coefficients for compression and generation conditions, respectively, and their detailed expressions are presented in literature [29].

As CAES usually uses the working pressure of the gas storage chamber to indicate its charge and discharge capacity, the equivalent state of charge (SOC) can be described as Equation (10).

$$p_{\text{CAES},t} = p_{\text{CAES},t-1} + u_{\text{CAES\_C},t} k_{\text{CAES\_C}} P_{\text{CAES\_C},t} \Delta t - u_{\text{CAES\_G},t} k_{\text{CAES\_G}} P_{\text{CAES\_G},t} \Delta t \tag{10}$$

where $u_{\text{CAES\_C},t} \in [0,1]$ and $u_{\text{CAES\_G},t} \in [0,1]$ are the charge and discharge status at time t, respectively. Thus, the constraints of CAES are shown in Equations (11)–(13).

$$\begin{cases} P_{\text{CASE\_G},min} \leq P_{\text{CASE\_G},t} \leq P_{\text{CAES\_G},max} \\ P_{\text{CAES\_C},min} \leq P_{\text{CAES\_C},t} \leq P_{\text{CAES\_C},max} \end{cases} \tag{11}$$

$$p_{\text{CAES},min} \leq p_{\text{CAES},t} \leq p_{\text{CAES},max} \tag{12}$$

$$u_{\text{CAES\_C},t} + u_{\text{CAES\_G},t} \leq 1 \tag{13}$$

$$p_{\text{CAES}}(T) = p_{\text{CAES}}(0) \tag{14}$$

Equation (11) is the power constraint of CAES for the charge and discharge stages. Equation (12) is the equivalent SOC constraint of CAES. Equation (13) is the charge and discharge status constraint of CAES; this constraint ensures that CAES can only be charging, discharging or in a stop state at any given time. Equation (14) is the equivalent SOC balance constraint of CAES; this constraint ensures that CAES maintains the pressure of the gas storage chamber before and after a scheduling cycle.

The remaining constraints are shown in Equations (15)–(17).

$$\sum_{i_{PV} \in N_{PV}} P_{PV,i_{PV},t} + \sum_{i_{WT} \in N_{WT}} P_{WT,i_{WT},t} + \sum_{i_{CAES} \in N_{CAES}} P_{CAES,i_{CAES},t} + \sum_{i_{FC} \in N_{FC}} P_{FC,i_{FC},t} + P_{\text{ILC\_DC},t} = P_{\text{L\_DC},t} \tag{15}$$

$$P_{\text{FCmin},i_{FC}} \leq P_{FC,i_{FC},t} \leq P_{\text{FCmax},i_{FC}} \tag{16}$$

$$P_{\text{ILC\_DC},min} \leq P_{\text{ILC\_DC},t} \leq P_{\text{ILC\_DC},max} \tag{17}$$

where Equation (15) is the power balance constraint of the DC section. Equation (16) is the output power constraint of the FC. Equation (17) is the output power constraint of the ILC.

3.2.2. Optimization Model of the AC Section

Similarly, the objective function of the AC section is shown in Equation (18).

$$\min F_{ll\_AC} = \sum_{t=1}^{T} (c_{g,t}P_{UPG,t} + c_{MT}P_{MT,i_{FC,t}} + c_{MT\_o}P_{MT,i_{MT,t}} + c_{g,t}P_{ILC\_AC,t})\Delta t \tag{18}$$

The constraints are shown in Equations (19)–(22).

$$P_{UPG,t} + \sum_{i_{MT} \in N_{MT}} P_{MT,i_{MT,t}} + P_{ILC\_AC,t} = P_{L\_AC,t} \tag{19}$$

$$P_{UPG,min} \leq P_{UPG,t} \leq P_{UPG,max} \tag{20}$$

$$P_{MTmin,i_{MT}} \leq P_{MT,i_{MT,t}} \leq P_{MTmax,i_{MT}} \tag{21}$$

$$P_{ILC\_AC,t} = -P_{ILC\_DC,t} \tag{22}$$

Equation (19) is the power balance constraint of the AC section, Equations (20) and (21) are the power constraints of UPG and MT respectively. According to the positive power direction shown in Figure 1, Equation (22) ensures the power balance of both the DC and AC sections.

Thus, the hierarchical scheduling model is established. It is summarized in Equation (23).

$$\begin{cases} the\ upper-level: \\ \min F_{ul} = \sum_{t=1}^{T} (c_g P_{loss\_l,t} + c_{loss\_ILC}P_{loss\_ILC,t}) \\ s.t.\ (4),(5) \\ the\ lower-level: \\ \min F_{ll\_DC} \qquad \min F_{ll\_AC} \\ s.t.\ (11) \sim (17) \quad s.t.\ (19) \sim (22) \end{cases} \tag{23}$$

4. Model Solution

The optimization model of the AC/DC hybrid ADN is a bi-level optimization problem. The'network loss information in the upper-level optimization needs to be obtained according to the lower-level optimization result. Conversely, the lower-level optimization requires the output information of flexible loads determined in the upper-level optimization. Based on existing research, this chapter will present a DWDO algorithm, and combine it with the YALMIP toolkit to solve the bi-level optimization problem.

4.1. Discrete Wind-Driven Optimization Algorithm

Through the treatment of flexible loads in this paper, the upper-level optimization is a nonlinear integer optimization problem. At present, most nonlinear optimization problems are solved by intelligent algorithms, such as particle swarm optimization (PSO), genetic algorithm (GA), and so on. However, most of them have the problems that they find it easy to fall into local optima and are computationally inefficient. Thus, a wind-driven optimization (WDO) technique was proposed by [30], which is also an iterative heuristic global optimization algorithm. WDO is used to describe the motion of air parcels within the earth's atmosphere, and Newton's second law of motion is used to describe the N-dimensional search space. As a conclusion, [31] points out that WDO is well-suited for problems with both discrete and continuous-valued parameters. The algorithm process of WDO is as follows.

(1) Initialize the population size, set the maximum number of iterations, related parameters, and search boundaries, and define pressure functions.
(2) Initialize the air mass points and assign the starting speeds and locations randomly.

(3) Calculate the pressure value of the air particles in the current iteration and rearrange the population according to the pressure values.

(4) Update the speeds of the air particles using Equation (24).

$$u_{new} = (1-\alpha)u_{cur} - gx_{cur} + [RT\left|\frac{1}{i}-1\right|(x_{opt}-x_{cur})] + (\frac{cu_{cur}^{otherdim}}{i}) \tag{24}$$

where u_{new} is the updated speed vector of air mass points, u_{cur} is current speed vector, x_{cur} is current location vector, x_{opt} is the global optimum location so far, and $u_{cur}^{otherdim}$ is the speed vector of any other dimension. i is the arrangement number of the air mass points in step 3. α, g, RT and c are parameters which will be changed according to the optimization problems.

(5) Update the positions of the air particles using Equation (25).

$$x_{new} = x_{cur} + (u_{new}\Delta t) \tag{25}$$

where x_{new} is the updated location vector of the air mass points.

(6) If the termination condition is not reached, go to step 3.

The purpose of the upper-level optimization is to determine the output state of the flexible loads at each time. This is a discrete nonlinear optimization problem, which cannot be solved directly using the WDO algorithm. Therefore, the original WDO algorithm needs to be discretized, which is called DWDO in this paper. In DWDO, the starting position parameters should be randomly initialized as 0–1 integer variables. The speed update formula is the same as Equation (24), but the positions need to be updated based on the probability of taking 1. Now, introduce the sigmoid function, which is shown in Equation (26).

$$s(u_{new}) = \frac{1}{1+\exp(-u_{new})} \tag{26}$$

In this place, $s(u_{new})$ indicates the probability of taking 1 for x_{new}. So the locations of the air particles can be updated by Equation (27).

$$x_{new} = \begin{cases} 1 & if\ rand() \leq s(u_{new}) \\ 0 & otherwise \end{cases} \tag{27}$$

The remaining optimization steps of DWDO are similar to the original WDO algorithm. Using the DWDO algorithm, the upper-level optimization can obtain the optimized output for flexible loads.

4.2. Solution Process of the Hierarchical Scheduling Model

The lower-level optimization is a mixed-integer linear programming (MILP) problem. As a mature optimization toolkit, YALMIP has great advantages in solving MILP problems, and has been widely used [32–34]. Furthermore, it can easily interact with MATLAB. Therefore, YALMIP is used to solve lower-level optimization and iterates with the DWDO algorithm in the upper-level optimization. The solution process is shown in Figure 3.

As this paper focuses on active economic dispatch of ADN, only active power is allowed to be exchanged between the ADN and the UPG. The DC and AC sections exchange active power through ILC. In addition to this, the AC section also exchanges active power with the UPG through the boundary line. Reactive power is generated locally by suitable means to satisfy the reactive demand and losses for both DC and AC sections. Thus, the solution process of the hierarchical scheduling model is as follows.

(1) Initialize the values of P_{CAES}, P_{FC}, P_{MT}, P_{ILC} using operation data of another similar day, and preset the planned output of the flexible loads.

(2) Solve the upper-level optimization using DWDO, in order to obtain the optimized output of the flexible loads ($P_{TL\_DC}^*, P_{TL\_AC}^*$). In this process, we expect that PV and wind power can be

absorbed as much as possible. In addition, as the AC side is connected to the upper-level power grid, its internal power adjustment ability is strong. Therefore, the optimization starts from the DC side in this step.

(3)　Solve the optimization of the DC and AC sections using YALMIP, in order to obtain P_{CAES}^*, P_{FC}^*, P_{ILC}^*, P_{MT}^*.

(4)　Determine whether the algorithm reaches a defined number of iterations or convergence accuracy. If the change of the system cost is lower than a certain constant between two iterations, the algorithm converges and then the result is presented. If not, go to step 2.

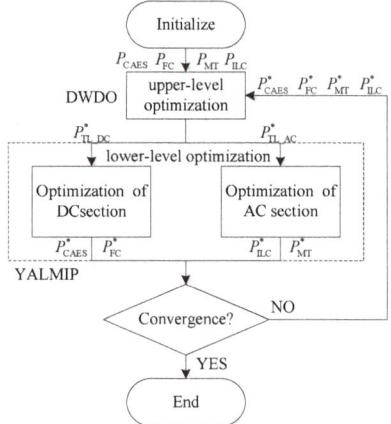

Figure 3. Solution process of the hierarchical scheduling model.

5. Case Study and Discussion

The hierarchical scheduling model is tested in a modified 38-node test system, which is shown in Figure 4. The modified test system is divided into a DC section and an AC section. The DC section includes two PVs, one WT, one FC and one CAES, while the AC section includes one MT and is connected to the upper-level power grid. The configuration of loads is also shown in Figure 4. The line parameters are given in literature [35].

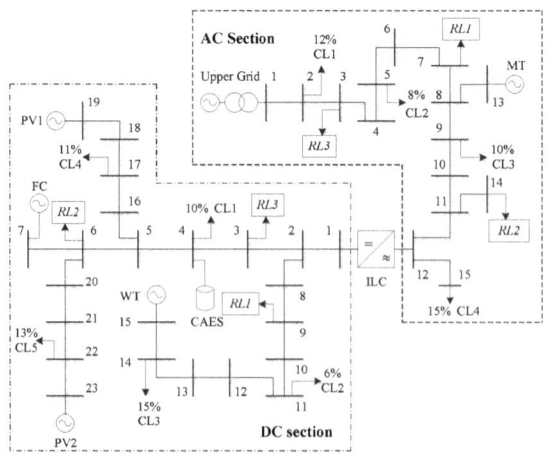

Figure 4. Modified 38-node test system.

In this system, the forecast output of the PVs and WT is shown in Figure 5, while the total rigid load prediction curve of the whole system is shown in Figure 6. The data for transferable loads is shown in Table 1. The data for CAES is shown in Table 2. The time-of-use price (TOU) of UPG is shown in Table 3. The maintenance cost of the PVs, WT, FC and MT are 0.0096 CNY/kWh, 0.0296 CNY/kWh, 0.088 CNY/kWh and 0.088 CNY/kWh, respectively. The fuel cost of the FC and MT are 0.525 CNY/kWh and 0.641 CNY/kWh, respectively. The penalty cost coefficients of PV and wind power curtailment is 10 CNY/kWh. The power limits of the ILC and the UPG are ±1000 kW and ±1500 kW, respectively.

The parameters of DWDO are $\alpha = 0.4$, $g = 0.2$, $RT = 3$, $c = 0.4$. The number of air mass points is 30, the number of iterations is 500. All programs are coded and tested in MATLAB, and the lower-level optimization is solved by YALMIP. The programs run on an Intel(R) Core(TM) i7-4710MQ @ 2.50 GHz, RAM 8 GB system.

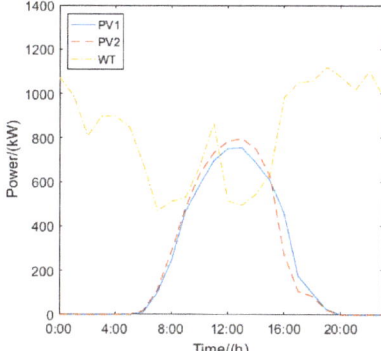

Figure 5. Prediction curve of PVs and WT.

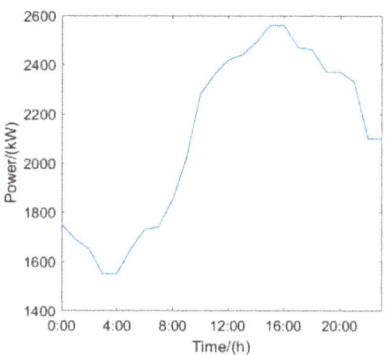

Figure 6. Prediction curve of rigid load.

Table 1. Data of transferable loads.

Transferable Loads	Rated Power (Kw)	Planned Operation Time (h)	Planned Outage Time (h)
RL1 of AC section	180	9–12, 15–18, 22–24	0–8, 13–17, 19–21
RL2 of AC section	200	11–14, 19–21	0–10, 15–18, 22–24
RL3 of AC section	160	0–18	19–24
RL1 of DC section	180	8–17	0–7, 18–24
RL2 of DC section	240	6–8, 11–13, 18–22	0–5, 9–10, 14–17, 23–24
RL3 of DC section	200	5–22	0–4, 23–24

Table 2. Data for CAES.

Item	Value
Rated discharge power (kW)	1000
Rated charge power (kW)	700
Initial pressure (bar)	55
Cavern operational pressure range (bar)	46–66
Operation coefficient for compression (bar/kW)	0.0071
Operation coefficient for generation (bar/kW)	0.0033
Power generation cost (CNY/kW)	0.16
Maintenance cost (CNY/kW)	0.0013

Table 3. Time-of-use price of UPG.

Time (h)	Price (CNY/kWh)
0–7	0.3
7–12	0.6
12–14	1
14–18	0.6
18–22	1
22–24	0.3

To analyze the effectiveness of the proposed hierarchical scheduling strategy, we consider the following two case studies:

CASE 1: Hierarchical optimization of two levels, but the lower-level is optimized as a whole.

CASE 2: Hierarchical optimization of two levels, and the lower-level is divided into two stakeholders.

5.1. Economic Comparison

CASE 1: This case is the conventional scheduling strategy that already exists. In this case, the AC/DC hybrid ADN is regarded as one stakeholder, and the ILC power limitation is not considered. The scheduling result is shown in Figure 7. In the depth of the night, the CAES charges, as both the system load and the electricity price of the UPG are low. At noon, from 12:00 to 14:00, the outputs of the PVs are high, and the power supply for the ADN is sufficient. The electricity price of the UPG reaches its maximum, which is higher than the MT and FC. Therefore, all DGs run in order to sell electricity to the UPG in order to gain substantial profits. At 15:00–17:00, although the rigid load of the system reaches peak value, the power of the PVs and WT is able to meet most load requirements. At 18:00–21:00, the electricity price of the UPG reaches maximum again. Now the outputs of the PVs are almost zero, and the FC and MT are running to sell electricity to the UPG. However, at this time, the sales power is less than that at noon. As a result, the operating costs of the AC and DC sections are 16,764 CNY and −2967 CNY respectively, and the cost of system loss is 4651 CNY.

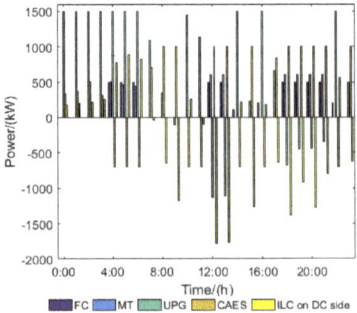

Figure 7. Scheduling result of CASE 1.

CASE 2: This case is the proposed hierarchical scheduling strategy in this paper. The lower level is divided into two stakeholders, and the transmission power limit between the two stakeholders is taken into account. The scheduling result is shown in Figure 8. Compared to CASE 1, the ADN will also sell electricity to the UPG at 12:00–14:00 and 18:00–21:00 in CASE 2, but the sales power is less than CASE 1. As a result, the operating costs of the AC and DC sections are 16,673 CNY and 2580 CNY, respectively, and the cost of the system loss is 4643 CNY. Obviously, the total cost in CASE 2 is more than that in CASE 1, due to the transmission power limitation of ILC. From Figure 7, we know that the maximum transmission power of ILC is about 1800 kW, which is far more than the 1000 kW in CASE 2. This means more investment is needed in the construction of the distribution network. Thus, it is inappropriate to regard the AC/DC hybrid ADN as one stakeholder without considering the constraints between the AC and DC sub-networks.

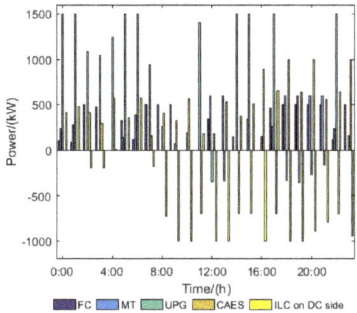

Figure 8. Scheduling result of CASE 2.

5.2. Scheduling Result of Loads

The scheduling result of the loads in CASE 2 is shown in Figure 9. The planned loads are low at 0:00–6:00 and high at 15:00–17:00. This may lead to excess or insufficient power supply for the ADN. By means of the optimization scheduling, flexible loads are transferred to achieve peak load shifting. The scheduled load curve is more stable, and the load peak-to-valley difference is smaller.

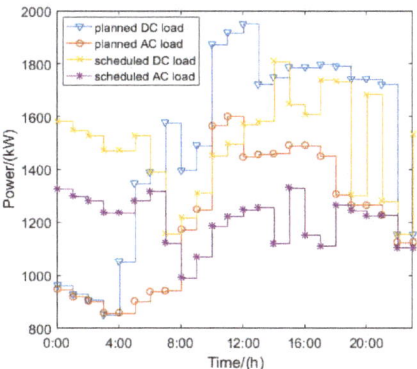

Figure 9. Load scheduling result of CASE 2.

5.3. ILC Power Analysis

The power balance of ILC is important in the optimization process. Usually, in optimization scheduling, decision-makers mainly focus on the active power. The power profiles of ILC on the DC side and the AC side are shown in Figure 10. According to the previous analysis, we know that a large amount of load was transferred to 0:00–2:00. Now, the electricity price of UPG is low and the

AC sub-network buys electricity from the UPG. Part of the purchased electricity is transmitted to the DC sub-network through the ILC. Between around 10:00 and 12:00, the outputs of WT and PVs are high, and ILC delivers power from the DC sub-network to the AC sub-network at maximum power. At 18:00–21:00, the electricity price of UPG is high, and the output of WT is high. Thus, power is also delivered from the DC sub-network to the AC sub-network during this time. It can be seen from Figure 10 that the power of the ILC on the AC and DC side is balanced in a scheduling period.

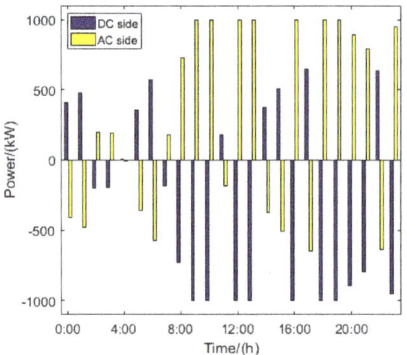

Figure 10. Power of ILC on both the DC and the AC side.

5.4. Rejected Power Analysis

Due to the existence of system constraints, the active power may be rejected by other stakeholders. The actual output powers of WT and PVs are shown in Figure 11. The actual outputs of PVs are the same as the prediction curve, but there is a reduction in the power of WT. In a scheduling period, the maximum total output of wind power is 19,811 kWh, but the total output is actually 19,501 kWh. The transmission power constraints of ILC and UPG will affect it, as well as the capacity of CAES.

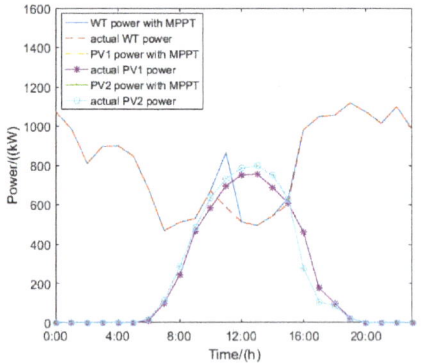

Figure 11. Actual output power and MPPT power of WT, PV1 and PV2.

For the DC section, the rejected power of WT and PVs is mainly related to the ILC power limitation and the capacity of the energy storage system. Meanwhile, for the whole AC/DC hybrid ADN, the rejected power is mainly affected by the power limitation of the boundary line with UPG. For instance, when the transmission power limit of the UPG is increased to 3000 kW, no rejected power will appear. Detailed quantitative analysis will be carried out in future research.

6. Conclusions and Prospects

As an AC/DC hybrid, ADN is becoming an important implementation forms of the future distribution network, its global optimization scheduling is of great significance. For this issue, a hierarchical multi-stage scheduling model is established in this paper. The upper level optimizes the total system losses by scheduling the flexible loads, while the lower level minimizes the scheduling costs of the AC and DC sections by scheduling the controllable DGs and CASE. The DWDO algorithm and the YALMIP toolbox are combined to solve the scheduling model. The case study shows that the proposed scheduling scheme can achieve lower system loss with low operating costs of both AC and DC sections. Thus, not only is the overall system optimized, but the profits of multi-stakeholders are also taken into account. The research in this paper can provide several theoretical references for the scheduling of the AC/DC hybrid distribution network.

Author Contributions: C.Y. and S.M. conceived and designed the study. C.Y., C.L. and Y.L. carried out the simulation. C.Y., S.M. and L.L. analyzed simulation results. C.Y., C.L. and Y.L. wrote the paper. C.Y., S.M. and L.L. reviewed and edited the manuscript. All authors read and approved the manuscript.

Funding: This research was funded by National Key Research and Development Program of China (2017YFB0903601), The National Natural Science Foundation of China (51777088) and 2018 scientific research project of State Grid Corporation of China (SGNXDK00DWJS1800016).

Conflicts of Interest: The authors declare no conflict of interest.

Nomenclature

F_{ul}	the upper-level optimization objective
$P_{loss,l,t}$	the line loss of the entire network at time t
$P_{loss\_ILC,t}$	the active power loss of ILC at time t
$c_{g,t}$	the time-of-use price of UPG at time t
c_{loss_ILC}	the cost of the ILC loss
$P_{loss,l_1,t}, P_{loss,l_2,t}$	line losses of AC and DC network respectively at time t
$P_{l_1,t}, Q_{l_1,t}$	active and reactive power of line l_1 at time t
$P_{l_2,t}$	active power of line l_2 at time t
R_{l_1}, R_{l_2}	line resistances of l_1 and l_2
U_{e_AC}, U_{e_DC}	rated voltages of AC and DC buses
$P_{ILC,t}$	transmission power of ILC at time t
η_{ILC}	effectiveness of ILC
$V_{s,t}$	voltage of node s
$V_{s,max}, V_{s,min}$	maximum and minimum allowable voltage of node s
X_j	total operating time in a scheduling period of transferable load j
$P_{TL,j}$	rated power of transferable load
$P_{TLj,t}$	actual power of transferable load at time t
$P_{L\_DC,t}, P_{L\_AC,t}$	total load of DC and AC section at time t
$P_{CL\_DC,t}, P_{CL\_AC,t}$	constant load of DC and AC section at time t
$P_{TL\_DC,t}, P_{TL\_AC,t}$	transferable load of DC and AC section at time t
N_{TL_DC}, N_{TL_AC}	total number of transferable loads in DC and AC section
$C_{PV}, C_{WT}, C_{CAES}, C_{FC}$	operating cost of PV, WT, CAES and FC
$c_{aban,PV}, c_{aban,WT}$	penalty cost coefficients of PV and wind power curtailment
$\overline{P}_{PV,i_{PV},t}, \overline{P}_{WT,i_{WT},t}$	output power of PV and WT with maximum power point tracking (MPPT)
C_{ILC_DC}	electricity transaction cost which is carried out through ILC on DC side
$c_{o,PV}, c_{o,PV}, c_{o,CAES}, c_{o,FC}$	maintenance cost coefficient of PV, WT, CAES and FC
c_{CAES}, c_{FC}	power generation cost coefficient of CAES and FC
$P_{PV,t}, P_{WT,t}, P_{CAES,t}, P_{FC,t}$	actual output power of PV, WT, CAES and FC
$P_{ILC\_DC,t}$	power of ILC on DC side
$\dot{p}_{CAES\_C,t}, \dot{p}_{CAES\_G,t}$	rate of pressure change for compression and generation conditions

$P_{CAES\_C,t}$, $P_{CAES\_G,t}$	power of CAES for compression and generation conditions
$p_{CAES,t}$, $p_{CAES,t-1}$	pressure of the gas storage chamber at time t and time $t-1$
Δt	interval between two scheduling time
$P_{CASE\_G,min}$, $P_{CAES\_G,max}$	minimum and maximum allowable discharge power of CAES
$P_{CAES\_C,min}$, $P_{CAES\_C,max}$	minimum and maximum allowable charge power of CAES
$p_{CAES,min}$, $p_{CAES,max}$	minimum and maximum allowable pressure of the gas storage chamber
$p_{CAES}(0)$, $p_{CAES}(T)$	pressure of the gas storage chamber at time 0 and time T
$P_{FCmin,i_{FC}}$, $P_{FCmax,i_{FC}}$	minimum and maximum allowable output power of FC
$P_{ILC\_DC,min}$, $P_{ILC\_DC,max}$	minimum and maximum allowable transmission power of ILC
$P_{UPG,t}$, $P_{MT,t}$	output power of UPG and MT at time t
$P_{ILC\_AC,t}$	power of ILC on AC side
$P_{UPG,min}$, $P_{UPG,max}$	minimum and maximum allowable transmission power Of UPG
$P_{MTmin,i_{MT}}$, $P_{MTmax,i_{MT}}$	minimum and maximum allowable output power of MT

References

1. Huang, S.; Wu, Q.; Liu, Z.; Nielsen, A.H. Review of congestion management methods for distribution networks with high penetration of distributed energy resources. In Proceedings of the 5th IEEE PES Innovative Smart Grid Technologies Europe, Istanbul, Turkey, 12–15 October 2014.

2. Cipcigan, L.M.; Taylor, P.C. Investigation of the reverse power flow requirements of high penetrations of small-scale embedded generation. *IET Renew. Power Gener.* **2007**, *1*, 160–166. [CrossRef]

3. Gabash, A.; Xie, D.; Li, P. Analysis of influence factors on rejected active power from active distribution networks. In Proceedings of the IEEE Power and Energy Student Summit, Ilmenau, Germany, 19–20 January 2012.

4. Imani, M.H.; Zalzar, S.; Mosavi, A.; Shamshirband, S. Strategic behavior of retailers for risk reduction and profit increment via distributed generators and demand response programs. *Energies* **2018**, *11*, 1602. [CrossRef]

5. Liu, X.; Wang, P.; Loh, P.C. A hybrid AC/DC microgrid and its coordination control. *IEEE Trans. Smart Grid* **2011**, *2*, 278–286.

6. Kurohane, K.; Senjyu, T.; Uehara, A.; Yona, A.; Funabashi, T.; Kim, C.-H. A hybrid smart AC/DC power system. In Proceedings of the 5th IEEE Conference on Industrial Electronics and Applications, Taichung, Taiwan, 15–17 June 2010.

7. Eghtedarpour, N.; Farjah, E. Power control and management in a hybrid AC/DC microgrid. *IEEE Trans. Smart Grid* **2014**, *5*, 1494–1505. [CrossRef]

8. Wang, P.; Goel, L.; Liu, X.; Choo, F.H. Harmonizing AC and DC: A hybrid AC/DC future grid solution. *IEEE Power Energy Mag.* **2013**, *11*, 76–83. [CrossRef]

9. Unamuno, E.; Barrena, J.A. Hybrid ac/dc microgrids—Part I: Review and classification of topologies. *Renew. Sustain. Energy Rev.* **2015**, *52*, 1251–1259. [CrossRef]

10. Yang, Z.; Zhong, H.; Bose, A.; Xia, Q.; Kang, C. Optimal power flow in AC-DC grids with discrete control devices. *IEEE Trans. Power Syst.* **2018**, *33*, 1461–1472. [CrossRef]

11. Cao, J.; Du, W.; Wang, H.F. An improved corrective security constrained OPF for meshed AC/DC grids with multi-terminal VSC-HVDC. *IEEE Trans. Power Syst.* **2016**, *31*, 485–495. [CrossRef]

12. Baradar, M.; Hesamzadeh, M.R.; Ghandhari, M. Second-Order cone programming for optimal power flow in VSC-type AC-DC grids. *IEEE Trans. Power Syst.* **2013**, *28*, 4282–4291. [CrossRef]

13. Malik, S.M.; Ai, X.; Sun, Y.; Chen, Z.; Zhou, S. Voltage and frequency control strategies of hybrid AC/DC microgrid: A review. *IET Gener. Transm. Distrib.* **2017**, *11*, 303–313. [CrossRef]

14. Shanthi, P.; Govindarajan, U.; Parvathyshankar, D. Instantaneous power-based current control scheme for VAR compensation in hybrid AC/DC networks for smart grid applications. *IET Power Electron.* **2014**, *7*, 1216–1226. [CrossRef]

15. Dong, L.; Ming, J.; Yu, T.; Fan, S.; Pu, T. Voltage division control in AC / DC hybrid distribution network based on model predictive control. In Proceedings of the 2016 IEEE PES Asia-Pacific Power and Energy Engineering Conference (APPEEC), Xi'an, China, 25–28 October 2016.

16. Zhu, M.; Hang, L.; Li, G.; Jiang, X. Protected control method for power conversion interface under unbalanced operating conditions in AC/DC hybrid distributed grid. *IEEE Trans. Energy Convers.* **2016**, *31*, 57–68. [CrossRef]

17. Nejabatkhah, F.; Li, Y. Overview of power management strategies of hybrid AC/DC microgrid. *IEEE Trans. Power Electron.* **2015**, *30*, 7072–7089. [CrossRef]

18. Liu, S.; Ding, L.; Miao, Y. Research of coordinated control strategy for multi-UHVDC in AC/DC hybrid power grid. *Energy Procedia* **2011**, *12*, 443–449. [CrossRef]

19. Hu, J.; Shan, Y.; Xu, Y.; Guerrero, J.M. A coordinated control of hybrid ac/dc microgrids with PV-wind-battery under variable generation and load conditions. *Int. J. Electr. Power Energy Syst.* **2019**, *104*, 583–592. [CrossRef]

20. Li, Y.; Li, Y.; Li, G.; Zhao, D.; Chen, C. Two-stage multi-objective OPF for AC/DC grids with VSC-HVDC: Incorporating decisions analysis into optimization process. *Energy* **2018**, *147*, 286–296. [CrossRef]

21. Ghadiri, A.; Haghifam, M.R.; Larimi, S.M.M. Comprehensive approach for hybrid AC/DC distribution network planning using genetic algorithm. *IET Gener. Transm. Distrib.* **2017**, *11*, 3892–3902. [CrossRef]

22. Liang, H.; Lin, J.; Ba, L.; Li, H. Research on optimal dispatch method for AC/DC hybrid active distribution network. In Proceedings of the International Conference on Renewable Power Generation (RPG), Beijing, China, 17–18 October 2015.

23. Indragandhi, V.; Logesh, R.; Subramaniyaswamy, V.; Vijayakumar, V.; Siarry, P.; Uden, L. Multi-objective optimization and energy management in renewable based AC/DC microgrid. *Comput. Electr. Eng.* **2018**, *70*, 179–198.

24. Qi, C.; Wang, K.; Li, G.; Han, B.; Xu, S.; Wei, Z. Hierarchical and distributed optimal scheduling of AC/DC hybrid active distribution network. *Proc. CSEE* **2017**, *37*, 1909–1917.

25. Eajal, A.A.; Shaaban, M.F.; Ponnambalam, K.; El-Saadany, E.F. Stochastic centralized dispatch scheme for AC/DC hybrid smart distribution systems. *IEEE Trans. Sustain. Energy* **2017**, *7*, 1046–1059. [CrossRef]

26. Qiu, H.; Zhao, B.; Gu, W.; Bo, R. Bi-Level two-stage robust optimal scheduling for AC/DC hybrid multi-microgrids. *IEEE Trans. Smart Grid* **2018**, *9*, 5455–5466. [CrossRef]

27. Qiu, H.; Gu, W.; Pan, J.; Xu, B.; Xu, Y.; Fan, M.; Wu, Z. Multi-interval-uncertainty constrained robust dispatch for AC/DC hybrid microgrids with dynamic energy storage degradation. *Appl. Energy* **2018**, *228*, 205–214. [CrossRef]

28. Gabash, A.; Li, P. On variable reverse power flow-part I: Active-reactive optimal power flow with reactive power of wind stations. *Energies* **2016**, *9*, 121. [CrossRef]

29. Li, Y.; Miao, S.; Luo, X.; Wang, J. Optimization model for the power system scheduling with wind generation and compressed air energy storage combination. In Proceedings of the 22nd International Conference on Automation and Computing (ICAC), Colchester, UK, 7–8 September 2016.

30. Bayraktar, Z.; Komurcu, M.; Werner, D.H. Wind Driven Optimization (WDO): A novel nature-inspired optimization algorithm and its application to electromagnetics. In Proceedings of the 2011 IEEE Antennas and Propagation Society International Symposium, Toronto, ON, Canada, 11–17 July 2010.

31. Bayraktar, Z.; Komurcu, M.; Bossard, J.A.; Werner, D.H. The wind driven optimization technique and its application in electromagnetics. *IEEE Trans. Antennas Propag.* **2013**, *61*, 2745–2757. [CrossRef]

32. Lofberg, J. YALMIP: A toolbox for modeling and optimization in MATLAB. In Proceedings of the 2004 IEEE International Conference on Robotics and Automation, New Orleans, LA, USA, 2–4 September 2004.

33. Kekatos, V.; Giannakis, G.B.; Wollenberg, B. Optimal placement of phasor measurement units via convex relaxation. *IEEE Trans. Power Syst.* **2012**, *27*, 1521–1530. [CrossRef]

34. Korres, G.N.; Manousakis, N.M.; Xygkis, T.C.; Löfberg, J. Optimal phasor measurement unit placement for numerical observability in the presence of conventional measurements using semi-definite programming. *IET Gener. Transm. Distrib.* **2015**, *9*, 2427–2436. [CrossRef]

35. Singh, D.; Misra, R.; Singh, D. Effect of load models in distributed generation planning. *IEEE Trans. Power Syst.* **2007**, *22*, 2204–2212. [CrossRef]

MDPI

St. Alban-Anlage 66

4052 Basel

Switzerland

Tel. +41 61 683 77 34

Fax +41 61 302 89 18

www.mdpi.com

Energies Editorial Office

E-mail: energies@mdpi.com

www.mdpi.com/journal/energies

CPSIA information can be obtained
at www.ICGtesting.com
Printed in the USA
BVHW021456040819
555055BV00011B/233/P